Bayesian Analysis of Time Series and Dynamic Models

STATISTICS: Textbooks and Monographs

A Series Edited by

Vol. 1: The Generalized Jackknife Statistic, *H. L. Gray and W. R. Schucany*
Vol. 2: Multivariate Analysis, *Anant M. Kshirsagar*
Vol. 3: Statistics and Society, *Walter T. Federer*
Vol. 4: Multivariate Analysis: A Selected and Abstracted Bibliography, 1957-1972, *Kocherlakota Subrahmaniam and Kathleen Subrahmaniam* (out of print)
Vol. 5: Design of Experiments: A Realistic Approach, *Virgil L. Anderson and Robert A. McLean*
Vol. 6: Statistical and Mathematical Aspects of Pollution Problems, *John W. Pratt*
Vol. 7: Introduction to Probability and Statistics (in two parts), Part I: Probability; Part II: Statistics, *Narayan C. Giri*
Vol. 8: Statistical Theory of the Analysis of Experimental Designs, *J. Ogawa*
Vol. 9: Statistical Techniques in Simulation (in two parts), *Jack P. C. Kleijnen*
Vol. 10: Data Quality Control and Editing, *Joseph I. Naus* (out of print)
Vol. 11: Cost of Living Index Numbers: Practice, Precision, and Theory, *Kali S. Banerjee*
Vol. 12: Weighing Designs: For Chemistry, Medicine, Economics, Operations Research, Statistics, *Kali S. Banerjee*
Vol. 13: The Search for Oil: Some Statistical Methods and Techniques, *edited by D. B. Owen*
Vol. 14: Sample Size Choice: Charts for Experiments with Linear Models, *Robert E. Odeh and Martin Fox*
Vol. 15: Statistical Methods for Engineers and Scientists, *Robert M. Bethea, Benjamin S. Duran, and Thomas L. Boullion*
Vol. 16: Statistical Quality Control Methods, *Irving W. Burr*
Vol. 17: On the History of Statistics and Probability, *edited by D. B. Owen*
Vol. 18: Econometrics, *Peter Schmidt*
Vol. 19: Sufficient Statistics: Selected Contributions, *Vasant S. Huzurbazar (edited by Anant M. Kshirsagar)*
Vol. 20: Handbook of Statistical Distributions, *Jagdish K. Patel, C. H. Kapadia, and D. B. Owen*
Vol. 21: Case Studies in Sample Design, *A. C. Rosander*
Vol. 22: Pocket Book of Statistical Tables, *compiled by R. E. Odeh, D. B. Owen, Z. W. Birnbaum, and L. Fisher*
Vol. 23: The Information in Contingency Tables, *D. V. Gokhale and Solomon Kullback*

Vol. 24: Statistical Analysis of Reliability and Life-Testing Models: Theory and Methods, *Lee J. Bain*
Vol. 25: Elementary Statistical Quality Control, *Irving W. Burr*
Vol. 26: An Introduction to Probability and Statistics Using BASIC, *Richard A. Groeneveld*
Vol. 27: Basic Applied Statistics, *B. L. Raktoe and J. J. Hubert*
Vol. 28: A Primer in Probability, *Kathleen Subrahmaniam*
Vol. 29: Random Processes: A First Look, *R. Syski*
Vol. 30: Regression Methods: A Tool for Data Analysis, *Rudolf J. Freund and Paul D. Minton*
Vol. 31: Randomization Tests, *Eugene S. Edgington*
Vol. 32: Tables for Normal Tolerance Limits, Sampling Plans, and Screening, *Robert E. Odeh and D. B. Owen*
Vol. 33: Statistical Computing, *William J. Kennedy, Jr. and James E. Gentle*
Vol. 34: Regression Analysis and Its Application: A Data-Oriented Approach, *Richard F. Gunst and Robert L. Mason*
Vol. 35: Scientific Strategies to Save Your Life, *I. D. J. Bross*
Vol. 36: Statistics in the Pharmaceutical Industry, *edited by C. Ralph Buncher and Jia-Yeong Tsay*
Vol. 37: Sampling from a Finite Population, *J. Hajek*
Vol. 38: Statistical Modeling Techniques, *S. S. Shapiro*
Vol. 39: Statistical Theory and Inference in Research, *T. A. Bancroft and C.-P. Han*
Vol. 40: Handbook of the Normal Distribution, *Jagdish K. Patel and Campbell B. Read*
Vol. 41: Recent Advances in Regression Methods, *Hrishikesh D. Vinod and Aman Ullah*
Vol. 42: Acceptance Sampling in Quality Control, *Edward G. Schilling*
Vol. 43: The Randomized Clinical Trial and Therapeutic Decisions, *edited by Niels Tygstrup, John M. Lachin, and Erik Juhl*
Vol. 44: Regression Analysis of Survival Data in Cancer Chemotherapy, *Walter H. Carter, Jr., Galen L. Wampler, and Donald M. Stablein*
Vol. 45: A Course in Linear Models, *Anant M. Kshirsagar*
Vol. 46: Clinical Trials: Issues and Approaches, *edited by Stanley H. Shapiro and Thomas H. Louis*
Vol. 47: Statistical Analysis of DNA Sequence Data, *edited by B. S. Weir*
Vol. 48: Nonlinear Regression Modeling: A Unified Practical Approach, *David A. Ratkowsky*
Vol. 49: Attribute Sampling Plans, Tables of Tests and Confidence Limits for Proportions, *Robert E. Odeh and D. B. Owen*
Vol. 50: Experimental Design, Statistical Models, and Genetic Statistics, *edited by Klaus Hinkelmann*
Vol. 51: Statistical Methods for Cancer Studies, *edited by Richard G. Cornell*
Vol. 52: Practical Statistical Sampling for Auditors, *Arthur J. Wilburn*
Vol. 53: Statistical Signal Processing, *edited by Edward J. Wegman and James G. Smith*
Vol. 54: Self-Organizing Methods in Modeling: GMDH Type Algorithms, *edited by Stanley J. Farlow*
Vol. 55: Applied Factorial and Fractional Designs, *Robert A. McLean and Virgil L. Anderson*
Vol. 56: Design of Experiments: Ranking and Selection, *edited by Thomas J. Santner and Ajit C. Tamhane*
Vol. 57: Statistical Methods for Engineers and Scientists. Second Edition, Revised and Expanded, *Robert M. Bethea, Benjamin S. Duran, and Thomas L. Boullion*
Vol. 58: Ensemble Modeling: Inference from Small-Scale Properties to Large-Scale Systems, *Alan E. Gelfand and Crayton C. Walker*
Vol. 59: Computer Modeling for Business and Industry, *Bruce L. Bowerman and Richard T. O'Connell*
Vol. 60: Bayesian Analysis of Linear Models, *Lyle D. Broemeling*

Vol. 61: Methodological Issues for Health Care Surveys, *Brenda Cox and Steven Cohen*

Vol. 62: Applied Regression Analysis and Experimental Design, *Richard J. Brook and Gregory C. Arnold*

Vol. 63: Statpal: A Statistical Package for Microcomputers – PC-DOS Version for the IBM PC and Compatibles, *Bruce J. Chalmer and David G. Whitmore*

Vol. 64: Statpal: A Statistical Package for Microcomputers – Apple Version for the II, II+, and IIe, *David G. Whitmore and Bruce J. Chalmer*

Vol. 65: Nonparametric Statistical Inference, Second Edition, Revised and Expanded, *Jean Dickinson Gibbons*

Vol. 66: Design and Analysis of Experiments, *Roger G. Petersen*

Vol. 67: Statistical Methods for Pharmaceutical Research Planning, *Sten W. Bergman and John C. Gittins*

Vol. 68: Goodness-of-Fit Techniques, *edited by Ralph B. D'Agostino and Michael A. Stephens*

Vol. 69: Statistical Methods in Discrimination Litigation, *edited by D. H. Kaye and Mikel Aickin*

Vol. 70: Truncated and Censored Samples from Normal Populations, *Helmut Schneider*

Vol. 71: Robust Inference, *M. L. Tiku, W. Y. Tan, and N. Balakrishnan*

Vol. 72: Statistical Image Processing and Graphics, *edited by Edward J. Wegman and Douglas J. DePriest*

Vol. 73: Assignment Methods in Combinatorial Data Analysis, *Lawrence J. Hubert*

Vol. 74: Econometrics and Structural Change, *Lyle D. Broemeling and Hiroki Tsurumi*

Vol. 75: Multivariate Interpretation of Clinical Laboratory Data, *Adelin Albert and Eugene K. Harris*

Vol. 76: Statistical Tools for Simulation Practitioners, *Jack P. C. Kleijnen*

Vol. 77: Randomization Tests, Second Edition, *Eugene S. Edgington*

Vol. 78: A Folio of Distributions: A Collection of Theoretical Quantile-Quantile Plots, *Edward B. Fowlkes*

Vol. 79: Applied Categorical Data Analysis, *Daniel H. Freeman, Jr.*

Vol. 80: Seemingly Unrelated Regression Equations Models: Estimation and Inference, *Virendra K. Srivastava and David E. A. Giles*

Vol. 81: Response Surfaces: Designs and Analyses, *Andre I. Khuri and John A. Cornell*

Vol. 82: Nonlinear Parameter Estimation: An Integrated System in BASIC, *John C. Nash and Mary Walker-Smith*

Vol. 83: Cancer Modeling, *edited by James R. Thompson and Barry W. Brown*

Vol. 84: Mixture Models: Inference and Applications to Clustering, *Geoffrey J. McLachlan and Kaye E. Basford*

Vol. 85: Randomized Response: Theory and Techniques, *Arijit Chaudhuri and Rahul Mukerjee*

Vol. 86: Biopharmaceutical Statistics for Drug Development, *edited by Karl E. Peace*

Vol. 87: Parts per Million Values for Estimating Quality Levels, *Robert E. Odeh and D. B. Owen*

Vol. 88: Lognormal Distributions: Theory and Applications, *edited by Edwin L. Crow and Kunio Shimizu*

Vol. 89: Properties of Estimators for the Gamma Distribution, *K. O. Bowman and L. R. Shenton*

Vol. 90: Spline Smoothing and Nonparametric Regression, *Randall L. Eubank*

Vol. 91: Linear Least Squares Computations, *R. W. Farebrother*

Vol. 92: Exploring Statistics, *Damaraju Raghavarao*

Vol. 93: Applied Time Series Analysis for Business and Economic Forecasting, *Sufi M. Nazem*

Vol. 94: Bayesian Analysis of Time Series and Dynamic Models, *edited by James C. Spall*

ADDITIONAL VOLUMES IN PREPARATION

Bayesian Analysis of Time Series and Dynamic Models

Edited by

JAMES C. SPALL
The Johns Hopkins University
Applied Physics Laboratory
Laurel, Maryland

MARCEL DEKKER, INC. New York and Basel

Library of Congress Cataloging-in-Publication Data

Bayesian analysis of time series and dynamic models/edited by James
C. Spall.
p. cm. -- (Statistics, textbooks and monographs; vol. 94)
Includes bibliographies and index.
ISBN 0-8247-7936-3
1. Bayesian statistical decision theory. 2. Time-series analysis.
3. System analysis. I. Spall, James C. II. Series: Statistics,
textbooks and monographs ; v. 94.
QA279.5.B385 1988
519.5′5--dc 19
88-14165
CIP

MARCEL DEKKER, INC.
270 Madison Avenue, New York, New York 10016

Current printing (last digit):
10 9 8 7 6 5 4 3 2 1

PRINTED IN THE UNITED STATES OF AMERICA

Preface

In perhaps every branch of the physical and social sciences, engineering, and medicine, one encounters problems involving the study of systems that change over time. When, for any number of reasons, the data collected on such dynamic systems are treated as random quantities, statistical techniques of the type considered here are used to make inferences and gain understanding. The techniques presented in this book for analyzing dynamic systems ultimately rest on one famous mathematical expression—Bayes' rule.

Although the Bayesian approach is sometimes controversial (owing primarily to issues surrounding the prior distribution), it has much to recommend it for analyzing dynamic systems. As evidence of this, one might consider the celebrated "Kalman filter," which is a Bayesian estimator that has enjoyed success in a myriad of applications.

There is a considerable amount of literature on Bayesian methods for studying dynamic systems. Much of the work is published in the fairly disparate fields of statistics, engineering, and econometrics. Unfortunately, individuals working in one field of study are often not aware of relevant results published in another field. This interdisciplinary book brings together individuals who are known primarily in either statistics, control systems engineering, or econometrics with the goal of increasing awareness of available methodologies in the scientific and technical communities at large.

Broadly speaking, the chapters in this volume deal with perhaps the two most commonly used types of time series models for studying dynamic systems:

(i) Traditional models such as ARMA (autoregressive moving average) or its variants
(ii) Linear and nonlinear state-space models

Although it is often possible to transfer from one type of model to the other (especially in the linear case), each type of model has its own body of theory as well as applications for which it is particularly well suited. As a general rule, most of the developments associated with models of type (i) have occurred in the statistics and econometrics literature while those for models of type (ii) have been concentrated in the control and systems engineering literature. A goal of this book is to increase communication among the statistics, engineering, and econometrics communities.

To help achieve the interdisciplinary goal of this book, each chapter underwent a thorough review, which generally included an examination by at least one specialist, as well as one referee not in the chapter author's main field of interest. All chapters were revised at least once after the initial submission. For almost every chapter, part of the revision involved the addition of discussion and references to related work published outside of the authors' field. This should help put the chapter in proper perspective and encourage less duplication of effort by alerting readers to a wider body of related results.

This book begins with an overview of some key historical developments in the field of dynamic modeling and estimation. This overview is followed by 18 self-contained chapters that span many areas of current (and future, no doubt) research interest. Although certain chapters treat models of both the ARMA type and state-space type, Chapters 1, 2, 3, 5, 12, and 13 focus mainly on models in the ARMA (or AR) form while Chapters 4, 6–11, and 14–18 emphasize the state-space approach. A brief review of the chapters is given below.

Chapters 1 (L. Broemeling and S. Shaarawy), 2 (M. J. Schervish and R. S. Tsay), and 3 (D. Poskitt) present Bayesian approaches for constructing AR and ARMA models, including procedures for model order determination, parameter estimation, and diagnostic checking. Chapter 4, by H. Tsurumi, reviews Bayesian and non-Bayesian methods for determining the time and magnitude of changes in dynamic regression models. The fifth chapter, by P. Cook, presents a procedure for estimating the frequency domain characteristics of AR models when a limited amount of data is available. Chapters 6 (H. W. Sorenson), 7 (A. Pole, M. West, P. J. Harrison), 8 (F. E. Daum), 9 (D. Peña and I. Guttman), and 10 (J.V. White) focus on the problem of state-space modeling and estimation in the presence of nonlinear system behavior and/or nonnormally distributed data. These chapters demonstrate how the Bayesian approach can effectively deal with the difficulties encountered when the usual assumptions of linearity and normality are relaxed.

Chapter 11, by W. W. Davis, discusses some of the problems encountered in the practical implementation of a Kalman filter in the context of a problem in image processing based on satellite radiometer measurements. Chapters 12

(M. Kárný and R. Kulhavý) and 13 (J. Diaz) consider problems in parameter estimation and order determination for models that are generalizations of the usual linear regression model, including, for example, AR models. Chapter 14 (K. Gordon and A. F. M. Smith) considers the problem of state-space modeling and estimation for systems that change abruptly.

R. Kohn and C. F. Ansley in Chapter 15 demonstrate that an important deterministic signal estimation procedure (optimal smoothing) for a signal-plus-noise problem can be interpreted as a Bayesian estimator, which leads to a number of significant new insights. Chapter 16, by W. Gersch and G. Kitagawa, considers stationary and nonstationary linear time series models with Gaussian disturbances, and also considers non-Gaussian/nonlinear models via a numerical integration of conditional densities. Chapter 17, by W. M. Sallas and D. A. Harville, discusses the problem of parameter estimation of state-space models when some of the states in the state vector have diffuse priors. Last, Chapter 18, by S. D. Hill and J. C. Spall, presents a tractable procedure for specifying "noninformative" prior distributions for parameters in state-space models that is based on Shannon information theory.

I would like to gratefully acknowledge the financial support of the Johns Hopkins University, Applied Physics Laboratory, through a Stuart S. Janney Fellowship. Additional support was received through U.S. Navy contract N00039-87-C-5301. I would also like to give special thanks to Dr. John Maryak for manifold assistance in settling issues that arose during the chapter review process, to Ms. Diane Harrison for secretarial support, and to Ms. Vickie Kearn at Marcel Dekker, Inc., for her able assistance in smoothing the way for the preparation of this book.

James C. Spall

Contents

Contributors

CRAIG F. ANSLEY Australian Graduate School of Management, University of New South Wales, Kensington, New South Wales, Australia

LYLE BROEMELING* Mathematical Sciences Division, Office of Naval Research, Arlington, Virginia

PEYTON COOK Department of Mathematical and Computer Sciences, The University of Tulsa, Tulsa, Oklahoma

FREDERICK E. DAUM Raytheon Company, Wayland, Massachusetts

WILLIAM W. DAVIS† Lockheed Missiles & Space Company, Inc., Austin, Texas

JOAQUIN DIAZ Department of Decision and Information Sciences, University of Houston, Houston, Texas

WILL GERSCH Department of Information and Computer Sciences, University of Hawaii, Honolulu, Hawaii

K. GORDON‡ Department of Mathematics, University of Nottingham, Nottingham, England

*Present address: Office of Academic Computing and Biostatistics, University of Texas Medical Branch, Galveston, Texas.

†Present address: Statistical Research Division, Bureau of the Census, Washington, D.C.

‡Present address: Mathematical Applications, CIBA-GEIGY AG, Basel, Switzerland.

IRWIN GUTTMAN Statistics Department, University of Toronto, Toronto, Ontario, Canada

P. JEFF HARRISON Department of Statistics, University of Warwick, Coventry, England

DAVID A. HARVILLE Department of Statistics, Iowa State University, Ames, Iowa

STACY D. HILL The Johns Hopkins University, Applied Physics Laboratory, Laurel, Maryland

MIROSLAV KÁRNÝ Institute of Information Theory and Automation, Czechoslovak Academy of Sciences, Prague, Czechoslovakia

GENSHIRO KITAGAWA The Institute of Statistical Mathematics, Tokyo, Japan

RUDOLF KULHAVÝ Institute of Information Theory and Automation, Czechoslovak Academy of Sciences, Prague, Czechoslovakia

ROBERT KOHN Australian Graduate School of Management, University of New South Wales, Kensington, New South Wales, Australia

DANIEL PEÑA Universidad Politecnica de Madrid, Madrid, Spain

ANDY POLE Department of Statistics, University of Warwick, Coventry, England

DONALD S. POSKITT Department of Statistics, IAS, Australian National University, Canberra, Australia

WILLIAM M. SALLAS IMSL Inc., Houston, Texas

MARK J. SCHERVISH Department of Statistics, Carnegie-Mellon University, Pittsburgh, Pennsylvania

SAMIR SHAARAWY Faculty of Economics and Political Science, Cairo University, Cairo, Egypt

ADRIAN F. M. SMITH Department of Mathematics, University of Nottingham, Nottingham, England

HAROLD W. SORENSON Chief Scientist, United States Air Force, Headquarters, USAF/CCN, Washington, District of Columbia

JAMES C. SPALL The Johns Hopkins University, Applied Physics Laboratory, Laurel, Maryland

RUEY S. TSAY Department of Statistics, Carnegie-Mellon University, Pittsburgh, Pennsylvania

HIROKI TSURUMI Department of Economics, Rutgers University, New Brunswick, New Jersey

MIKE WEST Department of Statistics, University of Warwick, Coventry, England

JAMES V. WHITE The Analytic Sciences Corporation, Reading, Massachusetts

An Overview of Key Developments in Dynamic Modeling and Estimation

1. INTRODUCTION

This overview discusses some key historical contributions that have led to important modern methods for analyzing dynamic systems. The emphasis here will be on the state-space (and associated Kalman filter) approach to time series modeling. (Chapter 1 of this volume, by L. Broemeling and S. Shaarawy, provides some historical background for the ARMA approach.) Aside from historical information, some discussion is also included on the form of the Kalman filter algorithm and on its associated properties.

Although this book is Bayesian in nature, most of the key historical developments in the field are non-Bayesian. This overview reflects that fact. However, since the Bayesian approach is becoming increasingly important (e.g., as motivated by problems related to the non-Gaussian/nonlinear models encountered in statistics and econometrics), an overview of the field written a decade or so from now would almost certainly feature more Bayesian historical developments than are included here.

Throughout, the focus will be on the discrete-time formulation for time-series problems, that is, a formulation relying on difference equations and data observed at discrete time points. This contrasts with the continuous time formulation, which relies on differential equations. Such discrete-time settings are the dominant modus operandi for time-series problems in statistics, econometrics, and biometrics, and are also widely used in engineering and other scientific applications. They are particularly appropriate for digital computer implementation.

The remainder of this overview is divided into the following sections:

Some Early History
The Wiener-Kolmogorov Approach

Developments Subsequent to Wiener-Kolmogorov
State-Space Models and Kalman Filtering
Contributions of the Aerospace Community
State-Space Models and the Kalman Filter in Econometrics and Statistics
Conclusions

The above organization has been chosen to highlight major milestones that have led to modern methods for dynamic modeling and estimation.

Obviously, much information had to be omitted in covering the above range of topics. The reader is invited to investigate pertinent references for further information on any of the various topics.

2. SOME EARLY HISTORY

Making inference from a time-series realization of data is a problem with a long history. Much of the early work in time series, such as that by Bernoulli, Gauss, and Legendre, grew out of problems in astronomy. The problem they addressed involved making inference as to the location of a "heavenly body" from a sequence of imperfect observations. Least squares (e.g., Gauss and Legendre), and maximum likelihood (Gauss, Bernoulli) techniques were both used. A more complete account of the contributions made to time series in the late 1700s to early 1800s is given in Seal (1967), Sorenson (1970, 1980), and Stigler (1986).

Later in the 1800s the Danish astronomer T. N. Thiele developed a recursive* procedure resembling what is now referred to as the Kalman filter for the problem of determining the distance from Copenhagen to Lund (Thiele, 1880). As with Gauss and Legendre, Thiele approached the problem from a least-squares perspective. In the language of the Kalman filter description, given in Section 5, Thiele's algorithm is equivalent to the Kalman filter for the special case of scalar state and measurements, state transition and measurement "matrices" equaling unity, and nonrandom initial state. Interestingly, Thiele also anticipated the "system identification" problem—an area of much current interest—by proposing a clever procedure for estimating the variances of the state and measurement noise terms. Further discussion on the work of Thiele is given in Hald (1981) and Lauritzen (1981).

*By a recursive procedure we mean one in which the data are processed sequentially rather than in batch form. Recursive estimators have the advantage of not requiring that the complete data set be stored prior to calculating the estimate and of not requiring a reprocessing of existing data if a new measurement becomes available.

3. THE WIENER-KOLMOGOROV APPROACH

In the early 1940s, N. Wiener and A. N. Kolmogorov independently presented procedures for estimating the signal in a signal-plus-noise problem (documented in Wiener, 1949, and Kolmogorov, 1941). Kolmogorov's formulation is the more general of the two and is antecedent to the work of Wiener by several years. Wiener, however, focuses more on developing a practical solution, having been motivated by a World War II anti-aircraft fire control problem. As Wiener (1949, p. 59) states:

> The present investigation was initiated in the early winter of 1940 as an attempt to solve an engineering problem. At that time and until the last week of 1941, by which time the paper was substantially complete, the author was not aware of the results of Kolmogorov's work and scarcely aware of its existence . . . it would appear that the work of Kolmogorov and that of the present writer represent two entirely separate attacks on the problem of time series.

Let us discuss an important special case of the Wiener-Kolmogorov (W-K) theory, which will lead logically into a discussion of the Kalman filter. In particular, we focus on the discrete time formulation as described in Levinson (1947). Consider a sequence of vector-valued measurements $z_1, z_2, \ldots,$ with

$$z_i = x_i + v_i \tag{3.1}$$

and x_i representing the signal (the quantity of interest) and v_i representing the noise. W-K theory is oriented toward finding the minimum mean-square error estimate of the sequence of signals $x_1, x_2, \ldots$ among linear estimators. That is, for $1 \leqslant k < \infty$, the W-K estimator (filter) for x_k has the form

$$\hat{x}_k = \sum_{i=1}^{k} A_{ki} z_i \tag{3.2}$$

where the A_{ki}'s are chosen to satisfy the minimum mean-square error criterion. To have a hope of determining the A_{ki}'s, it must be assumed that second-moment information on the quantities in Eq. (3.1) is available. In particular W-K theory requires covariance knowledge such as $E(x_i x_j^T)$ and $E(z_i x_j^T)$.

Since the W-K approach is oriented toward finding the minimum mean-square error estimator, the orthogonal projection theorem applies. This leads to a system of equations (in so-called Wiener-Hopf form) from which the A_{ki}'s can, in principle, be determined. However, since it is generally intractable to determine the A_{ki}'s in closed form, W-K theory introduces two assumptions that facilitate the computations: (i) the measurements (z_i's) and signal (x_i's) are stationary processes and (ii) the measurements are assumed to have been generated since the infinite past. Under (i) and (ii) the problem of estimating the signal vector x_k at any given time through the computation of the A_{ki}'s becomes feasible.

4. DEVELOPMENTS SUBSEQUENT TO WIENER-KOLMOGOROV

During the 1940s and 1950s much work continued on signal estimation problems of the type described above. In particular, it was possible to obtain tractable estimation algorithms for certain special cases when one or both of assumptions (i) or (ii) above were relaxed (e.g., Booton, 1952, for the nonstationary case and Zadeh and Ragazzini, 1950, for the finite initial time case). Unfortunately such procedures were still often difficult to implement, requiring significant modifications or new derivations (e.g., solution of an auxiliary differential equation) to apply to any particular problem. Furthermore, many of the procedures did not readily extend to the multivariate measurement case.

To address the shortcomings of the W-K and related approaches, a significantly different formulation of the signal estimation problem was begun in the mid 1950s. J. W. Follin and his colleagues at the Johns Hopkins University, Applied Physics Laboratory (Carlton and Follin, 1956, Bucy, 1958, Hanson, 1957a,b) and P. Swerling at the RAND Corp. (Swerling, 1958, 1959, 1971) replaced the assumption of knowledge of the auto- and cross-covariances for the measurements, signal, and noise with a specific mathematical model for how the signal evolves over time. They were then able to derive recursive estimators for the signal that are similar to the celebrated Kalman filter discussed below. The Follin group's work was motivated by a problem in developing antiballistic missile defense systems, which required that transient (i.e., nonstationary) effects be considered. Swerling's work derived from a problem in satellite orbit estimation.

Let us briefly discuss Swerling's (1958, 1959) discrete-time formulation. Suppose the signal evolves according to

$$x_k = f_{k-1}(x_{k-1}) \tag{4.1}$$

and measurements are given by

$$z_k = h_k(x_k) + v_k$$

with $f(\cdot)$ and $h(\cdot)$ generally nonlinear functions and v_k (as before) the measurement noise. Swerling uses a *least-squares* approach to develop recursive linear estimators for the signal. In his 1971 paper (Eqs. (58), (62)–(64)) he points out that it is straightforward to modify these recursive formulas to account for an additive noise term in the signal evolution equation, (4.1). In the case of linear $f(\cdot)$ and $h(\cdot)$, the Swerling recursions are then essentially equivalent to the Kalman filter.

In the statistics and econometrics literature, Plackett (1950) and Thiel and Goldberger (1961) present a least-squares procedure for recursively updating a regression estimate when additional data become available. In particular, for a constant signal

$$x_k = x_{k-1}$$

and measurements of the form

$$z_k = H_k x_k + v_k \tag{4.2}$$

they present a procedure that updates the estimate of the signal as an additional measurement z becomes available (H_k represents the design matrix).* Note that their procedure is a special case of the Swerling method outlined above.

5. STATE-SPACE MODELS AND KALMAN FILTERING

It was not until the famous papers of Kalman (1960) (discrete time) and Kalman and Bucy (1961) (continuous time), that an explicit connection was made between the problem of signal estimation and *state-space* models. Based on this connection, it was then possible to exploit a wealth of powerful theory that had been developed over the previous decade in the field of systems and control theory. It was (and is) also true that state-space models were suitable for describing many real-world dynamic systems.

In the discrete-time version of the model considered in Kalman (1960) and Kalman and Bucy (1961), the signal is assumed to evolve according to the linear equation

$$x_k = F_{k-1} x_{k-1} + w_k \tag{5.1a}$$

where F_k denotes the "state transition" matrix and w_k denotes a random noise. Consistent with the state-space vernacular, we henceforth refer to x_k as the *state* vector (instead of the signal vector). In practical applications, the state noise w_k may represent some inherent randomness in how the state of the system evolves; alternatively, it may serve as a "fudge factor" to account for the fact that the state vector does not evolve precisely according to the

*Interesting discussions of how the Plackett and Thiel/Goldberger formulation relates to the Kalman filter are given in Diderrich (1985) and Welch (1987).

simple linear propagation $x_k = F_{k-1}x_{k-1}$. The second equation in the state-space model relates the state to the observed quantity z_k, namely,

$$z_k = H_k x_k + v_k \tag{5.1b}$$

where H_k is the measurement (or design) matrix (as in Eq. (4.2)). Note from Eq. (5.1b) that for each k, x_k is not directly recoverable from z_k, because of the presence of the measurement noise as well as the fact that H_k is typically noninvertible. Kalman used the orthogonal projection theorem to develop a recursive expression for calculating the minimum mean-square error estimate $\hat{x}_k$ of the state vector x_k for $k = 1, 2, \ldots,$ based on the measurements $z_1, z_2, \ldots, z_k$ and prior information on the initial state x_0.

Several comments are in order on the linear state-space model of (5.1a, b). The state vector x_k contains all relevant terms that describe the system under study (e.g., terms related to position, velocity, orientation, and the inertial guidance mechanism for a ballistic missile system; terms related to the flow of money, goods, and services for an economic system). Obviously, for complex systems the dimension of x_k may be high, and in fact there will often be questions about whether certain terms should or should not be included. The measurement vector z_k, which is generally of lower dimension than the state vector, represents the observed quantities related to the elements of the state vector.

The Kalman filter (KF)* has the basic form

$$\hat{x}_k = T_{k-1}\hat{x}_{k-1} + K_k z_k \tag{5.2}$$

with the matrices T and K being functions of the underlying parameters associated with the state-space model, Eqs. (5.1a, b). Associated with $\hat{x}_k$ is an error-covariance matrix

$$P_k \equiv E[(\hat{x}_k - x_k)(\hat{x}_k - x_k)^T],$$

which can be obtained recursively in terms of P_{k-1} and the parameters of the state-space model.† The matrix P_k is an indicator of the accuracy to which $\hat{x}_k$ is tracking x_k. The filter is initialized with an a priori estimate of the initial state x_0 together with an associated error covariance matrix P_0. There is an abundance of papers and books that include the specific equations for calculating T_k, K_k, and P_k (see, e.g., Jazwinski, 1970, Rhodes, 1971, Sorenson, 1980, or Broemeling, 1985).

The KF estimate has several important statistical properties. In the case where the measurements are normally distributed and the noise sequences

*The continuous time analogue is generally called the Kalman-Bucy filter.

†Note that P_k is not the same as the covariance matrix of $\hat{x}_k$. In particular, $\text{cov}(\hat{x}_k) = \text{cov}(x_k) - P_k$.

$\{w_k\}$ and $\{v_k\}$ are mutually independent and independent of x_0, the KF estimate is the minimum mean-square error estimate of the state vector among *all* (linear and nonlinear) estimators, which implies that it is also the conditional mean, that is, $\hat{x}_k = E(x_k | z_1, z_2, \ldots, z_k)$. The KF can also be derived from first principles as a recursive Bayesian maximum a posteriori estimator, as shown in Ho and Lee (1964) (this can also be seen by using the fact that $\hat{x}_k$ is normally distributed together with the fact that it is the conditional mean). In the general non-normally distributed measurement case, the KF estimate is the generalized least squares and minimum mean-squared error estimate among *linear* estimators. As in the normal measurement case, the KF estimate is unbiased in the sense that $E(\hat{x}_k - x_k) = 0$. However, unlike the normal measurement case, $\hat{x}_k$ will not be normally distributed for finite k (it may, however, be normally distributed as $k \to \infty$; see Spall and Wall, 1984).

Aside from the above-mentioned statistical properties, there are a number of important implementation-oriented properties associated with the KF. Since the algorithm is recursive, there is no need to store the whole measurement sequence, and, in particular, the KF can be used for "real-time" data processing. Unlike the W-K approach there is no need for the measurement and signal (state) processes to be stationary, and the estimator is not based on the assumption that data have been generated from the infinite past. Finally, unlike W-K theory and other frequency domain techniques, the state-space approach is ideally suited to handle multivariate processes. Some additional implementation-oriented properties are given in Anderson and Moore (1979, pp. 41–43).

Since several other individuals had preceded Kalman in developing the recursion given in Eq. (5.2), why did the algorithm come to be called the Kalman filter? It seems that several factors contributed to this:

1. Kalman's accessible notation and general statement, formulation, and solution of the estimation problem (in contrast to being discussed in the context of a specific application, which would tend to obscure the discussion for those not familiar with the application).
2. The important connection of the signal estimation problem to the concept of state-space models, which were becoming increasingly prominent in the burgeoning field of control theory.
3. Related to (2), the duality of filtering to an optimal control problem, which allowed the specification of important technical conditions ("observability" and "controllability") under which the filter algorithm would be well behaved (in particular, being such that initial condition errors would damp out over time and such that the error covariance matrix P_k would be uniformly bounded in magnitude). Aside from the 1960 and 1961 papers, see Kalman (1963) for further work in this area.

4. The explosion in popularity of digital computers, for which the recursive Kalman filter was ideally suited, in the early 1960s.
5. A massive scientific program (Apollo) with a need in the early 1960s for just such an algorithm (see Section 6 below).

Evidently, Kalman was responsible for (1), (2), and (3), while (4) and (5) were more fortuitous.

6. CONTRIBUTIONS OF THE AEROSPACE COMMUNITY

Let us now elaborate on the last of the five items mentioned above, which played such a critical role in the development and popularity of the Kalman filter. Almost immediately after Kalman (1960) appeared, there was interest from the aerospace community in applications of the procedure. Kalman met with individuals at the NASA Ames Research Center in the fall of 1960 to discuss the potential application of his work in the navigation and guidance problem for a possible lunar mission. Although there was some initial confusion, it became clear by early 1961 that the procedure would represent a significant improvement over other techniques that had been examined such as W-K filtering and polynomial fitting.

In May 1961, President J. F. Kennedy stated in a special message to Congress:

> I believe that this nation should commit itself to achieve the goal, before this decade is out, of landing a man on the moon and returning him safely to Earth.

With this admonishment, the pace of research activity on the Apollo lunar program picked up considerably. In August 1961, NASA extended the first major Apollo contract to the M.I.T. Instrumentation Laboratory (now independent of M.I.T. as the Charles S. Draper Laboratory) to develop an on-board navigation and guidance system. This system was to rely on state-space/Kalman filtering ideas.

As with virtually any practical application of the KF, the standard filter equations were not simply implemented in "cookbook" form. Rather, several difficult problems had to be addressed. These included having to account for the underlying nonlinear differential equations of motion (via the so-called extended Kalman filter) and developing an algorithm that was numerically stable for use in an *on-board*, limited-capacity computer (this 1960s-vintage processor was limited to one cubic foot of space in the capsule!).

As it became apparent that the Kalman filter approach was going to be successful, other sectors of the aerospace community took note. In particular, during the mid 1960s it was decided that the navigation systems for the C5A transport plane and Boeing 747 would also be based on a KF approach. Today essentially all transoceanic commercial and military aircraft have an on-board KF-like algorithm for navigation and guidance.

The reader interested in applications of the KF in aerospace problems may wish to consult Battin (1982), Schmidt (1981), or Hutchinson (1984).

7. STATE-SPACE MODELS AND THE KALMAN FILTER IN ECONOMETRICS AND STATISTICS

It was not until the early 1970s that the KF/state-space approach achieved much recognition outside of the engineering community. Perhaps the first such area in which a significant amount of work was performed was in econometrics. The principal reason for much of the early interest in the KF among econometricians arose from the problem of modeling and estimation in a regression model when the coefficients are treated as random variables (although a different type of problem [inventory control] was addressed in Taylor, 1970). The need for this random-coefficient regression framework arises, for example, when the relative contribution of microeconomic effects on macroeconomic observables change randomly over time. Some of the early results on the application of KF ideas to the random coefficient problem were presented at a symposium sponsored by the National Bureau of Economic Research in January 1973; see, especially, Rosenberg (1973a, b) and Sarris (1973). The bibliography of Johnson (1977) also includes a number of the early contributions in this area.

The first treatment of state-space models and the KF that appeared in the statistics literature seems to be that given in the textbook by Hannan (1970, pp. 180–186). The first paper on the KF that appeared in the statistics literature is apparently Duncan and Horn (1972), which essentially rederives the KF equations from the viewpoint of least-squares regression theory (similar to Swerling, 1959). By the mid to late 1970s, a sprinkling of papers containing methodological contributions began appearing regularly in the statistics literature (for example, Harrison and Stevens 1976), while by the mid-1980s the general area of state-space modeling and Kalman filtering had become a mature area in statistics and econometrics. Several important areas of application for state-space and KF ideas have been spawned by problems in these areas. These include (among others) seasonal adjustment, spline smoothing and interpolation, time series hypothesis testing, and econometric intervention analysis.

8. CONCLUSIONS

The fact that state-space models and Kalman filtering represent a mature area does not signify a paucity of open problems. In particular, as indicated by the range of topics in this book, areas of active research interest include:

Nonlinear and/or non-Gaussian modeling and estimation, including robust methods
System identification (order determination, parameter estimation, model validation)
Fault detection and isolation (detecting abrupt or gradual changes in models, and, if necessary, modifying the models accordingly) and the closely related area of intervention analysis.

State-space modeling and Kalman filtering is now the approach of choice for a wide range of time series problems. Their flexibility in handling multivariate data and nonstationary processes provides a significant advantage over traditional time series techniques for many "real-world" applications.* In fact, applications of state-space models and Kalman filtering now appear in areas as diverse as geophysical exploration, biomedicine, demography, nuclear power plant failure detection, and macroeconomic forecasting, in addition to the more traditional aerospace applications.

ACKNOWLEDGMENTS

Partial financial support for the preparation of this overview was provided by a JHU/APL Janney Fellowship and U.S. Navy Contract N00039-87-C-5301. I have benefited from conversations with C. F. Ansley, J. W. Follin, J. E. Hanson, and H. Tsurumi.

REFERENCES

Aasnaes, H. B. and Kailath, T. (1973), An innovations approach to least squares estimation—Part VII: Some applications of vector autoregressive moving average models, *IEEE Trans. Auto Control*, AC-18, 601-607.
Anderson, B. D. O. and Moore, J. B. (1979), *Optimal Filtering*, Prentice-Hall, Englewood Cliffs, N.J.

*There are, however, still many instances in which other time series techniques are superior. Aasnaes and Kailath (1973), for example, discuss a situation involving sequentially correlated measurement noise where the ARMA approach is superior to the state-space approach.

Bayesian Analysis of Time Series and Dynamic Models

1

Time Series: A Bayesian Analysis in the Time Domain

LYLE BROEMELING* Mathematical Sciences Division, Office of Naval Research, Arlington, Virginia

SAMIR SHAARAWY Faculty of Economics and Political Science, Cairo University, Cairo, Egypt

1. INTRODUCTION

This chapter consists of two components. After a brief review of Bayesian contributions, we describe a Bayesian methodology to implement a complete time series analysis. Important Bayesian studies done since 1939 to the present are surveyed. The proposed Bayesian methodology is based on ARMA processes and is implemented by referring to an appropriate version of Bayes' theorem. Future trends in time series analysis are also outlined briefly.

2. REVIEW OF THE LITERATURE

Why a Bayesian approach to the analysis of time series? Is one needed? These questions are appropriate in view of the fact that there are many non-Bayesian methods. As will be seen, the Bayesian analysis proposed in this chapter implements in the time domain a complete analysis and uses only one

*Present address: Office of Academic Computing and Biostatistics, University of Texas Medical Branch, Galveston, Texas

theorem, Bayes' theorem. This approach has several advantages and the most obvious is pedagogical. It is much easier to learn the Bayes methodology once one has mastered the inferential interpretation of Bayes' theorem. On the other hand, with a non-Bayesian analysis one must learn a large variety of sampling theory techniques. Thus the Bayes approach unifies and simplifies the analysis of time series.

The statistical analysis of time series is a fascinating topic and is found in all areas of scientific endeavor. The review of the literature presented here is limited to those Bayesian analyses that use autoregressive moving average (ARMA) processes as a foundation and will not cover other important approaches, such as those based on linear dynamic (i.e., state-space) models. The literature on time series is vast and most of it is non-Bayesian (see, e.g., the *Journal of Time Series Analysis*, first published in 1980). The reader is referred to Box and Jenkins (1970) and to Priestley (1981) for the "standard" theory and methodology. There are no textbooks that are devoted exclusively to Bayesian time series, although Zellner (1971) and Broemeling (1985) do introduce the subject for some special ARMA processes and econometric models.

Most of the Bayesian contributions have occurred since 1970, but it seems that Jeffreys (1939) was the first to study spectral theory, and it was not until Box and Jenkins (1970) and then Zellner (1971) that a systematic study of time series began. Zellner derived the posterior and predictive distributions for first- and second-order autoregressive processes using a vague prior density and gave a complete analysis for regression models with autocorrelated errors and for distributed lag models. It is interesting to note that earlier Box and Jenkins (1970) had introduced a Bayesian analysis for the autoregressive process; however, their book focused on non-Bayesian methods.

In 1973, Newbold made an important contribution by his analysis of ARMA-type transfer function models. This represented a significant step forward because in less than two years the field had advanced from some special cases of ARMA processes to models that contain a single ARMA process as a special case. Newbold's results were based on a t-approximation for the posterior analysis, as did the later work of Zellner and Reynolds (1978). During this period, Bayesian forecasting was advanced by Chow (1975), who found the moments of the predictive distribution of vector autoregressive processes. In a related study, Litterman (1980) proposed a Bayesian forecasting technique that was based on a clever choice of the prior distribution. At the end of the 1970s, very little or no work had appeared on vector ARMA processes and on the frequency domain (spectral analysis) approach to time series.

It has been since 1980 that Bayesian techniques have begun to show that they did offer an attractive alternative to the popular Box-Jenkins

Battin, R. H. (1982), Space guidance evolution—A personal narrative, *J. Guidance*, 5, 97–110.

Booton, R. C. (1952), An optimization theory for time-varying linear systems with nonstationary statistical inputs, *Proc. IRE*, 40, 977–981.

Broemeling, L. D. (1985), *Bayesian Analysis of Linear Models*, Marcel Dekker, Inc., New York.

Bucy, R. S. (1958), Notes on finite time filtering with a derivation of the variance equations, The Johns Hopkins University, Applied Physics Laboratory (JHU/APL) memorandum BBD-500 (29 September) (declassified, formerly confidential).

Carlton, A. G. and Follin, J. W. (1956), Recent developments in fixed and adaptive filtering, *Proc. Second AGARD Guided Missiles Seminar (Guidance and Control)*, AGARDograph 21, pp. 285–300.

Diderrich, G. T. (1985), The Kalman filter from the perspective of Goldberger-Thiel estimators, *Am. Statistician*, 39, 193–198.

Duncan, B. and Horn, S. D. (1972), Linear dynamic recursive estimation from the viewpoint of regression analysis, *J. Am. Statist. Assoc.*, 67, 815–821.

Hald, A. (1981), T. N. Thiele's contributions to statistics, *Int. Statist. Rev.*, 49, 1–20.

Hannan, E. J. (1970), *Multiple Time Series*, Wiley, New York.

Hanson, J. E. (1957a), Some theoretical aspects of optimum track-while-scan smoothing and prediction for both the transient and steady-state cases, JHU/APL memorandum BBD-285 (1 July).

Hanson, J. E. (1957b), Some notes on the application of calculus of variations to smoothing for finite time, JHU/APL memorandum BBD-346 (31 October).

Harrison, P. J. and Stevens, C. F. (1976), Bayesian forecasting, *J. Roy. Statist. Soc.*, Ser. B, 38, 205–228.

Ho, Y. C. and Lee, R. C. K. (1964), A Bayesian approach to problems in stochastic estimation and control, *IEEE Trans. Auto. Control*, AC-9, 333–339.

Hutchinson, C. E. (1984), The Kalman filter applied to aerospace and electronic systems, *IEEE Trans. Aero. Elect. Syst.*, AES-20, 500–504.

Jazwinski, A. H. (1970), *Stochastic Processes and Filtering Theory*, Academic, New York.

Johnson, L. W. (1977), Stochastic parameter regression: An annotated bibliography, *Int. Statist. Rev.*, 45, 257–272.

Kalman, R. E. (1960), A new approach to linear filtering and prediction problems, *Trans. ASME, J. Basic Engineering*, 82, 35–45.

Kalman, R. E. (1963), New methods in Wiener filtering theory, *Proc. 1st Symp. on Engineering Applications of Random Function Theory*, Wiley, New York, 270–388.

Kalman, R. E. and Bucy, R. S. (1961), New results in linear filtering and prediction theory, *Trans. ASME, J. Basic Engineering*, 83, 95–108.

Kolmogorov, A. N. (1941), Interpolation and extrapolation of stationary random sequences, *Bull. de l'academie des sciences de U.S.S.R.*, Ser. Math., 5, 3-14 (see also translation: RAND Corp. report RM-3090-PR, 1962).

Lauritzen, S. L. (1981), Time series analysis in 1880: A discussion of contributions made by T. N. Thiele, *Int. Statist. Rev.*, 49, 319–331.

Levinson, N. (1947), The Wiener RMS error criterion in filter design and prediction, *J. Math. and Physics*, 25, 261–278.

Pagan, A. (1975), A note on the extraction of components from time series, *Econometrica*, 43, 163–168.

Plackett, R. L. (1950), Some theorems in least squares, *Biometrika*, 37, 149–157.

Rhodes, I. B. (1971), A tutorial introduction to estimation and filtering, *IEEE Trans. Auto. Control*, AC-16, 688–706.

Rosenberg, B. (1973a), A survey of stochastic parameter regression, *Ann. Econ. Soc. Meas.*, 2, 381–397.

Rosenberg, B. (1973b), The analysis of a cross section of time series by stochastically convergent parameter regression, *Ann. Econ. Soc. Meas.*, 2, 399–428.

Sarris, A. H. (1973), Kalman filter models: A Bayesian approach to estimation of time-varying regression coefficients, *Ann. Econ. Soc. Meas.*, 2, 501–523.

Schmidt, S. F. (1981), The Kalman filter: Its recognition and development for aerospace applications, *J. Guid. Control*, 4, 4–7.

Seal, H. L. (1967), The historical development of the Gauss linear model, *Biometrika*, 54, 1–24.

Sorenson, H. W. (1970), Least squares estimation: From Gauss to Kalman, *IEEE Spectrum*, 7, 63–68.

Sorenson, H. W. (1980), *Parameter Estimation*, Marcel Dekker, New York.

Spall, J. C. and Wall, K. D. (1984), Asymptotic distribution theory for the Kalman filter state estimator, *Commun. Statist.—Theory Meth.*, 13, 1981–2003.

Stigler, S. M. (1986), *The History of Statistics: The Measurement of Uncertainty Before 1900*, Harvard, Cambridge.

Swerling, P. (1958), A proposed stagewise differential correction procedure for satellite observations, RAND Corp. Paper P-1292 (8 January).

Swerling, P. (1959), First order error propagation in a stagewise smoothing procedure for satellite observations, *J. Astronaut. Sci.*, 6, 46–52.

Swerling, P. (1971), Modern state estimation methods from the viewpoint of the method of least squares, *IEEE Trans. Auto. Control*, AC-16, 707–719.

Taylor, L. (1970), The existence of optimal distributed lags, *Rev. Econ. Stud.*, 37, 95–106.

Thiel, H. and Goldberger, A. S. (1961), On pure and mixed statistical estimation in economics, *Int. Econ. Rev.*, 2, 65–78.

Thiele, T. N. (1880), *Sur la compensation de quelques erreurs quasi-systematiques par la methode des moindres carrees*, Reitzel, Copenhagan.

Welch, M. E. (1987), A Kalman filtering perspective, *Am. Statistician*, 41, 90–91 (letter to editor).

Wiener, N. (1949), *Extrapolation, Interpolation, and Smoothing of Stationary Time Series*, MIT Press, Cambridge, and Wiley, New York (jointly published).

Zadeh, L. A. and Ragazzini, J. R. (1950), An extension of Wiener's theory of prediction, *J. Appl. Phys.*, 21, 645–655.

James C. Spall

methodology. This was initiated by Monahan (1983), who used numerical integration to implement a complete time series analysis, including identification, diagnostic checking, estimation, and prediction. This was the first Bayesian attempt to perform a comprehensive analysis and was a valuable contribution.

On the analytical side (as compared to purely numerical), it was well known that the autoregressive process is easily analyzed with the normal-gamma posterior analysis. The challenging problem during the early 1980s was to find a purely analytical analysis of moving average and mixed models. If one considers a moving average process, the likelihood function is nonlinear in the coefficients of the process, and one is forced to use either a numerical procedure (which Monahan did) or an analytical approximation, as was done by Newbold (1973). These earlier approximations had not been checked for their accuracy and it was unknown how they would perform and how they would compare to non-Bayesian competitors. With this in mind, Broemeling and Shaarawy (1984, 1986) and Shaarawy and Broemeling (1985) proposed an approximation for the analysis of MA and ARMA processes. The likelihood function was approximated by a normal-gamma form and thus easily combined with a normal-gamma prior density. The approximation provided accurate inferences when using artificial data; however, a theoretical study of the accuracy of the approximation has not been undertaken.

There have been a few studies on structural change in time series, including the work of Salazar (1982), who worked with AR(1) and AR(2) models and found the posterior mass function of the change point. He reported similar findings for the regression model with autocorrelated errors. Smith (1980) devised a Bayesian test to detect a change in the parameters of an ARMA process, and Tsumuri (1982) developed Bayesian methods to study structural shifts in the simultaneous equation models of econometrics. These are important studies because Bayesian inferences for structural change aid one in checking the adequacy of the assumed model and serve as diagnostic checks for the analysis. Some recent articles in this area are in the special issue of the *Journal of Econometrics* (1982) and the book by Broemeling and Tsumuri (1986).

The traditional way to model a time series is with an additive model of trend, seasonal, and error components, and Gersch (Chapter 16, this volume), using a smoothness prior, has given a Bayesian analysis. This review has not emphasized the Bayesian contributions to time series given in the economics literature; thus the reader should consult Zellner (1977, 1983) for additional information.

As the reader may have noticed, most Bayesian approaches to time series are restricted to the time domain, but a notable exception is Shore (1980).

Using an approximate likelihood function, he provided Bayesian inferences for the spectral density of covariance stationary stochastic process. Although his results are approximate, they appear to be accurate and apply to a large class of time series models. More recently, Cook (1985) has performed a spectral analysis for autoregressive processes. He constructs credible regions for the spectral density at any frequency and the inferences are exact. The classical approach gives approximate confidence intervals because they are based on asymptotic arguments.

A review of the literature is not complete without a reference to the future trends in the field. Linear processes such as ARMA models have been very successful in representing a wide variety of series; however, as pointed out by Tong (1983), they are inadequate when one studies such nonlinear behavior as amplitude-frequency dependence, limit cycle, and jump phenomena. Models such as the bilinear, threshold autoregressive, and exponential autoregressive have recently been proposed to explain nonlinear behavior, and as far as this author knows, no Bayesian analyses have been made. The "new" era of time series analysis offers us many exciting possibilities.

3. ARMA PROCESSES AND THE BOX-JENKINS METHODS

A Bayesian analysis of time series is based on some special parametric models, including regression models with autocorrelated errors, distributed lag models, autoregressive processes, and others. Box and Jenkins (1970) developed a methodology for ARMA(p, q) processes

$$\phi(B)Y(t) = \theta(B)e(t), \tag{3.1}$$

where $Y(t)$ is the tth observation, $t = \ldots, -1, 0, 1, \ldots$, $e(t)$ is a sequence of independent normally distributed random variables with $e(t) \sim N(0, \tau^{-1})$, $\tau > 0$, where τ is unknown. The polynomial operators are

$$\phi(B) = 1 - \phi_1 B - \cdots - \phi_p B^p, \tag{3.2}$$

where the ϕ_i are real unknown parameters, p is a nonnegative integer, and

$$\theta(B) = 1 - \theta_1 B - \cdots - \theta_q B^q, \tag{3.3}$$

where θ_j are real unknown coefficients and B is a backshift operator defined by

$$B^s Y(t) = Y(t - s), \qquad s = 0, 1, 2, \ldots.$$

A discrete parameter stochastic process is said to be stationary in the strict sense if for all positive integers K and K integers $t_1, t_2, \ldots, t_k$, the distribution of $Y(t_1), \ldots, Y(t_k)$ is the same as the distribution of $Y(t_1 + h), \ldots, Y(t_k + h)$, where $h > 0$.

For example, the ARMA(1, 1) process is stationary if $|\phi_1| < 1$ and invertible if $|\phi_1| < 1$. Invertibility is imposed so that one may identify a unique model. If invertibility is not imposed, there are two ARMA(1, 1) models that give rise to the same autocorrelation function. The autocorrelation function of a stationary ARMA process is known and is compared to the sample autocorrelation function; thus stationarity plays an important role in the Box-Jenkins methodology.

A significant contribution of Box and Jenkins (1970) to time series analysis was to assume that the series (or a transformation of the series) could be represented by a parsimonious stationary and invertible ARMA processes and to delineate the four phases of an analysis: identification (order determination), estimation, diagnostic checking, and forecasting.

The identification of p and q is done by computing the sample autocorrelation function and the sample partial autocorrelation function and comparing them to their theoretical counterparts, which are known for low-order processes. After the model is tentatively identified, say as an ARMA (p, q), the moving average parameters $\theta = (\theta_1, \theta_2, \ldots, \theta_q)^T$, the autoregressive parameters $\phi = (\phi_1, \phi_2, \ldots, \phi_p)^T$, and the residual variance τ^{-1} are estimated by maximum likelihood (MLE) or by nonlinear least squares. The MLE and the least squares estimates may be based on either the full likelihood function or a conditional likelihood. These techniques are given by Priestley (1981, pp. 359–364) and explained in the following Bayesian analysis.

The next phase of the analysis is to check the model to see if it gives a reasonable fit to the data. This is accomplished by a series of diagnostic checks using the estimated residuals. The last phase, which is explained by Priestley (1981, pp. 762–773), is to forecast future observations where the predicted observations are computed recursively from an estimated conditional expectation, namely, the conditional expectation of a future observation given the past data.

The Box-Jenkins (BJ) method implements the four phases of a time series analysis by using a large variety of sampling-theory techniques. For example, the identification of an ARMA process is based on the method of moments by comparing theoretical and sample autocorrelation functions. The last three phases are implemented by using what is essentially the regression techniques of estimation, diagnostic residual checks, and prediction.

Our proposed Bayesian analysis of time series will adopt the BJ approach to the extent that the analysis is to be based on the ARMA class of processes and the four phases of identification, estimation, diagnostic checking, and forecasting. The Bayesian analysis will differ from the BJ methodology in the way the four phases are implemented. They will be implemented by the prior, posterior, and predictive analyses of an ARMA process, given below. In this way, many ad hoc techniques are replaced by one coherent methodology.

4. BAYESIAN ANALYSIS OF ARMA PROCESSES

The Bayesian analysis of a parametric model consists of three stages: the prior, posterior, and predictive stages. As will be shown, these three stages will be used to implement the four phases of a time series analysis.

Consider an ARMA(p, q) process (3.1) which is invertible and stationary and suppose there are n observations $S_n = [Y(1), Y(2), \ldots, Y(n)]^T$; then the residuals are given by

$$e(t) = Y(t) - \sum_{1}^{p} \phi_i Y(t-i) + \sum_{1}^{q} \theta_j e(t-j), \tag{4.1}$$

where $t = p, p+1, \ldots, n$.

By conditioning on the first p observations and letting $e(1) = \cdots = e(p) = e(0) = \cdots = e(p-q-1) = 0$, where $q > p+1$ (see Tiao and Box, 1981, p. 809; Priestley, 1981, p. 360), one may approximate the likelihood function by

$$L(\phi, \theta, \tau \mid S_n) \propto \tau^{(n-p)/2} \exp\left(-\frac{\tau}{2}\right) \sum_{p+1}^{n} e^2(t), \tag{4.2}$$

where $\phi \in R^p$, $\theta \in R^q$, and $\tau > 0$, and the $e(t)$ are given by (4.1). The conditional likelihood function (4.2) is nonlinear in the parameters ϕ and θ, because the residuals $e(t)$ are nonlinear functions of the ϕ_i and ϕ_j; however, the residuals can be estimated by

$$\hat{e}(t) = Y(t) - \sum_{1}^{p} \hat{\phi}_i Y(t-i) + \sum_{1}^{q} \hat{\theta}_j \hat{e}(t-j), \tag{4.3}$$

where $t = p+1, \ldots, n$, $\hat{e}(p-1) = \cdots = \hat{e}(p-q-1) = 0$, and $\hat{\phi}_i$ and $\hat{\theta}_j$ are the nonlinear least squares estimates of $\hat{\phi}_i$ and $\hat{\theta}_j$, respectively, and are found by minimizing the conditional sum of squares

$$\mathrm{SS}(\phi, \theta) = \sum_{p+1}^{n} e^2(t) \tag{4.4}$$

with respect to ϕ and θ over the region of invertibility and stationarity. The conditional least square estimates are found by a nonlinear regression algorithm, which is explained by Harvey (1981, p. 126).

The exponent of (4.2) is made linear in the parameters by approximating (4.2) with

$$L^*(\phi, \theta, \tau | S_n) \propto \tau^{(n-p)/2} \exp\left(-\frac{\tau}{2}\right)$$
$$\times \sum_{p+1}^{n} \left[Y(t) - \sum_{1}^{p} \phi_i Y(t-i) + \sum_{1}^{q} \theta_j \hat{e}(t-j) \right]^2, \qquad (4.5)$$

where $\tau > 0$, $\phi \in R^p$, $\theta \in R^q$, and $\hat{e}(p-1) = \ldots, -\hat{e}(p-q-1) = 0$.

But (4.5) as a function of ϕ and θ is a normal-gamma density; thus an appropriate choice of prior density function is

$$\xi(\phi, \theta, \tau) = \xi_1(\phi, \theta | \tau)\xi_2(\tau), \qquad \tau > 0, \quad \phi \in R^p, \quad \theta \in R^q, \qquad (4.6)$$

where ξ_1 is a normal density $N(\mu, \tau^{-1}, Q^{-1})$ and ξ_2 is a gamma density with parameters $\alpha > 0$ and $\beta > 0$. Of course, ξ is a normal-gamma density function with parameters μ, α, β, and Q which is a positive definite matrix of order $p + q$. When (4.6) is combined with (4.5), the product is a normal-gamma posterior distribution for the parameters. If one is confident of one's prior information, one would specify the prior parameters μ, Q, α, and β, but on the other hand, if one has "little" prior information, one may use Jeffreys' improper prior density

$$\xi(\phi, \theta, \tau) \propto \tau^{-1}, \qquad \tau > 0, \quad \phi \in R^p, \quad \theta \in R^q \qquad (4.7)$$

Since the posterior distribution of ϕ, θ, and τ will be normal-gamma, the joint posterior distribution of ϕ, θ, and τ will factor the same way as in (4.6).

See DeGroot (1970, p. 249) for the Bayesian analysis of linear models. These posterior distributions are approximate in the sense that they are based on the approximate conditional likelihood function (4.5). It remains to specify the parameters of the posterior distribution.

THEOREM 4.1 If the approximate conditional likelihood function (4.5) is combined with the prior density (4.6), the marginal density of ϕ and θ is a multivariate t with $n - p + 2\alpha$ degrees of freedom, location vector

$$E\left[\begin{matrix}\phi\\ \theta\end{matrix}\middle| S_n\right] = (A + Q)^{-1}(B + Q\mu), \qquad (4.8)$$

and covariance matrix

$$(n - p + 2\alpha)(n - p)^{-1} P^{-1} \left[\begin{matrix}\phi\\ \theta\end{matrix}\middle| S_n\right],$$

where the precision matrix is

$$P\left[\begin{matrix}\phi\\ \theta\end{matrix}\middle| S_n\right] = \frac{(n - p + 2\alpha)(A + Q)}{C - (B + Q\mu)^T (A + Q)^{-1}(B + Q\mu)}, \qquad (4.9)$$

where A is symmetric and of order $p + q$ and

$$A = \begin{bmatrix} A_{11} & A_{12} \\ A_{21} & A_{22} \end{bmatrix}. \tag{4.10}$$

For the properties of the multivariate t-distribution, see DeGroot (1970, p. 59).

Furthermore, A_{11} is of order p and has ith diagonal element $\sum_{p+1}^{n} Y^2(t-i)$ and ikth off-diagonal element $\sum_{p+1}^{n} Y(t-i)Y(t-k)$. A_{22} is of order q and has jkth element $\sum_{p+1}^{n} \hat{e}(t-j)\hat{e}(t-k)$. The $p \times q$ matrix A_{12} has ijth element $\sum_{p+1}^{n} Y(t-i)\hat{e}(t-j)$. The $p+q$ column vector B is

$$B = \begin{bmatrix} B_1 \\ B_2 \end{bmatrix}, \tag{4.11}$$

where B_1 is of order p and has ith element $\sum_{p+1}^{n} Y(t-i)Y(t)$, and B_2 is of order q and has jth element $\sum_{p+1}^{n} \hat{e}(t-j)Y(t)$. Finally, the scalar C is

$$C = 2\beta + \mu^T Q\mu + \sum_{p+1}^{n} Y^2(t). \tag{4.12}$$

Also, the marginal density of τ is gamma with parameters α' and β', where

$$\alpha' = (n - p + 2\alpha)|2 \tag{4.13}$$

and

$$2\beta' = C - (B + Q^{\mu})^T(A + Q)^{-1}(B + Q\mu). \tag{4.14}$$

The parameters of the prior density are μ, Q, α, and β, but if one is not willing or unable to determine these parameters and wants to employ the improper density (4.7), Theorem 4.1 can be modified by letting $\beta \to 0$, $Q \to 0$, $\alpha \to -(p+q)/2$ in the joint posterior distribution of all the parameters, which yields:

COROLLARY 4.1 If $n > 2p + q + 1$ and if the approximate conditional likelihood function is combined with Jeffreys' prior density (4.7), the marginal distribution of ϕ and θ is a multivariate t with $n - 2p - q$ degrees of freedom, location vector

$$E\left[\begin{matrix}\phi \\ \theta\end{matrix} \middle| S_n\right] = A^{-1}B, \tag{4.15}$$

and precision matrix

$$P\left[\begin{matrix}\phi \\ \theta\end{matrix} \middle| S_n\right] = \frac{(n - 2p - q)A}{\sum_{p+1}^{n} Y^2(t) - B^T A^{-1} B}. \tag{4.16}$$

Also, the marginal density of τ is gamma with parameters

$$\alpha' = \frac{n - 2p - q}{2}$$

and β' where

$$2\beta' = \sum_{p+1}^{n} Y^2(t) - B^T A^{-1} B.$$

Posterior inferential procedures of estimation and testing are well known and are illustrated in Box and Tiao (1973), Broemeling (1985), DeGroot (1970), and Zellner (1971). These inferential procedures will be employed in the identification and estimation phases of a time series analysis.

The Bayesian predictive density is the foundation from which future observations are forecasted and is the conditional density of a future observation $W_1 = Y(n+1)$ given the past observations S_n. If one employs the approximate conditional likelihood function (4.5) and the normal-gamma prior density, one may show

THEOREM 4.2 The approximate Bayesian predictive distribution of W_1 is a univariate t with $n - p + 2\alpha$ degrees of freedom, location vector

$$E(W_1 \mid S_n) = (1 - B_1^T A^{*-1} B_1)^{-1} B_1^T A^{*-1}(B + Q\mu), \tag{4.17}$$

and precision matrix

$$P(W_1 \mid S_n) = \frac{(n - p + 2\alpha)(1 - B_1^T A^{*-1} B_1)}{C - (B + Q\mu)^T A^{*-1}(B + Q\mu)}, \tag{4.18}$$

where B_1 is of order $p + q$,

$$B_1 = \begin{bmatrix} B_1^{(1)} \\ B_2^{(1)} \end{bmatrix} \tag{4.19}$$

and $B_{(1)}$ is a $p \times 1$ vector with ith element $Y(n + 1 - i)$ and $B_{(2)}^{(1)}$ is a $q \times 1$ vector with jth element $-\hat{e}(n + 1 - j)$. In addition, the matrix A^* is

$$A^* = A + A_1 + Q, \tag{4.20}$$

where A and Q are defined in Theorem 4.1, and A_1 is the symmetric matrix

$$A_1 = \begin{bmatrix} A_{11}^{(1)} & A_{12}^{(1)} \\ A_{21}^{(1)} & A_{22}^{(1)} \end{bmatrix}, \tag{4.21}$$

where $A_{11}^{(1)}$ is of order p and has ijth element $Y(n + 1 - i)Y(n + 1 - j)$. $A_{12}^{(1)}$ is $p \times q$ and ijth element $-Y(n + 1 - i)\hat{e}(n + 1 - j)$, and $A_{22}^{(1)}$ is of order q and has jkth element $\hat{e}(n + 1 - j)\hat{e}(n + 1 - k)$.

Suppose that the improper density (4.7) is used; then one may modify Theorem 4.2 to

COROLLARY 4.2 The approximate Bayesian predictive density of W_1 is a univariate t with $n - 2p - q$ degrees of freedom, location

$$E(W_1 | S_n) = [1 - B_1^T(A + A_1)^{-1}B_1]^{-1}B_1^T(A + A_1)^{-1}B, \tag{4.22}$$

and precision

$$P(W_1 | S_n) = \frac{(n - 2p - q)[1 - B_1^T(A + A_1)^{-1}B_1]}{\sum_{p+1}^{n} Y^2(t) - B_1^T(A + A_1)^{-1}B[1 - B_1^T(A + A_1)^{-1}B_1]B^T(A + A_1)^{-1}B_1 - D}, \tag{4.23}$$

$$D = B^T(A + A_1)^{-1}B.$$

Corollary 4.2 is obtained from Theorem 4.2 by modifying the parameters of the predictive distribution with $\alpha \to -(p + q)/2$, $Q \to 0$, and $\beta \to 0$.

The Bayesian predictive distribution will be used in the forecasting and diagnostic checking phases of the time series analysis. Theorem 4.2 and Corollary 4.2 apply when predicting only one lag ahead; however, it is possible to obtain the predictive distribution of any finite number of observations, say, k lags ahead, but the joint distribution is not a multivariate t. This was shown by Broemeling and Land (1984) for an autoregressive process with a normal-gamma prior, where it was shown the marginal distribution of $Y(n + 1)$ is a t, that of $Y(n + 2)$ given $Y(n + 1)$ is a t, and so on.

In order to forecast two steps ahead with Theorem 4.2 and Corollary 4.2, one may use the following:

COROLLARY 4.3 Assuming the proper prior density (4.6), the approximate conditional predictive distribution of $Y(n + 2)$ given $Y(n + 1) = E(W_1 | S_n)$, (4.17), is a univariate t with $n + 1 - p + 2\alpha$ ($\alpha > 0$) degrees of freedom, where the location and precision parameters are given by (4.17) and (4.18), respectively; however, the quantities A^*, A, B, B_1, and C are modified by letting $n = n + 1$, $Y(n + 1) = E(W_1 | S_n)$, (4.17), and

$$\hat{e}(n + 1) = Y(n + 1) - \sum_1^p \hat{\phi}_i Y(n + 1 - i) + \sum_1^q \hat{\theta}_j \hat{e}(n + 1 - j). \tag{4.24}$$

With the improper prior density (4.7), the approximate conditional predictive distribution of $Y(n + 2)$ given $Y(n + 1) = E(W_1 | S_n)$, (4.22), is a univariate t with $n + 1 - 2p - q$ degrees of freedom, and the location and precision parameters are given by (4.22) and (4.23), respectively, but the quantities A, A_1, B, and B_1 are modified by letting $n = n + 1$, $Y(n+1) = E(W_1 | S_n)$, (4.22), and $\hat{e}(n+1)$ is computed from (4.24). Thus, by a sequence of conditional expectations, one may forecast k steps ahead. If this is done, it is important to remember, for example, that if one is predicting two lags ahead, one is using the mean of the conditional density of $Y(n+2)$ given $Y(n+1) = E(W_1 | S_n)$ and not the mean of the marginal distribution of $Y(n+2)$. See Broemeling and Land (1984) and Chow (1975) for further details.

Theorems 4.1 and 4.2 and their corollaries will now be used to implement the four stages of a time series analysis.

5. TIME SERIES ANALYSIS

Identification, estimation, diagnostic checking, and prediction are the four phases of a time series analysis which are to be implemented by the posterior and predictive distributions of the preceding section. These theorems and corollaries will be sufficient to allow one to perform a complete time series analysis on a series generated by a univariate ARMA process.

Suppose that the data or some transformation of the data are denoted by $S_n = [Y(1), Y(2), \ldots, Y(n)]$ and has been generated by an ARMA(p, q) process, where p and q are unknown $(p, q = 0, 1, 2, \ldots)$ and that the prior information about the parameters is expressed by Jeffreys' prior density (4.7); then the first problem is to estimate p and q. If one uses a proper conjugate density, the hyperparameters μ, ϕ, α, and β of (4.6) must be assigned.

5.1 Identification

Suppose that one knows the maximum value of p is M and that of q, N; then:

1. Determine the marginal posterior distribution of $(\phi_1, \phi_2, \ldots, \phi_M)$ and $(\theta_1, \theta_2, \ldots, \theta_N)$, which is a $M + N$ multivariate t with the parameters given by (4.15) and (4.16) of Corollary 4.1.
2. Determine the orders of p and q of the process by a series of univariate t-tests:
   a. First test H_0: $\theta_N = 0$ versus H_1: $\theta_N \neq 0$, using the marginal posterior distribution of θ_N, which is a univariate t.
   b. If H_0 is not rejective above, test H_0: $\theta_{N-1} = 0$ versus H_1: $\theta_{N-1} \neq 0$ with the conditional distribution of θ_{N-1} given $\theta_N = 0$.
   c. Continue in this fashion until one arrives at the stage when one rejects H_0: $\theta_q = 0$ (versus H_1: $\theta_q \neq 0$) and conclude that the moving average order is q.

d. On the other hand, if the H_0 of (a) is rejected, test H_0: $\phi_M = 0$ versus H_1: $\phi_M \neq 0$.
e. And so on.

Continuing in this fashion, one will arrive at the stage where one will reject $\phi_p = 0$ and $\theta_q = 0$ and conclude that the data were generated by an ARMA(p, q) process, where $0 \leqslant p \leqslant M$ and $0 \leqslant q \leqslant M$. Often, M and N are not in excess of two (see Box and Tiao, 1970). There are many ARMA processes that could have generated the data. This identification technique is illustrated in the next section. Having tentatively identified the model as an ARMA(p, q), one begins the estimation stage.

5.2 Estimation

If the model ARMA(p, q) is tentatively identified, how should one estimate $\phi = (\phi_1, \phi_2, \ldots, \phi_p)^T$, $\theta = (\theta_1, \theta_2, \ldots, \theta_q)^T$, and τ? Of course, Corollary 4.1 will give a complete answer. Since ϕ and θ have a $p + q$ multivariate-t marginal posterior distribution, the moving average parameters can be separated from the autoregressive coefficients, and one can make inferences about ϕ, for example, using a p-dimensional multivariate t for the marginal posterior distribution. One would estimate ϕ with the mean of the marginal posterior distribution and compute confidence regions for ϕ as follows. If the posterior mean of ϕ is μ^* and the posterior precision is P^*, then

$$\{\theta:(\phi - \mu^*)^T P^*(\phi - \mu^*) < pF_{\alpha/2;p,n-2p-q}\} \tag{5.1}$$

is a $1 - \gamma$ HPD region for ϕ, where $F_{\alpha/2;n_1,n_2}$ is the upper $(100\alpha/2)\%$ point of the F-distribution with n_1 and n_2 degrees of freedom.

Another use of the credible region is to test H_0: $\phi = \phi_0$ versus H_1: $\phi \neq \phi_0$, where H_0 is rejected at level γ if ϕ_0 is not a member of the region (5.1). The marginal posterior mean and precision of ϕ can be computed easily, since the marginal posterior distribution of ϕ is a p-dimensional t-distribution with $n - 2p - q$ degrees of freedom. Any subset of the parameter vector (ϕ, θ) has a t-distribution. See DeGroot (1970, Chap. 5) for information about the multivariate-t.

To make inferences about τ or τ^{-1}, the residual variance, one would use the marginal posterior distribution of τ, which from Corollary 4.1 is a gamma with parameters α' and β', where $\alpha' = (n - 2p - q)/2$ and $2\beta' = \sum_{p+1}^{n} Y^2(t) - B^T A^{-1} B$, and one could estimate τ^{-1} by the posterior mean

$$E(\tau^{-1} \mid S_n) = \beta'(\alpha' - 1)^{-1}. \tag{5.2}$$

Regions of highest posterior density for τ must be found numerically because the gamma density is asymmetric.

We have also seen that estimation of the parameters is heavily involved in the identification and diagnostic checking phases of the analysis. Estimation, in the classical sense, implies point and interval estimation of the model parameters, where one is interested primarily in the sampling properties of the estimators. In the Bayesian approach, estimation is given a broader interpretation and means roughly that one inspects the marginal posterior distributions of the parameters to find the more plausible values of the parameters. To this end, the marginal posterior distributions should be plotted and the posterior characteristics (mean, variance, mode, etc.) computed.

This identification procedure is somewhat arbitrary in two aspects: the choice of M and N for the maximum value of the autoregressive and moving average orders, and the order by which the parameters θ and ϕ are tested. This problem is somewhat similar to choosing a subset of regressor variables. Of course, there are many ways to identify the ARMA model, but the one here does not behave in an unreasonable manner. Its sensitivity to M and N as well as the order of testing has not been investigated.

Diaz and Farah (1980) developed an identification for AR processes. With pure processes, either MA or AR, a step-down or backward strategy from high to lower order is quite "natural." However, with mixed processes this naturalness disappears, and one is forced to choose one of many possible paths.

5.3 Diagnostic Checking

An important phase of any analysis is to investigate the adequacy of the model, which was tentatively identified as an ARMA(p, q) process.

As suggested by Box and Jenkins (1970), one may overfit the models by fitting the series to an ARMA$(p + 1, q)$ process and check if $\phi_{p+1} = 0$. This is easily done since ϕ_{p+1} will have a univariate posterior t-distribution and one can test H_0: $\phi_{p+1} = 0$ versus H_1: $\phi_{p+1} \neq 0$. In a similar way, one can increase the moving average order by one and test H_0: $\theta_{q+1} = 0$ versus H_1: $\theta_{q+1} \neq 0$, after fitting an ARMA$(p, q + 1)$. The result will be a model identified as an ARMA(p', q'), where $p' = p$ or $p + 1$ and $q' = q$ or $q + 1$.

5.4 Forecasting

The last phase of a time series analysis is to predict future observations. Often this is the main purpose of a time series study. Thus, after passing through the first three phases, confident that an ARMA(p', q') process has generated the series, one would like to predict $Y(n + 1)$, $Y(n + 2)$, ... Theorem 4.2 and Corollary 4.2 give the approximate Bayesian predictive distribution for one step ahead, and it was shown that $Y(n+1)$ has a univariate t-distribution with mean and precision calculated from (4.22) and (4.23) [if one has an improper prior density (4.7)]. Thus one should plot the predictive density and compute its mean and precision. The mean will provide a point forecast, one step ahead, and the mean along with the variance will yield an interval forecast of

$$E(W_1 | S_n) \pm t_{\gamma/2; n-2p'-q'} \{P(W_1 | S_n)\}^{-1/2}, \tag{5.3}$$

with a confidence of $1-\gamma$, $0<\gamma<1$.

To forecast into the future, one may use Corollary 4.3, which explains how to develop a sequence of conditional (conditional on previous forecasts given by the conditional predictive means) forecasts based on the t-distribution. For example, for two steps ahead, an interval forecast of $Y(n+2)$ is

$$E[Y(n+2)|E(W_1 | S_n), S_n] \pm t_{\gamma/2; n+1-2p'-q'} \{P[Y(n+2)| E(W_1 | S_n), S_n]\}^{-1/2}, \tag{5.4}$$

where the conditional mean and precision of $Y(n+2)$ given $Y(n+1) = E(W_1 | S_n)$ are computed according to Corollary 4.3. The one-step-ahead predictions (5.3) are illustrated in the next section.

6. TIME SERIES ANALYSIS: SERIES A

To illustrate the four phases of a time series analysis, we consider series A of 197 observations given on page 525 of Box and Jenkins (1970), who analyzed this set of data on pages 85–86, 93–94, 145–146, 196, and 224 of their book. The series consists of chemical process concentration readings taken every 2 hours and Box and Jenkins identify the model, using the original observations, as an ARMA(1, 1) and as a MA(1) if the first difference is taken. Both models produce nearly identical forecasts.

6.1 Identification

Suppose that the sample mean of series A is subtracted from each observation and the resulting data $Y(t)$, $t = 1, 2, \ldots, 197$, is assumed to be generated by some ARMA(p, q) process with $p, q = 0, 1, 2$. As outlined in Section 5.1, suppose that the maximum value of p and q are $M = 2$ and $N = 2$, respectively; then the identification procedure is given by the binary decision tree of Figure 1.

First the series is fitted to an ARMA(2, 2) process

$$Y(t) - \phi_1 Y(t-1) - \phi_2 Y(t-2) = e(t) - \theta_1 e(t-1) - \theta_2 e(t-2), \tag{6.1}$$

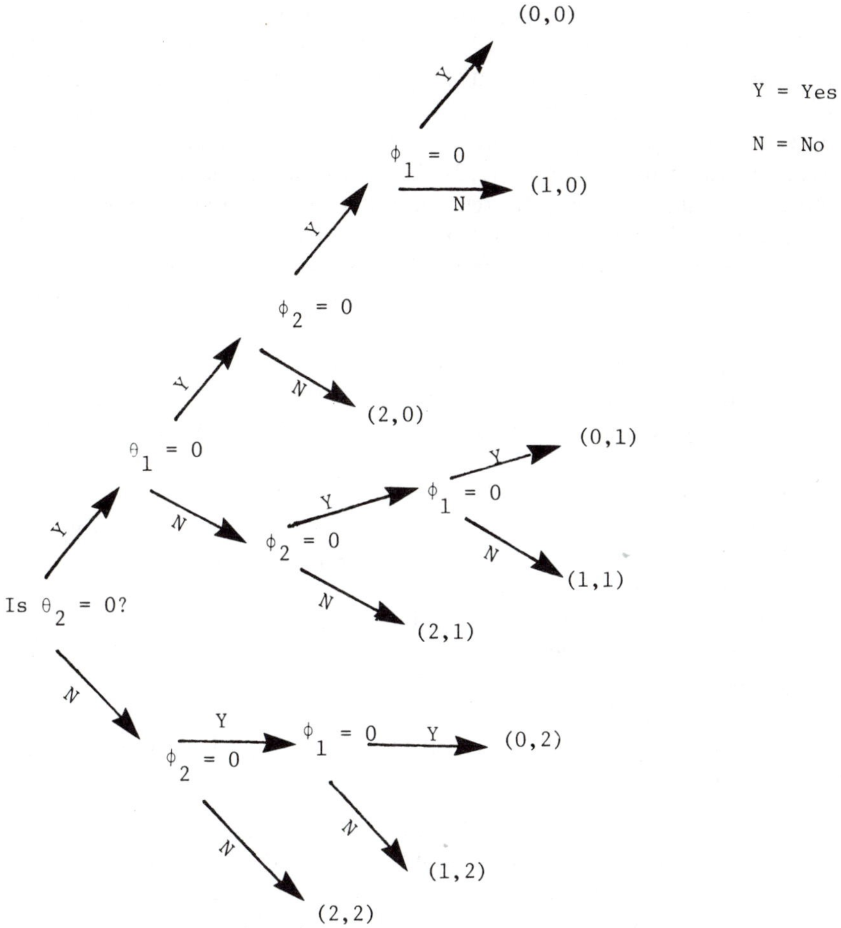

Figure 1. Identification of ARMA process.

where $\phi_i \in R$, $\theta_i \in R$, and the $e(t)$ are independent normal random variables with mean zero and variance τ^{-1}, $\tau > 0$. The first hypothesis to be tested is H: $\theta_2 = 0$ and is to be based on the marginal posterior distribution of θ_2, which is a univariate t distribution. Referring to Figure 1, if this hypothesis is rejected, one tests H: $\phi_2 = 0$, and if it is rejected, one concludes that the data were generated by an ARMA (2, 2). The decision tree depicts the way the other eight decisions may be reached. Note it gives the moving average coefficient θ_2 the first chance of being eliminated.

We now trace the path taken in the decision tree by Bayes' identification of the series A data. The techniques select an ARMA (1, 1) model as follows:

1. Test H_1: $\theta_2 = 0$ versus A_1: $\theta_2 \neq 0$.
   Use the marginal posterior distribution of θ_2, which is t with 191 degrees of freedom, and the 95% HPD region for θ_2 is $(-1.6439, 0.3678)$ and H_1 is not rejected.
2. Test H_2: $\theta_1 = 0$ versus A_2: $\theta_1 \neq 0$.
   Using the conditional posterior distribution of θ_1 given $\theta_2 = 0$ gives a 95% HPD region of (0.1106, 1.0591); thus H_2 is rejected.
3. Test H_3: $\phi_2 = 0$ versus A_3: $\phi_2 \neq 0$.
   The 95% HPD region for ϕ_2 based on the conditional posterior distribution of ϕ_2 given $\theta_2 = 0$ is $(-0.3349, 0.2271)$ and H_3 is not rejected.
4. Test H_4: $\phi_1 = 0$ versus A_4: $\phi_1 \neq 0$.
   This gives a 95% HPD region of (0.7025, 1.0592) based on the conditional posterior distribution of ϕ_1 given $\phi_2 = 0$ and $\theta_1 = 0$.

Referring to Figure 1, the decision path taken above leads to an ARMA (1, 1) process for tentative identification of the model and thus is in agreement with Box and Jenkins (pp. 189, 196). If the first difference between observations is taken, the Box-Jenkins analysis leads to an MA (1) model, which was the process identified by us; however, the details are not reported here.

6.2 Estimation of Parameters

Assuming that an ARMA (1, 1) model did indeed generate the data, how does one estimate the parameters ϕ_1, θ_1 and τ of the model

$$Y(t) - \phi_1 Y(t-1) = e(t) - \theta_1 e(t-1), \tag{6.2}$$

where $\phi_1 \in R$, $\theta_1 \in R$, and $\tau > 0$, and $t = \ldots, -1, 0, 1, \ldots$?

We appeal to Corollary 4.1, which describes the posterior analysis of the ARMA (p, q) process. First a nonlinear least squares algorithm is employed to determine the least squares estimates of ϕ_1 and θ_1 and the estimates are checked for stationarity and invertibility. The nonlinear least squares estimates were $\hat{\phi}_1 = 0.9$ and $\hat{\theta}_1 = 0.5$ and were computed by the ETS package of SAS (Statistical Analysis System). These estimates were then used to estimate the residuals via equation (4.3), after which we used Corollary 4.2 to compute the characteristics of the posterior distribution of ϕ_1, θ_1, and τ, with the following results:

$$\begin{aligned} E(\phi_1 | S_n) &= 0.87 \\ E(\theta_1 | S_n) &= 0.48 \\ P(\phi_1 | S_n) &= 119.69 \\ P(\theta_1 | S_n) &= 74.35 \\ \rho(\phi_1, \theta_1 | S_n) &= 0.78 \\ E(\tau^{-1} | S_n) &= 0.09 \\ \operatorname{var}(\tau^{-1} | S_n) &= 0.000104. \end{aligned} \tag{6.3}$$

For example, the posterior marginal distribution of ϕ_1 is a univariate t with $n - 2p - q = 196 - 2(1) - 1 = 193$ degrees of freedom, mean of 0.87, and precision 119.69 (variance of 0.008). The marginal posterior distribution of ϕ_1 and ϕ_1 is a bivariate t with 196 degrees of freedom and the correlation between them is 0.7871. The marginal posterior distribution of τ is a gamma and the posterior mean of the residual variance τ^{-1} is 0.09 and the posterior variance of τ^{-1} is 0.0001.

Consider a 95% HPD estimate of ϕ_1, which is given by

$$E(\phi_1 | S_n) + t_{0.025,193}(\sqrt{P(\phi_1 | S_n)})^{-1} = (0.69, 1.05) \tag{6.4}$$

which implies that one would reject the hypothesis $\phi_1 = 0$ at the 5% level. The results of the posterior analysis can be computed with the estimation of Box and Jenkins (1970, p. 239).

6.3 Diagnostic Checking

Can one confirm that the data were indeed generated by an ARMA(1, 1) model? How should the initial identification be checked? These and other questions are part of the diagnostic checking phase of the analysis.

We demonstrate an overfitting procedure that fits an ARMA(2, 1) to the data and tests the hypothesis H: $\phi_2 = 0$ versus A: $\phi_2 \neq 0$, where the fitted model is

$$Y(t) - 0.98Y(t-1) + 0.06Y(t-2) = e(t) - 0.60e(t-1) \tag{6.5}$$

for $t = \ldots, -1, 0, 1, \ldots$. The posterior properties of $(\phi_1, \phi_2, \theta_1)$ are given by

$$\begin{aligned} E(\phi_1 \mid S_n) &= 0.98 \\ E(\phi_2 \mid S_n) &= -0.065 \\ E(\theta_1 \mid S_n) &= 0.60 \\ \operatorname{var}(\phi_1 \mid S_n) &= 0.05 \\ \operatorname{var}(\phi_2 \mid S_n) &= 0.02 \\ \operatorname{var}(\theta_1 \mid S_n) &= 0.05 \end{aligned} \tag{6.6}$$

The test statistic for H versus A is based on the marginal posterior distribution of ϕ_2 (which is a t with 192 degrees of freedom, location -0.06, and variance 0.02) and has a value of -0.45737; thus H is not rejected, and the tentative identification is confirmed with this particular diagnostic check.

An overfitting procedure in the other direction increases the moving average order by one and fits an ARMA(1, 2) model to the data and tests H: $\theta_2 = 0$ versus A: $\theta_2 \neq 0$, where

$$Y(t) - \phi_1 Y(t-1) = e(t) - \theta_1 e(t-1) - \theta_2 e(t-2),$$

$$t = \ldots, -1, 0, 1, \ldots. \tag{6.7}$$

We found that H could not be rejected at the 5% level. Thus the overfitting methods confirms the initial identification of an ARMA(1, 1) model.

6.4 Forecasting

The basis for forecasting one step ahead is Corollary 4.2, which described the Bayesian predictive density $Y(n+1)$. Using $n = 196$ observations of series A (with the mean subtracted), the predictive distribution of $Y(197)$ is a univariate t with 193 degrees of freedom predictive mean

$$E[Y(197) \mid S_n] = 0.30$$

and predictive precision

$$P[Y(197) \mid S_n] = 9.74.$$

A 95% prediction interval for $Y(197)$ is

$$0.30 + (1.96)(9.74)^{-1/2} = (-0.31, 0.93)$$

and the 197 observation of series A is 0.2; thus the actual observation is in the 95% prediction interval and the precision or prediction error is 0.1088.

7. SUMMARY AND CONCLUSIONS

The aim of this chapter has been to show the reader that a Bayesian approach to time series is easily accomplished. The four stages of identification, estimation, diagnostic checking, and forecasting are each implemented by an appropriate version of Bayes' theorem, given by Theorem 4.1 and 4.2 and their corollaries. Thus one advantage of our method is pedagogical—it is much easier to learn Bayesian methodology than the large variety of sampling-theory techniques of other analyses.

Bayesian methods of time series are being developed and the methods presented here will be revised and expanded. There are many opportunites for research. For example, in the area of diagnostic checking, overfitting and underfitting were the only two ways proposed to check the model; however, the main assumption of the analysis is that of stationarity. The original observations or a transformation (usually by differencing) of the observations are assumed to be a realization of a stationary ARMA process, but the diagnostic tests do not explicitly test for stationarity.

Stationarity could be tested by using locally stationary processes as alternatives. For example, an ARMA process with a changing mean but a constant autocorrelation would be one alternative to stationarity, and a diagnostic test is possible using this approach. Some results in this direction have been made by Broemeling and Tsurumi (1986), but much remains to be done. This is just one of many areas where future research will produce improved methods for time series.

ACKNOWLEDGMENT

The Office of Naval Research supported the authors under Contract N00014-82-K-0292.

REFERENCES

Box, G. E. P., and G. Jenkins (1970). *Time Series Analysis, Forecasting and Control*, Holden-Day, San Francisco.

Box, G. E. P., and G. C. Tiao (1973). *Bayesian Inference in Statistical Analysis*, Addison-Wesley, Reading, Mass.

Broemeling, L. D. (1985). *The Bayesian Analysis of Linear Models*, Marcel Dekker, New York.

Broemeling, L. D., and M. Land (1984). On forecasting with univariate autoregressive processes: a Bayesian approach, *Commun. Stat. Theory Methods*, 13(11), 1305–1320.

Broemeling, L. D., and S. Shaarawy (1984). Bayesian inferences and forecasts with moving average processes, *Commun. Stat., Theory Methods*, 13(5), 1871–1888.

Broemeling, L. D., and S. Shaarawy (1986). A Bayesian Analysis of Time Series, in *Bayesian Inference and Decision Techniques* (P. Goel and A. Zellner, eds.), Elsevier, New York.

Broemeling, L. D., and H. Tsurumi (1986). *Econometrics and Structural Change*, Marcel Dekker, New York.

Chow, G. C. (1975). Multiperiod Predictions from Stochastic Difference Equations, in *Studies in Bayesian Econometrics and Statistics* (S. Fienberg and A. Zellner, eds.), North-Holland, Amsterdam.

Cook, P. (1985). Bayesian autoregressive spectral analysis, *Commun. Stat. Theory Methods*, 14(5), 1001–1018.

DeGroot, M. H. (1970). *Optimal Statistical Decision*, McGraw-Hill, New York.

Diaz, J., and J. L. Farah (1981). On the Identification of Autoregressive Processes, presented at the *NBER-NSF Seminar on Bayesian Inference in Econometrics*, May, Cornell University, Ithaca, N.Y.

Harvey, A. C. (1981). *Time Series Models*, Wiley, New York.

Jeffreys, H. (1939). *Theory of Probability*, Oxford University Press, Oxford.

Litterman, R. B. (1980). *A Bayesian Procedure for Forecasting with Informative Prior Distribution*, manuscript of Department of Economics, Massachusetts Institute of Technology, Cambridge, Mass.

Monohan, J. F. (1983). Fully Bayesian analysis of ARMA time series models, *J. Econometrics*, 21, 307–331.

Newbold, P. (1973). Bayesian estimation of Box-Jenkins transfer function models, *J. R. Stat. Soc. Ser. B* 35(2), 323–336.

Priestley, M. B. (1981). *Spectral Analysis and Time Series*, Academic Press, London.

Salazar, D. (1982). Structural change in time series, *J. Econometrics* 19(1), 147–164.

Shaarawy, S., and L. Broemeling (1985). Inferences and prediction with ARMA processes, *Commun. Stat. Theory Methods* 14(10), 2029–2040.

Shore, R. W. (1980). Bayesian Approach to the Spectral Analysis of Stationary Time Series, Part II of *Bayesian Analysis in Economic Theory and Time Series Analysis* (A. Zellner and J. B. Kadane, eds.), North-Holland, Amsterdam.

Smith, A. F. M. (1980). Change Point Problems: Approaches and Applications, in *Bayesian Statistics, Proceedings of the First International Meeting,*

Valencia (J. Bernardo, M. DeGroot, D. Lindley, and A. Smith, eds.) University Press, Valencia, Spain.

Tiao, G. C., and G. E. P. Box (1981). Modeling multiple time series with applications, *J. Am. Stat. Assoc.* 76(376), 802–816.

Tong, H. (1983). *Threshold Models in Non-linear Time Series Analysis*, Vol. 21 of *Lecture Notes in Statistics* (K. Krickegerg, ed.), Springer-Verlag, New York.

Tsurumi, H. (1982). A Bayesian maximum likelihood analysis of a gradual switching regression in a simultaneous equation framework, *J. Econometrics* 19(1), 165–182.

Zellner, A. (1971). *An Introduction to Bayesian Inference in Econometrics*, Wiley, New York.

Zellner, A. (1977). Statistical Analysis of Econometric Models, in *American Statistical Association Proceedings of Business of Economics*, pp. 51–53.

Zellner, A. (1983). Application of Bayesian analysis in econometrics, *Statistician* 32, 23–34.

Zellner, A. and R. A. Reynolds (1978). Bayesian Analysis of ARMA Models, presented at the *16th Seminar on Bayesian Inference in Econometrics.*

2

Bayesian Modeling and Forecasting in Autoregressive Models

MARK J. SCHERVISH and RUEY S. TSAY Department of Statistics, Carnegie-Mellon University, Pittsburgh, Pennsylvania

1. INTRODUCTION

Bayesian analysis has enjoyed success in many branches of statistics. Its potential for time series analysis, however, is underdeveloped, because time series analysis often involves highly nonlinear dynamic structure, which causes difficulties in prior specification and in posterior evaluation. The situation is changing, however. There is a growing interest in Bayesian time series analysis in recent years. The advance in statistical computation makes possible a fully Bayesian analysis in time series. Monahan (1983) gave a fully Bayesian treatment of autoregressive moving average (ARMA) models. Although he considered only six simple models, he successfully showed that Bayesian techniques are feasible and could be advantageous in time series analysis. For example, he designed a framework through which prior information can be incorporated and suggested a method for practical computation. Broemeling (1985) and Broemeling and Shaarawy (Chapter 1, this volume) also discuss fully Bayesian analysis of stationary linear time series. Zellner (1972, Chap. 7) discusses a Bayesian analysis of autoregressive models using diffuse priors.

There is Bayesian work in other aspects of time series analysis, too. In the statistics literature, Akaike (1980) proposed a Bayesian version of his famous criterion AIC for selecting hyperparameters involved in a Bayesian setting. Harrison and Stevens (1976), Litterman (1986), and West et al. (1985) showed that Bayesian forecasting is a useful alternative to the traditional non-Bayesian forecasting methods. West (1986) used Bayesian techniques to monitor a time series. Liu and Tiao (1980) adopted a Bayesian method to handle random coefficient autoregressive models. Abraham and Box (1978) used a Bayesian formulation to detect outliers in a univariate time series. More recently, Thompson and Miller (1986) proposed a Bayesian approach to forecasting and showed that better probability intervals can result by using the Bayesian method. These authors correctly pointed out that Bayesian analysis can take into account the uncertainty involved in model changes as well as in parameter estimation. In the control literature, Ho and Lee (1964) considered a Bayesian approach to stochastic estimation and control. Pierce and Sworder (1971) and Sworder (1972) studied Bayes' controller in a linear system with jump parameters. Sanyal and Shen (1974) used Bayes' decision rules in detection and adaptive estimation.

In this chapter we give a fully Bayesian analysis of autoregressive models with or without exogenous variables. We believe that there is no true model for a real-world time series and treat all autoregressive models as candidates, each of which provides an approximation to the underlying series. The adequacy of each autoregressive approximation is measured by its predictive likelihood. Consequently, a better way to forecasting is a weighted combination of forecasts produced by each candidate model and weighted by the corresponding predictive likelihood. Furthermore, we allow for observations to be missing and the process to have random model changes and/or outliers. The Bayesian time series analysis considered, therefore, is different from those available in the time series literature. For instance, the use of mixture models is a difference between the approach used here and that of West et al. (1985). In terms of fully Bayesian analysis, Monahan confined his work to stationary ARMA models and investigated model identification via the AIC criterion. On the other hand, we make no assumption on stationarity of the underlying process and we do not rely on or develop model selection criteria for the models we propose.

The problem of model changes considered in this chapter is related to fault detection in the control literature. A summary of fault detection methods can be found in Willsky (1976) and Basseville and Benveniste (1986). The Bayesian treatment of random model changes adopted here, however, is different from those fault detection methods. For instance, the use of mixture models weighted by predictive likelihoods has not been considered in the control literature, and we use predictive ability instead of likelihood ratio tests or residual-based methods as the criterion in the detection. Most of the likelihood ratio tests and residual-based methods are based on asymptotic results, whereas predictive likelihood is not.

The chapter is organized as follows. Section 2 gives the general Bayesian setting for autoregressive models. Normal distribution with normal prior for location parameters and gamma prior for scale parameters is used. This is a choice of simplicity, not necessity. Other distributions can be used at the cost of heavy numerical integration. Section 3 provides a means by which results of lower-order autoregressive models can be obtained from those of higher-order ones. Section 4 gives the predictive likelihood, which is a natural criterion for comparing predictive models. Section 5 handles the missing value problem. One and multiple missing observations are carefully treated. An algorithm is given for practical analysis. In Section 6 we describe how to handle outliers and model changes. Section 7 includes the results of numerical calculations done with the methods described in this chapter. The proposed methods are compared with non-Bayesian autoregressive models. Akaike's information criterion is used to select the best order in non-Bayesian analysis. Also, we discuss parallel computation for the proposed method. Parallel computation is shown to be highly efficient in handling large time series models.

2. SINGLE AR MODEL WITH EXOGENOUS VARIABLES

2.1 Statement of the Model

We consider a scalar linear model which includes both an autoregressive component and a regression component. In this chapter we use uppercase letters to denote the names of random variables, and lowercase letters to denote their observed values. Let Y_t denote the tth observation of the dependent variable in the time series and let $\mathbf{x}_t$ denote a vector of length q of independent (or exogenous) variables related to Y_t. If there are no exogenous variables, it may be useful to let $q = 1$ so as to fit an AR (0) coefficient, which is not included elsewhere in the model. We assume that one has observed the first l observations before formulating the model. For convenience, let

$$m = n - l.$$

Just before observing Y_t, denote by

$$\mathbf{y}_{t,p} = (y_{t-1}, y_{t-2}, \ldots, y_{t-p})^T$$

the vector of the previous p observed values of the dependent variable. The model takes the following form:

$$Y_t = \mathbf{x}_t^T \gamma + \mathbf{y}_{t,p}^T \beta + \varepsilon_t \qquad \text{for } t > l, \tag{2.1}$$

where ε_t is independent of the past observations with normal distribution $N(0, (\tau v_t)^{-1})$. The vector β contains the p AR coefficients and the vector γ

contains the q regression coefficients. We can write this model in matrix form as

$$\mathbf{Y} = X\boldsymbol{\mu} + \varepsilon,$$

where $\mathbf{Y}$ is the vector of data length m, $\boldsymbol{\mu}$ is the vector of length $q + p$,

$$\boldsymbol{\mu} = \begin{bmatrix} \gamma \\ \beta \end{bmatrix},$$

X is a matrix of dimension $m \times q + p$, and the conditional distribution of ε, given parameters $\boldsymbol{\mu}$ and τ, is multivariate normal

$$N_m(\mathbf{0}, \tau^{-1}V^{-1}). \tag{2.2}$$

In (2.2), V is a known $m \times m$ diagonal matrix with the numbers v_t on the diagonal. The symbol $N_k(\mathbf{a}, B)$ stands for the k-variate normal distribution with mean vector $\mathbf{a}$ and covariance matrix B. The matrix X is partitioned as follows:

$$X = [X_1 \mid X_2],$$

where X_1 is $m \times q$, X_2 is $m \times p$. The $t - l$ row of X_1 is $\mathbf{x}_t^T$, and the corresponding row of X_2 is $\mathbf{y}_{t,p}^T$.

In this chapter the parameter $\boldsymbol{\mu}$ depends on a second-level hyperparameter $\boldsymbol{\theta}$ and the prior distribution is specified hierarchically as

$$\begin{aligned} \boldsymbol{\mu} \mid \boldsymbol{\theta}, \tau &\sim N_{p+q}(A_0\boldsymbol{\theta} + C_0, \tau^{-1}W_0^{-1}) \\ \boldsymbol{\theta} \mid \tau &\sim N_s(\boldsymbol{\theta}_0, \tau^{-1}B_0^{-1}) \\ \tau &> \Gamma\left(\frac{a_0}{2}, \frac{b_0}{2}\right), \end{aligned} \tag{2.3}$$

where A_0, C_0, W_0, $\boldsymbol{\theta}_0$, a_0, $b_{\bar{0}}$ are known hyperparameters, and the symbol $\Gamma(a, b)$ stands for the *gamma distribution*, which has probability density function

$$f(t) = \frac{b^a}{\Gamma(a)} t^{a-1} \exp(-bt) \qquad \text{for } 0 \leqslant t < \infty,$$

and $\Gamma(\cdot)$ stands for the *gamma function*. The dimension s of $\boldsymbol{\theta}$ is the number of hyperparameters one wishes to introduce in the second level of the hierarchy of the distribution of $\boldsymbol{\mu}$. For example, if $q = 1$ and one wishes to model the AR coefficients $\beta_1, \ldots, \beta_p$ as a priori exchangeable, then one could use $s = 1$ hyperparameter and set $A_0^T = (0, 1, 1, \ldots, 1)$. The zero in the first coordinate of A_0 is to allow for the regression parameter γ_1 to not be exchangeable with the AR coefficients. Setting $A_0 = \mathbf{0}$ accomplishes the same goal as removing this level from the hierarchy altogether. Notice that we have treated the first l

of the Y_i observations as known. That is, the distributions above are all conditioned on the first l observations. If one wishes, one can use the first l observations to help formulate the prior distribution, that is, to choose the hyperparameters. The joint density of $\mathbf{Y}$, $\boldsymbol{\mu}$, $\boldsymbol{\theta}$, τ is

$$\frac{(b_0/2)^{a_0/2}|V|^{1/2}|W_0|^{1/2}|B_0|^{1/2}}{(2\pi)^{(m+p+q+s)/2}\Gamma(a_0/2)}\tau^{\delta-1}\exp\left(-\frac{\tau}{2}\{T_1+T_2+T_3+b_0\}\right), \qquad (2.4)$$

where $\delta = (a_0 + m + p + q + s)/2$,

$$T_1 = (\mathbf{Y} - X\boldsymbol{\mu})^T V(\mathbf{Y} - X\boldsymbol{\mu}),$$

$$T_2 = (\boldsymbol{\mu} - A_0\boldsymbol{\theta} - C_0)^T W_0(\boldsymbol{\mu} - A_0\boldsymbol{\theta} - C_0),$$

$$T_3 = (\boldsymbol{\theta} - \boldsymbol{\theta}_0)^T B_0(\boldsymbol{\theta} - \boldsymbol{\theta}_0).$$

One drawback to this model is the prior dependence between the precision τ of the innovation process ε_t and the regression parameters $\boldsymbol{\mu}$. This dependence is assumed for two reasons. The first is computational tractability. The assumption allows calculations to proceed in a manner analogous to least squares. Second, there will be posterior dependence of much the same form as the hypothesized prior dependence after data are observed, so we cannot keep τ and $\boldsymbol{\mu}$ independent for long anyway.

2.2 Posterior Distributions

This section is devoted to finding the posterior distribution of $(\boldsymbol{\mu}, \boldsymbol{\theta}, \tau)$ in the same hierarchical form as the prior and to find the predictive density of the observed data. We begin by simplifying T_1 from (2.4).

$$T_1 = \mathrm{RSS}_0 + (\boldsymbol{\mu} - \hat{\boldsymbol{\mu}})^T F(\boldsymbol{\mu} - \hat{\boldsymbol{\mu}}) = \mathrm{RSS}_0 + T_4, \qquad (2.5)$$

where $F = X^T V X$, $\hat{\boldsymbol{\mu}} = F^{-1}X^T V\mathbf{Y}$, and $\mathrm{RSS}_0 = \mathbf{Y}^T(V - VXF^{-1}X^T V)\mathbf{Y}$ is the weighted least squares residual sum of squares (the weights being the diagonal elements of V). Next we simplify $T_2 + T_4$.

$$T_2 + T_4 = [\boldsymbol{\mu} - \boldsymbol{\mu}^*(\boldsymbol{\theta})]^T W_1[\boldsymbol{\mu} - \boldsymbol{\mu}^*(\boldsymbol{\theta})] + (\boldsymbol{\mu} - A_0\theta - \mathbf{C}_0)^T D(\hat{\boldsymbol{\mu}} - A_0\boldsymbol{\theta} - \mathbf{C}_0), \qquad (2.6)$$

where $W_1 = W_0 + F$, $D = W_0 W_1^{-1} F$, and

$$\boldsymbol{\mu}^*(\theta) = W_1^{-1}[W_0(A_0\boldsymbol{\theta} + \mathbf{C}_0) + F\hat{\boldsymbol{\mu}}].$$

Next, we simplify the last term in (2.6).

$$(\hat{\boldsymbol{\mu}} - A_0\boldsymbol{\theta} - \mathbf{C}_0)^T D(\hat{\boldsymbol{\mu}} - A_0\boldsymbol{\theta} - \mathbf{C}_0) = (\boldsymbol{\theta} - \boldsymbol{\theta}^*)^T E(\boldsymbol{\theta} - \boldsymbol{\theta}^*) + (\mathbf{C}_0 - \hat{\boldsymbol{\mu}})^T[D - DA_0E^{-1}A_0^T D](\mathbf{C}_0 - \hat{\boldsymbol{\mu}}) = T_5 + \mathrm{RSS}_1,$$

where $E = A_0^T D A_0$ and $\boldsymbol{\theta}^* = E^{-1}A_0^T D(\hat{\boldsymbol{\mu}} - \mathbf{C}_0)$.

Next, we simplify $T_3 + T_5$.

$$T_3 + T_5 = (\boldsymbol{\theta} - \boldsymbol{\theta}_1)^T B_1(\boldsymbol{\theta} - \boldsymbol{\theta}_1) + \text{RSS}_2,$$

where $B_1 = B_0 + E$, $\boldsymbol{\theta}_1 = B_1^{-1}(B_0\boldsymbol{\theta}_0 + E\boldsymbol{\theta}^*)$, and

$$\text{RSS}_2 = (\boldsymbol{\theta}^* - \boldsymbol{\theta}_0)^T G(\boldsymbol{\theta}^* - \boldsymbol{\theta}_0)$$

with $G = B_0 B_1^{-1} E$. The exponent of the joint density (2.4) can now be written

$$-\frac{\tau}{2}\{b_1 + [\boldsymbol{\mu} - \boldsymbol{\mu}^*(\boldsymbol{\theta})]^T W_1[\boldsymbol{\mu} - \boldsymbol{\mu}^*(\boldsymbol{\theta})] + (\boldsymbol{\theta} - \boldsymbol{\theta}_1)^T B_1(\boldsymbol{\theta} - \boldsymbol{\theta}_1)\},$$

where

$$b_1 = b_0 + \text{RSS}_0 + \text{RSS}_1 + \text{RSS}_2. \tag{2.7}$$

It follows that the posterior distribution can be written

$$\begin{aligned} &\boldsymbol{\mu}|\mathbf{Y}, \boldsymbol{\theta}, \tau \sim N_{p+q}(A_1\boldsymbol{\theta} + \mathbf{C}_1, \tau^{-1}W_1^{-1}) \\ \boldsymbol{\theta}| \quad &\boldsymbol{\theta}|\tau, \mathbf{Y} \sim N_s(\boldsymbol{\theta}_1, \tau^{-1}B_1^{-1}) \\ &\tau|\mathbf{Y} \sim \Gamma\left(\frac{a_1}{2}, \frac{b_1}{2}\right), \end{aligned} \tag{2.8}$$

where $A_1 = W_1^{-1}W_0A_0$, $\mathbf{C}_1 = W_1^{-1}[W_0\mathbf{C}_0 + F\hat{\boldsymbol{\mu}}]$, and $a_1 = a_0 + m$.

The marginal density of the observed data will be needed later, and it is obtained by integrating the parameters $(\boldsymbol{\theta}, \boldsymbol{\mu}, \tau)$ out of (2.4). The result is

$$f(\mathbf{Y}) = (2\pi)^{-m/2}\frac{(b_0/2)^{a_0/2}\Gamma(a_1/2)|V|^{1/2}|W_0|^{1/2}|B_0|^{1/2}}{(b_1/2)^{a_1/2}\Gamma(a_0/2)|W_1|^{1/2}|B_1|^{1/2}}. \tag{2.9}$$

A slight variation arises when the matrix E is singular. For example, if $s = 1$ so that A_0 is a single column, it may be that $A_0^T D A_0 = 0$ because $A_0 = \mathbf{0}$. In this case, $\text{RSS}_1 = (\mathbf{C}_0 - \hat{\boldsymbol{\mu}})^T D(\mathbf{C}_0 - \hat{\boldsymbol{\mu}})$, $B_1 = B_0$, $\boldsymbol{\theta}_1 = \boldsymbol{\theta}_0$, $\text{RSS}_2 = 0$, and $\boldsymbol{\theta}$ plays no role in the inference since $A_1 = \mathbf{0}$ also. In general, F^{-1} is needed only for calculating $\boldsymbol{\theta}^*$ and RSS_1. When E^{-1} does not exist, one need only guarantee that $D^{1/2}\boldsymbol{\theta}^*$ is the projection of $D^{1/2}(\hat{\boldsymbol{\mu}} - \mathbf{C}_0)$ onto the linear space spanned by the columns of $D^{1/2}A_0$ and that RSS_1 is the squared length of $D^{1/2}(\hat{\boldsymbol{\mu}} - \mathbf{C}_0 - \boldsymbol{\theta}^*)$, where $D^{1/2}$ is any symmetric square root of D. This can be done using generalized least squares with a design matrix not of full rank as described by Seber (1977, Secs. 3.6 and 3.8).

2.3 Updating One Observation at a Time

It may be desirable to fit an autoregression from a large data set and then update the fit as new observations arrive. A naive way to do that is to recalculate the entire fit as each new observation arrives. This is rather wasteful of resources. An alternative is to wait until enough new observations arrive to make any difference.

A third approach is to use Bayes' theorem to update the model as each observation arrives. All one needs is the previous fit (2.8) and a model for the new observation:

$$Y_{n+1} \mid \mathbf{Y}, \boldsymbol{\mu}, \boldsymbol{\theta}, \tau \sim N(\mathbf{x}_{n+1}^T \boldsymbol{\mu}, (\tau v_{n+1})^{-1}),$$

where $\mathbf{x}_{n+1}$ is $q + p$ dimensional and the last p coordinates are the previous p of the Y_i observations. Here, v_{n+1} is a known constant. This model is now the same as that of Section 2.2 with $m = 1$ instead of $m = n - l$. The only part of the model fitting of Section 2.2 that cannot be done in identical fashion with only one observation is the part that requires F^{-1} where $F = X^T V X$. In the case of a single observation,

$$F = v_{n+1} \mathbf{x}_{n+1} \mathbf{x}_{n+1}^T,$$

a $(p + q) \times (p + q)$ matrix of rank 1. But all we really need is to ensure that $T_1 = \mathrm{RSS}_0 + T_4$ as in (2.5). This will be true if we set

$$\hat{\boldsymbol{\mu}} = \frac{\mathbf{x}_{n+1} y_{n+1}}{\mathbf{x}_{n+1}^T \mathbf{x}_{n+1}}$$

and $\mathrm{RSS}_0 = 0$. The remainder of the model fitting is identical with what was done in Section 2.2, using 1 in place of $n - l$ this time. Some time can be saved in calculating the new W_1^{-1} by using the following identity:

$$(W_0 + v_{n+1} \mathbf{x}\mathbf{x}^T)^{-1} = W_0^{-1} - \frac{W_0^{-1} \mathbf{x}\mathbf{x}^T W_0^{-1} v_{n+1}}{1 + v_{n+1} \mathbf{x}^T W_0^{-1} \mathbf{x}}. \tag{2.10}$$

The overall predictive density for all the data is the product of the previous predictive density for $Y_{l+1}, \ldots, Y_n$ times the one just calculated for Y_{n+1} given $Y_{l+1}, \ldots, Y_n$.

3. REDUCING A MODEL

Because one is seldom sure of what the best order is for an AR model, it is often desirable to fit several different order models. This allows one to make use of whatever predictive power each model has. Given a fit for an order p model, it is easy to generate the fit for an order $p - 1$ model. The purpose of this section is to give recursive formulas which can be used to reduce the AR order of a model successively without refitting the lower-order model. In Section 3.1 we derive the fit of an order $p - 1$ model from the existing fit of an order p model. In Section 3.2 we give a simple algorithm for performing the reduction recursively, based on the results from Section 3.1.

3.1 Derivation of Reduction Formulas

Let P_M be the highest-order model of interest. Assume that the highest-order AR coefficient is the last coordinate of $\boldsymbol{\mu}$. For any matrix (or vector) H, let Bot(H) denote the last row of H. Similarly, let Top(H) denote all but the last row of H. For matrices, partition them to isolate that last row and column and denote the upper left corner ULC(H), the upper right corner URC(H), and the lower right corner LRC(H). We will denote all densities $f(\cdot\,|\,\cdot)$, where the items to the left of the vertical bar are the random variables and those to the right have been conditioned on. When no vertical bar exists, $f(\cdot)$ is a marginal density. For example, (2.4) is $f(\mathbf{Y}, \boldsymbol{\mu}, \boldsymbol{\theta}, \tau)$. We prefer to write this as $f(\mathbf{Y}, \text{Top}(\boldsymbol{\mu}), \text{Bot}(\boldsymbol{\mu}), \boldsymbol{\theta}, \tau)$. The marginal of the data $f(\mathbf{Y})$ is (2.9). When we wish to condition on a specific value of a vector, or we wish to evaluate a density at a specific value, we will specify that the vector equals that value. For example, the conditional density of $\boldsymbol{\theta}$ and Bot$(\boldsymbol{\mu})$ given $\mathbf{Y}$ and τ evaluated at Bot$(\boldsymbol{\mu}) = 0$ is denoted $f(\boldsymbol{\theta}, \text{Bot}(\boldsymbol{\mu}) = 0\,|\,\mathbf{Y}, \tau)$. When no confusion will result, we will sometimes substitute the name of the distribution for the density.

Since the highest-order AR coefficient is the last coordinate of $\boldsymbol{\mu}$, we can reduce the order of a model by conditioning on Bot$(\boldsymbol{\mu}) = 0$, since this means that the highest-order AR coefficient is 0, hence the model is of one lower order. To do this conditioning, we need to express the joint conditional distribution

$$f(\text{Top}(\boldsymbol{\mu}), \boldsymbol{\theta}, \tau\,|\,\text{Bot}(\boldsymbol{\mu}) = 0, \mathbf{Y})$$

in hierarchical form, just as we stated the original prior (2.3) and the full model posterior (2.8). That is, we need the following distributions:

$$f(\text{Top}(\boldsymbol{\mu})\,|\,\boldsymbol{\theta}, \tau, \text{Bot}(\boldsymbol{\mu}) = 0, \mathbf{Y}), \tag{3.1}$$

$$f(\boldsymbol{\theta}\,|\,\tau, \text{Bot}(\boldsymbol{\mu}) = 0, \mathbf{Y}), \tag{3.2}$$

$$f(\tau\,|\,\text{Bot}(\boldsymbol{\theta}) = 0, \mathbf{Y}). \tag{3.3}$$

These can all be obtained from what we have already calculated with some small changes. We also need

$$f(\mathbf{Y}\,|\,\text{Bot}(\boldsymbol{\mu}) = 0),$$

which is the predictive density for the reduced model.

We begin by finding (3.1), which can be obtained from the first row of (2.8) using well-known results in multivariate normal distribution theory (see Morrison, 1976, Sect. 3.4). Straightforward application of these results gives

$$\text{Top}(\boldsymbol{\mu})|\text{Bot}(\boldsymbol{\mu}) = 0, \mathbf{Y}, \boldsymbol{\theta}, \tau \sim N_{p+q-1}(A_1^*\boldsymbol{\theta} + \mathbf{C}_1^*, \tau^{-1}W_1^{*-1}),$$

where

$$A_1^* = \text{Top}(A_1) - \frac{\text{URC}(W_1^{-1})\,\text{Bot}(A_1)}{\text{LRC}(W_1^{-1})},$$

$$\mathbf{C}_1^* = \text{Top}(\mathbf{C}_1) - \frac{\text{URC}(W_1^{-1})\,\text{Bot}(\mathbf{C}_1)}{\text{LRC}(W_1^{-1})},$$

$$W_1^{*-1} = \text{ULC}(W_1^{-1}) - \frac{\text{URC}(W_1^{-1})\,\text{URC}(W_1^{-1})^T}{\text{LRC}(W_1^{-1})}.$$

Hence we must replace A_1, $\mathbf{C}_1$, and W_1^{-1} by A_1^*, $\mathbf{C}^*$ and W_1^{*-1}, respectively, in (2.8) in order to specify the conditional distribution of Top$(\boldsymbol{\mu})$ given $\mathbf{Y}, \boldsymbol{\theta}, \tau$ in the reduced model.

We next obtain (3.2). The top row of (2.8) gives $f(\text{Top}(\boldsymbol{\mu}), \text{Bot}(\boldsymbol{\mu})|\boldsymbol{\theta}, \tau, \mathbf{Y})$, from which we can obtain $f(\text{Bot}(\boldsymbol{\mu})|\boldsymbol{\theta}, \tau, \mathbf{Y})$ by integrating out Top$(\boldsymbol{\mu})$. The result is

$$\text{Bot}(\boldsymbol{\mu})|\boldsymbol{\theta}, \tau, \mathbf{Y} \sim N(\text{Bot}(A_1)\boldsymbol{\theta} + \text{Bot}(\mathbf{C}_1), \tau^{-1}\,\text{LRC}(W_1^{-1})). \tag{3.4}$$

The second row of (2.8) is $f(\boldsymbol{\theta}|\mathbf{Y}, \tau)$. We can use this together with (3.4) and Bayes' theorem to obtain $f(\boldsymbol{\theta}|\text{Bot}(\boldsymbol{\mu}) = 0, \mathbf{Y}, \tau)$. When Bot$(\boldsymbol{\mu}) = 0$, the exponent of the density of (3.4) is

$$-\frac{\tau}{2}\frac{[-\text{Bot}(A_1)\boldsymbol{\theta} - \text{Bot}(\mathbf{C}_1)]^2}{\text{LRC}(W_1^{-1})} = -\frac{\tau}{2}(\boldsymbol{\theta} - \boldsymbol{\theta}')^T E'(\boldsymbol{\theta} - \boldsymbol{\theta}'),$$

where

$$E' = \frac{\text{Bot}(A_1)^T\,\text{Bot}(A_1)}{\text{LRC}(W_1^{-1})},$$

$$\boldsymbol{\theta}' = -\frac{\text{Bot}(A_1)^T\,\text{Bot}(\mathbf{C}_1)}{\text{Bot}(A_1)\,\text{Bot}(A_1)^T}.$$

Hence

$$f(\boldsymbol{\theta}, \text{Bot}(\boldsymbol{\mu}) = 0|\mathbf{Y}, \tau)$$
$$= \tau^{(s+1)/2}\frac{|B_1|^{1/2}\exp\{-0.5\tau[(\boldsymbol{\theta}-\boldsymbol{\theta}')^T E'(\boldsymbol{\theta}-\boldsymbol{\theta}') + (\boldsymbol{\theta}-\boldsymbol{\theta}_1)^T B_1(\boldsymbol{\theta}-\boldsymbol{\theta}_1)]\}}{(2\pi)^{(s+1)/2}|\text{LRC}(W_1^{-1})|^{1/2}}. \tag{3.5}$$

Rewriting the exponent in (3.5) yields

$$-\frac{\tau}{2}\{(\boldsymbol{\theta} - \boldsymbol{\theta}_1^*)^T B_1^*(\boldsymbol{\theta} - \boldsymbol{\theta}_1^*) + (\boldsymbol{\theta}_1 - \boldsymbol{\theta}')^T E' B_1^{*-1} B_1(\boldsymbol{\theta}_1 - \boldsymbol{\theta}')\}, \tag{3.6}$$

where

$$B_1^* = B_1 + E',$$

$$\boldsymbol{\theta}_1^* = B_1^{*-1}(B_1\boldsymbol{\theta}_1 + E'\boldsymbol{\theta}').$$

Hence, in order to get the distribution of $\boldsymbol{\theta}$ given τ in the reduced model, we must replace B_1 by B_1^* and replace $\boldsymbol{\theta}_1$ by $\boldsymbol{\theta}_1^*$ in (2.8).

Next, we obtain $f(\tau|\mathrm{Bot}(\boldsymbol{\mu}) = 0, \mathbf{Y})$ by Bayes' theorem. Integrate $\boldsymbol{\theta}$ out of (3.5), after rewriting the exponent as (3.6), to obtain

$$f(\mathrm{Bot}(\boldsymbol{\mu}) = 0|\mathbf{Y}, \tau)$$
$$= \frac{|B_1|^{1/2}\tau^{1/2}}{(2\pi)^{1/2}|B_1^*|^{1/2}|\mathrm{LRC}(W_1^{-1})|^{1/2}} \exp\left\{-\frac{\tau}{2}(\boldsymbol{\theta}_1 - \boldsymbol{\theta}')^T E' B_1^{*-1} B_1(\boldsymbol{\theta}_1 - \boldsymbol{\theta}')\right\}. \quad (3.7)$$

Multiply (3.7) by $f(\tau|\mathbf{Y})$ [the density corresponding to the last line of (2.8)] and renormalize to get

$$f(\tau|\mathbf{Y}, \mathrm{Bot}(\boldsymbol{\mu}) = 0) = \frac{(b_1^*/2)^{a_1^*/2}}{\Gamma(a_1^*/2)} \tau^{.5a_1^* - 1} \exp(-b_1^*\tau/2),$$

where

$$b_1^* = b_1 + (\boldsymbol{\theta}_1 - \boldsymbol{\theta}')^T E' B_1^{*-1} B_1(\boldsymbol{\theta}_1 - \boldsymbol{\theta}')\}, \qquad a_1^* = a_1 + 1.$$

This distribution is a gamma distribution, namely $\Gamma(a_1^*/2, b_1^*/2)$. Hence, we must replace a_1 by a_1^* and b_1 by b_1^* in (2.8), in order to get the posterior distribution of τ in the reduced model.

The only part we have not yet found is the predictive density, $f(\mathbf{Y}|\mathrm{Bot}(\boldsymbol{\mu}) = 0)$. This equals

$$\frac{f(\mathrm{Bot}(\boldsymbol{\mu}) = 0|\mathbf{Y})f(\mathbf{Y})}{f(\mathrm{Bot}(\boldsymbol{\mu}) = 0)}.$$

We can obtain $f(\mathrm{Bot}(\boldsymbol{\mu}) = 0|\mathbf{Y})$ as

$$\int_0^\infty f(\mathrm{Bot}(\boldsymbol{\mu}) = 0|\mathbf{Y}, \tau) f(\tau|\mathbf{Y})\, d\tau$$

using (3.7) and (2.8) and get that

$$f(\mathrm{Bot}(\boldsymbol{\mu}) = 0|\mathbf{Y})$$
$$= (2\pi)^{-1/2}|B_1|^{1/2}|B_1^*|^{-1/2}|\mathrm{LRC}(W_1^{-1})|^{-1/2}\left(\frac{b_1}{2}\right)^{a_1/2}\left(\frac{b_1^*}{2}\right)^{-a_1^*/2}. \quad (3.8)$$

Similarly,

$$f(\mathrm{Bot}(\boldsymbol{\mu}) = 0) = \int_0^\infty f(\mathrm{Bot}(\boldsymbol{\mu}) = 0|\tau) f(\tau)\, d\tau,$$

which is the same as (3.8) but with all the subscripts 1 changed to 0. The result for $f(\mathbf{Y}|\mathrm{Bot}(\boldsymbol{\mu}) = 0)$ is just (2.9) with all the required substitutions made.

3.2 Reduction Algorithm

In this section we summarize the model reduction as an algorithm, which is simple to implement. Assuming that one has fitted a model of a specific order, greater than 0, and that one wishes to fit a model of one lower order, one should perform the following steps:

Create the four vectors

$$\mathbf{E}' = \frac{\mathrm{Bot}(A_1)^T \,\mathrm{Bot}(A_1)}{\mathrm{LRC}(W_1^{-1})},$$

$$\boldsymbol{\theta}' = -\frac{\mathrm{Bot}(A_1)^T \,\mathrm{Bot}(C_1)}{\mathrm{Bot}(A_1)\,\mathrm{Bot}(A_1)^T},$$

$$\boldsymbol{\psi} = B_1(\boldsymbol{\theta}_1 - \boldsymbol{\theta}'),$$

$$\boldsymbol{\eta} = B_1\boldsymbol{\theta}_1 + \mathrm{Bot}(A_1)^T \,\mathrm{Bot}(C_1)\,\mathrm{LRC}(W_1^{-1}).$$

Replace $|B_1|$ by $|B_1|[1 + \mathrm{Bot}(A_1)B_1^{-1}\,\mathrm{Bot}(A_1)^T/\mathrm{LRC}(W_1^{-1})]$.
Replace B_1^{-1} by $B_1^{-1} - B_1^{-1}\mathbf{E}'B_1^{-1}/[1 + \mathrm{tr}(B_1^{-1}\mathbf{E}')]$, and use this replacement value wherever B_1^{-1} appears below.
Replace B_1 by $B_1 + \mathbf{E}'$.
Replace b_1 by $b_1 + (\boldsymbol{\theta}_1 - \boldsymbol{\theta}')^T\mathbf{E}'B_1^{-1}\boldsymbol{\psi}$, a_1 by $a_1 + 1$
Replace $\boldsymbol{\theta}_1$ by $B_1^{-1}\boldsymbol{\eta}$.
Replace A_1 by $\mathrm{Top}(A_1) - \mathrm{URC}(W_1^{-1})\,\mathrm{Bot}(A_1)/\mathrm{LRC}(W_1^{-1})$.
Replace C_1 by $\mathrm{Top}(C_1) - \mathrm{URC}(W_1^{-1})\,\mathrm{Bot}(C_1)/\mathrm{LRC}(W_1^{-1})$.
Replace $|W_1|$ by $|W_1|\,\mathrm{LRC}(W_1^{-1})$.
Replace W_1^{-1} by $\mathrm{ULC}(W_1^{-1}) - \mathrm{URC}(W_1^{-1})\,\mathrm{URC}(W_1^{-1})^T/\mathrm{LRC}(W_1^{-1})$.
Replace W_1 by $\mathrm{ULC}(W_1)$.

Finally, do all the replacements above for subscript 1 changed to 0 also. Iterate the foregoing procedure down to an AR (1) or AR (0) model as desired.

4. PREDICTIVE MODEL COMPARISON

4.1 Computing Predictive Densities for Multiple Models

One natural way to compare two (or more) predictive models is through predictive likelihood. Suppose that we have K models and model j assigns observation Y_i density $f_{j|i-1}(y)$ after seeing all observations $Y_1, \ldots, Y_{i-1}$. Then it is natural to judge models according to how much likelihood they assigned to the observation actually obtained. The predictive likelihood (or density) of $Y_{n+1}, \ldots, Y_{n+t}$ for model j is

$$\prod_{i=n+1}^{n+t} f_{j|i-1}(Y_i) = F_{j;n+1,t}. \tag{4.1}$$

Each factor on the left-hand side of (4.1) is the conditional density of observation i given the observations prior to i, so that the product is the joint conditional distribution of observations $n + 1, \ldots, n + t$ given observations $1, \ldots, n$. The higher $F_{j;n+1,t}$ is (for fixed n and t), the better model j was at predicting the t observations. This criterion is more sensible than a simple sum of squared deviations between the Y_i and the point predictions. It takes both the center of the distribution and the scale into account, as well as the entire shape. A close prediction with a high standard error may not be as impressive as one with a low standard error. Also, a poor prediction with a small standard error may be much worse than one with a larger standard error.

Using the results of Sections 2.3 and 3, it is easy to compare all the AR models of order 0 up to P_M by using the following scheme. First decide on how many points to condition on, $l \geqslant P_M$. Form a prior distribution based on the first l observations and what one knows and/or believes about the series. Next, fit the AR (P_M) model based on observations up to Y_n, conditioning on the first l. If $n = l$, this will be the prior. Begin the evaluation process with observation $n + 1$ by calculating $F_{j;n+1,t}$ for $j = 0, \ldots, P_M$ and any $t > 0$ one wishes. One may wish to observe the sequence of vectors $(F_{0;n+1,t}, \ldots, F_{P_M;n+1,t})$ as t goes from 1 on up in order to see if certain models are getting better as time elapses.

4.2 Two Algorithms for Computing Predictive Densities

To avoid storing the posterior distributions for many models or recalculating various reduced models, one may wish to proceed as follows. After fitting AR (P_M) based on observations up to Y_n, save the posterior distribution of the parameters. Assume that the next t observations are now available. Add one observation at a time as in Section 2.3 until time t. If t is not the end of time, one may wish to save this "new" posterior distribution for future reference. Now go back and apply the model reduction of Section 3 to both the prior and the posterior (replacing each one with the reduced model prior and posterior) in order to fit the order $P_M - 1$ model. Then add one observation at a time up to observation t. Continue going back to the reduced prior and posterior until they have been reduced as far as one wishes, say to AR (0). One now has calculated $(F_{0;n+1,t}, \ldots, F_{P_M;n+1,t})$, and is ready to continue on. This time, just copy the "new" posterior into the storage location of the old prior, continue updating one observation at a time, and so on. This scheme, of course, assumes that the observations from $n + 1$ to $n + t$ are all available at once.

If the observations become available only one at a time, and one needs predictions and posterior distributions at each time point, one must update the order P_M model each time a new observation arrives. The algorithm now requires that the reductions be performed right away, making sure to save the posterior for the AR (P_M) model at each time. In this case, updates need only be performed to the AR (P_M) model. This scheme has the added advantage of allowing one to calculate a distribution over the different models at each time point. We consider this point in more detail in Section 4.3.

4.3 Mixing Multiple Models Using Predictive Densities

Since one is rarely completely confident in what order AR model one should fit, imagine that one begins with a prior distribution over the order of the AR model. One needs values $q_{0,l}, \ldots, q_{P_M,l}$ such that $\sum_{i=0}^{P_M} q_{i,l} = 1$. Notice that the second subscript is l for each q, indicating that one need not calculate these numbers until after seeing the first l observations. After seeing n observations, one computes $q_{j,n}$, for $j = 0 \ldots, P_M$ as $cq_{j,l}F_{j;l+1,n-l}$, where c is whatever constant it takes to make the $q_{j,n}$ add to 1 over j. To update one observation at a time,

$$q_{j,n+1} = cq_{j,n}F_{j;n+1,1} = cq_{j,n}f_{j|n}(Y_{n+1}),$$

with a new c, of course.

At this point, let us see what our model looks like. At each time n, we have a distribution $(q_{0,n}, \ldots, q_{P_M,n})$ over the various AR models. Conditional on model j, we have a predictive density for Y_{n+1} given the past, $f_{j|n}(y)$. The overall (marginal) predictive density for Y_{n+1} is then

$$\sum_{j=0}^{P_M} q_{j,n}f_{j|n}(y) = f(Y_{n+1} = y \mid y_1, \ldots, y_n).$$

This marginal predictive density can be evaluated through predictive likelihood in the same manner that each of the individual models was evaluated. This model, however, has the property that it adapts as time goes on to whatever order AR model appears to be performing best. At the same time, it continues to hedge itself against the possibility that the best model may change. In particular, as long as several models do nearly equally well at predicting, the overall model will continue to be a weighted average of those several models.

5. HANDLING MISSING OBSERVATIONS

5.1 One Missing Observation

Missing observations in autoregressive processes are particularly troublesome because the predictive distribution of each new observation is so heavily dependent on the past observations. Consider the pth-order AR model with exogenous variables (or covariates):

$$Y_t = \beta^T \mathbf{y}_{t,p} + \gamma^T \mathbf{x}_t + \varepsilon_t,$$

where

$$\mathbf{y}_{t,p} = (y_{t-1}, y_{t-2}, \ldots, y_{t-p}),$$

ε_t is independent of $\mathbf{y}_{t,p}$ with normal distribution $N(0, (\tau v_t)^{-1})$, and $\mathbf{x}_t$ is a vector of exogenous variables. So given the past data, Y_t has normal distribution $N(\beta^T \mathbf{y}_{t,p} + \gamma^T \mathbf{x}_t, (\tau v_t)^{-1})$.

If Y_t is missing, each observation after time t up to time $t + p$ must be treated specially. Let us introduce some notation to make the discussion flow better. Let $\mathbf{Y}(s)$ denote $(Y_s, Y_{s-1}, \ldots)$ and let $\mathbf{y}(s)$ denote the observed values $(y_s, y_{s-1}, \ldots)$. Let $\mathbf{Y}(s|i_1, \ldots, i_r)$ be $\mathbf{Y}(s)$ with $Y_{i_1}, \ldots, Y_{i_r}$ removed, and let $\mathbf{y}(s|i_1, \ldots, i_r)$ be defined similarly for the observed values. We will often denote the conditional distribution of a future observation Y_{t+k} given a subset of the past, say $\mathbf{Y}(t + k - 1|t) = \mathbf{y}(t + k - 1|t)$, simply as the distribution of Y_{t+k} given $\mathbf{y}(t + k - 1|t)$ since no confusion should result. To get the distribution of Y_{t+k} given $\mathbf{y}(t + k - 1|t)$, we must integrate y_t out of the distribution of Y_{t+k} given $\mathbf{y}(t + k - 1)$. This will be possible as long as we know the distribution of Y_t given $\mathbf{y}(t + k - 1|t)$. The latter distribution can be obtained from Bayes' theorem as follows.

To get the distribution of Y_t given $\mathbf{y}(t + m|t)$ for $m \geqslant 1$, assume that Y_t given $\mathbf{y}(t + m - 1|t)$ has $N(\hat{y}_t(m - 1), (\tau\omega_t(m - 1))^{-1})$ distribution and that $\mathbf{Y}_{t+m}$ given $\mathbf{y}(t + m - 1)$ has the usual $N(\beta^T \mathbf{y}_{t+m,p} + \gamma^T \mathbf{x}_{t+m}, (\tau v_{t+m})^{-1})$ distribution. It follows that $\mathbf{Y}_t$ given $\mathbf{y}(t + m|t)$ has $N(\hat{y}_t(m), \{\tau\omega_t(m)\}^{-1})$ distribution, where

$$\hat{y}_t(j) = \omega_t(j)^{-1} \left[\omega_t(j - 1)\hat{y}_t(j - 1) - v_{t+j}\beta_j \left\{ \sum_{i \neq j} \beta_i y_{t+j-i} + \gamma^T \mathbf{x}_{t+j} \right\} \right] \quad (5.1)$$

and $\omega_t(j) = \omega_t(j - 1) + \beta_j^2 v_{t+j}$. Here we have invented an "AR (0)" coefficient β_0 and set it equal to -1 to make (5.1) easier to write. It is easy to see that $\hat{y}_t(0) = \beta^T \mathbf{y}_{t,p} + \gamma^T \mathbf{x}_t$ and $\omega_t(0) = v_t$; hence we get the distribution of Y_t given $\mathbf{y}(t + m|t)$ by recursion. It then follows that Y_{t+k} given $\mathbf{y}(t + k - 1|t)$ has normal distribution with mean and variance, respectively,

$$\beta^T \mathbf{y}_{t+k,p} + \gamma^T \mathbf{x}_{t+k} + \beta_k[\hat{y}_t(k - 1) - y_t] \quad \text{and}$$

$$\tau^{-1}[v_{t+k}^{-1} + \beta_k^2 \omega_t(k - 1)^{-1}].$$

5.2 Several Missing Observations

If more than one observation is missing in a stretch of p consecutive time units, the argument above must be applied recursively. There are two possible recursive formulations that come quickly to mind. One could construct the marginal distribution of the first missing observation and the conditional distributions of the later ones given the earlier ones. Or one could construct the marginal distribution of the last missing observation and the conditional distributions of the earlier ones given the later ones. Although this second formulation sounds backward, one should note two features which it has. First, we already calculate the distribution of the first missing observation $\mathbf{Y}$ given the future data $\mathbf{Y}(t+k|t)$. Hence future missing observations do not change the form of this distribution. Second, once p time periods have passed after a missing observation, we can ignore it without adjusting any of the distributions of later missing observations. For these reasons, we choose to implement this backwards formulation.

To handle multiple missing observations via the backwards formulation, we need the coefficient of each later missing observation in the formula of $\hat{y}_t(m)$ for $m \geqslant 1$. Let $C_t(k, m)$ be the coefficient of y_{t+k} in $\hat{y}_t(m)$ for $m \geqslant k$. From (5.1) we see that this is

$$C_t(k, m) = \omega_t(m)^{-1}[\omega_t(m-1)C_t(k, m-1) - v_{t+m}\beta_m\beta_{m-k}] \qquad \text{for } m > k$$

and

$$C_t(k, k) = \frac{\omega_{t+k}\beta_k}{\omega_t(k)}.$$

This same algorithm applies to later missing observations conditional on the future. Suppose that the current time is $s > t$ and t is the time point of the earliest missing observation within $\{s-p, \ldots, s\}$. The distribution of Y_s given $\mathbf{y}(s-1|t)$ was already calculated as normal with mean $\beta^T\mathbf{y}_{s,p} + \gamma^T\mathbf{x}_s$ with y_t replaced by $\hat{y}_t(s-t-1)$ and variance $\tau^{-1}[v_s^{-1} + \beta_{s-t}^2\omega_t(s-t-1)^{-1}]$. If we simply replace v_s^{-1} by $v_s^{-1} + \beta_{s-t}^2\omega_t(s-t-1)$ and replace y_t by $\hat{y}_t(s-t-1)$, this distribution is of the same form as that of $\mathbf{Y}_s$ given $\mathbf{y}(s-1)$. The time series now progresses just as before, as if Y_t were not missing, until the next missing observation occurs. The only changes are that the v_s precision scale factors must be made smaller to reflect the decrease in precision due to Y_t being missing, and Y_t must be replaced by its current best estimate $\hat{y}_t(s-t-1)$ at time s.

When the next missing observation occurs at time $t+l$ with $l \leqslant p$, we proceed as before. The distribution of Y_t given $\mathbf{y}(s|t)$ has precisely the same form as before, except that Y_{t+l} is now missing. We need the distribution of Y_{t+l} given $\mathbf{y}(t+l+m|t, t+l)$ for $m \geqslant 0$. To handle the general case, let t_i, $\ldots$, t_r be the indices of the missing observations in $(s-p, \ldots, s-1)$ in increasing order and Y_s is not missing. We will derive the distribution of Y_{t_r} given $\mathbf{y}(s|t_1, \ldots, t_{r-1})$ first and then the distribution of Y_{t_r} given $\mathbf{y}(s|t_1, \ldots, t_r)$. As before, assume that Y_{t_r} given $\mathbf{y}(s-1|t_1, \ldots, t_{r-1})$ has $N(\hat{y}_{t_r}(s-t_r-1), \{\tau\omega_t(s-t_r-1)\}^{-1})$ distribution. Note that this does not depend on whether or not Y_s is missing. Also assume that Y_s given $\mathbf{y}(s-1|t_1, \ldots, t_{r-1})$ has $N(\beta^T\mathbf{y}_{s,p} + \gamma^T\mathbf{x}_s, (\tau v_s^*)^{-1})$ distribution, where $y_{t_1}, \ldots, y_{t_{r-1}}$ in $\beta^T\mathbf{y}_s$ are replaced by $\hat{y}_{t_i}(s-1-t_i)$ for $i = 1, \ldots, r-1$. Each of these replacements $\hat{y}_{t_i}(s-1-t_i)$ has terms involving y_{t_j} for $j > i$ with coefficients $C_{t_i}(t_j - t_i, s-1-t_i)$ since these y_{t_j}'s are not in the conditioning set they must in turn be replaced by their best estimates. This is done recursively by substituting $\hat{y}_{t_r}(s-1-t_r)$ for y_{t_r} in $\hat{y}_{t_{r-1}}(s-1-t_{r-1})$ and then substituting both of these for y_{t_r} and $y_{t_{r-1}}$, respectively, in $\hat{y}_{t_{r-2}}(s-1-t_{r-2})$, and so on. After the recursive substitution is finished, let $d_{t_i}(t_r - t_i, s-1-t_i)$ be the net coefficient of $\hat{y}_{t_r}$ in y_{t_i}. We assume that the additional variance due to these substitutions is already reflected in v_s^* and $\omega_{t_r}(s-t_r-1)$. Bayes' theorem then can be used to find the distribution of Y_{t_r} given $\mathbf{y}(s|t_i, \ldots, t_r)$ as $N(\hat{y}_{t_r}(s-t_r), \{\tau\omega_{t_r}(s-t_r)\}^{-1})$, where

$$\hat{y}_{t_r}(s-t_r) = \omega_{t_r}(s-t_r)^{-1}\left[\omega_{t_r}(s-t_r-1)\hat{y}_{t_r}(s-t_r-1) - v_s^* b\left\{\sum_{i \neq s-t_r} \beta_i y_{s-1} - \gamma^T\mathbf{x}_s\right\}\right],$$

$$b = \beta_{s-t_r} + \sum_{i=1}^{r-1} \beta_{s-t_i} d_{t_i}(t_r - t_i, s-t_i-1)$$

and $\omega_{t_r}(s-t_r) = \omega_{t_r}(s-t_r-1) + b^2 v_s^*$. The distribution of Y_{s+1} given $\mathbf{y}(s|t_i, \ldots, t_{r-1})$ is $N(\beta^T\mathbf{y}_{s+1,p} + \gamma^T\mathbf{x}_{s+1}, (\tau v_{s+1}^*)^{-1})$ with $y_{t_1}, \ldots, y_{t_{r-1}}$ recursively replaced by $\hat{y}_{t_i}(s-t_i)$, $i = j, \ldots, r-1$ as above so that only y_{t_r} appears explicitly. It follows then that Y_{s+1} given $\mathbf{y}(s|t_i, \ldots, t_r)$ has $N(\beta^T\mathbf{y}_{s+1,p} + \gamma^T\mathbf{x}_{s+1}, \tau^{-1}v_{s+1}^{*-1} + b^2\omega_{t_r}(s-t_r)^{-1})$ distribution, where even y_{t_r} has now been replaced by $\hat{\mathbf{y}}_{t_r}(s-t_r)$. All that remains is to find v_s^*, $\omega_{t_r}(0)$, and $\hat{y}_{t_r}(0)$. By definition, Y_{t_r} given $\mathbf{y}(t_r - 1|t_1, \ldots, t_{r-1})$ has a $N(\hat{y}_{t_r}(0), \{\tau\omega_{t_r}(0)\}^{-1})$ distribution. So $\hat{y}_{t_r}(0) = \beta^T\mathbf{y}_{t_r,p} + \gamma^T\mathbf{x}_{t_r}$, where $y_{t_r}, \ldots, y_{t_{r-1}}$ have been recursively replaced by $\hat{y}_{t_i}(t_r - t_i - 1)$, so that no missing values appear explicitly in $\hat{y}_{t_r}(0)$. Also $\omega_{t_r}(0)^{-1} = v_{t_r}^{*-1} + F^2\omega_{t_{r-1}}(t_r - t_{r-1} - 1)^{-1}$ with

$$F = \beta_{t_r - t_{r-1}} + \beta_{t_r - t_{r-2}} d_{t_{r-2}}(t_{r-1} - t_{r-2}, t_r - t_{r-2} - 1) + \cdots + \beta_{t_r - t} d_{t_1}(t_{r-1} - t_1, t_r - t_1 - 1).$$

It is also clear that if $(\tau v_s)^{-1}$ is the variance of Y_s given $y(s-1|t_1, \ldots, t_{r-2})$, then $v_s^{*-1} = v_s^{-1} + g^2\omega_{t_{r-1}}(s - t_{r-1} - 1)^{-1}$ with

$$g = \beta_{s-t_{r-1}} + \beta_{s-t_{r-2}} d_{t_{r-2}}(t_{r-1} - t_{r-2}, s - t_{r-2} - 1) + \cdots + \beta_{s-t_1} d_t(t_{s-t_1} - t_1, s - t_1 - 1).$$

Finally, we need to calculate the d_{t_i}'s recursively, and all distributions will then be defined recursively.

To obtain d_{t_i}, first note that as long as Y_s is not missing, the same formula that updates $C_{t_i}(k, s - t_i - 1)$ to $C_{t_i}(k, s - t_i)$ updates d_{t_i}. That is, if Y_s is observed,

$$d_{t_i}(k, s - t_i) = \omega_{t_i}(s - t_i)^{-1}[\omega_{t_i}(s - t_i - 1)d_{t_i}(k, s - t_i - 1) - v_s\beta_{s-t_i}\beta_{s-t_i-k}].$$

If Y_s is missing (i.e., $s = t_r$), the formula above works for $t_i + k < t_r$, but $d_{t_i}(t_r - t_i, t_r - t_i)$ must be calculated as follows. For $i < r$,

$$d_{t_i}(t_r - t_i, t_r - t_i) = \frac{\omega_s\beta_{s-t_i}}{\omega_{t_i}(s - t_i)} + \sum_{j=1}^{r-i-1} d_{t_{i+j}}(t_r - t_{i+j}, t_r - t_{i+j})d_{t_i}(t_{i+j} - t_i, s - t_i). \tag{5.2}$$

The recursion begins at $i = r - 1$ and proceeds down to $i = 1$.

5.3 Algorithm for Updating Distributions of Missing Observations

Since all the means and variances in the preceding section are defined recursively, there exists a recursive algorithm for calculating them.The algorithm tells one how to proceed each new time period. Let s be the new time period. Assume that $t_i, \ldots, t_r$ are the indices of missing observations in $(s-p, \ldots, s-1)$. We must find the predictive density of Y_s and the posterior distributions of $Y_{t_1}, \ldots, Y_{t_r}$.

1. First, calculate the mean of Y_s given $\mathbf{y}(s - 1|t_1, \ldots, t_r)$ by substituting $\hat{y}_{t_r}(s - 1)$ for y_{t_r} in the formula for $\hat{y}_{t_{r-1}}(s - 1)$ and substituting both of these into the formula for $\hat{y}_{t_{r-2}}(s - 1)$, and so on, down to $\hat{y}_{t_r}(s - 1)$. Then substitute all of these results, together with the observed data in $\{s - 1, \ldots, s - p\}$, into $\beta^T\mathbf{y}_{s,p} + \gamma^T\mathbf{x}_s$ to get the mean of Y_s. The vector of independent variables for the regression is $(x_s^T, y_{s,p}^T)^T$ with the noted substitutions made.
2. Update the distributions of $Y_{t_1}, \ldots, Y_{t_r}$ beginning with t_1. That is, start with $i = 1$ and use (5.1) to update $\hat{y}_{t_i}$ and then set $\omega_{t_i}(s - t_i) = \omega_{t_i}(s - t_i - 1) + \beta_{s-t}^2 v_s$. Also update $C_{t_i}(t_j - t_i, s - t_i)$ and $d_{t_i}(t_j - t_i, s - t_i)$ for $0 = i + 1, \ldots, r$. Then replace v_s^{-1} by $v_s^{-1} + \beta_{s-t_i}^2\omega_{t_i}(s - t_i - 1)$ *before* updating the distribution of $Y_{t_{i+1}}$. The final v_s will be part of the predictive density of Y_s given $\mathbf{y}(s - 1|t_1, \ldots, t_r)$. Now set $i = i + 1$ and go back to the beginning of this step until $i = r$ is finished. The usual one observation at a time updating formulas for the posterior now apply if Y_s is observed.
3. If Y_s is missing, we still update the distributions of $Y_{i_1}, \ldots, Y_{t_r}$ exactly as in step 2. Now $s = t_{r+1}$ and we must calculate $d_{t_i}(t_{r+1} - t_i, t_{r+1} - t_i)$ as in (5.2).

5.4 Implementation of the Algorithm for Actual Data Analysis

The major drawback to the approach described is that when there is missing data, the precision scale factor v_s for an observation given past data depends on the coefficients β which are unknown. The predictive density of Y_s involves an integral over β. This integral is now complicated by the fact that the integrand has a factor of $v_s^{1/2}$. We propose to ignore this problem and merely substitute the current best estimate of β for β in the formula for v_s and treat it as constant. We hope that this provides a reasonable approximation to what one would obtain if one did the actual integration, which is too complicated to do in closed form and too time consuming to do numerically for $p > 1$.

Since so many of the aspects of the distributions of missing observations are computed recursively, we can save a great deal of time and storage by keeping only the current values of such quantities as $\omega_{t_i}(\cdot)$, $C_{t_i}(t_j - t_i, \cdot)$, and $d_{t_i}(t_j - t_i, \cdot)$. At each new time point, we plug the current best estimate of β into the recursive formula, but we should remember that the current values also depend on the previous best estimates of β. This is another degree of approximation.

The other aspects of model fitting remain unaffected by missing observations under the approximations described above. We can still update the posterior distribution of the model parameters one observation at a time if Y_s is observed. We can also reduce the models one observation at a time after the posterior distribution for the largest model is formed. This provides a further level of approximation because a missing observation for an order 10 model, say, is not missing for an order 5 model if it occurred seven time periods ago.

6. MODEL GENERALIZATIONS

In this section we consider two possible generalizations to the AR models which have been considered so far. First, we allow the possibility that some observations are outliers in Section 6.1. Second, we allow certain types of changes in the model to occur in Section 6.2. Smith and West (1983) also consider model generalizations of the type described here. Whereas we allow the number of candidate models to increase over time, Smith and West choose to approximate the mixture of several models by a single model after each observation is observed.

6.1 Handling Outliers

If we know that the observation that arrives at time s is unrelated to the rest of the series, it makes sense to treat Y_s as missing. However, it could be that we just observed an observation from the tail of the distribution or the model has somehow changed at time s. The latter possibility is discussed further in Section 6.2. Since each observation is either part of the same series as the rest of the data or not, we leave each of these possibilities open for each observation. If an observation is part of the same series, we should continue the usual updating. If it is not, we should treat it as missing as we did in Section 5.

The problem with outliers is that we do not know which of the two possibilities has occurred. In the Bayesian framework, the natural thing to do is to have a prior probability for each and calculate a posterior probability for each. To do this, we also need a distribution for Y_s given that it is an outlier. Call the density of this distribution $f_O(y)$. If p_0 is the prior probability that Y_s is an outlier and p_1 is the posterior probability, then

$$\frac{p_1}{1-p_1} = \frac{p_0}{1-p_0} \frac{f_O(y_s)}{f_P(y_s|\text{past})},$$

where $f_P(y_s|\text{past})$ is the predictive density of y_s given the past and given that it is not an outlier.

For each observation, we consider the possibility that it is an outlier. We need not reject the observations immediately because, after several more observations arrive, the probability that Y_s is an outlier may decrease. For example, if y_s is in the tail of the predictive distribution of Y_s given the past, but y_{s+1} is in the middle of the predictive distribution of Y_{s+1} given the past including $Y_s = y_s$, then the probability that Y_s is an outlier could go down after observing $Y_{s+1} = y_{s+1}$. For this reason (and others described in next section), we intend to hold open the possibility that Y_s is an outlier for as long as possible. Of course, if we hold open the possibility that each observation is an outlier, we have 2^n possibilities after n observarions. There are obvious limitations to carrying out this method. These are discussed later in more detail.

6.2 Random Model Changes

As we mentioned earlier, it is possible that one may wish to allow the AR model to change occasionally. The types of changes that can be modeled are quite general. In fact, any feature that has a distribution can be allowed to change at any time period.

For example, one might believe that there is a chance that the order of the AR model has increased. If so, one can mix two models with different distributions over the order of the AR model. Perhaps one believes that there is a chance that some of the coefficients have changed. If so, one can mix two models where one of them has higher variances for the coefficients which might have changed. Similarly, one can mix in a model in which the degrees of freedom of the error precision τ is much smaller than in the posterior based on the previous observations, to allow that the error precision has changed. To keep the estimated precision the same, one would also decrease b_1 [see (2.7)] by the same factor.

In our implementation of each case above, the new model introduced will give a large variance to the predictive distribution of the next observation, but the predictive median of the next observation will be the same as in the original model. Hence, if the next observation is sufficiently far from the median of the predictive distribution, it will have higher predictive density under the new model than under the original one. Similarly, if the next observation is close to the median, it will have higher density under the original model. If the density under the new model is substantially larger than the density under the original model, the posterior probability of the new model will go up from its prior value.

Just as with outliers, the number of models one needs to consider increases geometrically as time goes on. Storage and time limitations require that one discard many models. One should however, keep models for as long as possible, because the vindication of a model may not occur for several time periods. For example, if a level shift occurs at time t, the predictive densities of the next several observations $t + 1, \ldots, t + k$ should all be higher under the model which says that a shift occurred at time t than under the original model. After the k observations, the posterior probability of the model with a level shift at time t may have been built up to the point where it is a model to be taken seriously. If a level shift did not occur, that model should rapidly disappear due to low posterior probability. Also, it takes a few observations to decide whether a level shift occurred or if an observation was an outlier.

7. NUMERICAL RESULTS

7.1 Comparisons of Procedures

To illustrate the proposed Bayesian method for analyzing time series data, we compare it with a non-Bayesian autoregressive method. Akaike's information criterion (AIC; Akaike, 1974), is used to select the best order for the non-Bayesian method. This criterion is commonly used for stationary autoregressive models. A justification for the use of AIC to specify the order of a nonstationary autoregressive model is given in Tsay (1985). We used 19 real time series in the comparison. The series are listed in the Appendix. These series are often analyzed in the literature. They have various sample sizes and represent the type of time series data commonly encountered in application.

None of these series have exogenous variables, but the AR(0) coefficient is actually treated as an exogenous variable in all of our examples, that is, $q = 1$.

For the sunspot data, we also employed a transformation $y^* = \log_e(1 + y)$, denoted by LNsunspot. For the Resex data, we employed a transformation $y^* = \log_e(y)$, denoted LNresex. These transformations were applied because these data sets appeared to be very skewed and because the data varied over several orders of magnitude. It has been our experience that such data are often better modeled as log-normal rather than normal. Retransformations were made before computing forecast errors, and the logs of the predictive densities include the Jacobian factors, so that the transformed and untransformed models are comparable.

In the comparison, we used $n - 20$ observations for model building, where n is the number of observations, and reserved the last 20 observations for forecast comparison. We referred to these two parts of data as modeling and forecasting samples, respectively. Four criteria are used to summarize the comparison results. They are the mean square forecast error, mean absolute forecast error, mean percentage forecast error, and the natural logarithm of the predictive likelihood of the last 20 observations.

For each series except TS02 the maximum autoregressive order entertained is 20. For the TS02, 10 is used because of the sample size limitation. For a given series, the non-Bayesian method first selects the best order based on the modeling sample, then estimates the parameters and computes the one-step-ahead forecast recursively for the forecasting sample. The computation is relatively easy for the non-Bayesian method. In Bayesian analysis, the prior distribution of (2.3) used is

$$a_0 = b_0 = 0.01, \quad s = 1, \quad \theta_0 = 1, \quad A_0 = (0, 1, \ldots, 1)^T,$$
$$B_0 = 0.01, \quad C_0 = \mathbf{0}, \quad V = I,$$

W_0^{-1} is diagonal with $W_0^{-1}(1, 1) = 100$ and (i, i)th element equal to $100\sqrt{i - 1}$, for $i > 1$.

Prior of AR(i) model q_i proportional to $(i + 1)^{-0.1}$ for $i = 0, \ldots, P_M$.

This prior distribution is somewhat flat, but still proper.

For a given series, the original estimates for 21 AR models, including the AR(0) model, are computed from the modeling sample. Then, for each observation in the forecasting sample we considered three possibilities: a level shift, an outlier, and a regular observation. The prior probabilities for these three cases are 0.025, 0.025, and 0.95, respectively. Consequently, we obtained a tree structure for Bayesian models as new observations became available. The number of models considered, therefore, grows rapidly as more and more new observations become available. To keep the number of models manageable, we retain only 60 models for each new observation. By a "model," now, we mean a statement as to which observations were outliers and when level shifts occurred. Each of the 60 models is still a mixture of 21 AR submodels. These 60 models are the ones that have the highest cumulative predictive densities among all the possible models at a given time. Before fitting the next observation, we discard the 40 worst models and create 3 models for each of the 20 that remain, one for level shift, one for outlier, and one for regular observation. The outlier density f_O was a t density with degrees of freedom equal to the prior hyperparameter $a_0 = 0.01$, median equal to the prediction based on the overall Bayesian model, and scale factor equal to $2\sqrt{b_1/a_1}$ [see (2.7)]. This is twice the scale factor for a regular observation, with considerably fewer degrees of freedom, so that the tails of the outlier distribution are much heavier than those of a regular observation. The way a level shift was modeled was to change the multiple of τ^{-1} in the variance of the level parameter to four times its current value. Since $\tau^{-1}W_0^{-1}$ is the covariance matrix of the AR and regression parameters given τ, this can be accomplished by multiplying the $(1, 1)$th entry of the matrix W_0^{-1} by 4.

At each observation in the forecasting sample, we computed point forecasts and predictive densities for each Bayesian model. (The total number of Bayesian models is 60 for each observation.) These results were then used in two ways. First, they were used to obtain a combined forecast for that observation (a weighted average with weights equal to posterior probabilities for the individual models.) Second, the posterior probabilities were used to identify the "best" Bayesian model (i.e., the one that has the maximum posterior probability). This best Bayesian model is the model used to forecast the next observation if one is merely interested in a single model. The results of the combined forecast are referred to as the overall results of Bayesian analysis, while those of the single best Bayesian model as the maximum posterior probability (MAP) results. Both results are used in the comparison below. In addition, we also provide results for Bayesian analysis without considering outliers or level shifts, labeled "regular" in the tables.

Tables 1 to 3 summarize the comparison results based on the mean square error, mean absolute error, and mean percentage error for the 20 series. For most of the data, the results of "regular" are the same as those of MAP, indicating that the Bayesian model of highest posterior probability is stable for the forecasting period. This is expected because most of the series used are stable. A difference between MAP and "regular" signifies the possible existence of outliers or level shifts. This is clearly seen in the Resex series, which is known to contain outliers (see Martin et al., 1983). Table 4 gives the natural logarithms of the predictive likelihoods of non-Bayesian and Bayesian models.

From the comparison, we made the following observations. First, no

Table 1 Forecast Comparison Based on Mean Square Error

Data	Non-Bayesian	Overall	MAP	Regular if different from MAP
SeriesA	0.137	0.153	0.200	0.136
SeriesB	60.9	58.0	57.6	
SeriesC	0.014	0.014	0.014	
SeriesD	0.098	0.099	0.098	
SeriesF	227.4	138.68	136.69	
Sunspot	378.1	431.0	432.0	
LNsunspot	198.5	205.6	205.4	
TS01	1.05	2.99	2.03	1.08
TS02	0.006	0.006	0.006	
TS03	0.075	0.050	0.048	
TS05	0.057	0.043	0.043	
TS08	13.8	12.40	12.40	
Gasrate	0.009	0.008	0.008	
CO2	0.110	0.102	0.101	
Flourp	64.4	53.6	53.8	
CanaGNP	0.0005	0.0005	0.0005	
CanaM1	0.0001	0.0001	0.0001	
CanaM2	0.0003	0.0004	0.0004	
Resex	4.77×10^8	2.62×10^8	3.48×10^8	2.12×10^8
LNresex	2.08×10^8	2.50×10^8	2.48×10^8	1.74×10^8
Crest	0.0026	0.0030	0.0030	

method is uniformly better than the others, regardless of the criterion used. This suggests that Bayesian time series analysis deserves further investigation. Second, for Bayesian analysis the results of "overall" and MAP are close, suggesting that one may use the "overall" model without being concerned with model selection in Bayesian analysis. In reading the "overall" column of Table 4, one expects to see slightly smaller numbers unless there are outliers or level shifts. The log of the predictive likelihood for each observation which is not an outlier and at which a level shift does not occur is $-\log(0.95) = 0.051$ smaller than it would be if we did not consider the possibility of level shifts. Over 20 observations, this adds up to 1.026. That is, if we wish to allow the possibility of outliers or level shifts in our predictions, we must be willing to suffer a loss of log-likelihood amounting to 0.051 for each observation. The "regular" and MAP predictions do not suffer this loss

Table 2 Forecast Comparison Based on Mean Absolute Error

Data	Non-Bayesian	Overall	MAP	Regular if different from MAP
SeriesA	0.299	0.317	0.344	0.298
SeriesB	6.71	6.55	6.53	
SeriesC	0.098	0.098	0.098	
SeriesD	0.243	0.244	0.245	
SeriesF	12.6	9.64	9.58	
Sunspot	16.3	16.9	16.9	
LNsunspot	10.8	11.3	11.3	
TS01	0.750	0.816	0.848	0.752
TS02	0.062	0.057	0.058	
TS03	0.193	0.151	0.150	
TS05	0.203	0.151	0.151	
TS08	3.16	3.00	3.00	
Gasrate	0.071	0.070	0.070	
CO2	0.259	0.249	0.250	
Flourp	5.68	5.34	5.34	
CanaGNP	0.016	0.018	0.018	
CanaM1	0.010	0.009	0.009	
CanaM2	0.013	0.016	0.017	
Resex	7.50×10^3	9.64×10^3	1.07×10^4	7.89×10^3
LNresex	7.50×10^3	7.68×10^3	7.78×10^3	5.90×10^3
Crest	0.042	0.045	0.045	

because they use a single model, determined before the next observation is seen, for each prediction.

The most variation between the methods occurs with the Resex data. This data set has several observations which are candidates for being called outliers. In fact, based on the overall Bayesian analysis of the last 20 observations, the probability is approximately 1.0 that a particular set of three observations $\{77, 83, 84\}$ are all outliers. And the probability is about 0.93 that observation number 76 is also an outlier. These four observations alone account for 76% of the sum of squared prediction errors for the Bayesian model.

Table 3 Forecast Comparison Based on Mean Percentage Error

Data	Non-Bayesian	Overall	MAP	Regular if different from MAP
SeriesA	1.70	1.81	1.96	1.70
SeriesB	1.92	1.87	1.87	
SeriesC	0.467	0.455	0.457	
SeriesD	2.77	2.79	2.79	
SeriesF	30.3	24.9	24.8	
Sunspot	47.9	40.9	41.0	
LNsunspot	28.5	29.5	29.1	
TS01	13.5	13.1	13.4	13.2
TS02	1.14	1.06	1.07	
TS03	2.31	1.80	1.79	
TS05	4.67	3.62	3.61	
TS08	13.8	12.44	12.44	
Gasrate	1.98	1.95	1.95	
CO2	0.470	0.447	0.449	
Flourp	3.27	3.08	3.08	
CanaGNP	0.097	0.108	0.108	
CanaM1	0.056	0.049	0.050	
CanaM2	0.110	0.138	0.139	
Resex	23.9	29.8	36.4	21.9
LNresex	23.9	24.6	25.2	16.29
Crest	10.5	11.4	11.3	

7.2 Parallel Computation

Due to the large number of models that must be considered at each time period, one must either discard most models very quickly, find a very fast computer, or do a little bit of both. We have chosen this third option. Since high-speed personal workstations have become relatively inexpensive in recent years compared to large mainframes, their popularity has grown correspondingly. Communication between these workstations is possible at several levels of sophistication. Even small personal computers are capable of communicating with each other and with larger computers. Since the total computing capacity of the machines available in a given research environ-

Table 4 Forecast Comparison Based on Natural Logarithm of Predictive Likelihood

Data	Non-Bayesian	Overall	MAP	Regular if different from MAP
SeriesA	−8.751	−9.989	−13.551	−8.788
SeriesB	−69.55	−70.036	−68.979	
SeriesC	13.816	12.748	13.789	
SeriesD	−5.128	−6.292	−5.242	
SeriesF	−84.108	−79.334	−78.269	
Sunspot	−88.580	−91.133	−90.036	
LNsunspot	−86.119	−87.582	−86.342	
TS01	−31.611	−30.237	−42.819	−34.130
TS02	21.811	21.018	22.192	
TS03	−39.891	0.316	2.060	
TS05	0.276	0.176	1.212	
TS08	−55.092	−54.866	−53.837	
Gasrate	12.378	10.930	11.976	
CO2	−6.212	−6.693	−5.609	
Flourp	−70.054	−70.682	−69.621	
CanaGNP	47.081	45.820	47.283	
CanaM1	60.135	51.134	52.608	
CanaM2	54.116	47.971	49.176	
Resex	−214.586	−210.989	−355.378	−262.141
LNresex	−207.919	−216.679	−262.325	−229.506
Crest	33.824	28.377	29.426	

ment is rarely used continuously, a system has been developed that allows a single program to run simultaneously on several computers in a network. This system is described by Eddy and Schervish (1986). What the system does is to break the computational task into smaller pieces, which then run in parallel on as many different machines as are available. The results of the individual tasks are accumulated by one of the machines acting as a master who assigns work to the slaves.

In the case of Bayesian time series analysis, the task of calculating posterior distributions and predictions for all the different models for each successive time period is quite large. The parallel system, sends each model out to a different computer and collects the results when they are done. This technique can substantially shorten the computation time in analyzing time series with large numbers of observations.

Table 5 Timing Comparison of Parallel and Serial Computations (Times in Seconds)

Data	Serial	Parallel
SeriesA	1391	360
SeriesB	1396	390
SeriesC	1393	382
SeriesD	1391	374
SeriesF	1308	410
Sunspot	1395	458
TS01	1393	433
TS02	283	270
TS03	1388	415
TS05	1391	369
TS08	1388	371
Gasrate	1378	337
CO2	1392	363
Flourp	1396	380
CanaGNP	1417	386
CanaM1	1401	381
CanaM2	1450	369
Resex	1394	400
Crest	1406	366

Table 5 gives comparative timings for serial and parallel computations. The serial timings are CPU time in order to reduce the effect of variations in system load. The parallel timings are elapsed time and are the minimum of three runs, in order to reduce the effect of variations in system loads. The parallel runs were made on six micro VAX II computers with differing amounts of random access memory. A good deal of information is contained in the line of Table 5 for data set TS02, which had a maximum AR order of 10 instead of 20. There is a significant amount of overhead involved in starting up the parallel computation. The AR (10) model case appears to be very close to the break-even point below which parallel computation is not helpful.

8. CONCLUSION

In this chapter we have illustrated a Bayesian approach to autoregressive time series modeling. The approach includes a hierarchical model for the parameters, a way to handle outliers and missing observations within the model, and a distribution over various competing models which obviates the need for ad hoc model selection criteria. Numerical comparisons with a popular model selection criterion show that the two methods can perform very closely. One thing that would help to make more effective use of Bayesian methods is a way to choose the prior distribution sensibly. Hill and Spall (Chapter 18, this volume) consider the problem of producing noninformative priors for linear dynamic models. One should be aware that the prior distributions used in the Bayesian calculations described in Tables 1 to 4 were chosen without looking at any of the data ahead of time, while the non-Bayesian method examined the beginning $n - 20$ observations first in order to choose the order of the AR model. The Bayesian method automatically adapted its modeling based on the first $n - 20$ observations before making the predictions for the last 20. Furthermore, the data used are those available in the literature which were thought to give good results for non-Bayesian methods.

In none of the sample calculations did we include exogenous variables, since there were none in any of the sample series. Further investigation is needed to study the performance of the Bayesian model with exogenous variables.

ACKNOWLEDGMENTS

Mark Schervish would like to acknowledge the support of the Bayesian Statistics Study Year 1986, University of Warwick, Coventry, United Kingdom, for support during the preparation of this chapter. Ruey Tsay wishes to acknowledge partial support from the ALCOA Foundation while preparing this chapter. Both authors would like to acknowledge the help provided by two referees and the editor in making the chapter easier to read.

APPENDIX: DATA DESCRIPTION

Seriesa to Seriesd and Seriesf: Series A to D and Series F of Box and Jenkins (1976) with observations 197, 369, 226, 310, and 70, respectively

Sunspot: annual sunspot data with 231 observations from 1749 to 1924

TS01–TS03, TS05, and TS08: U.S. economic time series with observations 104, 52, 90, 80, and 73, respectively

Gasrate and CO2: input and output processes of Series J of Box and Jenkins (1976), both with 296 observations

Flourp: monthly flour price index of Buffalo, New York, with 119 observations from January 1971 to March 1982

CanaGNP, CanaM1, CanaM2: Canadian quarterly GNP, M1, M2 in logarithms with 92 observations given in Hsiao (1979).

Resex: Monthly inward movement of residential telephone extensions with 89 observations given in Martin et al. (1983)

Crest: weekly market share of Crest toothpaste with 276 observations given in Wichern and Jones (1977)

Total number of data sets = 19

REFERENCES

Abraham, B., and G. E. P. Box (1979). Bayesian analysis of some outlier problems in time series, *Biometrika* 66, 229–236.

Akaike, H. (1974). A new look at the statistical model identification, *IEEE Trans. Autom. Control* AC-19, 716–722.

Akaike, H. (1980). Seasonal adjustment by a Bayesian modeling, *J. Time Ser. Anal.* 1, 1–13.

Basseville, M., and A. Benveniste eds. (1986). *Detection of Abrupt Changes in Signals and Dynamical Systems*, Springer-Verlag, New York.

Box, G. E. P., and G. M. Jenkins (1976). *Time Series Analysis: Forecasting and Control*, Holden-Day, San Francisco.

Broemeling, L. D. (1985). *Bayesian Analysis of Linear Models*, Marcel Dekker, New York.

Eddy, W. F., and M. J. Schervish (1986). Discrete-Finite Inference on a Network of VAXes, in *Computer Science and Statistics, Proceedings of the 18th Symposium on the Interface* (T. J. Boardman, ed.), American Statistical Association, Washington, D.C., pp. 30–36.

Harrison, P. J., and C. F. Stevens (1976). Bayesian forecasting, *J. R. Stat. Soc. Ser. B* 38, 205–247.

Ho, Y. C., and R. C. K. Lee (1964). A Bayesian Approach to Problems in Stochastic Estimation and Control, in *Proceedings of the Joint Automatic Control Conference*, Stanford University, Stanford, Calif., pp. 382–387.

Hsiao, C. (1979). Autoregressive modeling of Canadian money and income data, *J. Am. Stat. Assoc.* 74, 553–560.

Litterman, R. B. (1986). Forecasting with Bayesian vector autoregressions—five years of experience, *J. Bus. Econ. Stat.* 4, 25–38.

Liu, L. M., and G. C. Tiao (1980). Random coefficient first-order autoregressive models, *J. Econometrics* 13, 305–325.

Martin, R. D., A. Samarov, and W. Vandaele (1983). Robust Methods for

ARIMA Models, in *Applied Time Series Analysis of Economic Data* (A. Zellner, ed.), Bureau of the Census, Washington, D.C.

Monahan, J. F. (1983). Fully Bayesian analysis of ARMA time series models, *J. Econometrics* 21, 307–331.

Morrison, D. F. (1976). *Multivariate Statistical Methods*, McGraw-Hill, New York.

Pierce, B. D., and D. D. Sworder (1971). Bayes and minimax controllers for a linear system with stochastic jump parameters, *IEEE Trans. Autom. Control* AC-16, 300–307.

Sanyal, P., and C. N. Shen (1974). Bayes' decision rule for rapid detection and adaptive estimation scheme with space applications, *IEEE Trans. Autom. Control* AC-19, 228–231.

Seber, G. A. F. (1977). *Linear Regression Analysis*, Wiley, New York.

Smith, A. F. M., and M. West (1983). Monitoring renal transplants: an application of the multiprocess Kalman filter, *Biometrics* 39, 867–878.

Sworder, D. D. (1972). Bayes' controllers with memory for a linear system with jump parameters, *IEEE Trans. Autom. Control* AC-17, 118–121.

Thompson, P. A., and R. B. Miller (1986). Sampling the future: a Bayesian approach to forecasting from univariate time series models, *J. Bus. Econ. Stat.* 4, 427–436.

Tsay, R. S. (1985). Order selection in nonstationary autoregressive models, *Ann. Stat.* 12, 1425–1433.

West, M. (1986). Bayesian model monitoring, *J. R. Stat. Soc. Ser. B* 48, 70–78.

West, M., P. J. Harrison, and H. S. Migon (1985). Dynamic generalized linear models and Bayesian forecasting, *J. Am. Stat. Assoc.* 80, 73–97.

Wichern, D. W., and R. H. Jones, (1977). Assessing the impact of market disturbances using intervention analysis, *Manage. Sci.* 24, 329–337.

Willsky, A. S. (1976). A survey of design methods for failure detection in dynamic systems, *Automatica* 12, 601–611.

Zellner, A. (1972). *An Introduction to Bayesian Inference in Econometrics*, Wiley, New York.

3

Decision-Theoretic Approach to Order Determination and Structure Selection in Vector Linear Processes

DONALD S. POSKITT Department of Statistics, IAS, Australian National University, Canberra, Australia

A recurrent problem in the statistical analysis of time series data is that of model specification. In the context of vector processes this problem is confounded by the need not only to determine an appropriate order for the process but also to select a corresponding set of structural indices. In this chapter the concepts of order and structure are explained and an asymptotic Bayes rule for order determination and structure selection is derived. This is achieved by setting up a utility function appropriate for the observational decision problem of model choice and analyzing the large-sample behavior of the posterior expected utility. The operational characteristics of the model selection criterion so obtained are also investigated.

1. INTRODUCTION

Let $X_T = \mathrm{vec}(x(1):x(2):\cdots:x(T))$ denote a data set where for each time point t, $t = 1, \ldots, T$, $x(t)$ is a vector of v variables or measurements which is to be used to build a mathematical model of some phenomenon of interest. Although different prior considerations will influence the general character and complexity of the specifications employed, determination of the most appropriate model will rarely be achieved entirely on the basis of prior deliberations. The problem of ascertaining the dynamic structure of the system generating the observations arises in many areas of study (economics, engineering, hydrology, and medicine), for example. The practitioner is likely to try a range of different models, and will seek for the best one by reference to an analysis of X_T, having defined an appropriate criterion for assessing the relative merits of the alternative specifications. Such a process is fundamental in a statistical procedure for model determination and the use of model selection criteria such as AIC (Akaike, 1974) and BIC (Rissanen, 1978; Schwarz, 1978) to select an appropriate specification is now almost commonplace. In this chapter a decision-theoretic approach to the observational decision problem of selecting models for multivariate linear dynamic systems is investigated and an alternative model selection criterion, similar in spirit to those current in vogue, is proposed.

Assume that X_T is a realization of a stationary and ergodic stochastic process $x(t)$, $t \in \mathbf{Z}$. Assume also that any deterministic components have been removed, so that, without loss of generality, $x(t)$ is a zero mean, regular process and that there is no nonnull vector a such that $a^T x(t) = 0$ almost surely, (a.s.), for otherwise the dimension of $x(t)$ could be reduced. To avoid excessive notation $x(t)$ is employed to denote both a given process and a realized value of that process. Wold's decomposition (Wiener and Masani, 1957) implies that $x(t)$ may be expressed as

$$x(t) = \sum_{j=0}^{\infty} \psi(j)\xi(t-j), \tag{1.1}$$

where the transfer function

$$\Psi(z) = \sum_{j=0}^{\infty} \psi(j) z^{-j}$$

is finite at infinity and holomorphic outside the unit disk in the complex plane. The innovation $\xi(t)$ is a standardized noise process, that is, $E[\xi(t)] = 0$ and $E[\xi(t)\xi(s)^T] = I\delta_{t,s}$, $t, s \in \mathbf{Z}$, where $\delta_{t,s}$ denotes Kronecker's delta and I the vth-order identity matrix. Using an asterisk for the complex conjugate transpose, the power spectrum of $x(t)$ is given by

$$2\pi S(\omega) = \Psi(e^{i\omega})\Psi(e^{i\omega})^*.$$

It will be assumed that the squared norm

$$\begin{aligned} \|\Psi\| &= (2\pi)^{-1} \int_{-\pi}^{\pi} \operatorname{tr} \Psi(e^{i\omega})\Psi(e^{i\omega})^* \, d\omega \\ &= \sum_{j=0}^{\infty} \operatorname{tr} \psi(j)\psi(j)^T \end{aligned} \tag{1.2}$$

is bounded so that $x(t)$ has finite variance. Thus modulo postmultiplication by unitary matrices, there exists a unique correspondence between the transfer function Ψ and the power spectrum S, and hence the second moment properties of the process. In principle, knowledge of any one of these quantities allows for the evaluation of the other two. The process $x(t)$ may therefore be completely characterized by the sequence of $v \times v$ matrix coefficients $\{\psi(0), \ldots, \psi(i), \ldots\}$, referred to as the impulse response. Consequently, when modeling $x(t)$ it will be necessary to capture the salient features of the impulse response associated with the linear process representation (1.1).

In the following section a very brief account of some properties of linear systems is given to clarify concepts and terminology and motivate the consideration of a general class of parametric specifications. The Gaussian likelihood of a model structure and its associated parameters is discussed in Section 3. The large-sample behavior of the likelihood function and related quantities is considered, recognizing that in the context of model determination and structure selection it cannot be assumed that there exists a particular member from within the model set that corresponds to the true data-generating mechanism. This gives rise to the concept of pseudo true model parameters and the notion of an asymptotically best approximating model. Sections 2 and 3 may be viewed as providing the background and theoretical framework necessary for the subsequent discussion. Section 4 considers the structure and rationale underlying a utility function that may be employed to quantify the adequacy of a model relative to the system being analyzed. This utility function yields a stable decision problem and forms the basis for the practical implementation of an optimal statistical decision rule. In Section 5 the principle of precise measurement is pursued and a Bayes criterion for model determination derived. Sections 4 and 5 provide natural generalizations to the multivariate case of results previously obtained in Poskitt (1986) for univariate processes. For an application of Bayesian methodology in a similar context but from a rather different perspective, see also Shore (1980). The chapter concludes with a brief summary. As an *aide-mémoire*, an index of the more important notational conventions adopted is presented before the references.

2. ORDER, STRUCTURE, AND PARAMETRIC MODELS

When modeling dynamic systems it is common practice to model the linear process representation indirectly. Consider, for example, an autoregressive moving average (ARMA) specification of the form

$$\sum_{j=0}^{p} \alpha(j)x(t-j) = \sum_{j=0}^{p} \mu(j)\eta(t-j), \tag{2.1}$$

where the observed process $x(t)$ is expressed as a linear transformation of an unobservable disturbance process $\eta(t)$ assumed to be standardized noise. A particular ARMA form is obtained by fixing a value for the integer p and allotting numerical values to the elements of the parameter matrices $\alpha(j)$ and $\mu(j)$, $j = 0, 1, \ldots, p$, which may be chosen so that (2.1) is stationary and invertible. The transfer function of the model, Φ, is obtained from the matrix polynomials

$$A(z) = \sum_{j=0}^{p} \alpha(j)z^j \quad \text{and} \quad M(z) = \sum_{j=0}^{p} \mu(j)z^j$$

via the matrix fraction description, MFD, $\Phi(z) = A(z)^{-1}M(z)$. Restricting attention to irreducible left coprime pairs to avoid redundancy of description, the degree of $\det A(z)$, the determinant of $A(z)$, will be minimal among all MFDs of the same transfer function and is an invariant of Φ called the order or McMillan degree of Φ or the system.

Two structures are said to be observationally equivalent if they give rise to the same second-order moments for $x(t)$, and a model is said to be identifiable or estimable if there are no two structures belonging to the same model set that are observationally equivalent. If restrictions are not imposed on $A(z)$ and $M(z)$, the model will contain observationally equivalent structures given by polynomial pairs, $\{A_1(z):M_1(z)\}$ and $\{A_2(z):M_2(z)\}$, say, which are related by the equation

$$\{A_1(z):M_1(z)\} = U(z)\{A_2(z):M_2(z)\},$$

where $U(z)$ is unimodular. The existence of such equivalence classes necessitates the consideration of subsets of canonical ARMA forms in which the parameters are subject to appropriate a priori conditions that provide a bijective relationship between structures within a model set and transfer functions of given order.

Similar considerations arise in the context of a Markovian or state-space innovation representation. Each structure in a model set is obtained by assigning numerical values to the matrices A, B, C, and D, of dimension $(n \times n)$, $(n \times v)$, $(v \times n)$, and $(v \times v)$, respectively, in the equations

$$z(t+1) = Az(t) + \beta\eta(t)$$

$$x(t) = Cz(t) + D\eta(t),$$

where n is the dimension of the state vector $z(t)$ and $\eta(t)$ is again assumed to be a standardized noise process. Redundancy of description is avoided in this case by taking the dimension of the state vector as small as possible, which implies that the model will be reachable, $\text{rank}\,[B, AB, A^2B, \ldots, A^{n-1}B] = n$, and observable, $\text{rank}\,[C^T, (CA)^T, (CA^2)^T, \ldots, (CA^{n-1})^T] = n$, with n equal to the McMillan degree. The impulse response of the model is given by $\{\phi(0),$

$\phi(1), \ldots, \phi(i), \ldots\} = \{D, CB, \ldots, CA^{i-1}B, \ldots\}$. However, because of the nonuniqueness of representations of minimal state dimension, if (A, B, C, D) determines a minimal representation, then so does $(TAT^{-1}, TB, CT^{-1}, D)$ for any nonsingular T, it is necessary as with MFDs to consider additional identifying conditions leading to canonical forms. For a more detailed exposition of the algebraic properties of state-space and matrix fraction representations of linear systems, see Barnett (1971) and Kailath (1980).

From the preceding discussion it is apparent that just as the stochastic properties of the process $x(t)$ are characterized by Ψ, so the characteristics of a model are given by Φ. Furthermore, MFDs $(A(z); M(z))$ and state-space structures (A, B, C, D) give rise to rational transfer functions of the form

$$\Phi(z) = \sum_{j=0}^{\infty} \phi(j)z^{-j} = \frac{N_\Phi(z)}{d_\Phi(z)},$$

where the numerator polynomials of $N_\Phi(z)$ and the denominator polynomial $d_\Phi(z)$ are determined by

$$N_\Phi(z) = \operatorname{adj}\{A(z)\}M(z), \qquad d_\Phi(z) = \det A(z)$$

and

$$N_\Phi(z) = C\operatorname{adj}\{(zI - A)\}B + Dd_\Phi(z), \qquad d_\Phi(z) = \det(zI - A)$$

respectively.

Let H denote a Hankel matrix having $\phi(i + j - 1), i, j = 1, 2, \ldots$, in its ijth block of v^2 elements. Regarding the rows of H as elements of the Hilbert space of square summable series, it is known (Kalman et al., 1969; Rissanen, 1974) that if Φ is rational, H will have finite rank and the maximum number of linearly independent rows equals the order of Φ. Let $\mathcal{M}(n)$ denote the set of all proper rational transfer functions $\Phi(z)$, $\Phi(z)$ analytic and $\det \Phi(z) \neq 0 |z| \geqslant 1$, of McMillan degree n. Given any $\Phi \in \mathcal{M}(n)$, a state-space realization (A, B, C, D) of Φ can be derived from H in the following way. Let h_i denote the ith row of H and assume that $b = \{i(1), \ldots, i(n)\}$, $i(j) < i(j+1)$, $j = 1, \ldots, n-1$, defines an index set of n rows of H that form a basis. Set $H_b = (H_b(1) \colon H_b(2) \colon \cdots)$, where H_b contains the n rows $h_{i(j)}$, $j = 1, \ldots, n$, arranged in the same order as in H and partitioned into the first block of v columns, the second block of v columns, and so on. Writing the first v rows of H in terms of this basis gives $\phi(j) = CH_b(j)$, $j = 1, 2, \ldots$, for some $v \times n$ matrix C and hence

$$\begin{aligned} x(t) &= \sum_{j=0}^{\infty} \phi(j)\eta(t - j) \\ &= C \sum_{j=1}^{\infty} H_b(j)\eta(t - j) + \phi(0)\eta(t). \end{aligned} \tag{2.2a}$$

If $h_{i,j}$ denotes the ijth element of H, expressing rows $h_{i(j)+v}$, $j = 1, \ldots, n$, in terms of H_b gives coefficients a_{jk}, $k = 1, \ldots, n$, such that

$$h_{i(j)+v,r} = \sum_{k=1}^{n} a_{jk} h_{i(k),r}, \qquad r = 1, 2, \ldots. \tag{2.2b}$$

From the orthosymmetry of H, $h_{i,j+v} = h_{i+v,j}$, which generates the recursion $H_b(j) = AH_b(j-1)$, $j = 2, 3, \ldots$. It follows that

$$\sum_{j=1}^{\infty} H_b(j)\eta(t-j) = A \sum_{j=1}^{\infty} H_b(j)\eta(t-1-j) + H_b(1)\eta(t-1) \tag{2.2c}$$

and setting

$$z(t) = \sum_{j=1}^{\infty} H_b(j)\eta(t-j) \tag{2.2d}$$

gives the required realization. A completely analogous construction can be obtained for a corresponding MFD (see Rissanen, 1974; Deistler and Hannan, 1981; Guidorzi, 1981). Canonical forms can now be obtained by choosing the basis b in a canonical way.

There are, of course, many ways to define a canonical form. For instance, when seeking for the first n linearly independent rows of H in natural order, from the property of H that if h_i is in the span of $h_{i(1)}, \ldots, h_{i(k)}$, then h_{i+v} is in the span of $h_{i(1)+v}, \ldots, h_{i(k)+v}$ (Rissanen, 1974), it follows that if the ith row is in the basis, so will be the $(i-v)$th and any row not in the basis may be expressed in terms of the preceding basic rows. The search defines a multi-index $\{n_1, \ldots, n_v\}$, $\sum n_i = n$, $n_i > 0$, such that the rows i, $i+v$, $\ldots$, $i+(n_i-1)v$, $i = 1, \ldots, v$ form a basis for H. The integers n_i, $i = 1, \ldots, v$, are referred to as the dynamic or Kronecker indices of the system and the state-space realization or MFD associated with such a multi-index is said to be in echelon form. Hermite's canonical form may be obtained by selecting the basis rows in a lexicographical manner. These canonical forms are discussed in Denham (1974), Dickinson, Kailath and Morf (1974), and Deistler (1983).

More generally, let $\{n_1, \ldots, n_v\}$ denote a multi-index where the structural indices $n_i > 0$, $i = 1, \ldots, v$, which are not necessarily the Kronecker indices, are such that rows i, $i+v, \ldots, i+(n_i-1)v$, $i = 1, \ldots, v$ form a basis for H. Such a basis is said to be "nice." Denote by $\mathcal{M}_n(n_1, \ldots, n_v)$ the set of $\Phi \in \mathcal{M}(n)$ for which these rows form a basis. Each $\Phi \in \mathcal{M}_n(n_1, \ldots, n_v)$ may be parameterized using (2.2) or its matrix fraction equivalent, and thereby a coordinate system for $\mathcal{M}_n(n_1, \ldots, n_v)$ is obtained. If $\pi_{\{n_1,\ldots,n_v\}}$ gives the parameterization, then Clark (1976) has shown that as Φ varies over $\mathcal{M}_n(n_1, \ldots, n_v)$, $\pi_{\{n_1,\ldots,n_v\}}$ maps $\mathcal{M}_n(n_1, \ldots, n_v)$ into an open set of dimension $2nv$ in $\mathbf{R}^{2nv}$, Euclidean space of that dimension, and that the family $\{\mathcal{M}_n(n_1, \ldots, n_v), \pi_{\{n_1,\ldots,n_v\}}\}$, $\sum n_i = n$, yields the charts of an analytic manifold for $\mathcal{M}(n)$.

Differential geometry will not be used in this chapter. It is sufficient for present purposes to note that, roughly speaking, an analytic manifold is a geometrical configuration in which each point has a neighborhood which is homeomorphic to a neighborhood of the origin in a finite-dimensional Euclidean space. Since these neighborhoods are not disjoint, that is, $\mathscr{M}_n(n_1, \ldots, n_v) \cap \mathscr{M}_n(n'_1, \ldots, n'_v)$, $\sum n_i = \sum n'_i = n$, is in general not empty, the coordination of $\mathscr{M}(n)$ via multi-indices has been termed the overlapping parameterization.

The canonical forms as determined above are based on the matrices $\phi(i)$, $i = 1, 2, \ldots$, and serve to identify the system apart from the instantaneous response $\phi(0)$. A complete set of canonical forms can be obtained by observing that the additional freedom may be taken up by requiring $\phi(0)$ to be lower triangular with positive diagonal elements, thus ensuring a unique spectral factorization

$$2\pi g(\omega) = \Phi(e^{i\omega})\Phi(e^{i\omega})^* \tag{2.3a}$$

$$= K(e^{i\omega})\Omega K(e^{i\omega})^*, \tag{2.3b}$$

$K(z) = \sum_{j=0}^{\infty} k(j)z^{-j} = \Phi(z)\Phi(\infty)^{-1}$, $\Omega = \phi(0)\phi(0)^T$, for the power spectrum $g(\omega)$ of the model (see Denham, 1974, Sect. 4; Hannan, 1979, Theorem 2.2). Henceforth it will be assumed that this condition has been imposed. This brings the specification of the system down to the unique description of $K(z)$ and Ω, Ω being coordinatized by the $\frac{1}{2}v(v+1)$ elements of $\phi(0)$, which lie in an open set of $\mathbf{R}^{(1/2)v(v+1)}$, and $K(z)$ as described above.

A model of the process $x(t)$ will thus in general be composed of a set of canonical forms or structures that define a transfer function $\Phi \in \mathscr{M}(n)$, where n is the assumed order of the system or process being observed. These structures are determined by a finite-dimensional time-invariant parameter vector that can vary freely over a parameter set that may be taken as a subset of a Euclidean space. In an attempt to encapsulate these features, consider the class of models of the form

$$x(t) = \sum_{j=0}^{\infty} \phi(j, \theta)\eta(t-j),$$

where

$$\Phi(z, \theta) = \sum_{j=0}^{\infty} \phi(j, \theta)z^{-j}$$

is an element of $\mathscr{M}(n)$, determined by a parameter vector $\theta \in \Theta \subset \mathbf{R}^{d(n)}$, and $\eta(t)$ is a standardized noise process. When specifying such a model, a value of n and an appropriate element from within $\mathscr{M}(n)$ will have to be chosen, and therefore the integer $d(n)$, referred to as the model dimensionality, will depend on the order and the particular canonical form or structure used to represent the system. For example, if a state-space representation is being employed then θ would correspond to the freely varying elements of the quadruple (A, B, C, D) not constrained to be zero or one. The parameter space Θ is assumed to be an open subset of $\mathbf{R}^{d(n)}$ and as θ varies in Θ $\Phi(z, \theta)$ delineates an open subset $U \subset \mathscr{M}(n)$ such that the parameterization $\pi_M: U \to \Theta$ defines a homeomorphism between Θ and U. For the parameter space Θ the natural Euclidean topology is taken and for sets of transfer functions the pointwise topology associated with the norm (1.2) (Hannan, 1979, Theorem 2.6). For technical reasons it is necessary to assume that the elements of $\Phi(z, \theta)$ are twice differentiable with respect to θ and that these derivatives are continuous in θ and z.

A more detailed discussion of the ideas considered in this section is contained in Hannan (1979), Hannan and Deistler (1981), and Hannan and Kavalieris (1984). Deistler (1983) provides a thorough review with extensive references. The connection between the true theoretical structure of the process $x(t)$ as given in (1.1) and a model $M = \{\Phi(z, \theta) \in \mathscr{M}(n) : \theta \in \Theta \subseteq \mathbf{R}^{d(n)}\}$ is provided by the following definition, which concludes this section. A model M is said to be true or equivalently, to hold or to obtain, if there exists a parameter value $\theta_0 \in \Theta$ such that $\Phi(z, \theta_0) = \Psi(z)$, implying that

$$\begin{aligned} 2\pi g(\omega, \theta_0) &= \Phi(e^{i\omega}, \theta_0)\Phi(e^{i\omega}, \theta_0)^* \\ &= K(e^{i\omega}, \theta_0)\Omega(\theta_0)K(e^{i\omega}, \theta_0)^* \\ &= S(\omega) \quad \text{a.s.} \end{aligned}$$

3. LIKELIHOOD AND PSEUDO TRUE PARAMETERS

The likelihood of a model is proportional to $p(X_T \mid \theta, M)$, the probability of the data when the model is taken as true. Assuming that $x(t)$ is Gaussian, this gives rise to the support or scaled log likelihood function

$$\begin{aligned} -T^{-1} 2 \log p(X_T \mid \theta, M(\Theta)) - v \log 2\pi &= L_T(\theta, M) \\ &= \frac{\log \det \Gamma_T(\theta) + X_T^T \Gamma_T(\theta)^{-1} X_T}{T} \end{aligned} \tag{3.1a}$$

where the $Tv \times Tv$ block Toeplitz covariance matrix

$$\Gamma_T(\theta) = \int_{-\pi}^{\pi} \zeta\zeta^* \otimes g(\omega, \theta)\, d\omega,$$

the vector $\zeta = (e^{i\omega}, \ldots, e^{i\omega T})^T$, $g(\omega, \theta)$ being defined as in (2.3). Estimation methods based on a normal likelihood but applicable to more general processes are discussed in Dunsmuir and Hannan (1976) (see also Shaman, 1975). Closely related papers are those by Caines (1978), Ljung (1978), Ljung and Caines (1979), and Kabaila and Goodwin (1980). These authors consider estimating the parameters of a model by minimizing the prediction error. For Gaussian processes, of course, the best predictor in a mean square error sense is the conditional expectation, which is linear. Recognition of this property allows the likelihood to be constructed by a process of orthogonalization. If $\hat{x}(t)$ denotes the orthogonal projection of $x(t)$ onto the space spanned by $x(s)$, $1 \leqslant s < t$, then since $x(t)$ is Gaussian, $\hat{x}(t) = E[x(t) \mid x(s), 1 \leqslant s < t]$ and the error sequence defined by

$$e(t) = x(t) - \hat{x}(t)$$

yields an orthogonal process with covariance $\Sigma(t) = E[e(t)e(t)^T]$. Then following Schweppe (1965), an alternative expression for the support is given by

$$L_T(\theta, M) = T^{-1} \sum_{t=1}^{T} \log \det \Sigma(t, \theta) + e(t, \theta)^T \Sigma(t, \theta)^{-1} e(t, \theta), \tag{3.1b}$$

where the errors $e(t, \theta)$ and their covariances $\Sigma(t, \theta)$ may be obtained from the state-space representation using the Kalman filter and its associated Riccati equation or equivalently, since $\Phi(z, \theta)$ is rational, via the prediction algorithm discussed in Rissanen and Caines (1979).

Whatever version of the likelihood is employed in practice, it is important to observe that several alternative specifications are likely to be contemplated when choosing between models, and while it is necessary to adopt the premise that a model holds in order to construct the likelihood, it cannot be assumed unconditionally that any model actually obtains. Therefore, consideration is given to the asymptotic properties of the likelihood when it is explicitly recognized that the model under consideration need not encompass the data-generating mechanism.

For the purposes of theoretical developments, and in the formulation of appropriate numerical procedures, it is desirable to consider the behavior of $L_T(\theta, M)$ as θ varies over a compact set. In general, Θ will be an open subset of $\mathbf{R}^{d(n)}$ and points within the boundary $\bar{\Theta} \backslash \Theta$, the relative complement of Θ with respect to its closure $\bar{\Theta}$, will violate one or more of the model assumptions, such as the stationarity, miniphase, or rank condition. Such violations can lead to discontinuities in the likelihood (Deistler and Pötscher, 1983). What is required is a compact set $\Theta_\delta \subset \Theta$ determined such that the content of $\Theta \backslash \Theta_\delta$ is negligible, in the sense that the likelihood of any points omitted from Θ_δ can be made arbitrarily small. This may be achieved by adopting an idea due to Dunsmuir and Hannan (1976) and developed in Deistler and Pötscher

(1983). Let

$$U_\delta = \left\{\Phi \in U : \delta_1 \leqslant \min_{i \in \{1,\dots,v\}} \phi_{ii}(0) \leqslant \max_{i \in \{1,\dots,v\}} \phi_{ii}(0) \leqslant \delta_2,\ \|\Phi^{-1}\| \leqslant \delta_3\right\}.$$

LEMMA 3.1 If $\Theta_\delta := \pi_M(U_\delta)$, then Θ_δ is compact and

$$\Theta\backslash\Theta_\delta \subseteq \{\theta \in \Theta = \pi_M(U) : p(X_T \mid \theta, M) < \exp(-Ta/2)\},$$

for some $\delta = (\delta_1, \delta_2, \delta_3) = \delta(a)$, $0 < a < \infty$. Furthermore, $p(X_T \mid \theta, M)$ is uniformly continuous on Θ_δ.

Proof: Given any arbitrary sequence Φ_i in U_δ, it is possible to determine a subsequence $\Phi_{i(j)}$ such that $\Phi_{i(j)}^{-1}$ is Cauchy. The compactness of Θ_δ now follows essentially from the Riesz-Fischer theorem plus the fact that $\Phi_{i(j)}^{-1} \to \Phi^{-1}$ implies that $\Phi_{i(j)} \to \Phi$ and the parameterization π_M is, by assumption, a homeomorphism. The remaining statements of the lemma follow immediately from Deistler and Pötscher (1983): Theorem 2.3 and the results in Section 3.

Lemma 3.1 implies that there exists a $\hat{\theta}$ minimizing $L_T(\theta, M)$ and making a measurable selection for each T yields a sequence of maximum likelihood or Gaussian estimators $\hat{\theta}_T$. The asymptotic properties of this estimator are then determined by the large-sample behavior of $L_T(\theta, M)$, which is given in the following lemma.

LEMMA 3.2 Let $L_T(\theta, M)$ be defined as in (3.1a) or (3.1b). Then almost surely and uniformly in $\theta \in \Theta_\delta$, $L_T(\theta, M)$ converges to the function

$$L(\theta, M) = \log\det \Omega(\theta) + \operatorname{tr} \int_{-\pi}^{\pi} \Phi(e^{i\omega}, \theta)^{-1} S(\omega) \Phi(e^{i\omega}, \theta)^{-*}\, d\omega,$$

where $\Omega(\theta) = \phi(0, \theta)\phi(0, \theta)^T$ and $\Phi(e^{i\omega}, \theta)^{-*} = \{\Phi(e^{-i\omega}, \theta)^T\}^{-1}$.

Proof: For $L_T(\theta, M)$ defined as in (3.1a), the proof is given in Lemmas 2 and 3 of Dunsmuir and Hannan (1976) (see also Deistler et al., 1978, Sec. 3). For (3.1b) the proof proceeds exactly as for Lemma 3 of Rissanen and Caines (1979) to give

$$L_T(\theta, M) \to \log\det \Omega(\theta) + E[\operatorname{tr} \eta(t, \theta)\eta(t, \theta)^T] \quad \text{a.s.}$$

as $T \to \infty$, uniformly in $\theta \in \Theta_\delta$, where the process

$$\begin{aligned} \eta(t, \theta) &= \Phi(z, \theta)^{-1} x(t) \\ &= \phi(0, \theta)^{-1} K(z, \theta)^{-1} x(t) \\ &= \phi(0, \theta)^{-1} \varepsilon(t, \theta), \end{aligned}$$

interpreting z^{-1} as the lag operator. The result is then immediate.

The motivation behind the terminology employed in Lemma 3.1 should be clear. As the sample size increases, the likelihood will evince the plausibility of the model via $L(\theta, M)$, a continuous function of θ on Θ_δ. Let $\theta\dagger$ denote a value of θ at which the minimum of $L(\theta, M)$ is achieved; that is, $L(\theta\dagger, M) \leqslant L(\theta, M)$ for all $\theta \in \Theta_\delta$. It can be verified that

$$L(\theta, M) \geqslant \log\det \int_{-\pi}^{\pi} K(e^{i\omega}, \theta\dagger)^{-1} S(\omega) K(e^{i\omega}, \theta\dagger)^{-*}\, d\omega + v$$
$$\geqslant \log\det \psi(0)\psi(0)^T + v, \tag{3.2}$$

since $\log\det Y + \operatorname{tr}(Y^{-1}A)$ is minimized at $Y = A$ for positive definite vth-order matrices Y and A, and the one-step-ahead prediction error variance of a model is never less than the innovation variance of the process. The vector $\theta\dagger$ may then be viewed as providing the asymptotically most likely member of the model family M, the lower bound in (3.2) occurring if and only if $\Phi(z, \theta\dagger) = \Psi(z)$. If the model obtains then $\theta\dagger = \theta_0$, the unique true value of theta, but in general $\theta\dagger$ need not be unique, Åström and Söderström (1974), and, as indicated previously, this equality cannot be assumed to hold for any specification. For this reason $\theta\dagger$ will be called the *pseudo true* parameter for the model. Using Lemmas 3.1 and 3.2, we may proceed as in Dunsmuir and Hannan (1976, Sec. 3) to establish the following corollary.

COROLLARY 3.3 The Gaussian estimator $\hat{\theta}_T$ will converge with probability 1 to the pseudo true parameter set $\{\theta\dagger \in \Theta_\delta : L(\theta\dagger, M) \leqslant L(\theta, M)$ for all $\theta \in \Theta_\delta\}$. If the model obtains, $\hat{\theta}_T$ is a strongly consistent estimator of θ_0.

To determine the expected utility of a model, it will also be necessary to evaluate Fisher's measure of information pertaining to θ, the matrix of second derivatives or the curvature of Kullback's mean information integral for discriminating between values of θ per observation from the process $x(t)$. From (3.1b) the ijth element of the Hessian matrix $H_T(\theta, M) = [\partial^2 L_T(\theta, M)/\partial\theta_i \partial\theta_j]$ $i, j, = 1, \ldots, d(n)$, is given by

$$\frac{\partial^2 L_T(\theta, M)}{\partial\theta_i \partial\theta_j} = \frac{1}{T} \sum \operatorname{tr}\left\{\frac{\partial^2 \Sigma}{\partial\theta_i \partial\theta_j}(\Sigma^{-1} - \Sigma^{-1}ee^T\Sigma^{-1})\right\}$$
$$-\operatorname{tr}\left\{\Sigma^{-1}\frac{\partial\Sigma}{\partial\theta_i}\left(\Sigma^{-1}\frac{\partial\Sigma}{\partial\theta_j} - \Sigma^{-1}ee^T\Sigma^{-1}\frac{\partial\Sigma}{\partial\theta_j} - \Sigma^{-1}\frac{\partial\Sigma}{\partial\theta_j}\Sigma^{-1}ee^T\right)\right\}$$
$$-2\left\{e^T\Sigma^{-1}\frac{\partial\Sigma}{\partial\theta_i}\Sigma^{-1}\frac{\partial e}{\partial\theta_j} + \frac{\partial e^T}{\partial\theta_i}\Sigma^{-1}\frac{\partial\Sigma}{\partial\theta_j}\Sigma^{-1}e\right\}$$
$$+2\left\{e^T\Sigma^{-1}\frac{\partial e}{\partial\theta_i\partial\theta_j} + \frac{\partial e^T}{\partial\theta_i}\Sigma^{-1}\frac{\partial e}{\partial\theta_j}\right\},$$

the arguments t and θ having been omitted for simplicity. Evaluating $E[\partial^2 L_T(\theta, M)/\partial\theta_i, \partial\theta_j]$ as a nested sequence of conditional expectations, noting that by construction $E[e(t,\theta)e(t,\theta)^T] = \Sigma(t,\theta)$ and $E[e(t,\theta)\,\partial e(t,\theta)^T/\delta\theta_i] = 0$ yields Fisher's information

$$i_T(\theta_i, \theta_j, M) = \frac{1}{T}\sum_{t=1}^{T} \frac{1}{2}\operatorname{tr}\left\{\Sigma(t,\theta)^{-1}\frac{\partial\Sigma(t,\theta)}{\partial\theta_i}\Sigma(t,\theta)^{-1}\frac{\partial\Sigma(t,\theta)}{\partial\theta_j}\right\} + \frac{\partial e(t,\theta)^T}{\partial\theta_i}\Sigma(t,\theta)^{-1}\frac{\partial e(t,\theta)}{\partial\theta_j},$$

(see Cooper and Wood, 1981).

LEMMA 3.4 Let $\mathscr{I}_T(\theta, M) = [i_T(\theta_i, \theta_j, M)]$, $i, j = 1, \ldots, d(n)$. Then $\mathscr{I}_T(\theta, M)$ and $\frac{1}{2}H_T(\theta, M)$ converge to the matrix $\mathscr{I}(\theta, M)$ with typical ijth element

$$\frac{1}{2}\operatorname{tr}\left\{\Sigma(\theta)^{-1}\frac{\partial\Sigma(\theta)}{\partial\theta_i}\Sigma(\theta)^{-1}\frac{\partial\Sigma(\theta)}{\partial\theta_j}\right\} + \int_{-\pi}^{\pi}\Phi(e^{i\omega},\theta)^{-1}\frac{\partial\Phi(e^{i\omega},\theta)}{\partial\theta_i}\Phi(e^{i\omega},\theta)^{-1}S(\omega)\Phi(e^{i\omega},\theta)^{-*}\frac{\partial\Phi(e^{i\omega},\theta)^*}{\partial\theta_j} \times \Phi(e^{i\omega},\theta)^{-*}\,d\omega.$$

Proof: It is easily shown that the derivative sequence $\partial e(t,\theta)/\partial\theta_i$ is related to the process $\zeta_i(t,\theta) = \partial\varepsilon(t,\theta)/\partial\theta_i$, given by

$$N_K(z,\theta)\zeta_i(t,\theta) = \left\{\frac{\partial d_K(z,\theta)}{\partial\theta_i}K(z,\theta) - \frac{\partial N_K(z,\theta)}{\partial\theta_i}\right\}\varepsilon(t,\theta),$$

in a way that exactly parallels the relationship between $e(t,\theta)$ and $\varepsilon(t,\theta)$. The proof is then obtained in a manner completely analogous to the developments in the Appendix of Rissanen and Caines (1979). The details, which are straightforward but tedious, are omitted.

Remark: Presume that the model obtains and partition θ as $(\sigma^T : \tau^T)^T$, where σ determines the scale parameters of Ω and is independent of the structural parameters of the system τ that determine $K(z)$. In this case $\mathscr{I}(\theta_0, M)$ simplifies to the usual expressions given for the information measure. See Dunsmuir and Hannan (1976, pp. 357–358, 361–362).

4. ASSESSING MODEL UTILITY

The utility associated with the action of choosing a particular model will depend in part on the intended use of the model, for prediction or interpolation, for example. Whatever the ultimate objective of the analysis, however, it is often the case that the performance of a model may be evaluated by reference to the discrepancy between $S(\omega)$ and $g(\omega, \theta)$. In practice, of course, the power spectrum $S(\omega)$ will not be known and given a collection of models M_i, $i = 1, \ldots, m$, what will be perceived as the best action will also depend on the extent of available knowledge concerning the true system. For practical purposes it will be necessary to consider a utility measure that depends upon and can be modified by the observations.

Define the sample power spectrum or periodogram by

$$P_T(\omega) = (2\pi T)^{-1} Z_T(\omega) Z_T(\omega)^*,$$

$$Z_T(\omega) = \sum_{t=1}^{T} x(t) \exp(i\omega t),$$

and let $a(\theta, M)$ denote the action of choosing from the family M the particular member given by the parameter value θ. For convenience the model subscript $i = 1, \ldots, m$ will be omitted where this raises no ambiguity. A natural quantity to consider when assessing the regret or loss involved in taking action $a(\theta, M)$ is the integrated squared error, or discrepancy,

$$\rho_T\{a(\theta, M)\} = \|P_T(\omega) - g(\omega, \theta)\|^2, \tag{4.1a}$$

or if it is the proportionate difference between $S(\omega)$ and $g(\omega, \theta) = (2\pi)^{-1}\Phi(e^{i\omega}, \theta)\Phi(e^{i\omega}, \theta)^*$ that is felt to be important, the mean square relative error

$$\rho_T\{a(\theta, M)\} = \|2\pi\Phi(e^{i\omega}, \theta)^{-1} P_T(\omega) \Phi(e^{i\omega}, \theta)^{-*} - I\|^2. \tag{4.1b}$$

Assuming that such regrets are measured in Naperian units and acknowledging that any value of the error (4.1) is likely to provide greater utility if it can be achieved with a parsimonious rather than a profligate model, the associated utility will be taken as

$$\exp[-\tfrac{1}{2}d(n)\rho_T\{a(\theta, M)\}]. \tag{4.2a}$$

Such a utility function may be regarded as providing an approximation to the true preference ordering obtained in the unrealistic situation where $S(\omega)$ is known, and if it is to be employed, it should not lead to decisions with substantially less utility than would have been obtained had the true preferences been available. In the terminology of Kadane and Chuang (1978), but with the added real-world interpretation that the index set over which sequences are defined corresponds to time, it is necessary for the utility function to yield a stable decision problem.

PROPOSITION 4.1 Preference orderings that are induced by $\exp[-\frac{1}{2}d(n)\rho_T\{a(\theta, M)\}]$ determine strongly stable decision problems when paired with any convergent sequence of posterior distributions or opinions.

Proof: Since $\exp[-\frac{1}{2}d(n)\rho_T\{a(\theta, M)\}]$ is continuous in θ and bounded for all T, by Theorem 2 of Kadane and Chuang (1978) it is sufficient to show that it converges uniformly in θ with probability 1. Consider

$$\rho_T\{a(\theta, M)\}$$
$$= \operatorname{tr}\left[\int_{-\pi}^{\pi} P_T(\omega)P_T(\omega)^* - 2\operatorname{Re}\{P_T(\omega)g(\omega, \theta)^*\} + g(\omega, \theta)g(\omega, \theta)^*\, d\omega\right], \tag{4.2b}$$

where $\operatorname{Re}\{\cdot\}$ denotes the real part of the argument. The last term in this expression is determined uniquely by $a(\theta, M)$. Given $a(\theta, M)$ the model assumptions imply that $g(\omega, \theta)$ is of bounded variation in ω, and by Theorem 7.6.1 of Brillinger (1975) the cumulants of the second term of order $k \geqslant 2$ are bounded by $T^{-k+1}N$ for some fixed $N < \infty$. Applying the method of proof employed in Brillinger (1975, pp. 405–407) using the cumulant generating function, Markov's inequality, and the Borel-Cantelli lemma, it follows that

$$\left|\operatorname{tr}\int \{S(\omega) - P_T(\omega)\}g(\omega, \theta)^*\, d\omega\right| = O\left\{\left(\frac{\log T}{T}\right)^{1/2}\right\} \quad \text{a.s.}$$

The convergence is uniform in θ because the mean value theorem implies that

$$|g(\omega, \theta_1)_{i,j} - g(\omega, \theta_2)_{i,j}| < \frac{\delta \sum_{h=1}^{d(n)} \sup \partial g(\omega, \theta)_{i,j}}{\partial \theta_h},$$

where $\|\theta_1 - \theta_2\| < \delta$ and the supremum is taken over $\omega \in [-\pi, \pi]$ and $\theta \in N_\delta(\theta_1) = \{\theta : \|\theta_1 - \theta\| < \delta\}$. Consequently, there exists a constant $N < \infty$ such that

$$\operatorname{tr}\int_{-\pi}^{\pi} P_T(\omega)\{g(\omega, \theta_1) - g(\omega, \theta_2)\}^*\, d\omega < \delta N \operatorname{tr}\int_{-\pi}^{\pi} P_T(\omega)\, d\omega,$$

and hence $\operatorname{tr}\int P_T(\omega)g(\omega, \theta)\, d\omega$ is equicontinuous and converges uniformly in θ. Turning to the first term, standard results (Brillinger, 1975, Corollaries 7.2.1 and 7.2.2) indicate that

$$\lim_{T\to\infty} E\left[\operatorname{tr}\int_{-\pi}^{\pi} P_T(\omega)P_T(\omega)^*\, d\omega\right] = \int_{-\pi}^{\pi} \operatorname{tr}(S(\omega)S(\omega)^*) + \{\operatorname{tr} S(\omega)\}^2\, d\omega,$$

and a similar argument to that described above indicates the almost sure convergence of this term to its asymptotic expectation. A parallel derivation shows that the mean squared relative error also converges in the manner required. For a detailed proof in the univariate case, see Poskitt (1986a).

Clearly, a utility function based solely on $\rho_T\{a(\theta, M)\}$ takes no account of other *desiderata* relating to different features of the model that the practitioner may regard as important. For example, it seems reasonable to suppose that other things being equal, a preference would be revealed for models that are simpler and/or whose parameters can be estimated with greater precision. A measure of the utility associated with the posterior precision and complexity of the model proposed by Poskitt (1987a) in the context of general statistical modeling is

$$\exp\left[-\tfrac{1}{2}d(n)\log\left\{\frac{\operatorname{tr} H_T(\theta, M)^{-1}}{d(n)}\right\}\right]. \tag{4.3}$$

Expression (4.3) reflects, first, the notion that the precision with which θ can be estimated is governed directly by the curvature of the likelihood function, and second, that the dispersion matrix of any random variable can be used to determine an information theoretic judgment as to its stochastic complexity (Van-Emden, 1971). In addition, Bernardo (1979) has argued that the utility to be derived from a model can be equated to expected gains in information, and (Poskitt 1987a) this can be characterized, locally at least, by the function

$$\exp\left[-\tfrac{1}{2}d(n)\frac{1 + \log 2\pi - \log\{p(\theta\,|\,M)/\det H_T(\theta, M)\}}{d(n)}\right], \tag{4.4}$$

where $p(\theta\,|\,M)$ denotes the prior distribution of the parameter θ given the model M. Combining these different components by taking the product of (4.2) with (4.3) and (4.4) results in an overall assessment of the utility of the action $a(\theta, M)$, $U_T\{a(\theta, M)\}$, say, that blends together different qualitative features of the model that the practitioner is likely to bear in mind when considering the relative merits of alternative specifications.

5. LARGE-SAMPLE BAYES' DECISION RULE

Given a set or range of models $M_i = \{\Phi_i(z, \theta_i) \in \mathscr{M}(n_i) : \theta_i \in \Theta_i \subset \mathbf{R}^{d_i(n_i)}\}$ $i = 1, \ldots, m$, a decision rule is a function specifying for any possible data set X_T the action $a(\theta, M)$ that is to be taken. The decision function that maximizes

$$v_T(M) = \int_\Theta U_T\{a(\theta, M)\}\frac{p(M, \theta)p(X_T\,|\,\theta, M)}{p(X_T)}\,d\theta, \tag{5.1}$$

where

$$p(X_T) = \sum_{i=1}^{m} p(M_i)p(X_T\,|\,M_i)$$

and

$$p(X_T\,|\,M) = \int_\Theta p(\theta\,|\,M)p(X_T\,|\,\theta, M)\,d\theta$$

(i.e., expected utility where the expectation is taken with respect to the posterior distribution of the model and its parameters) is termed the Bayes decision rule. Using this decision rule, a comparison of alternative decisions is made by reference to the data currently available, an examination of decisions that might have been taken given different realizations that might have been obtained but which in fact were not observed is not required. Contemplation of the Bayes decision rule in conjunction with the utility structure and results presented above leads to the following theorem.

THEOREM 5.1 Suppose that models are held to be equally likely a priori and that utility is measured by reference to expected gains in information, posterior precision and complexity, and mean squared discrepancy. Then the Bayes decision rule is given asymptotically by the decision function $D(X_T) = a(\hat{\theta}_K, M_K)$, where

$$M_K = \arg\min_{i=1,\ldots,K} \Delta_T(\hat{\theta}_{iT}, M_i)$$

$$\Delta_T(\hat{\theta}_T, M) = L_T(\hat{\theta}_T, M) + d(n)\frac{1 + \log\{T \operatorname{tr} H_T(\hat{\theta}_T, M)^{-1}/d(n)\} + \rho_T\{a(\hat{\theta}_T, M)\}}{T}$$

Proof: Using Lemmas 3.1, 3.2, and 3.4 together with Proposition 4.1, it follows that regularity conditions *A* of Poskitt (1987a) are satisfied, and using Lemma 2.1 of that paper yields the theorem.

Expressing the delta criterion in the form $L_T(\hat{\theta}_T, M) + d(n)C_T/T$, it is clear that it is of the same generic structure as AIC and BIC, and in practice, Δ_T is applied in a similar manner. Preference is given to the specification that minimizes the value of the criterion function over the range of models being considered. It also follows from the results of Hannan (1981) that Δ_T will be strongly consistent in the sense that it will ultimately reveal a preference for a true model of lowest dimensionality. As emphasized above, however, there is no assumption that the model set $\{M_i^i = 1, \ldots, m\}$ will encompass the data-generating mechanism. Indeed, all specifications under consideration are likely to be only approximations to reality and no true model will exist, reality will generally contain far more complicacy than any model. From this viewpoint the property of consistency may not be an overriding consideration and other optimality properties (Shibata, 1980) may be more important. Nevertheless, the fiction that there exists a true model will provide some guide to the behavior of the criterion in situations where $x(t)$ is near a process that can be so parameterized.

The feature that distinguishes Δ_T from model selection criteria such as AIC and BIC is that the penalty adjustment term depends not only on the dimensionality $d(n)$ and sample size T but also on other performance characteristics of the model. These features will influence Δ_T in particular ways. Assume that M_1 is a true specification and is to be compared with an alternative model M_2. Let

$$R_T(M_1 : M_2) = \exp[-\tfrac{1}{2}T\{\Delta_T(\hat{\theta}_{1T}, M_1) - \Delta_T(\hat{\theta}_{2T}, M_2)\}], \tag{5.2}$$

the ratio of the expected utility of the first model to that of the second. A second-order Taylor series expansion of the logarithm of the determinant of a matrix expressed as a function of the matrix elements shows that

$$\begin{aligned}\log\det 2\pi g_2(\omega, \theta_2) &+ \operatorname{tr}\{g_1(\omega, \theta_1)g_2(\omega, \theta_2)^{-1}\}\\ &= v + \log\det 2\pi g_1(\omega, \theta_1)\\ &\quad + \|\Phi_2(e^{i\omega}, \theta_2)^{-1}g_1(\omega, \theta_1)\Phi_2(e^{i\omega}, \theta_2)^{-*} - I\|^2\\ &\quad + E_2\|g_1(\omega, \theta_1) - g_2(\omega, \theta_2)\|,\end{aligned} \tag{5.3}$$

where the remainder $E_2 \to 0$ as $\|g_1(\omega, \theta_1) - g_2(\omega, \theta_2)\| \to 0$. Substituting in Δ_T, using Lemma 3.2, Corollary 3.3, Lemma 3.4, and Proposition 4.1, it follows that to this order of approximation

$$R_T(M_1 : M_2) = \exp[\tfrac{1}{2}Q]\exp[\tfrac{1}{2}(\alpha T + \beta)]T^{\gamma}\exp[\tfrac{1}{2}\log(P)] \quad \text{a.s.}, \tag{5.4}$$

where

$$Q = \frac{d_2(n_2)}{2\pi}\int_{-\pi}^{\pi}\{\operatorname{tr} g_1(\omega, \theta_1^{\dagger})g_2(\omega, \theta_2^{\dagger})^{-1}\}^2\,d\omega - v^2 d_1(n_1),$$

$$\alpha T + \beta = \{T + d_2(n_2)\}\cdot\|\Phi_2(e^{i\omega}, \theta_2^{\dagger})^{-1}g_1(\omega, \theta_1^{\dagger})\Phi_2(e^{i\omega}, \theta_2^{\dagger})^{-*} - I\|^2,$$

$$\gamma = \tfrac{1}{2}\{d_2(n_2) - d_1(n_1)\},$$

and

$$P = \frac{\{\operatorname{tr}\mathcal{I}(\theta_2^{\dagger}, M_2)^{-1}/d_2(n_2)\}^{d_n(n_2)}}{\{\operatorname{tr}\mathcal{I}(\theta_1^{\dagger}, M_1)^{-1}/d_1(n_1)\}^{d_1(n_1)}}.$$

Thus if $\|\Phi_2(e^{i\omega}, \theta_2^{\dagger})^{-1}g_1(\omega, \theta_1^{\dagger})\Phi_2(e^{i\omega}, \theta_2^{\dagger})^{-*} - I\| > 0$, as will happen if $n_2 < n_1$, then $\alpha, \beta > 0$ and $R_T(M_1 : M_2) \to \infty$ as $T \to \infty$, so that the true model M_1 will ultimately be determined. The dominant second term in (5.4) is, however, amplified or attenuated by $\exp[\tfrac{1}{2}Q]T^{\gamma}\exp[\tfrac{1}{2}\log(P)]$ according to the relative magnitudes of $g_1(\omega, \theta_1^{\dagger}) = S(\omega)$ and $g_2(\omega, \theta_2^{\dagger})$, $d_1(n_1)$ and $d_2(n_2)$, and the posterior precision and complexity of the two models. If $n_2 > n_1$ and none of the structural indices are underspecified, then M_2 will also obtain. Although now $\alpha = \beta = 0$, $\theta_2^{\dagger}$ will lie in an affine subspace of $\mathbf{R}^{d_2(n_2)}$ (Deistler and Hannan, 1981, Theorem 3), and $\mathcal{I}(\theta_2^{\dagger}, M_2)$ will be singular. This will manifest itself in the value of $\exp[\tfrac{1}{2}\log(P)]$, $\gamma > 0$, and R_T will again increase without bound, indicating a preference for the true model of smallest degree.

However, even if $n_2 = n_1$ and a correct set of structural indices are specified so that $\alpha = \beta = \gamma = Q = 0$, a similar effect may occur. As shown in Wertz et al. (1983, App.), the Jacobian of the transformation between overlapping parameterizations will have maximum and minimum eigenvalues approaching infinity and zero, respectively, near the edge of a coordinate system, implying that $\exp[\frac{1}{2}\log(P)] \to \infty$ if M_2 specifies a bad set of structural indices. These arguments indicate the possible efficacy of using the criterion Δ_T for order determination and structure selection.

As a simple numerical illustration, consider the scalar process with transfer function

$$\Psi(z) = \frac{z^2 + 0.285z}{z^2 - 1.463z + 0.616}.$$

In simulations of 250 realizations of this process with $T = 98$ observations, Δ_T selected the true process on 65.4% of occasions when choosing between the correct model and a pure autoregression of order 2. This is to be compared to 56.8% for BIC. These figures must be judged in the light of the fact that this example has been chosen deliberately to exemplify the problems that model selection criteria can encounter. Although the coefficient of z in the numerator of $\Psi(z)$ is not small, $\Psi(z)$ is well approximated by

$$\Phi(z, \theta^{\dagger}) = \frac{z^2}{z^2 - 1.580z + 0.725},$$

$$(2\pi)^{-1} \int_{-\pi}^{\pi} \left| \frac{\Psi(e^{i\omega})}{\Phi(e^{i\omega}, \theta^{\dagger})} \right|^2 d\omega = 1.045.$$

Thus little may be lost by adopting the more parsimonious parameterization. Both criteria will, of course, reveal a preference for the true process almost always as T increases, but the evidence from this example indicates that the delta criterion's dependence on a wider range of performance characteristics can reduce the risk of selecting an incorrect specification. For some additional evidence on the practical performance of Δ_T, see Poskitt (1986b, 1987b).

6. CONCLUSION

In this chapter it has been shown how decision-theoretic arguments can be used to determine a model selection criterion, Δ_T, which is similar in form to AIC and BIC but which takes account of performance characteristics of the model other than the prediction error variance and model dimensionality.

The criterion is employed in a manner parallel to the use of other criteria generally current. Theoretical arguments and some empirical evidence indicate that use of Δ_T for order determination and structure selection is likely to prove beneficial.

Finally, on the question of practical implementation, it should perhaps be pointed out that when assessing the relative merits of different models using a criterion such as Δ_T to contemplate searching over a range of multi-indices $\{n_1, \ldots, n_v\}$ such that $\sum n_i \leqslant N$, N fixed a priori, presents a daunting task. The large number of possible specifications that would have to be examined even when v and N are quite small makes such a strategy unattractive, if not prohibitive. For example, if $v = 2$ and $N = 10$, a total of 66 different models would have to be examined, some with over 40 parameters, and if $v = 3$, the corresponding figures would be 286 and 60. A possible way of proceeding might be to conduct a preliminary data analysis in an attempt to ascertain specifications that might warrant closer investigation. Alternatively, since it is not necessary to examine every overlapping parameterization for each order, attention could be focused initially on the generic neighborhood where the first n rows of H form a basis (i.e., $n_1 = n_2 = \cdots = n_q = n_{q+1} + 1 = \cdots = n_v + 1$), this corresponds to an echelon canonical form that is dense in $\mathscr{M}(n)$. Such a procedure has been suggested by Hannan and Kavalieris (1984), who also discuss algorithm construction. What is the most effective search strategy in the context of order determination and structure selection still seems, however, to be an open question.

INDEX OF NOTATION

$x(t)$	Vector process of v variables
$\Psi(z)$ and $S(\omega)$	Transfer function and power spectrum of process
$\Phi(z)$ and $g(\omega)$	Transfer function and power spectrum of model
$\mathscr{M}(n)$	Set of proper rational transfer functions of McMillan degree n with neither zeros nor poles outside the unit circle
$M = \{\Phi(z, \theta) \in \mathscr{M}(n) : \theta \in \Theta \subseteq \mathbf{R}^{d(n)}\}$	Statistical model set with parameter space theta of dimension $d(n)$
$L_T(\theta, M)$	Scaled log likelihood
$\hat{\theta}_T$	Maximum likelihood estimator
$H_T(\theta, M) = [\partial^2 L_T(\theta, M)/\partial\theta_i\partial\theta_j]$	Hessian matrix
$\mathscr{I}_T(\theta, M) = [i_T(\theta_i, \theta_j, M)]$	Fisher's measure of information

$a(\theta, M)$	Action of choosing from model M element characterized by parameter value θ
$\rho_T\{a(\theta, M)\}$	Squared error or discrepancy
$U_T\{a(\theta, M)\}$	Utility of action
$\Delta_T(\hat{\theta}_T, M)$	Model selection criterion

REFERENCES

Akaike, H. (1974). A new look at the statistical model identification, *IEEE Trans. Autom. Control* AC-19, 716–723.

Åström, K. J., and T. Söderström (1974). Uniqueness of the maximum likelihood estimates of the parameters of an ARMA model *IEEE Trans. Autom. Control* AC-19, 769–73.

Barnett, S. (1971). *Matrices in Control Theory*, Van Nostrand Rienhold, London.

Bernardo, J. M. (1979). Expected information as expected utility, *Ann. Stat.* 7, 686–690.

Brillinger, D. R. (1975). *Time Series: Data Analysis and Theory*, Holt, Rinehart and Winston, New York.

Caines, P. E. (1978). Stationary linear and nonlinear identification and predictor set completeness, *IEEE Trans. Autom. Control* AC-23, 583–594.

Clark, J. H. C. (1976). The Consistent Selection of Parameterizations in System Identification, presented at the *Joint Automatic Control Conference*, Purdue University, West Lafayette, Ind.

Cooper, D. M., and E. F. Wood (1981). Estimation of the parameters of the Markovian representation of the autoregressive-moving average model, *Biometrika* 68, 320–322.

Deistler, M. (1985). General Structures and Parameterization of ARMA and State Space Systems and Its Relation to Statistical Problems, in *Handbook of Statistics*, Vol. 5 (E. J. Hannan, P. R. Krishnaiah, and M. M. Rao, eds.), North-Holland, Amsterdam.

Deistler, M., and E. J. Hannan (1981). Some properties of the parameterisation of ARMA systems of unknown order, *J. Multivar. Anal.* 11, 474–484.

Deistler, M., and B. M. Pötscher (1983). The behaviour of the likelihood function for ARMA models, *Adv. Appl. Probab.* 16, 843–866.

Deistler, M., W. Dunsmuir, and E. J. Hannan (1978). Vector linear time series models: corrections and extensions, *Adv. Appl. Probab.* 10, 360–372.

Denham, M. J. (1974). Canonical forms for the identification of multivariate linear systems, *IEEE Trans. Autom. Control* AC-19, 646–656.

Dickinson, B. W., T. Kailath, and M. Morf (1974). Canonical matrix fraction and state space descriptions of deterministic and stochastic linear systems, *IEEE Trans. Autom. Control* AC-19, 656–666.

Dunsmuir, W., and E. J. Hannan (1976). Vector linear time series models, *Adv. Appl. Probab.* 8, 339–364.

Guidorzi, R. (1981). Invariants and canonical forms for systems structural and parametric identification, *Automatica* 17, 117–133.

Hannan, E. J. (1979). The Statistical Theory of Linear Systems, in *Developments in Statistics*, Vol. 2 (P. R. Krishnaiah, ed.), Academic Press, New York.

Hannan, E. J. (1981). Estimating the dimension of a linear system, *J. Multivar. Anal.* 11, 458–473.

Hannan, E. J., and L. Kavalieris (1984). Multivariate linear time series models, *Adv. Appl. Probab.* 16, 492–561.

Kabaila, P. V., and G. C. Goodwin (1980). On the estimation of the parameter of an optimal interpolator when the class of interpolators is restricted, *SIAM J. Control Optimization* 18, 121–144.

Kadane, J. B., and D. T. Chuang (1978). Stable decision problems, *Ann. Stat.* 6, 1095–1110.

Kailath, T. (1980). *Linear Systems*, Prentice-Hall, Englewood Cliffs, N.J.

Kalman, R. E., P. L. Falb, and M. A. Arbib (1969). *Topics in Mathematical System Theory*, McGraw-Hill, New York.

Ljung, L. (1978). Convergence analysis of parametric identification methods, *IEEE Trans. Autom. Control* . AC-23, 770–783.

Ljung, L., and P. E. Caines (1979). Asymptotic normality of prediction error estimation for approximate system models, *Stochastics* 3, 29–46.

Poskitt, D. S. (1986a). A Bayes procedure for the identification of univariate time series models, *Ann. Stat.* 14, 502–516.

Poskitt, D. S. (1986b). Bayes ARMA model determination: some empirical evidence, *Austral. J. Statist.*, to appear.

Poskitt, D. S. (1987a). Precision, complexity and Bayesian model determination, *J. R. Stat. Soc. Ser. B* 49, 199–208.

Poskitt, D. S. (1987b). A modified Hannan-Rissanen strategy for mixed autoregressive moving-average order determination, *Biometrika* 74, 781–790.

Rissanen, J. (1974). Bases of invariants and canonical forms for linear dynamic systems, *Automatica* 10, 175–182.

Rissanen, J. (1978). Modelling by shortest data description, *Automatica* 14, 465–471.

Rissanen, J., and P. E. Caines (1979). The strong consistency of maximum likelihood estimators for ARMA processes, *Ann. Stat.* 7, 297–315.

Schwarz, G. (1978). Estimating the dimension of a model, *Ann. Stat.* 6, 461–464.

Schweppe, F. C. (1965). Evaluation of likelihood functions for Gaussian signals, *IEEE Trans. Inf. Theory* 11, 61–70.
Shaman, P. (1975). An approximate inverse for the covariance matrix of moving-average and autoregressive processes, *Ann. Stat.* 3, 532–538.
Shibata, R. (1980). Asymptotically efficient selection of the order of the model for estimating parameters of a linear process, *Ann. Stat.* 8, 147–164.
Shore, R. W. (1980). A Bayesian Approach to the Spectral Analysis of Stationary Time Series, in *Bayesian Analysis in Economic Theory and Time Series Analysis*, Vol. II (A. Zellner and J. B. Kadane, eds.), North-Holland, Amsterdam.
Van Emden, M. H. (1971). *An Analysis of Complexity*, Mathematical Centre Tracts, Vol. 35, Amsterdam.
Wertz, V., M. Gevers, and E. J. Hannan (1982). The determination of optimal structures for the state space representation of multivariate stochastic processes, *IEEE Trans. Autom. Control.* AC-27, 1200–1210.
Wiener, N., and P. Masani (1957). The prediction theory of multivariate stochastic processes: 1. The regularity condition, *Acta Math.* 98, 111–150.

4

Survey of Bayesian and Non-Bayesian Testing of Model Stability in Econometrics

HIROKI TSURUMI Department of Economics, Rutgers University, New Brunswick, New Jersey

1. INTRODUCTION

It is recognized that a physical entity experiences structural change as it evolves over time. In economics the notion of structural change has been prevalent in various theories of economic development. Karl Marx, for example, presented different stages of economic development, from primitive to capitalistic and postcapitalistic society. A transition from one stage to another implies that economic relationship changes, and such a change had been explained often in descriptive fashion until the 1950s. With the introduction of regression analysis as the principal tools of economic data processing in the 1950s and 1960s, attempts were made to describe changes of economic relationship in regression framework. Sengupta and Tintner (1963), for example, used long-run per capita U.S. data (1869–1953) and fitted a logistic curve to each of roughly four equal subperiods to show that the upper asymptote of the income curve followed an increasing sequence from one subperiod to another. The upward shift of the upper asymptote is demonstrated in a descriptive manner, rather than testing whether such a shift is statistically significant.

Since econometric models are often specified in regression framework, model stability may be defined as a switch in a regression equation from one subsample period (or regime) to another. Chow (1960) proposed an *F* test for the case where there are two regression regimes and the join point to separate observations into two subsamples is known. By the 1970s the F criterion that was proposed by Chow had been used extensively in empirical studies to test model stability. As long as it is known to which regime each observation belongs, the F test can be applied either to time series or to cross-sectional data; a historical time point or some threshold value of a key variable may serve as the join point. For example, Hamermesh (1970) estimated wage equations that switch regimes at the threshold value of the price index. Ever since Goldfeld (1976) examined the stability of the U.S. demand-for-money equation, stability of regression equation has become one of the major issues of empirical studies of demand for money. The examination of stability of the money demand equation has been done in two different ways: examination of postsample forecast performances by descriptive or inferential statistical criteria, and hypothesis testing of parameter shifts.

Changes in regression parameters bring with them two separate issues: the one is on join points and the other is on the significance of parameter changes. In some cases researchers may have information on exactly when parameters changed and they need to test the significance of parameter shifts. In other cases researchers may have some vague information on the possible time interval in which parameters changed, and thus they need to make inferences on the exact join point as well as on the significance of parameter shifts.

In this chapter we present a survey of statistical tests of model stability in econometrics. The tests available in the literature and some new tests are presented, whenever possible, primarily from the Bayesian point of view, so that the reader will see that some non-Bayesian tests, such as the Chow test, have a Bayesian interpretation. First we present ways to test the significance of model stability when the join point is known exactly. This is done within the classical linear regression framework, and we present the U.S. demand for money as an example. Next, we discuss inference about the unknown join point and gradual switching regressions in which the join point is not known and it takes some time to adjust to the second regime. We then present tests of parameter stability without using the join point. This is done by specifying the model in state-space representation in which regression coefficients vary from one period to another. Gradual switching regressions and time-varying parameter models have many similarities to "fault detection" models in the control literature. In fault detection, linear stochastic models in state-space form are maintained in the null and alternative hypotheses, whereas in the tests of econometric model stability, state-space representation is employed in such a way that under the null hypothesis of no change of regression parameters, the model is reduced to the classical linear regression case. The main concern of fault detection models and time-varying regression models are the same. Both are concerned with the time and magnitude of an abrupt change in a model. More discussions on this are given later. Finally, we discuss tests of constancy of a structural equation within a simultaneous equation system.

2. TESTS OF CONSTANCY OF REGRESSION MODELS WITHIN THE CLASSICAL LINEAR REGRESSION FRAMEWORK

We shall present tests of constancy of regression models when the join point is known. Let a regression line have two regimes:

$$\begin{aligned} y_1 &= X_1\beta_1 + \varepsilon_1, \qquad \varepsilon_1 \sim N(0, \sigma_1^2 I_{n_1}), \\ y_2 &= X_2\beta_2 + \varepsilon_2, \qquad \varepsilon_2 \sim N(0, \sigma_2^2 I_{n_2}), \end{aligned} \tag{2.1}$$

where y_i and ε_i are $n_i \times 1$ vectors of dependent variables and of error terms $(i = 1, 2)$, respectively. X_i is a $n_i \times k$ matrix of explanatory variables $(i = 1, 2)$, and β_i is a $(k \times 1)$ vector of parameters in regime i $(i = 1, 2)$. Equation (2.1) can be rewritten as

$$z = W\zeta + \varepsilon, \tag{2.2}$$

where $z^T = (y_1^T,\ y_2^T)^T$, $W = (X,\ V)$, $X = (X_1^T,\ X_2^T)^T$, $V = (0^T,\ \omega X_2^T)^T$, $\omega = \sigma_1/\sigma_2$, $\zeta = (\beta_1^T, \delta^T)^T$, $\delta = \beta_2 - \beta_1$, and $\varepsilon = (\varepsilon_1^T, \varepsilon_2^T)^T$. If $\delta = 0$, then we have $\beta_1 = \beta_2$ in equations (2.1), implying that the single regression equation holds true for the entire sample periods of $n_1 + n_2$. If $\delta \neq 0$, on the other hand, equations in (2.1) mean that regression coefficients, β_1, in regime 1 switch abruptly at time $n_1 + 1$ to β_2, but in each regime they stay constant. Accordingly, the null hypothesis of constancy of two regressions is given by H: $\delta = 0$ against the alternative hypothesis K: $\delta \neq 0$.

We shall first formulate the likelihood ratio test, and see that it can be derived as a Bayes factor. The likelihood function under the null hypothesis H is given by

$$l(\beta_1, \sigma_1, \omega|\text{data}) = (2\pi)^{-n/2}\sigma_1^{-n}\omega^{n_2}\exp\left[-\frac{1}{2\sigma_1^2}(y - X\beta_1)^T(y - X\beta_1)\right] \tag{2.3}$$

where $n = n_1 + n_2$. On the other hand, the likelihood function under the alternative hypothesis K is

$$l(\beta_1, \delta, \sigma_1, \omega|\text{data}) = (2\pi)^{-n/2}\sigma_1^{n}\omega^{n_2}\exp\left[-\frac{1}{2\sigma_1^2}(z - W\zeta)^T(z - W\zeta)\right] \tag{2.4}$$

From equations (2.3) and (2.4), we obtain an expression for $-2\log\lambda$, where λ

is the likelihood ratio:

$$-2 \log \lambda = -2n_2 \log \frac{\hat{\omega}_H}{\hat{\omega}_K} + n \log \frac{z^T M_x z}{z^T M_w z} \tag{2.5}$$

where $M_w = I - W(W^T W)^{-1} W^T$, and $M_x = I - X(X^T X)^{-1} X^T$. The expressions $\hat{\omega}_H$ and $\hat{\omega}_K$ in (2.5) are the maximum likelihood estimates (MLEs) of ω under H and K, respectively; $z^T M_x z$ and $z^T M_w z$ are evaluated by $\hat{\omega}_H$ and $\hat{\omega}_K$, respectively. It is well known that $-2 \log \lambda$ is distributed as χ^2 with k degrees of freedom as $n_1, n_2 \to \infty$.

If we know the value of ω to be ω_0, we can form the likelihood ratio given the value of ω, $\omega = \omega_0$:

$$\begin{aligned} -2 \log \lambda &= n \log \frac{z^T M_x}{z^T M_w z} \\ &= n \log \left[1 + \frac{z^T M_x V(V^T M_x V)^{-1} V^T M_x z}{z^T M z} \right] \\ &= n \log \left[1 + \omega_0^2 \frac{k}{n - 2k} F(k, n - 2k | \omega_0) \right], \end{aligned} \tag{2.6}$$

where $F(k, n - 2k|\omega_0) = [(\text{SSR}_c - \text{SSR}_u)/k]/[\text{SSR}_u/(n - 2k)]$, $\text{SSR}_c = z^T M_x z$, $\text{SSR}_u = z^T M_w z = u_1^T u_1 + \omega_0^2 u_2^T u_2$, $u_i = [I - X_i(X_i^T X_i)^{-1} X_i^T] y_i$ $(i = 1, 2)$, and z and M_x are evaluated using ω_0.

Given ω_0, $F(k, n - 2k|\omega_0)$ has an F distribution with $(k, n - 2k)$ degrees of freedom. If we put $\omega_0 = 1$, $F(k, n - 2k|\omega_0 = 1)$ becomes Chow's (1960) test.

The test statistic, $F(k, n - 2k|\omega_0)$, can also be derived as a Bayes factor as follows:

$$B_{HK|\omega=\omega_0} = \frac{\int_0^\infty \int_{-\infty}^\infty p_H(\beta_1, \sigma_1) l(\beta_1, \sigma_1 | \text{data}, \omega = \omega_0, H) \, d\beta_1 \, d\sigma_1}{\int_0^\infty \int_{-\infty}^\infty \int_{-\infty}^\infty p_K(\beta_1, \delta, \sigma_1) l(\beta_1, \delta, \sigma_1 | \text{data}, \omega = \omega_0, K) \, d\beta_1 \, d\delta \, d\sigma_1}, \tag{2.7}$$

where $p(\beta_1, \sigma_1) \propto \sigma^{-1}$, $p(\beta_1, \delta, \sigma_1) \propto \sigma_1^{-(k+1)}$, and $l(\cdot|\text{data}, \omega = \omega_0, H)$ and $l(\cdot|\text{data}, \omega = \omega_0, K)$ are given by (2.3) and (2.4), respectively. Integrating out β_1, δ, and σ_1, we obtain

$$B_{HK|\omega=\omega_0} \propto \left[1 + \omega_0^2 \frac{k}{n - 2k} F(k, n - 2k|\omega_0) \right]^{-(n-k)/2}. \tag{2.8}$$

The prior under K is chosen to be proportional to $\sigma_1^{-(k+1)}$ instead of σ_1^{-1}, so that the Bayes factor can be expressed as a monotone function of $F(k, n - 2k|\omega_0)$.

The Bayes factor in (2.7) is derived conditionally on $\omega = \omega_0$. The unconditional Bayes factor is given by

$$B_{HK} = \frac{\int_0^\infty \int_0^\infty \int_{-\infty}^\infty p_H(\beta_1, \sigma_1, \omega) l(\beta_1, \sigma_1, \omega | \text{data}, H)\, d\beta_1\, d\sigma_1\, d\omega}{\int_0^\infty \int_0^\infty \int_{-\infty}^\infty \int_{-\infty}^\infty p_K(\beta_1, \delta, \sigma_1, \omega) l(\beta_1, \delta, \sigma_1, \omega | \text{data}, K)\, d\beta_1\, d\delta\, d\sigma_1\, d\omega}, \tag{2.9}$$

and if we set $p_H(\beta_1, \sigma_1, \omega) = c_1 \sigma_1^{-1} \omega^{-1}$ and $p_K(\beta_1, \delta, \sigma_1, \omega) = c_2 \sigma_1^{-(k+1)} \omega^{k-1}$, equation (2.9) becomes

$$B_{HK} = \pi^{-k/2} \frac{c_1 \int_0^\infty (v_1 s_1^2)^{-v_1} \omega^{n_2 - 1} |X^T X|^{-1/2}\, d\omega}{c_2 \int_0^\infty (v_2 s_2^2)^{-v_1/2} \omega^{n_2 + k - 1} |W^T W|^{-1/2}\, d\omega}, \tag{2.10}$$

where $v_1 s_1^2 = z^T M_x z$, $v_1 = n - k$, $v_2 s_2^2 = z^T M_w z$, $v_2 = n - 2k$, and c_1 and c_2 are normalizing constants.

The posterior odds ratio is given by the Bayes factor multiplied by the prior odds ratio:

$$K_{HK} = \frac{\Pi_H}{\Pi_K} B_{HK}, \tag{2.11}$$

where Π_H and Π_K are prior probabilities of the null and alternative hypotheses, H: $\delta = 0$ and K: $\delta \neq 0$, respectively. To use the posterior odds ratio for hypothesis testing, one may minimize the expected loss of the hypotheses. If one knows the values of Π_H, Π_K, c_1, and c_2 as well as the loss function, the Bayesian posterior odds analysis will give a stronger result than the significance test and it can avoid the difficulties that are associated with the latter. An obvious difficulty of significance tests is that there is no explicit way to adjust the significance level as the sample size increases. This is due to the fact that in significance testing the importance of a departure from the null hypothesis is absent. In general, there is no criterion to choose a suitable significance test statistic.

If the researchers do not possess enough information to specify both the loss function and Π_H, Π_K, c_1, and c_2, however, they will be at a loss to interpret the particular value of K_{HK} that the sample produces. In such a case, it may be better to interpret the Bayes factor as a sampling statistic. Conditioned on $\omega = \omega_0$, the Bayes factor (2.8) is a monotone function of the F statistic and hence we may use the F statistic for testing H: $\delta = 0$ against K: $\delta \neq 0$. In the case where ω is not known, Tsurumi and Sheflin (1985) suggest using ω_0 as the posterior mean of ω under the alternative hypothesis K, ω_{PK}, that is given by

$$\omega_{PK} = E(\omega | \text{data}, K) = \int_0^\infty \omega^{n_2 - k} (z^T M_w z)^{-(n-2k)/2}\, d\omega \tag{2.12}$$

As in equations (2.7), (2.9), and (2.10), equation (2.12) can easily be evaluated by a numerical integration procedure such as Simpson's rule.

In time series analysis, in addition to heteroscedasticity [i.e., $\sigma_1 \neq \sigma_2$ in (2.1)], we need to be concerned with possible autocorrelation in the error terms, ε_i in (2.1). Ilmakunnas and Tsurumi (1984) derived a highest posterior density interval (HPDI) test to see whether an individual parameter in the regression model shifted or not. Let us derive a Bayes factor test for a set of parameters, that is, the hypotheses H: $\delta = 0$ versus K: $\delta \neq 0$. We rewrite equation (2.1) to allow for autocorrelated errors:

$$\begin{aligned} y_0 - x_0\beta &= M + \varepsilon_0 \\ y_t &= x_t\beta_1 + u_t, \qquad t = 1, \ldots, n_1 \\ y_t &= x_t\beta_2 + u_t, \qquad t = n_1 + 1, \ldots, n \end{aligned} \tag{2.13}$$

where u_t obeys

$$\begin{aligned} u_t &= \rho_1 u_{t-1} + \varepsilon_{1t}, \qquad t = 1, \ldots, n_1 \\ u_t &= \rho_2 u_{t-1} + \varepsilon_{2t}, \qquad t = n_1 + 1, \ldots, n, \end{aligned}$$

where $x_t = (x_{1t}, \ldots, x_{kt})$ is a $1 \times k$ vector of explanatory variables (or regressors). The error terms, ε_{it}, are assumed to have independent normal distributions with zero mean and variance σ_i^2; the error of the initial observation, ε_0, has distribution $N(0, \sigma_1^2)$. The treatment of the first observation is the same as in Zellner and Tiao (1964): M is an unknown constant, which will be integrated out in the posterior analysis. This approach is chosen primarily to examine the probability of $|\rho_i| \geqslant 1$. Using the definitions of the error terms, u_t, we can write equation (2.13) in the form

$$\begin{aligned} y_0 - x_0\beta &= M + \varepsilon_0 \\ y_t - \rho_1 y_{t-1} &= (x_t - \rho_1 x_{t-1})\beta_1 + (u_t - \rho_1 u_{t-1}), \qquad t = 1, \ldots, n_1 \\ y_{n_1+1} - \rho_2 y_{n_1} &= (x_{n_1+1} - \rho_2 x_{n_1})\beta_1 + x_{n_1+1}\delta + (u_{n_1+1} - \rho_2 u_{n_1}) \\ y_t - \rho_2 y_{t-1} &= (x_t - \rho_2 x_{t-1})\beta_1 + (x_t - \rho_2 x_{t-1})\delta + (u_t - \rho_2 u_{t-1}), \\ &\qquad t = n_1 + 2, \ldots, n, \end{aligned} \tag{2.14}$$

where $\delta = \beta_2 - \beta_1$. The equation system (2.14) can be expressed as

$$\begin{aligned} Z &= X\beta_1 + V\delta + \varepsilon \\ &= W\zeta + \varepsilon, \end{aligned}$$

where $Z = (Y_1^T, \omega Y_2^T)^T$, $X = (X_1^T, \omega X_2^T)^T$, $V = (0^T, \omega \bar{X}_2^T)^T$, $\varepsilon = (\varepsilon_1^T, \omega\varepsilon_2^T)^T$, $W = (X, V)$, $\zeta = (\beta_1^T, \delta^T)^T$, $\omega = \sigma_1/\sigma_2$,

$$Y_1 = \begin{bmatrix} y_1 - \rho_1 y_0 \\ \vdots \\ y_{n_1} - \rho_1 y_{n_1-1} \end{bmatrix} \qquad Y_2 = \begin{bmatrix} y_{n_1+1} - \rho_2 y_{n_1} \\ \vdots \\ y_n - \rho_2 y_{n-1} \end{bmatrix}$$

$$X_1 = \begin{bmatrix} x_1 - \rho_1 x_0 \\ \vdots \\ x_{n_1} - \rho_1 x_{n_1 - 1} \end{bmatrix} \quad X_2 = \begin{bmatrix} x_{n_1+1} - \rho_2 x_{n_1} \\ \vdots \\ x_n - \rho_2 x_{n-1} \end{bmatrix}$$

$$\bar{X}_2 = \begin{bmatrix} x_{n_1+1} \\ x_{n_1+2} - \rho_2 x_{n_1+1} \\ \vdots \\ x_n - \rho_2 x_{n-1} \end{bmatrix} \quad \varepsilon_1 = \begin{bmatrix} u_1 - \rho_1 u_0 \\ \vdots \\ u_{n_1} - \rho_1 u_{n_1-1} \end{bmatrix} \quad \varepsilon_2 = \begin{bmatrix} u_{n_1+1} - \rho_2 u_{n_1} \\ \vdots \\ u_n - \rho_2 u_{n-1} \end{bmatrix}.$$

We shall derive the highest posterior density region (HPDR) for δ to test H: $\delta = 0$ against K: $\delta \neq 0$. The HPDR will be derived conditionally on ρ_1, ρ_2, and ω. The likelihood function can be written as

$$l(\beta_1, \delta, M, \rho_1, \rho_2, \sigma_1, \omega|\text{data}) \propto \sigma_1^{-(n+1)}\omega^{n_2} \exp\left[-\frac{1}{2\sigma_1^2}(Z - W\zeta)^T(Z - W\zeta) - \frac{1}{2\sigma_1^2}(y_0 - x_0\beta - M)^2 \right]. \tag{2.15}$$

Let us use the prior probability density function (p.d.f.)

$$p(\beta_1, \delta, M, \sigma_1, \rho_1, \rho_2, \omega) = p(\beta_1, \delta, M|\sigma_1, \rho_1, \rho_2, \omega)p(\rho_1, \rho_2|\sigma_1, \omega)p(\sigma_1, \omega). \tag{2.16}$$

We use Jeffrey's diffuse priors: $p(\beta_1, \delta, M|\sigma_1, \rho_1, \rho_2, \omega) \propto$ constant, $p(\sigma_1, \omega) \propto \sigma_1^{-1}\omega^{-1}$, and as for the prior for ρ_1 and ρ_2, we choose $p(\rho_1, \rho_2|\sigma_1, \omega) \propto |W^TW|^{1/2}$. This prior is chosen to avoid the singularity of W^TW when $\rho_1 = \rho_2 = 1$. For further discussions on this, see O'Brien (1970) and Ilmakunnas and Tsurumi (1984). Combining (2.15) and (2.16) and integrating out β_1, M, and σ_1, we obtain the conditional posterior p.d.f. for δ given ρ_1, ρ_2, and ω:

$$p(\delta|\rho_1, \rho_2, \omega, \text{data}) \propto \left[1 + \frac{(\delta - \hat{\delta})^T V^T M_x V(\delta - \hat{\delta})}{\nu s^2} \right]^{-(n-k)/2}, \tag{2.17}$$

where $\nu s^2 = Z^T[I - W(W^TW)^{-1}W^T]Z$, $\nu = n - 2k$, $\hat{\delta} = (V^TM_xV)^{-1}V^TM_xZ$. If we set $\delta = 0$, then $\hat{\delta}^TV^TM_xV\hat{\delta}/\nu s^2$ is distributed as $F(k, n-2k)$ a posteriori, and we may use this as the HPDR test statistic given ρ_1, ρ_2, and ω. As the values of ρ_1, ρ_2, and ω, we may use their posterior means, which can be computed numerically using the joint posterior p.d.f.:

$$p(\rho_1, \rho_2, \omega|\text{data}) \propto \omega^{n_2 - 1}(\nu s^2)^{-\nu/2} \tag{2.18}$$

Although the statistic, $\hat{\delta}^TV^TM_xV\hat{\delta}/\nu s^2$, is derived as an HPDR, we can derive it as the Bayes factor conditionally on ρ_1, ρ_2, and ω in the manner similar to the derivation of equation (2.8).

3. U.S. DEMAND FOR MONEY: AN EXAMPLE OF MODEL STABILITY TEST

In this section we apply the HPDR test of model stability [equation (2.17)] to the money demand function in the United States. Goldfeld (1976) examined the stability of the U.S. demand for money equation by post-sample-period forecasts for 10 quarters, from the first quarter of 1974 (1974.1) to the second quarter of 1976 (1976.2), and concluded that it showed no systematic tendency in forecast errors up to 1973, but starting in 1974, forecasts began to overpredict real money balances substantially. Goldfeld's study attracted much attention in the field of macro and monetary economics. A survey by Judd and Scadding (1982) summarizes the issue well. The subsequent research has taken three routes: (1) evaluation of various definitions of money, (2) examination of the specification of the money demand equation, and (3) examination of estimation procedures. In this section, rather than following any of these routes, we stay with Goldfeld's specification of demand for money and apply the HPDR test procedure that is developed in the preceding section to see whether the regression equation shifted or not.

The U.S. money demand equation is specified by

$$\log\left(\frac{M_t}{P}\right)_t = \beta_1 + \beta_2 \log \mathrm{GNP}_t + \beta_3 \log \mathrm{RMS}_t + \beta_4 \log \mathrm{RSAV}_t + \beta_5 \log\left(\frac{M_1}{P}\right)_{t-1} + \varepsilon_t, \tag{3.1}$$

where M_1 is the currency plus checkable deposits, P the aggregate price level (GNP deflator), GNP the real gross national product, RMS the 6 months treasury bill rate, and RSAV the interest rate on corporate bonds rated as Aaa. Goldfeld (1976) estimated equation (3.1) using seasonally adjusted data from the second quarter of 1952 (1952.2) to the fourth quarter of 1973 (1973.4) and obtained postsample simulation errors for 10 quarters beyond the end of the sample period (1974.1–1976.2). Since money data have been revised in recent years, let us use the revised data from 1959.2 to 1979.4 and choose the first quarter of 1974 (1974.1) as the join point. Since preliminary data analyses indicate that the coefficients of autocorrelation in two regimes, ρ_1 and ρ_2 in equation (2.13) appear to be the same, we modify the HPDR test, $\hat{\delta}^T V^T M_x V \hat{\delta}/vs^2$, and the posterior p.d.f. in (2.18) so that $\rho = \rho_1 = \rho_2$.

Table 1 presents the marginal posterior p.d.f.'s for the coefficient of autocorrelation, ρ, and for the ratio of the standard deviation of the error terms of the two regimes, $\omega = \sigma_1/\sigma_2$. Most of the mass of the posterior distribution for ω is between 0.5 and 0.9, indicating that heteroscedasticity exists. If we take the posterior mean of ω, $\omega_p = 0.7272$, as the point estimate,

Table 1 Marginal Posterior PDFs for ω and ρ: U.S. Demand for Money Equation[a]

Marginal posterior PDF for ω		Marginal posterior PDF for ρ	
ω	$p(\omega\|\cdot)$	ρ	$p(\rho\|\cdot)$
0.3	0.00004	0	0.08841
0.4	0.01309	0.1	0.15182
0.5	0.36635	0.2	1.01734
0.6	2.06113	0.3	2.70384
0.7	3.61720	0.4	3.23485
0.8	2.65285	0.5	2.00832
0.9	1.00193	0.6	0.71404
1.0	0.22735	0.7	0.14417
1.1	0.03487	0.8	0.01281
1.2	0.00395	0.9	0.00024
1.3	0.00035	1.0	0
1.4	0.00003		
Posterior		Posterior	
Mean	0.7272	Mean	0.3866
Standard deviation	0.1094	Standard deviation	0.1203

[a]The posterior probability density is normalized so that $\int p(x|\cdot)\,dx = 1$.

we see that the variance of the error term in the second regime increased by about 1.9 times. The posterior p.d.f. of the coefficient of autocorrelation shows that the error term is positively autocorrelated.

Table 2 presents the posterior means and standard deviations of the regression coefficients. These regression coefficient estimates are conditioned on the posterior means of ω and ρ. Comparing the estimated regression coefficients in two regimes, we notice that the constant term increased, while the coefficient of log GNP, log RSAV, and $\log(M/P)_{t-1}$ decreased. The coefficient of log RMS, on the other hand, changed signs, but this coefficient seems to be insignificant in both regimes. The HPDR test yields the value of 3.234, and this is significant at the 1% level, and thus we may conclude that the regression model shifted. The change of the coefficient of log RSAV from -0.0083 in regime 1 to -0.0710 in regime 2 shows that the demand for money became more sensitive to the long-term interest rate.

Table 2 Estimates of the U.S. Demand for Money: 1959.2–1979.4[a]

	Posterior means and standard deviations		
Coefficient of:	Regime 1 (1959.2–1973.4)		Regime 2 (1974.1–1979.4)
Constant term	0.716		1.718
	(0.374)		(0.989)
log GNP	0.1194		0.0226
	(0.0330)		(0.0405)
log RMS	−0.0066		0.0152
	(0.0058)		(0.0201)
log RSAV	−0.0083		−0.0710
	(0.0082)		(0.0380)
$\log \left(\dfrac{M}{P}\right)_{t-1}$	0.7245		0.6774
	(0.1056)		(0.1377)
Posteriorior mean of ω		0.7272	
Posterior mean of ρ		0.3866	
HPDR test		3.234	

[a]The figures in parentheses are posterior standard deviations. The posterior standard deviations are given here for diagnostic checks. Since the posterior p.d.f. for β is not Gaussian, one should avoid the precise use of a posterior standard deviation for evaluating whether or not β is significant.

4. INFERENCE ABOUT THE UNKNOWN JOIN POINT AND GRADUAL SWITCHING REGRESSION

In Section 3 we presented tests of constancy of the regression model assuming that the join point, $n_1 + 1$, is known. In many data analyses, the join point is not known exactly, although researchers may have some information on the time interval in which the regression switched regimes. Quandt (1958) presented a maximum likelihood procedure for estimating the join point, while Ferreira (1975) and Chin Choy and Broemeling (1980) gave Bayesian procedures. If we modify Ferreira's procedure to allow for heteroscedasticity and for two regimes of autocorrelation, we may derive the joint posterior p.d.f. for n_1, ω, ρ_1, and ρ_2, and this becomes the same equation as (2.18) if we use (2.16) as the prior p.d.f.:

$$p(n_1, \omega, \rho_1, \rho_2|\text{data}) \propto \omega^{n_2-1}(\nu s^2)^{-\nu/2} \qquad \text{for } k < n_1 < n-k \tag{4.1}$$

Quandt (1958), Ferreira (1975), and Chin Choy and Broemeling (1980) assumed an abrupt switch of regimes. Also, equation (4.1) is derived assuming an abrupt switch of regimes. In many cases, however, the shift from one regime to the other may be achieved gradually. Bacon and Watts (1971) introduced a class of transition functions with two parameters, one for the join point and the other for the speed of transition. Tsurumi (1980) modified Bacon and Watt's parametric transition function, and derived an estimation procedure for a simultaneous equation model. He also introduced a parametric temporary shift function that is to be used when a discontinuity in the parameter values is of a temporary nature rather than a permanent shift from an old to a new regime.

Let us present a parametric transition model allowing for heteroscedasticity and for autocorrelation. The regression with a transition function may be written as

$$\begin{aligned} Z &= X\beta_1 + \text{TRN}\cdot V\delta + \varepsilon \\ &= R\zeta + \varepsilon \end{aligned} \tag{4.2}$$

where $R = (X, \text{TRN}\cdot V)$; Z, X, and ζ are defined below (2.14) and $\text{TRN}\cdot V$ is given by

$$\text{TRN}\cdot V = \begin{bmatrix} \text{trn}\left(\dfrac{s_1}{\eta}\right) v_{11}, \ldots, \text{trn}\left(\dfrac{s_1}{\eta}\right) v_{1k} \\ \vdots \\ \text{trn}\left(\dfrac{s_n}{\eta}\right) v_{n1}, \ldots, \text{trn}\left(\dfrac{s_n}{\eta}\right) v_{nk} \end{bmatrix}$$

and v_{ij} is the ijth element of $V = (0^T, \omega \bar{X}_2^T)^T$. The transition function, $\text{trn}(s_t/\eta)$, satisfies

$$\begin{aligned} &\text{(i)} \quad \lim_{s_t \to \infty} \text{trn}\left(\frac{s_t}{\eta}\right) = 1 \\ &\text{(ii)} \quad \text{trn}(0) = 0 \\ &\text{(iii)} \quad \lim_{\eta \to 0} \text{trn}\left(\frac{s_t}{\eta}\right) = 1 \end{aligned} \tag{4.3}$$

and s_t is

$$s_t = \begin{cases} 0 & \text{for } t \leqslant n_1 \\ t - n_1 & \text{for } t > n_1. \end{cases}$$

An intuitive way to look at (4.3) is that trn (s_t/η) is a dummy variable that is zero for $t \leqslant n_1$ and approaches unity gradually, with the speed of adjustment depending on the value of η. Given the prior $p(n_1, \eta, \sigma_1, \rho_1 \cdot \rho_2, \omega) \propto \sigma_1^{-1}|R^TR|^{1/2}$, we obtain the joint posterior p.d.f. for $n_1, \eta, \rho_1, \rho_2$, and ω as

$$p(n_1, \eta, \omega, \rho_1, \rho_2|\text{data}) \propto \omega^{n_2-1}(vs_3^2)^{-(n-2k)/2} \tag{4.4}$$

where $vs_3^2 = Z^T[I - R(R^TR)^{-1}R^T]Z$. The marginal posterior p.d.f. of n_1 or of η may be obtained by numerical integration.

Let us apply (4.1) and (4.4) to the U.S. money demand equation. Again, we choose the autocorrelation parameters in two regimes to be the same ($\rho = \rho_1 = \rho_2$). Table 3 present the posterior means and standard deviations of n_1, ρ, and ω for the abrupt switching model of (4.1) and those of n_1, η, ρ, and ω for the gradual switching regression model of (4.2). We present the results for the hyperbolic tangent as the transition function since specification of the transition function (such as the Poisson distribution, a ratio of quadratic function, and an exponential function) yields similar results.

In Section 3 we used the join point, $n_1 + 1$, of 1974.1. This join point is taken from Goldfeld's study (1976). The estimations of the join point give 1973.1 for the abrupt switching model and 1972.4 for the gradual switching model, and they are close to Goldfeld's join point. The parameter for the speed of adjustment, η, has the posterior mean of 12.7 quarters. This implies that it takes roughly 26 quarters (or 6.5 years) to reach the second regime since tanh(26/12.7) 0.96. According to this gradual switching model, the

Table 3 Posterior Means and Standard Deviations of Abrupt and Gradual Switching Regression Models: U.S. Demand for Money Equation

Parameter[a]		Abrupt switch	Gradual switch
n_1	Mean	1972.4	1972.3
	(S.D.)	(1.87)	(2.16)
η	Mean	—	12.68
	(S.D.)		(4.72)
ρ	Mean	0.3579	0.4148
	(S.D.)	(0.1182)	(0.1261)
ω	Mean	0.7078	0.6031
	(S.D.)	(0.1009)	(0.1189)

[a]S.D., posterior standard deviation. S.D. is given as a descriptive statistic. See the footnote for Table 2.

period from 1972.4 to almost the end of the sample period (1979.4) is an adjustment period. The Bayes estimators of the regression coefficients, β_1 and δ of the gradual switching model can be estimated either unconditionally or conditionally on the posterior means of n_1, η, ρ, and ω. The conditional posterior means and standard deviations of the regression coefficients are given in the following equation:

$$\begin{aligned}
\log\left(\frac{M_1}{P}\right)_t = {} & \underset{(0.3654)}{0.6953} + \underset{(0.0325)}{0.1204}\log \text{GNP}_t - \underset{(0.0059)}{0.0066}\log \text{RMS}_t \quad (0.0059) \\
& - \underset{(0.0082)}{0.0105}\log \text{RSAV}_t + \underset{(0.1056)}{0.7245}\log\left(\frac{M_1}{P}\right)_{t-1} \\
& + \underset{(1.288)}{1.539}\,\text{trn} + \underset{(0.0588)}{0.0866}\,\text{trn}\log \text{GNP}_t \\
& + \underset{(0.0231)}{0.0025}\,\text{trn}\log \text{RMS}_t - \underset{(0.0441)}{0.0395}\,\text{trn}\log \text{RSAV}_t \\
& - \underset{(0.2243)}{0.3895}\,\text{trn}\log\left(\frac{M_1}{P}\right)_{t-1},
\end{aligned} \tag{4.5}$$

where trn denotes trn (s_t/η). The comparison of the regression coefficients in regimes 1 and 2 (regime 2 being attained by setting trn = 1) are given in Table

Table 4 Regression Coefficients of the U.S. Demand for Money in Regimes 1 and 2: Gradual Switching Regression

Coefficient of:	Regime 1	Regime 2[a]
Constant term	0.6953	2.2343
log GNP	0.1204	0.2070
log RMS	−0.0066	−0.0041
log RSAV	−0.0105	−0.0500
log $(M_1/P)_{t-1}$	0.7245	0.3350

[a]Coefficients in regime 2 are computed by setting trn in (4.5) to be 1 and adding each coefficient to the corresponding coefficient with trn. For example, the coefficient of log GNP in regime 2 is obtained by $0.1204 + 0.0866 = 0.2070$.

4. As in the case of Table 2, the coefficient of log RSAV increased in absolute value terms. The coefficient of log RMS is still negative in the second regime, although it became closer to zero than in the first regime. The coefficient of log GNP increased by 0.0860 from 0.1204 to 0.2070. The results of the gradual switching regression model seem to make more economic sense than those of the abrupt change model in Table 2.

5. TESTS OF MODEL STABILITY WITHOUT USING THE JOIN POINT

Inference on model stability that has been presented so far is based on the assumption that within each regime parameters stay the same from one observation to another, and tests of parameter stability are derived conditionally on the join point. There are some tests of model stability that are derived without assuming the join point. Let us present three of these tests: the Brown-Durbin-Evans (BDE) test, Lamotte-McWhorter (LM) test, and Cooley-Prescott (CP) test. These tests are presented here primarily to compare them with the Bayesian gradual switching regression model presented in the preceding section and to point out similarities with the fault detection models in the control literature.

The BDE and LM tests are based on the regression model, in which regression coefficients vary from one observation to another:

$$y_t = \beta_{t1} + \beta_{t2}x_{t2} + \cdots + \beta_{tk}x_{tk} + \varepsilon_t, \qquad t = 1, \ldots, n$$

or

$$y_t = x_t\beta_t + \varepsilon_t, \tag{5.1}$$

where $x_t = (1, x_{t2}, \ldots, x_{tk})$ and $\beta_t = (\beta_{t1}, \ldots, \beta_{tk})^T$. The null hypothesis of parameter stability is given by

$$H\colon\ \beta_1 = \beta_2 = \cdots = \beta_n \tag{5.2}$$

against the alternative hypothesis, K, that is given by "not all β_t's are the same." Equation (5.1) reduces to the gradual switching regression model of (4.2) if

$$\begin{aligned} \beta_1 = \cdots = \beta_{n_1} \qquad & \text{(regime 1)} \\ \beta_{n_1+1} \neq \cdots \neq \beta_{n_1+t^*} \qquad & \text{(transition periods)} \\ \beta_{n_1+t^*+1} = \cdots = \beta_n \qquad & \text{(regime 2)} \end{aligned}$$

where t^* is the period that makes trn at period $n_1 + t^* + 1$ equal to 1.

The null hypothesis (5.2) may be appealing to some researchers since one does not have to know the join point or specify the alternative hypothesis as clearly as in the case of the switching regression. On the issue of U.S. demand for money, many authors used procedures to test the null hypothesis (5.2), perhaps because the join point is not known with certainty or because the time-varying model is preferred to the abrupt or gradual switching regression model. Some used the BDE test (see, e.g., Heller and Khan, 1979). Others used the test procedure proposed by Cooley and Prescott (1973, 1976) (see, e.g., Roll, 1972; Laumas and Mehra, 1976; Rausser and Laumas, 1976; Cooley and Decanio, 1977; Laumas, 1978; Mullineaux, 1980).

5.1 Brown-Durbin-Evan Tests

The BDE tests are based on a series of normalized errors in one-period-ahead predictions from the sequentially extended sample periods, w_t:

$$w_t = \frac{y_t - x_t\hat{\beta}_{t-1}}{[1 + x_t(X_{t-1}^T X_{t-1})^{-1} x_t^T]^{1/2}}, \qquad t = k+1, \ldots, n,$$

where $\hat{\beta}_{t-1} = (X_{t-1}^T X_{t-1})^{-1} X_{t-1}^T Y_{t-1}$, $X_{t-1} = (x_1^T, \ \ldots, \ x_{t-1}^T)^T$, and $Y_{t-1} = (y_1, \ldots, y_{t-1})^T$. Using the normalized error in one-period predictions, w_t, two tests, the "cusum" and "cusum of squares" tests, are proposed. The cusum test is given by

$$W_t = \frac{\sum_{j=k+1}^{t} w_j}{s}, \qquad t = k+1, \ldots, n,$$

where $s^2 = Y_n^T[I - X_n(X_n^T X_n)^{-1} X_n^T] Y_n/(n-k)$. The null hypothesis (5.2) is rejected if $|W_t| > [a(n-k)^{1/2} + 2a(t-k)/(n-k)]$ occurs for any $t = k+1, \ldots, n$. The scalar, a, is set at 1.143 for a 1% significance level and at 0.948 for a 5% level. The cusum-of-squares test is given by

$$S_t = \frac{\sum_{j=k+1}^{t} w_j^2}{\sum_{j=k+1}^{n} w_j^2}, \qquad t = k+1, \ldots, n,$$

and the null hypothesis is rejected if

$$\left| S_t - \frac{t-k}{n-k} \right| > c_0 \qquad \text{for any } t = k+1, \ldots, n.$$

The critical value, c_0, is calculated from Table 1 of Durbin (1969).

The rejection criteria of the cusum and cusum-of-squares tests are multiple comparison procedures, and as such they may turn out to be too conservative; actual significance levels tend to be smaller than the normal significance levels. Garbade's (1977) sampling results show this conservative nature of the BDE tests. In surveying the literature on U.S. money demand,

Judd and Scadding (1982) note that the BDE tests were applied to the money demand equations, and they tended to accept the null hypothesis of parameter stability, whereas such stability was rejected by other tests. The BDE tests are obtained under the assumption that the error term is homoscedastic and is not autocorrelated. Since forecast errors and their variances tend to be greatly influenced by heteroscedasticity and autocorrelation, the BDE tests tend to be sensitive to heteroscedasticity and autocorrelation.

5.2 Lamotte and McWhorter's Test

Lamotte and McWhorter (1978) suggest an exact F test for the null hypothesis in (5.2). Writing the regression (5.1) in state-space form yields

$$\begin{aligned} y_t &= x_t\beta_t + \varepsilon_t \\ \beta_t &= \beta_{t-1} + u_t. \end{aligned} \tag{5.3}$$

Lamotte and McWhorter assume that $\varepsilon_t \sim N(0, \sigma_t^2)$ and $u_t \sim N(0, \sigma_u^2 D)$, where a $k \times k$ positive semidefinite matrix D is known. Given the model (5.3) and the null hypothesis (5.2), it is apparent that the null hypothesis is equivalent to H: $\sigma_u^2 = 0$. The model (5.3) can be written as

$$y = X\beta_0 + \psi w + \varepsilon \tag{5.4}$$

where $y = (y_1, \ldots, y_n)^T$, $X = (x_1^T, \ldots, x_n^T)^T$, $\psi = \operatorname{diag}(x_i)$, $w = (w_1^T, \ldots, w_n^T)^T$, $w_i = \sum_{j=1}^{i} u_j$, and $\varepsilon = (\varepsilon_1, \ldots, \varepsilon_n)^T$. Although β_0 is a random vector, Lamotte and McWhorter treat it as given. Then the covariance of y given β_0 is

$$\operatorname{cov}(y|\beta_0) = \sigma_\varepsilon^2 + \sigma_u^2 V, \tag{5.5}$$

where $V = (v_{ts})$, $v_{ts} = \min(t, s)x_t D x_s^T$. Let H be an $n \times (n-k)$ matrix with $H^T H = I_{n-k}$, and $HH^T = I - X(X^T X)^{-1} X^T$. Let $\lambda_1, \ldots, \lambda_q$ be the distinct characteristic roots of $H^T V H$ with multiplicity $r_1, \ldots, r_q$. Let P_i be a $(n-k) \times r_i$ matrix whose columns are orthogonal characteristic vectors of $H^T V H$. Then $H^T V H = \sum \lambda_i P_i P_i^T$, and $\sum P_i P_i^T = I_{n-k}$, and the sum of squared residuals, SSR, becomes

$$\begin{aligned} \text{SSR} &= y^T[I - X(X^T X)^{-1} X^T]y = y^T HH^T y \\ &= \sum_{i=1}^{q} y^T H P_i P_i^T H^T y = \sum_{i=1}^{q} Q_i, \end{aligned} \tag{5.6}$$

where $Q_i = y^T H P_i P_i^T H^T y$. Since $P_i^T H^T y \sim N[0, (\sigma_\varepsilon^2 + \lambda_i \sigma_u^2) I_{r_i}]$, the term $(\sigma_\varepsilon^2 + \lambda_i \sigma_u^2)^{-1} Q_i$ is distributed as chi-square with r_i degrees of freedom under the null hypothesis. Assume that $\lambda_1 > \lambda_2 > \cdots > \lambda_q$. If we partition $(Q_1, \ldots, Q_q)$ into $(Q_1, \ldots, Q_g)$ and $(Q_{g+1}, \ldots, Q_q)$ for $1 \leqslant g \leqslant k$, then defining

$s_g = \sum_{i=1}^{g} Q_i$, we see that under the null hypothesis,

$$F = \frac{s_g/n_g}{(\text{SSR} - s_g)/(n - k - n_g)} \tag{5.7}$$

is distributed as an F with $n_g = \sum_{i=1}^{g} r_i$ and $n - k - n_g$ degrees of freedom. Lamotte and McWhorter's statistic is appealing since it has an exact F distribution under the null hypothesis. Their statistic, however, may be sensitive to the choice of g as well as to the values of D.

5.3 Cooley and Prescott's Test

Cooley and Prescott (1973a, 1976) proposed a time-varying parameter model which is a variation of equation (5.3):

$$\begin{aligned} y_t &= x_t \beta_t \\ \beta_t &= \beta_{t-1}^p + u_t \\ \beta_t^p &= \beta_{t-1}^p + v_t. \end{aligned} \tag{5.8}$$

The $k \times 1$ error vectors u_t and v_t are independent normal variables with

$$\begin{aligned} E(u_t) &= E(v_t) = 0 \\ \operatorname{cov}(u_t) &= (1 - \gamma)\sigma^2 \Sigma_u \\ \operatorname{cov}(v_t) &= \gamma\sigma^2 \Sigma_v \\ E(u_t u_s^T) &= E(v_t v_s^T) = 0 \qquad \text{for } t \neq s. \end{aligned}$$

The $k \times 1$ regression coefficient vector, β_t^p, is called the permanent component of the parameters. Cooley and Prescott designed their model not to estimate β_t or β_t^p per se, but to test the stability of the regression coefficients, β_t. If $\gamma = 0$, then $v_t = 0$, and in this case the regression model becomes $y_t = x_t \beta + \varepsilon_t$ with $\beta = \beta_t^p = \beta_{t-1}^p$ and $\varepsilon_t = x_t u_t$. This is the classical regression model except that the error term is heteroscedastic. On the other hand, if $\gamma \neq 0$, equation (5.8) becomes the varying parameter model. The $k \times k$ matrices Σ_u and Σ_v are assumed to be known up to scale factors.

Model (5.8) can be rewritten as

$$y_t = x_t \beta_{n+1}^p + \mu_t, \tag{5.9}$$

where $\mu_t = x_t u_t - x_t \sum_{s=t+1}^{n+1} v_t$. Two testing procedures are suggested: One is a Bayesian highest posterior density interval (HPDI) test, and the other is a maximum likelihood estimation procedure. Using (5.9) and the prior p.d.f.,

$$p(\beta_{n+1}^p, \gamma, \sigma^2) \propto \sigma^{-1}.$$

Cooley and Prescott (1973b) derive the marginal posterior p.d.f. of γ as

$$p(\gamma|\Sigma_u, \Sigma_v, \text{data}) \propto |\Omega(\gamma)|^{-1/2}|X^T\Omega(\gamma)^{-1}X|^{-1/2}s^{-v}, \tag{5.10}$$

where $\Omega(\gamma) = (1 - \gamma)R + \gamma Q$, $vs^2 = (y - X\hat{\beta}^p_{n+1})^T\Omega(\gamma)^{-1}(y - X\hat{\beta}^p_{n+1})$, $\hat{\beta}^p_{n+1} = [X^T\Omega(\gamma)^{-1}X]^{-1}X^T\Omega(\gamma)^{-1}y$, $v = n - k$, $R = \text{diag}(x_i\Sigma_u x_i^T)$. $Q = (q_{ij})$, $q_{ij} = \min[n - i + 1, n - j + 1]x_i\Sigma_v x_j^T$, and $X = (x_1^T, \ldots, x_n^T)^T$. Cooley and Prescott (1976) derived the maximum likelihood estimator of γ, $\hat{\gamma}$, using (5.9) and suggest to test $\gamma = 0$ with $\hat{\gamma}$ since they claim that $\sqrt{n}(\hat{\gamma} - \gamma)$ is asymptotically normal with mean zero and a constant variance.

There are two caveats for the user of the Cooley-Prescott test. The first is, as demonstrated in Tsurumi and Shiba (1981), the sensitivity of the inference on γ to the specification of the two covariances, Σ_u and Σ_v, whose elements are assumed to be known. The second pertains to the asymptotic properties of Cooley-Prescott's maximum likelihood estimators. Swamy and Tinsley (1980) and Pagan (1980) cast doubts on this point. Tsurumi and Shiba (1981) prove that their maximum likelihood estimators of γ and σ^2 are not consistent. Consequently, the inference on γ may yield a strange result since the asymptotic distribution of $\sqrt{n}\,\hat{\gamma}$ is not established to be normal. In the special case of $k = 1$, Tanaka (1983) shows that the test statistic of this type does not converge to normality under the null hypothesis of $\gamma = 0$.

Lamotte and McWhorter's test as well as Cooley and Prescott's test are based on dynamic systems. In the control literature model stability is examined by fault detection procedures, mainly using sequential probability tests (SPRT) and generalized likelihood ratio tests (GLRT). Willsky (1976) and Isermann (1984) survey fault detection methods. The dynamic systems, expressed in a regression framework, may be given by

$$\begin{aligned} y_t &= x_t\beta_t + \varepsilon_t \\ \beta_t &= \Phi_t\beta_{t-1} + v_t. \end{aligned} \tag{5.11}$$

The Cooley-Prescott model of (5.8), for example, becomes a special case of (5.11) with a heteroscedastic error term if we put $\beta_t = \beta_t^p$, $\Phi_t = I$, and $\varepsilon_t = x_t u_t$.

In the fault detection literature, abrupt faults may appear in the state variables or in output variables. If we express them in the dynamic systems of (5.11), we have

$$y_t = x_t\beta_t + \varepsilon_t + v_1\theta_{tr} \qquad \text{(sensor step)}$$

and (5.12)

$$\beta_t = \Phi_t\beta_{t-1} + v_t + v_2\sigma_{ts} \qquad \text{(dynamic step)}$$

where v_1 and v_2 are unknown k-vectors, and θ_{t1} and σ_{ts} are given by

$$\theta_{t1} = \begin{cases} 1, & t \geqslant r \\ 0, & t < r \end{cases} \quad \text{and} \quad \sigma_{ts} = \begin{cases} 1, & t \geqslant s \\ 0, & t < s \end{cases} \tag{5.13}$$

Several testing methods have been used, including the chi-square test of the residuals of the normal Kalman filter, sequential probability ratio tests (SPRTs), and generalized likelihood ratio tests (GLRTs). A good critique of the use of SPRT and GLRT techniques is given in Kerr (1983). Bayesian methods are given in Chow and Willsky (1984) and Spall (1988), but such Bayesian procedures seem to be more of an exception at present (the Spall method is not restricted to state space models).

If we change the unit step function in (5.13) to a transition function of (4.3), we may have a gradual fault detection model. In fault detection, linear dynamic models in state-space form are maintained whether v_1 and v_2 are zero or not, whereas in econometric models, model stability implies a regression model with constant parameters. It is a future task to evaluate whether an abrupt or gradual fault detection model explains economic behavior better than the current predominant use of a regression with constant parameters.

6. TESTS OF CONSTANCY OF A STRUCTURAL EQUATION IN SIMULTANEOUS EQUATION SYSTEM

There have been some attempts to develop tests of constancy of a structural equation in a simultaneous equation system. In econometric model building the specification of structural equations is often based on preliminary regression runs using least squares procedures. Insofar as the least squares procedures serve as the basis for specifying a structural equation, one may argue that we may use the testing procedures that are developed for the classical linear regression model as descriptive diagnostic checks at the stage of determining the specification of each equation of the model.

An attempt to establish a test of constancy of a structural equation almost always leads to an asymptotic or conditional inference. Within the limited information Bayesian (LIB) framework, one can derive a highest posterior density interval (HPDI) test of an individual parameter shift. The procedure is derived and applied to a macro savings function in Tsurumi (1978). The derivation of the posterior density for the difference of a regression coefficient in two subperiods is given more fully in Broemeling and Tsurumi (1986). Here let us consider some tests to examine whether a set of parameters, rather than individual parameter, in a structural equation has shifted or not.

Let a structural equation of interest for two subsample periods be given by

$$\begin{aligned} y_{1,\mathrm{I}} &= Y_{1,\mathrm{i}}\gamma_{1,\mathrm{i}} + X_{1,\mathrm{i}}\beta_{1,\mathrm{I}} + u_{1,\mathrm{I}} \\ y_{1,\mathrm{II}} &= Y_{1,\mathrm{II}}\gamma_{1,\mathrm{II}} + X_{1,\mathrm{II}}\beta_{1,\mathrm{II}} + u_{2,\mathrm{II}}, \end{aligned} \tag{6.1}$$

where

$y_{1,i} = (n_i \times 1)$ vector of dependent variables in regime i, $i = \mathrm{I, II}$
$Y_{1,i} = (n_i \times m_1)$ matrix of endogenous variables in the right-hand side of the structural equation in regime i, $i = \mathrm{I, II}$
$X_{1,i} = (n_i \times k_1)$ matrix of exogenous variables in the right-hand side of the structural equation in regime i, $i = \mathrm{I, II}$
$u_{1,i} = (n_i \times 1)$ vector of error terms in regime i, $i = \mathrm{I, II}$

and $\gamma_{1,i}$ and $\beta_{1,i}$ are, respectively, $m_1 \times 1$ and $k_1 \times 1$ vectors of parameters in regime i ($i = \mathrm{I, II}$). Let us assume that $u_{1,\mathrm{I}}$ and $u_{2,\mathrm{II}}$ are independent. The hypothesis to be tested is

$$H\colon \begin{bmatrix} \gamma_{1,\mathrm{II}} \\ \beta_{1,\mathrm{II}} \end{bmatrix} = \begin{bmatrix} \gamma_{1,\mathrm{I}} \\ \beta_{1,\mathrm{I}} \end{bmatrix} \quad \text{versus} \quad K\colon \begin{bmatrix} \gamma_{1,\mathrm{ii}} \\ \beta_{1,\mathrm{II}} \end{bmatrix} \neq \begin{bmatrix} \gamma_{1,\mathrm{i}} \\ \beta_{1,\mathrm{I}} \end{bmatrix} \tag{6.2}$$

Morimune (1983) suggests a limited information likelihood ratio test

$$\log \frac{(1+\lambda_0)^n}{(1+\lambda_{\mathrm{I}})^{n_1}(1+\lambda_{\mathrm{II}})^{n_2}} \tag{6.3a}$$

where λ_i, $i = 0, \mathrm{I, II}$, is the minimum characteristic root of

$$|W_i^T M_{1,i} W_i - \lambda_i W_i^T M_i W_i| = 0 \tag{6.3b}$$

and $W_i = (y_{1,i}, Y_{1,i})$, $i = \mathrm{I, II}$, $W_0 = (y_1, Y_1)$, $y_1 = (y_{1,\mathrm{I}}^T, y_{1,\mathrm{II}}^T)^T$, $Y_1 = (Y_{1,\mathrm{I}}^T, Y_{1,\mathrm{II}}^T)^T$, $M_{1,i} = \mathrm{I} - X_{1,i}(X_{1,i}^T X_{1,i})^{-1} X_{1,i}^T$, $i = 0, \mathrm{I, II}$, $X_{1,0} = (X_{1,\mathrm{I}}^T, X_{1,\mathrm{II}}^T)^T$, $M_i = \mathrm{I} - X_i(X_i^T X_i)^{-1} X_i^T$, $i = 0, \mathrm{I, II}$, $X_i = (X_{1,i}^T, X_{2,i}^T)^T$, and $X_{2,i}$ is a $(n_i \times k_2)$ matrix of observations on the exogenous variables excluded from equation of interest. Asymptotically, (6.3b) is distributed as chi-square with $k_1 + m_1$ degrees of freedom under the null hypothesis, H.

As an alternative limited-information Bayesian (LIB) approach, Tsurumi (1985) derives a marginal posterior p.d.f. of the coefficients of the structural equation conditioned on the coefficients of the reduced-form equation. If we apply this method to the test of constancy of structural coefficients, one can obtain an extension of the F test for the structural equation that allows for heteroscedasticity. For each regime the likelihood function can be written as

$$\begin{aligned} &l(\gamma_{1,i}, \beta_{1,i}, \zeta_i, \theta_i, \Pi_{2,i}, \Omega_{11.2,i}, \Omega_{22,i} | \text{data}) \\ &\quad \propto |\Omega_{11.2,i}|^{-n_i/2} |\Omega_{22,i}|^{n_i/2} \exp\left(-\frac{1}{2\Omega_{11.2,i}} Q_{1,i}\right) \exp\left(-\frac{1}{2} Q_{2,i}\right), \end{aligned} \tag{6.4}$$

where $Q_{1,i} = (y_{1,i} - Y_{1,i}\gamma_{1,i} - X_{1,i}\beta_{1,i} - X_{2,i}\zeta_i - V_{1,i}\theta_i)^T(y_{1,i} - Y_{1,i}\gamma_{1,i} - X_{1,i}\beta_{1,i} - X_{2,i}\zeta_i - V_{1,i}\theta_i)$, $Q_{2,i} = (Y_{1,i} - X_i\Pi_{2,i})^T(Y_{1,i} - X_i\Pi_{2,i})\Omega_{22,i}^{-1}$, $\Omega_{11,2,i} = \sigma_{11,i} - \zeta_i^T\Omega_{22,i}^{-1}\zeta_i$; $\sigma_{11,i}$ is the variance of the tth element of $u_{1,i}$ in regime i; $\Omega_{22,i}$ is the variance of the tth row of $Y_{1,i}$ in the reduced-form equation; ξ_i is the covariance between the tth row of $Y_{1,i}$ and the tth element of $u_{1,i}$ in regime i; ζ_i is the identification parameter; $\theta_i = \Omega_{22,i}^{-1}\xi_i$ and $\Pi_{2,i}$ is the reduced-form coefficients and $V_{1,i} = Y_{1,i} - X_i\Pi_{2,i}$. Equation (6.4) can be rewritten as, after imposing the identifiability condition, $\zeta_{\rm i} = \zeta_{\rm ii} = 0$:

$$
\begin{aligned}
&l(\mu_1, \delta_1, \Pi_{2,i}, \Pi_{2,\mathrm{II}}, \Omega_{11.2,i}, \omega, \Omega_{22,\mathrm{I}}, \Omega_{22,\mathrm{II}} | \zeta_{\mathrm{I}} = \zeta_{\mathrm{II}} = 0, \text{data})\\
&\quad \propto \Omega_{11.2,\mathrm{I}}^{-n/2}\omega^{n_2}|\Omega_{22,\mathrm{I}}|^{-n_1/2}|\Omega_{22,\mathrm{II}}|^{-n_2/2}\\
&\quad\quad \times \exp\left[-\frac{1}{2\Omega_{11.2,\mathrm{I}}}(y_1 - R\mu_1 - T_1\delta_1)^T(y_1 - R\mu_1 - T_1\delta_1)\right]\\
&\quad\quad \times \prod_{i=\mathrm{I}}^{\mathrm{II}} \exp\left[-\frac{1}{2}\mathrm{tr}\,(Y_{1,i} - X_i\Pi_{2,i})^T(Y_{1,i} - X_i\Pi_{2,i})\Omega_{22,i}^{-1}\right],
\end{aligned}
\tag{6.5}
$$

where $\omega = \Omega_{11.2,\mathrm{I}}/\Omega_{11.2,\mathrm{II}}$, $y_1 = (y_{1,\mathrm{I}}^T, y_{1,\mathrm{II}}^T)^T$,

$$
R = \begin{bmatrix} Z_{1,\mathrm{I}} & V_{1,\mathrm{I}} & 0 \\ \omega Z_{1,\mathrm{II}} & 0 & \omega V_{1,\mathrm{II}} \end{bmatrix}, \qquad Z_{1,i} = (Y_{1,i}, X_{1,i}), \qquad T_1 = \begin{bmatrix} 0 \\ \omega Z_{1,\mathrm{II}} \end{bmatrix},
$$

$$
\mu_1 = (\eta_{1,\mathrm{I}}^T, \theta_{\mathrm{I}}^T, \theta_{\mathrm{II}}^T)^T, \qquad \eta_{1,i} = (\gamma_{1,i}^T, \beta_{1,i}^T)^T, \qquad \delta_1 = \eta_{1,\mathrm{II}} - \eta_{1,\mathrm{I}}.
$$

Given the prior p.d.f.,

$$
p(\mu_1, \delta_1, \Pi_{2,\mathrm{I}}, \Pi_{2,\mathrm{II}}, \Omega_{11.2,\mathrm{I}}, \omega, \Omega_{22,\mathrm{I}}, \Omega_{22,\mathrm{II}}) \propto \Omega_{11.2,\mathrm{I}}^{-1/2}|\Omega_{22.\mathrm{I}}|^{-1/2}|\Omega_{22.\mathrm{II}}|^{-1/2},
$$

we obtain the joint posterior for μ_1, δ_1, $\Omega_{11.2,\mathrm{I}}$, and ω conditionally on $\hat{\Pi}_{2,i} = (X_i^TX_i)^{-1}X_i^TY_{1,i}$, $i = \mathrm{I}, \mathrm{II}$, and on $\hat{\Omega}_{22,i} = (Y_{1,i} - X_i\hat{\Pi}_{2,i})^T \times (Y_{1,i} - X_i\hat{\Pi}_{2,i})/n$:

$$
p(\mu_1, \delta_1, \Omega_{11.2,\mathrm{I}}, \omega | \zeta_i = 0, \hat{\Pi}_{22,i}, \hat{\Omega}_{2,i}, i = \mathrm{I}, \mathrm{II}, \text{data}) \propto \omega^{n_2}\Omega_{11.2,\mathrm{I}}^{-(n+1)/2}\exp\left(-\frac{1}{2\Omega_{11.2,\mathrm{I}}}Q_3\right), \tag{6.6}
$$

where $Q_3 = (y_1 - R\mu_1 - T_1\delta_1)^T(y_1 - R\mu_1 - T\delta_1)$, and $V_{1,\mathrm{I}}$ and $V_{1,\mathrm{II}}$ are now evaluated by $\hat{V}_{1,i} = Y_{1,i} - X_i\hat{\Pi}_{2,i}$, $i = \mathrm{I}, \mathrm{II}$. Integrating $\Omega_{11.2,\mathrm{I}}$ and μ_1 out of (6.6), we obtain

$$
p(\delta_1 | \omega, \hat{\Pi}_{2,i}, \hat{\Omega}_{22,i}, i = \mathrm{I}, \mathrm{II}, \text{data}) \propto \left[1 + (\delta_1 - \hat{\delta}_1)^T\frac{T_1^TM_RT_1}{vs^2}(\delta_1 - \hat{\delta}_1)\right]^{-(n-2m_1-k_1)/2}, \tag{6.7}
$$

where $\hat{\delta}_1 = (T_1^T M_R T)^{-1} T_1^T M_R y_1$, $M_R = I - R(R^T R)^{-1} R^T$, $\nu s^2 = y_1^T[I - E \times (E^T E)^{-1} E^T] y_1 E = (R, T_1)$, and $\nu = n - 4m_1 - 2k_1$. We may evaluate ω by its posterior mean:

$$\omega_p = E(\omega|\cdot) = \int_0^\infty \omega^{n_2+1} |E^T E|^{-1/2} (\nu s^2)^{-(\nu-2)/2} \tag{6.8}$$

From (6.7) we see that conditionally on ω_p, $\hat{\Pi}_{2,i}$, and $\hat{\Omega}_{22,i}$ (i = I, II),

$$\frac{1}{m_1 + k_1} \hat{\delta}^T T_1^T M_R T_1 \hat{\delta}_1 / s^2 \tag{6.9}$$

is distributed as $F(m_1 + k_1, n - 4m_1 - 2k_1)$. Expression (6.9) may be used to test the hypotheses given in (6.2).

7. CONCLUSION

In this chapter we presented tests of constancy of regression models. We focused our attention on the tests for a set of parameters in one regime versus that in another regime. Since heteroscedasticity and/or autocorrelation influence test statistics substantially, we derived the highest posterior density region test that allows for heteroscedasticity and autocorrelation, and applied it to the U.S. demand for money as an empirical example. The highest posterior density region test is derived assuming that the join point is known. If it is not known and if the switch from one regime to another is believed to be gradual, the Bayesian inference procedure for the gradual switching regression provides a convenient tool since it does not rely on asymptotic distributions. It provides an exact posterior distributional inference given a realized sample.

It would be interesting to compare the performances of various tests of parameter stability. We derived a Bayesian highest posterior density region test of constancy of a structural equation conditionally on certain reduced form parameters. Its performance via-à-vis other tests needs to be evaluated in terms of sizes and powers of tests.

REFERENCES

Bacon, D. W., and D. G. Watts (1971). Estimating the transition between two intersecting lines, *Biometrica* 58, 525–534.

Bromeling, L. D., and H. Tsurumi (1986). *Econometrics and Structural Change*, Marcel Dekker, New York.

Brown, R. L., J. Durbin, and J. M. Evans (1975). Techniques for testing the constancy of regression relationships over time, *J. R. Stat. Soc. Ser. B* 37, 149–163.

Chin Choy, J. H., and L. D. Broemeling (1980). Some Bayesian inferences for a changing linear model, *Technometrics* 22, 71–78.

Chow, G. (1960). Tests of the equality between two sets of coefficients in two linear regressions, *Econometrica* 28, 561–605.

Chow, E. Y., and A. S. Willsky (1984). Bayesian design of decision rules for failure detection, *IEEE Trans. Aerosp. Electron. Syst.* AES-20, 761–773.

Cooley, T., and J. DeCanio (1977). Rational expectations in American agriculture, 1867–1914, *Rev. Econ. Stat.* 59, 9–17.

Cooley, T., and E. Prescott (1973a). An adaptive regression model, *Int. Econ. Rev.* 14, 364–371.

Cooley, T., and E. Prescott (1973b). Tests of an adaptive regression model, *Rev. Econ. Stat.* 14, 364–371.

Cooley, T., and E. Prescott (1976). Estimation in the presence of stochastic parameter variation, *Econometrica* 44, 167–184.

Durbin, J. (1960). Tests for serial correlation in regression analysis based on the periodogram of least squares residuals, *Biometrika* 56, 1–15.

Ferreira, P. E. (1975). A Bayesian analysis of a switching regression model: known number of regimes, *J. Am. Stat. Assoc.* 70, 370–374.

Garbade, K. (1977). Two methods for examining the stability of regression coefficients, *J. Am. Stat. Assoc.* 72, 54–63.

Goldfeld, S. M. (1976). The case of the missing money, *Brookings Pap. Econ. Act.* 3, 588–635.

Hamermesh, D. S. (1970). Wage bargains, threshold effects and the Phillips curve. *Q. J. Econ.* 84, 501–517.

Heller, H. R., and M. S. Khan (1979). The demand for money and the term structure of interest rates, *J. Polit. Econ.* 87, 109–129.

Ilmakunnas, P., and H. Tsurumi (1984). Testing for parameter shifts in a regression model with two regimes of autocorrelated errors, *Econ. Stud. Q.* 35, 46–56.

Ilmakunnas, P., and H. Tsurumi (1985). Testing the Lucas hypothesis on output-inflation trade-offs, *J. Bus. Econ. Stat.* 3, 43–53.

Isermann, R. (1984). Process fault detection based on modeling and estimation—a survey, *Automatica* 20, 387–404.

Judd, J. P., and J. L. Scadding (1982). The search for a stable money demand function: a survey of the post-1973 literature, *J. Econ. Lit.* 20, 993–1023.

Kerr, T. H. (1983). The Controversy over Use of SPRT and GLR Techniques and Other Loose Ends in Failure Detection, in *Proceedings of the American Control Conference*, pp. 966–977.

LaMotte, L. R., and A. McWhorter, Jr. (1978). An exact test for the presence of random walk coefficients in a linear regression model, *J. Am. Stat. Assoc.* 73, 816–820.

Laumas, P. S. (1978). Monetarization, economic development and the demand for money, *Rev. Econ. Stat.* 60, 614–615.

Laumas, G. S., and Y. P. Mehra (1976). The stability of the demand for money function: the evidence from quarterly data, *Rev. Econ. Stat.* 58, 463–468.

Mullineaux, D. (1980). Inflation expectations and money growth in the United States, *Am. Econ. Rev.* 70, 149–161.

O'Brien, R. J. (1970). Serial Correlation in Econometric Models, in *Econometric Study of the United Kingdom* (K. Hilton and D. F. Heathfield, eds.), Kelley, New York, pp. 375–437.

Pagan, A. (1980). Some identification and estimation results for regression models with stochastically varying coefficients, *J. Econometrics* 13, 341–363.

Quandt, R. (1958). The estimation of the parameters of a linear regression system obeying two separate regimes, *J. Am. Stat. Assoc.* 53, 837–880.

Rausser, G. C., and P. S. Laumas (1976). The stability of the demand for money in Canada, *J. Monetary Econ.* 2, 367–380.

Roll, R. (1972). Interest rates on monetary assets and commodity price index changes, *J. Finance* 27, 241–277.

Sengupta, J. K., and G. Tintner (1963). On some aspects of trend in the aggregative models of economic growth, *Kyklos* 16, 47–61.

Spall, J. C. (1988). Bayesian error isolation for models of large-scale systems, *IEEE Trans. Autom. Control* AC-33, 341–347.

Swamy, P. A. V. B., and P. A. Tinsley (1980). Linear prediction and estimation methods for regression models with stationary stochastic coefficients, *J. Econometrics* 12, 103–142.

Tanaka, K. (1983). Non-normality of the Lagrangian multiplier statistic for testing the constancy of regression coefficients, *Econometrica* 51, 1577–1582.

Tsurumi, H. (1978). A Bayesian test of a parameter shift in a simultaneous equation with an application to a macro savings function in Japan, *Econ. Stud. Q.* 29, 216–230.

Tsurumi, H. (1980). Bayesian Estimation of Structural Shifts by Gradual Switching Regressions with an Application to the U.S. Gasoline Market, in *Bayesian Analysis in Econometrics and Statistics: Essays in Honor of Harold Jeffreys* (A. Zellner, ed.), North-Holland, Amsterdam, pp. 212–240.

Tsurumi, H. (1985). Limited information Bayesian analysis of a structural coefficient in a simultaneous equations system, *Commun. Stat. Theory Methods* 14, 1103–1120.

Tsurumi, H., and N. Sheflin (1985). Some tests for the constancy of regressions under heteroscedasticity, *J. Econometrics* 27, 221–234.

Tsurumi, H., and T. Shiba (1981). On Cooley and Prescott's time varying parameter model, *Econ. Stud. Q.* 32, 176–179.

Willsky, A. S. (1976). A survey of design methods for failure detection in dynamic systems, *Automatica* 12, 601–611.

Zellner, A., and G. C. Tiao (1964). Bayesian analysis of the regression model with autocorrelated errors, *J. Am. Stat. Assoc.* 59, 673–778.

5

Small-Sample Bayesian Frequency-Domain Analysis of Autoregressive Models

PEYTON COOK University of Tulsa, Department of Mathematics and Computer Science, The University of Tulsa, Tulsa, Oklahoma

1. INTRODUCTION

Spectral analysis has been developing for a considerable time (see Robinson, 1983, pp. 345–407). Perhaps the first of the modern researchers to popularize techniques for estimating the spectrum of a time series was Schuster (1898), who was interested in examining hidden periodicities in data. The periodogram developed by Schuster was hampered by the computational devices of the day. During the 1950s modern computers were able to perform the calculations which previously had been so tedious that spectrum estimation was performed only by those with extreme need for frequency domain information. The fast Fourier transform (FFT) became popular after its discovery (see Brigham, 1974, pp. 8–9 for a brief history). Upon entering the 1960s, the approach of Blackman and Tukey (1959) became widely used. Tukey's work with the FFT in the 1960s spurred the use of periodogram-based techniques for spectral analysis. In all, the late 1950s and early 1960s proved a very productive period for those perfecting general nonparametric techniques for spectrum estimation. Yet early in this century, parametric time series models were discovered to possess characteristics useful in modeling pseudoperiodic behavior (see Slutzky, 1937). Even so, it was not until the mid to late 1970s that general autoregressive moving average (ARMA) models were recommended as a routine class of models for dealing with pseudocyclic activity (see Akaike, 1978; Parzen, 1979).

Statistical approaches to the estimation of frequency-domain characteristics of time series have primarily been based on the frequentist or repeated-sampling approach. Also, the performance of the frequentist procedures has generally been justified using only asymptotic theory rather than small-sample theory. Very few techniques seem to be justified for small samples. That is, the primary justification for the popular techniques is consistency of the estimators. Also, asymptotic distributions of popular estimators are often used. However, the sample size required to justify the use of asymptotic distributions is not always obvious. Even today, many of the proponents of the frequentist approach still maintain that small-sample justifications are unlikely to be of much value because they are so cumbersome (see Dzhaparidze, 1986, pp. 2–3). Furthermore, in practice, one does not often encounter attempts to indicate the amount of uncertainty associated with point estimates.

By the beginning of the 1970s, it was known that the Bayesian approach could be very helpful for developing small-sample approaches to estimation of the time-domain characteristics of ARMA models. Zellner (1971, pp. 186–200) describes the posterior distributions of the parameters of some autoregressive models. Box and Jenkins (1976, pp. 250–258) briefly describe some general aspects of Bayesian estimation of the parameters of ARMA models. They also include a brief description of asymptotic approximations which can be used. The literature on spectrum estimation using a Bayesian approach is very sparse. Jeffreys (1961, pp. 442–453) was perhaps the first to describe a Bayesian implementation in the frequency domain. Shore (1980, pp. 83–180) described certain results on Bayesian spectral analysis exploiting the asymptotic behavior of the likelihood function. Cook (1983, 1985, 1986) appears to be the first Bayesian attempt to develop posterior results for autoregressive models which do not use asymptotic approximations.

In this chapter we describe Bayesian techniques that are computationally feasible for examining Gaussian univariate autoregressive processes in the frequency domain. The approach to be described will not introduce asymptotic approximations. Section 2 will give a brief review of the Bayesian time-domain analysis of Gaussian univariate autogressive processes. Much of the basic notation used in this chapter is developed in Section 2. In Section 3 we give a brief review of the notion of the spectrum of covariance stationary stochastic processes. In particular, the functional form of the spectrum of an autoregressive process will be given. In Section 4 we develop the posterior expectation and variance for the reciprocal of the autoregressive spectrum. Also, in Section 4 we describe how one can create a conservative posterior probability interval or credible interval for the spectrum and its reciprocal at any particular desired frequency. These interval estimates are based on Chebychev's inequality, which is used in the simplest manner possible. The goal is to make convenient use of the posterior expectation and variance of the reciprocal of the spectrum. Although it might be possible to obtain tighter bounds than are possible with Chebychev's inequality, tighter bounds often require that one know more than simply a mean and variance. In addition, a cleaner result that will apply to all frequencies simultaneously is described in a subsequent section. Some practitioners prefer to work with the squared gain rather than the spectrum. In Section 5 we extend the approach of Section 4 to handle the squared gain of the process and the reciprocal of the squared gain. In Section 6 we deal directly with the real and imaginary parts of the frequency-response function. The real and imaginary parts of the reciprocal of the frequency-response function of the autoregressive process can be expressed as linear combinations of the autoregressive coefficients. This will allow the derivation of the posterior distribution of the real and imaginary parts of the reciprocal of the frequency-response function of the process. The derivation is based on standard linear model theory, which allows the use of an adaptation of the Scheffé result for making inferences simultaneously at infinitely many frequencies. In Section 7 we extend the results on the real and imaginary parts of the reciprocal of the frequency-response function of the process to the squared gain and its reciprocal. In particular, we show how one can make inferences at infinitely many frequencies simultaneously for the squared gain and its reciprocal. These inferences do not require asymptotic justifications and are the main contribution of this chapter. In Section 8 we implement many of the techniques described in this chapter. The famous sunspot series of Wolfer is used. Finally, in Section 9 we provide a summary and a description of further work in progress.

2. AUTOREGRESSIVE MODEL

The univariate pth-order autoregressive model [AR (p)] examined in this chapter is

$$x(t) = \phi_0 + \sum_{j=1}^{p} \phi_j x(t-j) + e(t). \tag{2.1}$$

In expression (2.1), $t = 1, 2, \ldots, n$, $x(t)$ is an observable random variable, $e(t)$ is an unobservable random variable with a normal distribution where $E[e(t)] = 0$, $\operatorname{var}(e(t)) = \tau^{-1}$, $\operatorname{cov}(e(t_1), e(t_2)) = 0$, $t_1 \neq t_2$. Also, $x(0)$, $x(-1), \ldots, x(-p+1)$ are fixed initial starting values. The precision $\tau = \sigma^{-2}$, rather than the variance, of the unobservable random variables $e(t)$ will be used for convenience. The fixed unknown parameters of the model are ϕ_0, $\phi_1, \ldots, \phi_p$, and τ. Properties of the process (2.1) are described in Box and Jenkins (1976, pp. 53–66), Fuller (1976, pp. 18–75), and Broemeling (1985, pp.

183–186). It should be noted that (2.1) has fixed nonrandom starting values. The notation will be greatly simplified by adopting standard matrix notation. Column vectors will be indicated by boldface and matrices will be indicated with capital letters. This is intended to emphasize the dimensions of the various quantities. This produces a conflict with the notation used in probability where capital letters indicate a random variable. Since the Bayesian approach will be used, the parameters of the model described in (2.1) become random variables even though they do not result from a random selection. When it is useful to emphasize that a quantity is random, emphasis will be given. By using matrix notation, expression (2.1) can be simplified to

$$\begin{bmatrix} x(1) \\ x(2) \\ \vdots \\ x(n) \end{bmatrix} = \begin{bmatrix} 1 & x(0) & \cdots & x(-p+1) \\ 1 & x(1) & \cdots & x(-p+2) \\ \vdots & \vdots & \ddots & \vdots \\ 1 & x(n-1) & \cdots & x(n-p) \end{bmatrix} \begin{bmatrix} \phi_0 \\ \phi_1 \\ \vdots \\ \phi_p \end{bmatrix} + \begin{bmatrix} e(1) \\ e(2) \\ \vdots \\ e(n) \end{bmatrix}$$

or just

$$\mathbf{x} = Z\boldsymbol{\phi} + \mathbf{e}, \tag{2.2}$$

the probability density function (p.d.f.) of $\mathbf{x}$ given $(\boldsymbol{\phi}, \tau)$, and the starting values for the process is

$$f(\mathbf{x}|\boldsymbol{\phi}, \tau) \propto \tau^{n/2} \exp\left[-\frac{\tau}{2}(\mathbf{x} - Z\boldsymbol{\phi})^T(\mathbf{x} - Z\boldsymbol{\phi}) \right], \qquad \mathbf{x} \in R^n. \tag{2.3}$$

Note that dependence on the initial starting values $x(0), x(-1), \ldots, x(-p+1)$ is not indicated explicitly in (2.2) or (2.3). A standard Bayesian approach is to use the natural conjugate prior p.d.f. for the fixed unknown parameters of the model. For (2.3), the natural conjugate prior is the normal-gamma family. The parameterization to be used is

$$f(\boldsymbol{\phi}|\tau) \propto \tau^{(p+1)/2} \exp\left[-\frac{\tau(\boldsymbol{\phi} - \mathbf{b})^T S(\boldsymbol{\phi} - \mathbf{b})}{2} \right], \qquad \boldsymbol{\phi} \in R^{p+1},$$

$$f(\tau) \propto \tau^{\alpha/2-1} \exp\left(-\frac{\tau}{2}\beta \right), \qquad \tau > 0 \tag{2.4}$$

Expression (2.4) is well known and is simply the normal-gamma conjugate prior p.d.f. for $(\boldsymbol{\phi}, \tau)$. Note that expression (2.4) is similar to but not identical to expression (5.7) found in Broemeling (1985, p. 184). The hyperparameters of the prior p.d.f. for $(\boldsymbol{\phi}, \tau)$ are $\mathbf{b} \in R^{p+1}$, S a $(p+1)$ by $(p+1)$ symmetric positive definite matrix, $\alpha > 0$, and $\beta > 0$. By choosing values of these hyperparameters, one explicitly chooses a specific prior p.d.f. in the conjugate family. Since spectral analysis is the current concern, only covariance stationary processes are of interest. Actually, the initial value model (2.1) is nonstationary precisely because of the starting values. However, under certain conditions to be specified in Section 3, the initial value model (2.1) can be approximated by a corresponding covariance stationary process that is similar except that the starting values are moved infinitely far into the past. As a result, no constraints will be placed on the parameters $(\boldsymbol{\phi}, \tau)$ to guarantee stationarity. Since the Bayesian approach used here will not require asymptotic justifications to obtain posterior p.d.f.'s, the parameter space is the set of values for $(\boldsymbol{\phi}, \tau)$ such that (2.3) is a legitimate p.d.f. for $\mathbf{x}$. That is, $\boldsymbol{\phi} \in R^{p+1}$ and $\tau > 0$.

To simplify the time-domain notation, let

$$\begin{aligned} A_1 &= Z^T Z + S \\ \mathbf{a}_2 &= Z^T \mathbf{x} + S\mathbf{b} \\ \hat{\boldsymbol{\phi}} &= A_1^{-1} \mathbf{a}_2 \\ r &= \beta + \mathbf{b}^T S \mathbf{b} + \mathbf{x}^T \mathbf{x} - \hat{\boldsymbol{\phi}}^T A_1 \hat{\boldsymbol{\phi}} \\ v &= n + \alpha \\ \hat{P} &= v r^{-1} A_1. \end{aligned} \tag{2.5}$$

Then the posterior p.d.f. for $(\boldsymbol{\phi}, \tau)$ is

$$f(\boldsymbol{\phi}, \tau | \mathbf{x}) \propto \tau^{(v+p+1)/2-1} \exp\left\{ -\frac{\tau}{2} \left[(\boldsymbol{\phi} - \hat{\boldsymbol{\phi}})^T A_1 (\boldsymbol{\phi} - \hat{\boldsymbol{\phi}}) + r \right] \right\}. \tag{2.6}$$

Since the prior p.d.f. for the parameters was in the conjugate normal-gamma family, this posterior p.d.f. for the parameters will also be in the normal-gamma family. For more information about the derivation of (2.6), see Broemeling (1985, pp. 181–185). The posterior distribution of $\boldsymbol{\phi}$ given τ is a $(p + 1)$ variate normal with mean $\hat{\boldsymbol{\phi}}$ and variance $(\tau A_1)^{-1}$. The posterior distribution of $\boldsymbol{\phi}$ (not conditional on τ) is a $(p + 1)$ variate t-distribution with v degrees of freedom, mean $\hat{\boldsymbol{\phi}}$, and precision matrix $\hat{P}$. The posterior distribution of τ is gamma with parameters $v/2$ and $r/2$. The posterior expectation of τ is v/r and the variance is $2v/r^2$. This can be summarized with the notation

$$\begin{aligned} \boldsymbol{\phi} | \tau, \mathbf{x} &\sim N_{p+1}(\hat{\boldsymbol{\phi}}, (\tau A_1)^{-1}) \\ \boldsymbol{\phi} | \mathbf{x} &\sim T_{p+1}(v, \hat{\boldsymbol{\phi}}, \hat{P}) \\ \tau | \mathbf{x} &\sim \text{gamma}\left(\frac{v}{2}, \frac{r}{2}\right). \end{aligned} \tag{2.7}$$

The properties of these posterior p.d.f.'s will be used in subsequent sections. Additional properties that will be used will be introduced as needed.

However, general discussions of the properties of these distributions can be found in DeGroot (1970, Chap. 5), Box and Tiao (1973, pp. 117–118), and Press (1982, pp. 136–138).

Due to the specification of the hyperparameters, (2.6) will exist for all sample sizes where $\nu > 0$. However, the variance of $(\boldsymbol{\phi}|x)$ is $(\nu/(\nu-2))\hat{P}^{-1}$, which exists only when $\nu > 2$.

In Section 3 we give a brief review of the spectral density function and its interpretation. Also, the spectral density function of the AR (p) process will be reviewed.

3. AR (p) SPECTRUM

The notion of the spectrum of a covariance stationary process will now be reviewed briefly. If $\{x(t)\}$ is a discrete-time covariance stationary stochastic process with absolutely summable autocovariance function $\{\gamma(h)\}$, the power spectrum of the process is defined to be

$$S(f) = \sum_{h=-\infty}^{\infty} \gamma(h) \exp(-2\pi i f h), \qquad -\tfrac{1}{2} < f \leqslant \tfrac{1}{2} \tag{3.1}$$

(see Fuller, 1976, p. 127; Brillinger, 1975, p. 23; or Priestley, 1981, p. 225). The variable f denotes cycles per unit time and $i = \sqrt{-1}$. Also, one is generally interested only in the nonnegative frequencies. Implicitly, a unit sampling interval is assumed. As can be seen, $S(f)$ is simply the Fourier transform of the autocovariance function. The inverse Fourier transform of $S(f)$ is

$$\gamma(h) = \int_{-1/2}^{1/2} S(f) \exp(2\pi i f h)\, df. \tag{3.2}$$

The interpretation of $S(f)$ can be seen by substituting $h = 0$ into (3.2), giving

$$\sigma_X^2 = \gamma(0) = \int_{-1/2}^{1/2} S(f)\, df. \tag{3.3}$$

From (3.3), one sees that the spectral density function $S(f)$ (or just the spectrum) provides a decomposition, across the frequencies from $-\frac{1}{2}$ to $\frac{1}{2}$, of the variance of the process $\{x(t)\}$. If one were to examine the graph of the spectrum, frequency ranges that contain a definite peak are frequencies where variance is added to the process. As a result, such peaks indicate pseudo-periodic behavior. Expressions (3.1) through (3.3) apply to any covariance stationary process with absolutely summable autocovariance function. Until the late 1970s, most interest on the spectrum focused on the definitions above, which do not make use of specific parametric models such as the AR (p)

process. In the late 1970s (see Akaike, 1978, or Parzen, 1979) interest began to center on the use of models whose spectrum depended only on a finite number of parameters, such as the autoregressive processes. Not all processes defined by stochastic difference equations are covariance stationary. The usual way to demonstrate covariance stationarity of an AR(p) process requires that the process be expressed as an infinite-order moving average process (see Fuller, 1976, pp. 56–58). The AR(p) process (2.1) is not covariance stationary due to the initial nonrandom starting values, as indicated in Section 2. However, it can be approximated by an AR(p) process whose initial values are moved infinitely far into the past. For such an AR process, a different notation will give some insight.

Let B denote the backshift operator so that $Bx(t) = x(t-1)$ and $B^k x(t) = x(t-k)$. Then define the operator

$$\phi(B) = 1 - \phi_1 B - \phi_2 B^2 - \cdots - \phi_p B^p. \tag{3.4}$$

Expression (2.1) can then be rewritten as

$$\phi(B)x(t) = \phi_0 + e(t). \tag{3.5}$$

The deterministic homogeneous difference equation that corresponds tc (3.5) has a characteristic polynomial that appears similar to the polynomial operator (3.4). The appearance is deceptive. The argument of the polynomial operator is the backshift operator B. The argument of the characteristic polynomial is a complex variable which appears to be the reciprocal of B. When the zeros of the characteristic polynomial are inside the unit circle, the solution to the deterministic homogeneous difference equation will be asymptotically stable. As time goes to infinity, the solution to the homogeneous deterministic equation tends to zero for any possible starting values. This corresponds to the probabilistic notion of stationarity. Although some abuse of the notation occurs, it is sometimes convenient to act as if the autoregressive operator $\phi(B)$ were a polynomial whose argument is a complex number. Then a condition that will guarantee stationarity for the AR(p) process is that the zeros of $\phi(B)$ lie outside the unit circle. Then the process is covariance stationary with an absolutely summable autocovariance function. This corresponds to a process defined over time infinitely far into the past as well as the future. The theory of linear filters (used in Sections 5 and 6) suggests that the technicality regarding the time span is not so important. That is, the operator $\phi(B)$ is important from the point of view of linear filtering theory as well as from the point of view of stationary stochastic processes. The spectral density function of the covariance stationary AR(p) process will exist and is

$$S(f) = \sigma^2 |\phi(\exp(-2\pi i f))|^{-2}, \qquad -\tfrac{1}{2} < f \leqslant \tfrac{1}{2}. \tag{3.6}$$

Expression (3.6) in effect treats the symbol B as if it were the complex number $B = \exp(-2\pi i f)$ (an extension of the abuse of the notation described above).

Another way of viewing the condition producing stationarity (for the process defined over all time) simply requires that the operator $\phi(B)$ be invertible. This is possible when it zeros lie outside the unit circle. This condition also corresponds to an asymptotically stable linear filter. A derivation of (3.6) can be found in Fuller (1976, pp. 144–145) or Priestley (1981, p. 282). It is relevant to point out that (3.6) depends only on the $p + 1$ parameters $\phi_1, \phi_2, \ldots, \phi_p$, and τ.

It is clear that expression (3.6) requires that one know p, the order of the process. For the purposes of this chapter it will be assumed that it is known that forecasts from an AR(p) process will give adequate forecasts. That is, problems of order identification will be left to other authors. In Section 4 we describe a Bayesian approach to making inferences regarding the spectral density function $S(f)$ as well as its reciprocal $S^{-1}(f)$. Further, the use of $S^{-1}(f)$ is also developed in Section 4.

4. POSTERIOR MOMENTS FOR $S^{-1}(f)$

The most prevalent techniques for estimating the spectral density function $S(f)$ are based on the periodogram. Modern computational techniques for the periodogram (or a smoothed version of the periodogram) make use of the fast Fourier transform (FFT). Accounts of the FFT can be found in Brigham (1974, pp. 172–197) or Bloomfield (1976, pp. 61–76). In the last decade parametric approaches to estimating $S(f)$ have been based on ARMA models. For details on ARMA spectral analysis from the repeated sampling point of view, one may consult Akaike (1978), Parzen (1979), or Priestley (1981, pp. 600–607). The literature for the frequentist or repeated sampling estimators of the spectral density function usually describes asymptotic properties of the estimators. As will be shown shortly, the reciprocal of the spectral density function $S^{-1}(f)$ for AR models can be expressed as a quadratic form of the autoregressive coefficients. Since moments of quadratic forms are well known, and since the posterior distribution of the parameters of the AR(p) process is known, it will be possible to compute the exact posterior mean and variance of $S^{-1}(f)$ using the technique of Cook (1985). Using the posterior mean and variance of $S^{-1}(f)$, one can use Chebychev's well-known inequality to construct conservative (nonoptimal) credible intervals for $S^{-1}(f)$ at any particular frequency of interest. These intervals can be transformed to produce credible intervals for the spectral density function without requiring asymptotic approximations.

To simplify derivation of the posterior moments of $S^{-1}(f)$, the following notation will be helpful. Let

$$\begin{aligned}\boldsymbol{\phi}^* &= [-1, \phi_1, \phi_2, \ldots, \phi_p]^T \\ \mathbf{c}^*(f) &= [\cos 2\pi 0f, \cos 2\pi 1f, \ldots, \cos 2\pi pf]^T \\ \mathbf{s}^*(f) &= [\sin 2\pi 0f, \sin 2\pi 1f, \ldots, \sin 2\pi pf]^T \\ D(f) &= \mathbf{c}^*(f)\mathbf{c}^{*T}(f) + \mathbf{s}^*(f)\mathbf{s}^{*T}(f).\end{aligned}$$

Then the reciprocal of the spectral density function can be written as

$$\begin{aligned}S^{-1}(f) &= \tau|\phi(\exp(-2\pi i f))|^2 \\ &= \tau\boldsymbol{\phi}^{*T}D(f)\boldsymbol{\phi}^*, \qquad -\tfrac{1}{2} < f \leqslant \tfrac{1}{2},\end{aligned} \tag{4.1}$$

which is a quadratic form in $\boldsymbol{\phi}^*$. There are two reasons for choosing to work with the reciprocal of the spectral density function. First, one need only work with quadratic forms rather than the reciprocal of a quadratic form. Second, attempts to produce interval estimates for $S^{-1}(f)$ do not encounter the possibility of infinite limits. The relationship between the two functions, $S(f)$ and its reciprocal, is straightforward. Peaks of one function correspond to troughs of the other. This is mentioned explicitly because most practitioners are used to working primarily with the spectrum rather than its reciprocal.

Notice that posterior moments of (4.1) merely involve moments of quadratic forms of random vectors with a t-distribution. The posterior mean and variance can be obtained in a two-step fashion. Conditional upon τ, the quadratic form will involve a normally distributed random vector. The conditioning on τ can then be removed to obtain the exact posterior expectation of (4.1) without introducing asymptotic approximations. Recall that $E[\boldsymbol{\phi}|\tau, x] = \hat{\boldsymbol{\phi}}$, from (2.5) and (2.6). The posterior expectation of $\boldsymbol{\phi}^*$, conditional upon τ, is obtained by replacing the first component of $\hat{\boldsymbol{\phi}}$ with the number -1. Also, recall that the posterior variance of $\boldsymbol{\phi}$, conditional upon τ, is $\tau^{-1}A_1^{-1}$. Let $\hat{C}$ be a matrix obtained by replacing the first row and column of A_1^{-1} with zeros. Then

$$\begin{aligned}E[\boldsymbol{\phi}^*|\tau, \mathbf{x}] &= [-1, \hat{\phi}_1, \hat{\phi}_2, \ldots, \hat{\phi}_p]^T = \hat{\boldsymbol{\phi}}^* \\ \operatorname{var}(\boldsymbol{\phi}^*|\tau, \mathbf{x}) &= \tau^{-1}\begin{bmatrix} 0 & \mathbf{0}^T \\ \mathbf{0} & A_2^{-1}\end{bmatrix} \\ &= \tau^{-1}\hat{C},\end{aligned} \tag{4.2}$$

where A_2^{-1} is obtained by stripping off the first row and the first column of

A_1^{-1}. With (4.2) and the well-known expectation of quadratic forms, one sees that

$$\begin{aligned} E[S^{-1}(f)|\tau, \mathbf{x}] &= E[\tau\boldsymbol{\phi}^{*T}D(f)\boldsymbol{\phi}^*|\tau, \mathbf{x}] \\ &= \tau E[\boldsymbol{\phi}^{*T}D(f)\boldsymbol{\phi}^*|\tau, \mathbf{x}] \\ &= \tau\{\operatorname{tr}(D(f)\tau^{-1}\hat{C}) + \hat{\boldsymbol{\phi}}^{*T}D(f)\hat{\boldsymbol{\phi}}^*\} \\ &= \operatorname{tr}(D(f)\hat{C}) + \tau\hat{\boldsymbol{\phi}}^{*T}D(f)\hat{\boldsymbol{\phi}}^*. \end{aligned} \tag{4.3}$$

To remove the conditioning upon τ in expression (4.3), one need only take a final expectation using the posterior distribution of τ. Then the posterior expectation of $S^{-1}(f)$ is

$$\begin{aligned} E[S^{-1}(f)|\mathbf{x}] &= \operatorname{tr}(D(f)\hat{C}) + \mathrm{E}|\tau|\mathbf{x}]\hat{\boldsymbol{\phi}}^{*T}D(f)\hat{\boldsymbol{\phi}}^* \\ &= \operatorname{tr}(D(f)\hat{C}) + (v/r)\hat{\boldsymbol{\phi}}^{*T}D(f)\hat{\boldsymbol{\phi}}^*. \end{aligned} \tag{4.4}$$

Expression (4.4) is an exact result based on the initial value model (2.1) and can be computed whenever one uses the conjugate normal-gamma prior p.d.f. described in (2.4). That is, asymptotic approximations were not used to derive the result. As is well known, the asymptotic behavior of posterior distributions is similar to that of maximum likelihood (see Zellner, 1971, pp. 31–34, for a general discussion of the asymptotic behavior of posterior p.d.f.'s).

The posterior variance of $S^{-1}(f)$ has two parts which arise when one expresses a variance in terms of conditional means and variances (see DeGroot, 1970, p. 28). The posterior variance of $S^{-1}(f)$ can be expressed as

$$\operatorname{var}(S^{-1}(f)|\mathbf{x}) = E_{\tau|\mathbf{x}}[\operatorname{var}(S^{-1}(f)|\tau, \mathbf{x})] + \operatorname{var}_{\tau|\mathbf{x}}(E[S^{-1}(f)|\tau, \mathbf{x}]).$$

In the expression above, the subscript of $\tau|\mathbf{x}$ is intended to indicate explicitly that the expectations are to be computed with respect to the posterior distribution of τ.

Now

$$\begin{aligned} \operatorname{var}_{\tau|\mathbf{x}}(E[S^{-1}(f)|\tau, \mathbf{x}]) &= \operatorname{var}_{\tau|\mathbf{x}}(\tau\hat{\boldsymbol{\phi}}^{*T}D(f)\hat{\boldsymbol{\phi}}^*|\tau, \mathbf{x}) \\ &= (\hat{\boldsymbol{\phi}}^{*T}D(f)\hat{\boldsymbol{\phi}}^*)^2 \operatorname{var}(\tau|\mathbf{x}) \\ &= (\hat{\boldsymbol{\phi}}^{*T}D(f)\hat{\boldsymbol{\phi}}^*)^2(2v/r^2) \end{aligned} \tag{4.5}$$

and

$$\begin{aligned} E_{\tau|\mathbf{x}}[\operatorname{var}(S^{-1}(f)|\tau, \mathbf{x})] &= E_{\tau|\mathbf{x}}[\operatorname{var}(\tau\boldsymbol{\phi}^{*T}D(f)\boldsymbol{\phi}^*|\tau, \mathbf{x})] \\ &= E_{\tau|\mathbf{x}}[\tau^2\{2\operatorname{tr}(D(f)\tau^{-1}\hat{C}D(f)\tau^{-1}\hat{C}) \\ &\quad + 4\hat{\boldsymbol{\phi}}^{*T}D(f)\tau^{-1}\hat{C}D(f)\hat{\boldsymbol{\phi}}^*\}|\mathbf{x}] \end{aligned}$$

Then

$$\begin{aligned} E_{\tau|\mathbf{x}}[\operatorname{var}(S^{-1}(f)|\tau, \mathbf{x}] &= E_{\tau|\mathbf{x}}[2\operatorname{tr}(D(f)\hat{C}D(f)\hat{C}) + 4\tau\hat{\boldsymbol{\phi}}^{*T}D(f)\hat{C}D(f)\hat{\boldsymbol{\phi}}^*|\mathbf{x}] \\ &= 2\operatorname{tr}(D(f)\hat{C}D(f)\hat{C}) + 4(v/r)\hat{\boldsymbol{\phi}}^{*T}D(f)\hat{C}D(f)\hat{\boldsymbol{\phi}}^*. \end{aligned} \tag{4.6}$$

By putting (4.5) and (4.6) together, one obtains

$$\operatorname{var}(S^{-1}(f)|\mathbf{x}) = 2v(\hat{\boldsymbol{\phi}}^{*T}D(f)\hat{\boldsymbol{\phi}}^{*})^2/r^2 + 2\operatorname{tr}(D(f)\hat{C}D(f)\hat{C}) + 4(v/r)\hat{\boldsymbol{\phi}}^{*T}D(f)\hat{C}D(f)\hat{\boldsymbol{\phi}}^{*}. \tag{4.7}$$

Expressions (4.4) and (4.7) give formulas for the posterior mean and variance of $S^{-1}(f)$, $-\frac{1}{2} < f \leqslant \frac{1}{2}$ and apply to all sample sizes. These formulas are relatively easy to program and are computationally fast. Additional details may be found in Cook (1983, 1985).

The function $S^{-1}(f)$ has its own interpretation, which is related to the concept of filtering. Our model $\phi(B)x(t) = \phi_0 + e(t)$ is one where data are filtered to create white noise. $S^{-1}(f)$ is proportional to the spectral density function of the filter (or a hypothetical white noise input process) rather than the process $\{x(t)\}$. Of course, $S^{-1}(f)$ is simply the reciprocal of $S(f)$, so that peaks of $S^{-1}(f)$ correspond to troughs of $S(f)$, and vice versa. By plotting $S^{-1}(f)$ (a logarithmic scale is often needed to increase resolution) one can easily determine frequency bands where $S^{-1}(f)$ has its troughs.

Exact highest posterior density (HPD) credible intervals for $S^{-1}(f)$ require that one obtain its posterior distribution. Apparently, its posterior distribution is analytically intractable and one would be forced into a numerically intensive technique. However, one can use Chebychev's inequality knowing only the posterior mean and variance of the reciprocal of the spectral density function. In particular, if k is a number such that $k \geqslant 1$, then at a given frequency f, the posterior probability that $S^{-1}(f)$ is within k standard deviations of its mean is at least $1 - 1/k^2$. Since the goal of this chapter is to present only results that do not depend on asymptotic approximations, the use of Chebychev's inequality may be desirable. If one does use Chebychev's inequality to create a conservative credible interval for $S^{-1}(f)$ at a particular frequency f, one may easily transform the interval with a reciprocal transformation to produce a conservative credible interval for the spectrum itself. However, one should use caution. The use of the Chebychev inequality with $S^{-1}(f)$ produces a credible interval which is symmetric about its posterior expectation. Because of this and the fact that the Chebychev inequality is conservative, it is possible that such a credible interval could include negative numbers. Since the reciprocal of the spectral density function is nonnegative, one would truncate the lower endpoint of the credible interval at zero. If one tried to use the reciprocal transformation on such a credible interval, one would have a credible interval for $S(f)$ with an infinite upper bound. Consequently, one should not select too large a value for k.

Rather than attempting to create tighter credible intervals for the spectrum at a particular frequency, in a later section we derive a technique for constructing infinitely many credible intervals for the squared gain, which are simultaneously valid with a preselected amount of posterior probability.

Since the spectrum is a continuous function such a result may prove more desirable to many practitioners than tighter credible intervals at individual frequencies.

The reciprocal of the squared gain of an AR (p) process is proportional to (4.1). Since many researchers are used to working with the squared gain, in the next section we describe posterior moments for the reciprocal of the squared gain of the AR (p) process. The approach used will closely resemble the derivations of this section.

5. POSTERIOR MOMENTS FOR $G^{-2}(f)$

The frequency-response function for the AR (p) process is

$$\phi^{-1}(f) = \phi^{-1}(\exp(-2\pi i f))$$

(see Bloomfield, 1976, p. 124; Fuller, 1976, p. 152). This function exists for covariance stationary AR (p) processes whose operator in (3.5) can be inverted, giving

$$x(t) = \phi^{-1}(B)[\phi_0 + e(t)]. \tag{5.1}$$

As can be seen from (5.1), $\phi^{-1}(B)$ acts as a linear filter on the white noise input series to create the observable data. [It should be mentioned here that in Section 6 we deal with the frequency-response function, $\phi(f)$, of the AR (p) operator rather than the frequency-response function of the AR (p) process.] The squared norm of the frequency-response function is proportional to the spectral density function of the process and is called the squared gain of the process. Let the gain of the process, the norm of $\phi^{-1}\exp(-2\pi i f)$, be denoted by $G(f)$. Then the squared gain of the AR process is

$$\begin{aligned} G^2(f) &= |\phi^{-1}(\exp(-2\pi i f))^2 \\ &= |\phi(\exp(-2\pi i f))|^{-2} \\ &= (\boldsymbol{\phi}^{*T} D(f) \boldsymbol{\phi}^*)^{-1}, \end{aligned}$$

where $0 \leqslant f \leqslant \frac{1}{2}$. Note that $G(f)$ is real valued, whereas the frequency-response function is complex valued. The squared gain is the reciprocal of a quadratic form similar to that of the spectrum. The reciprocal of the squared gain of the process, denoted $G^{-2}(f)$, is a quadratic form, which, as in Section 4, is relatively easy to examine. The reciprocal of the squared gain is just

$$G^{-2}(f) = \boldsymbol{\phi}^{*T} D(f) \boldsymbol{\phi}^*, \qquad 0 \leqslant f \leqslant \tfrac{1}{2}. \tag{5.2}$$

Expression (5.2) is proportional to (4.1). $S^{-1}(f)$ contains τ, whereas $G^{-2}(f)$ does not. Frequencies where $G^2(f)$ exceeds the number 1 are frequencies where the filter adds variance to the white noise input to create the data. Alternatively, frequencies where $G^{-2}(f)$ is less than 1 are frequencies where variance is added to create the data. Once again, there is the obvious reciprocal relationship just as was seen with the spectral density function and its reciprocal. Perhaps it should be mentioned that a typical way to express the frequency-response function is with polar coordinates. The magnitude coordinate is called the gain, defined above, and the squared magnitude is called the squared gain. Although the phase coordinate is certainly of interest (see Robinson and Treital, 1981, pp. 55–56), most interest centers on the magnitude. Only the squared gain and its reciprocal will be examined in this section.

Posterior moments for $G^{-2}(f)$ can be derived in a manner similar to the posterior moments for $S^{-1}(f)$. By using the same approach as the preceding section, one can obtain the posterior expectation and variance for $G^{-2}(f)$. One first conditions on τ and then one removes the conditioning. Also, one uses the same results regarding the mean and variance of quadratic forms of normally distributed random vectors as in the preceding section. Then, conditional upon τ,

$$E[G^{-2}(f)|\tau, \mathbf{x}] = \operatorname{tr}(D(f)\tau^{-1}\hat{C}) + \hat{\boldsymbol{\phi}}^{*T}D(f)\hat{\boldsymbol{\phi}}^{*}.$$

One now computes the expectation with respect to the posterior distribution of τ, and the posterior expectation of $G^{-2}(f)$ at any frequency is

$$\begin{aligned} E[G^{-2}(f)|\mathbf{x}] &= E[\tau^{-1}|\mathbf{x}]\operatorname{tr}(D(f)\hat{C}) + \hat{\boldsymbol{\phi}}^{*T}D(f)\hat{\boldsymbol{\phi}}^{*}, \\ &= \frac{r}{v-2}\operatorname{tr}(D(f)\hat{C}) + \hat{\boldsymbol{\phi}}^{*T}D(f)\hat{\boldsymbol{\phi}}^{*}. \end{aligned} \tag{5.3}$$

The posterior expectation of τ^{-1} used in (5.3) merely involves a standard result regarding moments from the gamma distribution. As with $S^{-1}(f)$, the posterior variance of $G^{-2}(f)$ can be obtained in two steps, the first deriving expressions conditional on τ and the second removing the conditioning on τ. Then

$$\begin{aligned} \operatorname{var}_{\tau|\mathbf{x}}(E[G^{-2}(f)|\tau, \mathbf{x}]) &= \operatorname{var}_{\tau|\mathbf{x}}(\tau^{-1}\operatorname{tr}(D(f)\hat{C}) + \hat{\boldsymbol{\phi}}^{*T}D(f)\hat{\boldsymbol{\phi}}^{*}|\mathbf{x}) \\ &= (\operatorname{tr}(D(f)\hat{C}))^2\operatorname{var}(\tau^{-1}|\mathbf{x}) \\ &= (2r^2(v-2)^{-2}(v-4)^{-1})(\operatorname{tr}(D(f)\hat{C}))^2 \end{aligned} \tag{5.4}$$

$$\operatorname{var}(G^{-2}(f)|\tau, \mathbf{x}) = 2\tau^{-2}\operatorname{tr}(D(f)\hat{C}D(f)\hat{C}) + 4\tau^{-1}\hat{\boldsymbol{\phi}}^{*T}D(f)\hat{C}D(f)\hat{\boldsymbol{\phi}}^{*}.$$

The last conditional expression needed is

$$\begin{aligned} E_{\tau|\mathbf{x}}[\operatorname{var}(G^{-2}(f)|\tau, \mathbf{x})] &= 2r^2(v-2)^{-1}(v-4)^{-1}\operatorname{tr}(D(f)\hat{C}D(f)\hat{C}) \\ &\quad + 4r(v-2)^{-1}\hat{\boldsymbol{\phi}}^{*T}D(f)\hat{C}D(f)\hat{\boldsymbol{\phi}}^{*}. \end{aligned} \tag{5.5}$$

The conditional expressions (5.4) and (5.5) can be added together to obtain the exact posterior variance of the reciprocal of the squared gain and is

$$\operatorname{var}(G^{-2}(f)|\mathbf{x}) = E_{\tau|\mathbf{x}}[\operatorname{var}(G^{-2}(f)|\tau, \mathbf{x})] + \operatorname{var}_{\tau|\mathbf{x}}(E[G^{-2}(f)|\tau, \mathbf{x}]). \tag{5.6}$$

Expression (5.6) requires that $v - 4 > 0$, but otherwise exists for all sample sizes. As with the reciprocal of the spectrum, the posterior expectation and variance of the reciprocal of the squared gain do not involve the introduction of asymptotic approximations. As in Section 4, one can use Chebychev's inequality to create credible intervals for $G^{-2}(f)$ and $G^2(f)$ for any frequency in the interval form $-\frac{1}{2}$ to $\frac{1}{2}$. As before, these intervals are chosen for convenience rather than for optimality and make use only of the posterior mean and variance for the reciprocal of the squared gain. Such credible intervals can be computed by taking the posterior expectation of $G^{-2}(f) \pm k$ posterior standard deviations. The resulting credible interval for $G^{-2}(f)$ at a particular frequency f will have a posterior probability of at least $1 - 1/k^2$. Also, as before, the reciprocal relationship is so direct that one may prefer to work directly with the reciprocal of the squared gain so that the possibility of infinite limits is removed.

In Section 6 we focus more attention on the frequency-response function. In particular, the reciprocal of the frequency-response function is a complex-valued linear combination of the coefficients of the AR(p) process. One can adapt the Scheffé result (see Graybill, 1975, pp. 197–199, or Miller, 1981, pp. 110–114 for a description of the Scheffé result) to create an uncountably infinite set of credible intervals for the real or imaginary part of the reciprocal of the frequency-response function which will simultaneously be valid with a preselected amount of posterior probability. This adaptation of the Scheffé result for simultaneous inferences is described in Section 6. In Section 7 we describe how to transform these infinite sets of credible intervals to make inferences for the squared gain and its reciprocal.

6. FREQUENCY-RESPONSE FUNCTION

Recall that the operator $\phi(B)$ from (3.4) can be thought of as a filter. The function $\phi(f)$, defined in Section 5, is the reciprocal of the frequency-response function of the AR process. The function $\phi(f)$ can also be interpreted as the frequency-response function associated with the filter $\phi(B)$ and is the subject of this section. In this section we derive the exact posterior distribution of the real and imaginary parts of $\phi(f)$, HPD intervals for the real and imaginary parts at any individual frequency desired, and an infinite set of credible intervals that will hold simultaneously at all frequencies. These infinite sets of credible intervals are used in Section 7 to develop an infinite set of credible intervals for $G^{-2}(f)$ and $G(f)$ which hold simultaneously at all frequencies.

In Section 5 the complex-valued function $\phi(f)$ was defined by using the complex exponential $\exp(-2\pi i f)$. It will now be convenient to replace the complex exponential with trigonometric functions and explicitly examine the real and imaginary parts. Then

$$\begin{aligned}\phi(f) &= 1 - \sum_{j=1}^{p} \phi_j \exp(-2\pi i j f) \\ &= 1 - \sum_{j=1}^{p} \phi_j(\cos 2\pi j f - i \sin 2\pi j f) \\ &= \left(1 - \sum_{j=1}^{p} \phi_j \cos 2\pi j f\right) + i\left(\sum_{j=1}^{p} \phi_j \sin 2\pi j f\right) \\ &= \mathrm{Re}(f) + i\,\mathrm{Im}(f).\end{aligned} \tag{6.1}$$

The real and imaginary parts of $\phi(f)$, as shown in (6.1), are real-valued linear combinations of the AR coefficients $\phi_1, \phi_2, \ldots, \phi_p$. In Section 4 we used a convenient notation that allowed for degenerate distributions. Such a convention is now inconvenient. In particular, notice that the AR coefficient ϕ_0 is not needed.

Let

$$\begin{aligned} H &= [\mathbf{0}_p, I_{p\times p}] \\ \mathbf{c}(f) &= [\cos 2\pi f, \cos 2\pi 2f, \ldots, \cos 2\pi p f]^T \\ \mathbf{s}(f) &= [\sin 2\pi f, \sin 2\pi 2f, \ldots, \sin 2\pi p f]^T. \end{aligned}$$

Notice that $H\boldsymbol{\phi}$ simply picks out the coefficients $\phi_1, \phi_2, \ldots, \phi_p$. Then the function $\phi(f)$ of expression (6.1) can be reexpressed as

$$\phi(f) = (1 - \mathbf{c}^T(f)H\boldsymbol{\phi}) + i(\mathbf{s}^T(f)H\boldsymbol{\phi}). \tag{6.2}$$

In Section 2 it was seen that the posterior distribution of the vector of AR coefficients has a multivariate t-distribution with v degrees of freedom, posterior mean $\hat{\boldsymbol{\phi}}$, and precision matrix $\hat{P}$. Linear combinations of random variables with a multivariate t-distribution will also have a t-distribution. The real part of $\phi(f)$ is $\mathrm{Re}(f) = 1 - \mathbf{c}^T(f)H\boldsymbol{\phi}$, which is a shifted linear combination of the AR coefficients. The real part, at any frequency f, has a univariate t-distribution with v degrees of freedom, mean $1 - \mathbf{c}^T(f)H\hat{\boldsymbol{\phi}}$, and a precision matrix of $(\mathbf{c}^T(f)H\hat{P}^{-1}H^T\mathbf{c}(f))^{-1}$. One may consult Box and Tiao (1973, pp. 117–118) for additional properties of the multivariate t-distribution. The imaginary part of $\phi(f)$ is also a linear combination of the AR coefficients and has a posterior distribution that is a univariate t-

distribution. Using notation introduced in Section 2, this can be summarized as

$$\begin{aligned}\mathrm{Re}(f)|\mathbf{x} &\sim T_1(\nu,\ 1 - \mathbf{c}^T(f)H\hat{\boldsymbol{\phi}},\ (\mathbf{c}^T(f)H\hat{P}^{-1}H^T\mathbf{c}(f))^{-1})\\ \mathrm{Im}(f)|\mathbf{x} &\sim T_1(\nu,\ \mathbf{s}^T(f)H\hat{\boldsymbol{\phi}},\ (\mathbf{s}^T(f)H\hat{P}^{-1}H^T\mathbf{s}(f))^{-1}).\end{aligned} \tag{6.3}$$

From (6.3) one can construct HPD intervals for the real and imaginary parts of $\phi(f)$ at any desired frequency. (Although asymptotic approximations are not of central importance in this chapter, it is relevent to mention the well-known fact that with ν bigger than 30 or 40, the t-distribution will be very similar to a normal distribution.) The results above allow one to make use of the usual tables containing percentage points from Student's t-distribution to create HPD intervals for the real and imaginary parts of $\phi(f)$ at any desired frequency. However, since the real and the imaginary parts are continuous functions defined at an uncountably infinite number of frequencies, one may wish a stronger result that takes into account infinitely many frequencies simultaneously.

The Scheffé result will now be adapted to the Bayesian approach to make inferences simultaneously at all frequencies. A basic result needed is that if the posterior distribution of $\boldsymbol{\phi}$ is a $(p + 1)$-dimensional t-distribution with ν degrees of freedom, mean $\hat{\boldsymbol{\phi}}$, and precision matrix $\hat{P}$, the quadratic form

$$\frac{(\boldsymbol{\phi} - \hat{\boldsymbol{\phi}}^T)\hat{P}(\boldsymbol{\phi} - \hat{\boldsymbol{\phi}})}{p + 1}$$

has, a posteriori, a central F distribution with $p + 1$ numerator and ν denominator degrees of freedom (see Box and Tiao, 1973, p. 117). Recall that the real and imaginary parts of $\phi(f)$ at any given frequency f are linear combinations of the AR coefficients $\phi_1, \phi_2, \ldots, \phi_p$. As f ranges through the interval from 0 to $\frac{1}{2}$ (or $-\frac{1}{2}$ to $\frac{1}{2}$) infinitely, many linear combinations of the AR coefficients will be generated. Let $\mathbf{l} \in R^p \backslash \{\mathbf{0}\}$ be any vector except the zero vector, and let H be the matrix in (6.2). Consider $\mathbf{l}^T H\boldsymbol{\phi}$. As $\mathbf{l}$ ranges throughout R^p, all possible linear combinations of $\phi_1, \phi_2, \ldots, \phi_p$ will be generated. The usual way of deriving the Scheffé result requires the following intermediate result. The maximum value of

$$(\mathbf{l}^T H\boldsymbol{\phi} - \mathbf{l}^T H\hat{\boldsymbol{\phi}})^2(\mathbf{l}^T H\hat{P}^{-1}H^T\mathbf{l})^{-1}$$

over all $\mathbf{l}$ is

$$(\boldsymbol{\phi} - \hat{\boldsymbol{\phi}})^T H^T(H\hat{P}^{-1}H^T)^{-1}H(\boldsymbol{\phi} - \hat{\boldsymbol{\phi}}),$$

which is proportional to a quadratic form with a central F distribution with p numerator and ν denominator degrees of freedom. The numerator degrees of freedom is now p rather than $p + 1$ because the matrix H has rank p. Notice that $\mathbf{l}$ is not allowed to be the zero vector. If $\mathbf{l}$ were allowed to be zero, the ratio of quadratic forms maximized above would be indeterminate. That is, the numerator and denominator would be zero and would have degenerate posterior distributions.

One more intermediate result will be needed to construct a set of credible intervals for an uncountably infinite set of linear combinations of $\boldsymbol{\phi}$ which will all hold simultaneously with a preselected amount of posterior probability. Let W be a positive number; then

$$\begin{aligned}
&\Pr((\mathbf{l}^T H\boldsymbol{\phi} - \mathbf{l}^T H\hat{\boldsymbol{\phi}})^2(\mathbf{l}^T H\hat{P}^{-1}H^T\mathbf{l})^{-1} \leqslant W, \forall \mathbf{l} \in R^p\backslash\{\mathbf{0}\}|\mathbf{x}) \\
&\quad = \Pr\left(\max_{\mathbf{l}\in R^p\backslash\{\mathbf{0}\}} (\mathbf{l}^T H\boldsymbol{\phi} - \mathbf{l}^T H\hat{\boldsymbol{\phi}})^2(\mathbf{l}^T H\hat{P}^{-1}H^T\mathbf{l})^{-1} \leqslant W|\mathbf{x}\right) \\
&\quad = \Pr((\boldsymbol{\phi} - \hat{\boldsymbol{\phi}})^T H^T(H\hat{P}^{-1}H)^{-1}H(\boldsymbol{\phi} - \hat{\boldsymbol{\phi}}) \leqslant W|\mathbf{x}) \\
&\quad = \Pr((\boldsymbol{\phi} - \hat{\boldsymbol{\phi}})^T H^T(H\hat{P}^{-1}H)^{-1}H(\boldsymbol{\phi} - \hat{\boldsymbol{\phi}})/p \leqslant W/p|\mathbf{x}).
\end{aligned}$$

As indicated earlier, the quadratic form of the last probability statement has an F distribution with p numerator and v denominator degrees of freedom. Let $0 < \varepsilon < 1$, and let $F_{1-\varepsilon,p,v}$ be the $(1-\varepsilon)100\%$ point of the central F distribution. Then

$$\begin{aligned}
&\Pr((\mathbf{l}^T H\boldsymbol{\phi} - \mathbf{l}^T H\hat{\boldsymbol{\phi}})^2(\mathbf{l}^T H\hat{P}^{-1}H^T\mathbf{l})^{-1} \leqslant pF_{1-\varepsilon,p,v}, \forall \mathbf{l} \in R^p\backslash\{\mathbf{0}\}\ |\mathbf{x}) \\
&\quad = 1 - \varepsilon. \qquad (6.4)
\end{aligned}$$

By transforming the set indicated in the probability statement (6.4) with a square root and using a small amount of algebra, one finds that

$$\begin{aligned}
&\Pr(\mathbf{l}^T H\hat{\boldsymbol{\phi}} - (\mathbf{l}^T H\hat{P}^{-1}H^T\mathbf{l}pF_{1-\varepsilon,p,v})^{1/2} \leqslant \mathbf{l}^T H\boldsymbol{\phi} \\
&\qquad\qquad \leqslant \mathbf{l}^T H\hat{\boldsymbol{\phi}} + (\mathbf{l}^T H\hat{P}^{-1}H^T\mathbf{l}pF_{1-\varepsilon,p,v})^{1/2}, \forall \mathbf{l} \in R^p\backslash\{\mathbf{0}\}\ |\mathbf{x}) \\
&\quad = 1 - \varepsilon. \qquad (6.5)
\end{aligned}$$

It should be emphasized that (6.5) is a single probability statement that applies simultaneously to an uncountably infinite set of credible intervals. If one does not let $\mathbf{l}$ range completely throughout $R^p\backslash\{\mathbf{0}\}$, then the actual probability will be at least $1 - \varepsilon$. However, if in doing so, one still ranges through infinitely many values of $\mathbf{l}$, then $1 - \varepsilon$ may be a tight bound.

From (6.5) one can obtain an infinite set of credible intervals for $\mathrm{Re}(f)$ which hold simultaneously at all frequencies in the interval $0 < f \leqslant \frac{1}{2}$ with a posterior probability of at least $1 - \varepsilon$. Let $-\mathbf{c}(f)$ take the place of $\mathbf{l}$, shift the set of intervals by the number 1, and let f range continuously throughout the interval from 0 to $\frac{1}{2}$. In doing this, one traces out a continuous band surrounding $\mathrm{Re}(f)$. One can also create a similar set of credible intervals for $\mathrm{Im}(f)$ by replacing $\mathbf{l}$ with $\mathbf{s}(f)$ and letting f range through all positive frequencies. Since one does not generate all possible linear combinations by letting f range through the interval from 0 to $\frac{1}{2}$, $1 - \varepsilon$ will be a lower bound on the probability. So the probability associated with the continuous bands will be conservative. Explicit formulas for the bands for the real and imaginary parts, respectively, are given by

$$(1 - \mathbf{c}^T(f)H\hat{\boldsymbol{\phi}}) \pm (\mathbf{c}^T(f)H\hat{P}^{-1}H^T\mathbf{c}(f)pF_{1-\varepsilon,p,v})^{1/2},$$
$$(\mathbf{s}^T(f)H\hat{\boldsymbol{\phi}}) \pm (\mathbf{s}^T(f)H\hat{P}^{-1}H^T\mathbf{s}(f)pF_{1-\varepsilon,p,v})^{1/2} \tag{6.6}$$

as f ranges from 0 to $\frac{1}{2}$. One really needs to look only at the positive frequencies. But if one wishes, one may include negative frequencies and the probability bound will remain unchanged. Since the continuous bands defined in (6.6) will be used in the next section, it will be useful to give the upper and lower limits symbols of their own. Let the lower and upper limits of the real part at a given frequency be denoted by $\mathrm{Re}_{\mathrm{L}}(f)$ and $\mathrm{Re}_{\mathrm{U}}(f)$, respectively. Let the lower and upper limits of the imaginary part at a given frequency be denoted by $\mathrm{Im}_{\mathrm{L}}(f)$ and $\mathrm{Im}_{\mathrm{U}}(f)$, respectively.

The frequency-response function of this section has been expressed in rectangular coordinates. While rectangular coordinates are of interest, so are polar coordinates where the magnitude corresponds to the gain of the filter. The posterior mean and variance of the squared gain of the filter $\phi(B)$ were presented in Section 5.

Individual credible intervals were presented for the squared gain of the process and for the squared gain of the filter in Section 5. The credible intervals, at a given frequency f, were based on Chebychev's inequality and were conservative. Also, the interpretation of an infinite set of such intervals does not appear to be straightforward since $G^{-2}(f)$ at any two different frequencies will be dependent on each other. This is true because they are both functions of the same autoregressive coefficients. In Section 7 we describe how the infinite sets of credible intervals derived in this section can be transformed into an infinite set of credible intervals for the squared gain of the process (or for the reciprocal of the squared gain) which hold simultaneously at all frequencies.

7. SIMULTANEOUS POSTERIOR BANDS FOR $G^{-2}(f)$ AND $G^2(f)$

The primary goal of this chapter is to derive an infinite set of credible intervals for $G^2(f)$ and $G^{-2}(f)$ which will hold simultaneously at all frequencies. These continuous bands will now be developed without introducing asymptotic approximations. Attention will first center on the reciprocal of the squared gain of the process (or the squared gain of the filter) $G^{-2}(f)$. The infinite set of credible intervals created for $G^{-2}(f)$ can then be transformed with a reciprocal transformation to create an infinite set of credible intervals for the process which will hold simultaneously.

Expression (6.6) in the preceding section gives continuous bands for estimating $\mathrm{Re}(f)$ and one for $\mathrm{Im}(f)$. Recall that the squared gain of the AR filter is just $G^{-2}(f) = \mathrm{Re}^2(f) + \mathrm{Im}^2(f)$, $0 \leqslant f \leqslant \frac{1}{2}$. One way to produce a simultaneous band for $G^{-2}(f)$, which is convenient, though not optimal, is to square the bands for the real and for the imaginary parts. Adding the corresponding lower limits together produces $G_L^{-2}(f)$, the lower limit for the squared gain of the filter. Adding the corresponding upper limits of the real and imaginary bands will produce $G_U^{-2}(f)$, the upper limit for a simultaneous band for $G^{-2}(f)$. One must remember that one is transforming sets. The lower limit for a continuous band on $\mathrm{Re}^2(f)$ will be zero at a particular frequency if the simultaneous band for $\mathrm{Re}(f)$ contains zero at that frequency. A similar result holds for the imaginary part. As a result, at any frequency where continuous bands for the real and imaginary parts both contain zero the lower limit for the simultaneous band for $G^{-2}(f)$ will also be zero. To be explicit, at all other frequencies define

$$
\begin{aligned}
G_L^{-2}(f) &= \min\{\mathrm{Re}_{\mathrm{L}}^2(f) + \mathrm{Im}_{\mathrm{L}}^2(f), \mathrm{Re}_{\mathrm{U}}^2(f) + \mathrm{Im}_{\mathrm{U}}^2(f)\}, \\
G_U^{-2}(f) &= \max\{\mathrm{Re}_{\mathrm{L}}^2(f) + \mathrm{Im}_{\mathrm{L}}^2(f), \mathrm{Re}_{\mathrm{U}}^2(f) + \mathrm{Im}_{\mathrm{U}}^2(f)\},
\end{aligned} \tag{7.1}
$$

where $0 \leqslant f \leqslant \frac{1}{2}$. If the continuous bands for $\mathrm{Re}(f)$ and $\mathrm{Im}(f)$ each have a posterior probability of at least $1 - \varepsilon$, then the posterior probability for the continuous band on $G^{-2}(f)$, given in (7.1), will be at least $1 - 2\varepsilon$. This arises from a basic result in probability. Let A and B be arbitrary sets each with probability $1 - \varepsilon$. Then

$$
\begin{aligned}
\Pr(A \cap B) &= \Pr(A) + \Pr(B) - \Pr(A \cup B) \\
&\geqslant 2 - 2\varepsilon - 1 \\
&\geqslant 1 - 2\varepsilon.
\end{aligned}
$$

The result above is easily applied. Let A denote the points in the squared continuous band for the real part of $\phi(f)$, and let B denote the set of points in the squared continuous band for the imaginary part of $\phi(f)$. The probability is calculated using the posterior p.d.f. of $\boldsymbol{\phi}$. One is interested in those values of $\boldsymbol{\phi}$ that produce $\mathrm{Re}^2(f) \in A$ and $\mathrm{Im}^2(f) \in B$. This result indicates that the simultaneous posterior band for the squared gain of the filter is conservative.

It should be emphasized that at any frequency where the bands for both the real and imaginary parts contain zero, the lower limit for the simultaneous band for the squared gain of the filter will be zero. This has implications for the squared gain of the process $G^2(f)$. At all frequencies where $G_L^{-2}(f)$ is greater than zero, one can take the reciprocal transformation to make inferences for $G^2(f)$ simultaneously at all frequencies. This reciprocal will not exist at any frequency where $G_L^{-2}(f)$ is equal to zero. Most approaches for creating an infinite set of credible intervals, which hold simultaneously at all frequencies, for the squared gain of the AR process or for its spectral density function are subject to infinite upper limits (see Koslov and Jones, 1985; Newton and Pagano, 1983). The infinite upper limits can occur since optimal results are not used. In the Bayesian context, optimal results apparently require extensive numerical integrations. For this reason, the approach taken in this section has been based on convenience. One should remember that for the scalar process examined in this chapter, all relevant information about the squared gain of the process is summarized in the squared gain of the filter. So direct examination of the characteristics of the filter may be preferable. Nevertheless, it appears that the continuous band for $G^2(f)$ presented in this section will be less likely to have infinite upper limits than other approaches.

In Section 8 we describe some numerical illustrations of the techniques presented in this chapter.

8. NUMERICAL ILLUSTRATIONS

Several techniques have been described for estimating the frequency-domain characteristics of pure AR models. Illustrations will now be developed to show that the techniques described in this chapter are operational. Concerns such as the sensitivity of the posterior p.d.f.'s to specification of the prior p.d.f.'s and asymptotic properties will not be examined. Wolfer's sunspot numbers (see Pandit and Wu, 1985, p. 487) will be modeled using an AR (2) process, which seems to be a popular model for the sunspot numbers. For additional information about sunspots, one should consult Morris (1977) or Shove (1983). All figures presented use a logarithmic vertical scale.

As in all Bayesian applications, one must select a particular prior p.d.f. for the parameters of the model used. To give the readers a common reference point, Jeffreys' prior will be used because it will produce results that are similar to maximum likelihood. Jeffreys' prior p.d.f. for the AR (p) parameters is

$$f(\boldsymbol{\phi}, \tau) \propto \tau^{-1}, \qquad \boldsymbol{\phi} \in R^{p+1}, \quad \tau > 0 \tag{8.1}$$

This prior p.d.f. can be obtained as a limiting case of the conjugate prior p.d.f. described in Section 2. In expression (2.4), let S, $\mathbf{b}$, and β tend to 0, and let α tend to $-(p+1)$. Then one obtains expression (8.1). One should note that while Jeffreys' prior is an improper prior, it will produce a proper posterior p.d.f. in this situation.

Figure 1 contains credible intervals for the spectrum of Wolfer's sunspot

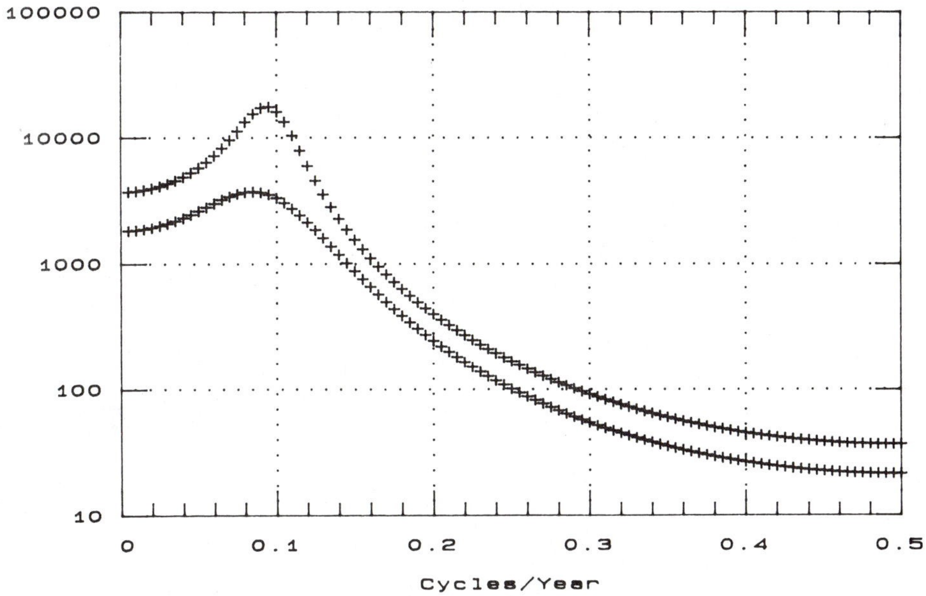

Figure 1. Sunspot spectrum—Chebychev ($k = 2$), Jeffreys prior.

data based on an AR(2) model. In estimating the parameters, the first two data values were used as the startup values for the initial value model of (2.1). Although there are actually 176 data values, there are 174 effective observations. Credible intervals were computed based on Chebychev's inequality with $k = 2$ producing a posterior probability of at least 0.75 at each individual frequency chosen. The credible intervals were computed for 100 equally spaced intervals from 0 to 0.5 cycle per year (excluding 0). The credible intervals are plotted using a "+" symbol. Figure 1 shows a definite peak at a frequency of about 0.09 cycle per year or about 11 years per cycle. This is consistent with most analyses of the sunspot data.

Figure 2 graphs the squared gain for Wolfer's sunspot data. The solid band is based on the Scheffé result for simultaneous inferences. As an intermediate step, bands were computed for the real and the imaginary parts of the AR operator, each with a posterior probability of at least 0.975. As a result, the solid outer band for the squared gain of the AR(2) process used to model the sunspot data has a posterior probability of at least 0.95. That is, the posterior probability is at least 0.95 that the squared gain of the process, at all frequencies, is inside the solid outer band. The frequencies actually chosen for plotting were the same as in Figure 1. To get an insight into the conservative

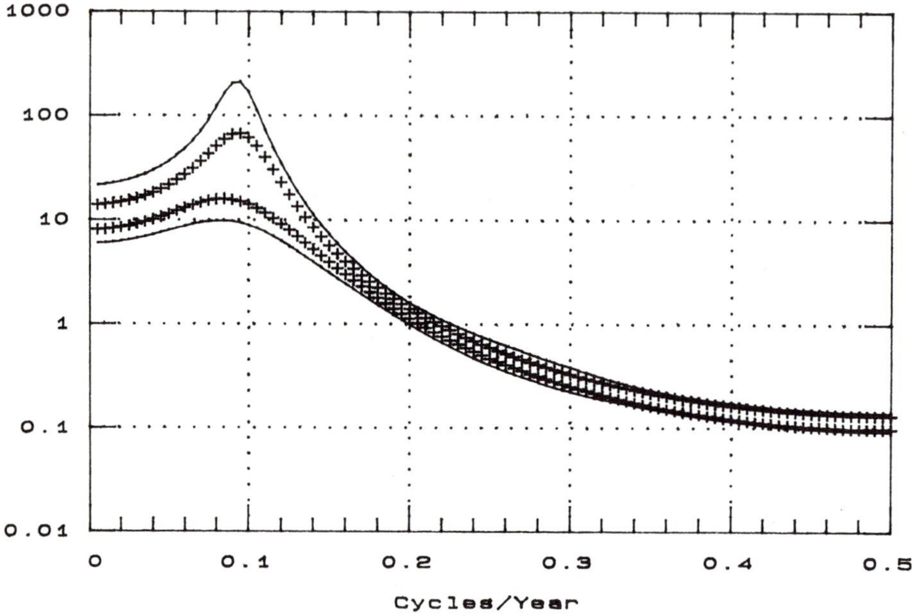

Figure 2. Sunspot squared gain, Jeffreys prior.

nature of Chebychev's inequality, credible intervals were computed and plotted with the symbol "+" at each of the 100 frequencies. Each credible interval has a posterior probability of at least 0.75 (as in Figure 1) and is based on Chebychev's inequality with $k=2$. As can be seen from Figure 2, at many frequencies, the individual credible intervals are almost as wide as the solid outer band. The advantage of examining the squared gain rather than the spectrum is the ease of making inferences simultaneously at all frequencies. Of course, Figure 2 shows the same definite peak corresponding to sunspot cycles of about 11 years.

Figure 3 examines the squared gain of the residuals obtained from fitting the AR(2) model to the sunspot numbers. It was decided to use an AR(2) process for the residuals for the purpose of diagnostic checking. If linear information is present in the residuals, the AR(2) model fitted above would be seen to be inadequate. The choice of an AR(2) model for the residuals was arbitrary. Jeffreys' prior [in expression (8.1)] was used in modeling the residuals. Since an AR(2) process was used for 174 residuals, there are 172 effective residuals for the analysis. The format for Figure 3 is the same as for Figure 2. The solid outer band is used for inferences at all frequencies simultaneously. The posterior probability is at least 0.95 that the squared

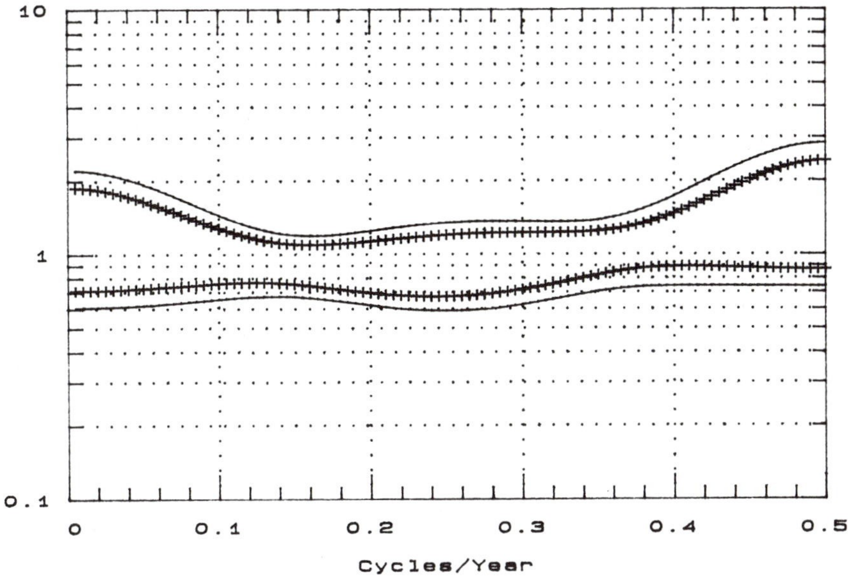

Figure 3. Square gain of sunspot residuals, Jeffreys prior.

gain of the residuals is contained inside the solid band. Note that a horizontal line of height 1 is contained inside the solid outer band. Since the squared gain of white noise is equal to 1 at all frequencies, the AR (2) model fit to the sunspot data produces residuals with properties that are similar to that of white noise. Also, credible intervals at each of the 100 frequencies chosen were computed using Chebychev's inequality based on $k = 2$. These credible intervals at each of the 100 frequencies have a posterior probability of at least 0.75 and are plotted with "+" symbols. As with Figure 2, the individual credible intervals are almost as wide as the band created with the Scheffé result.

All computations were performed on an IBM PC*/XT with a math coprocessor using STSC's APL*PLUS/PC and STATGRAPHICS software.

9. SUMMARY

Techniques for estimating the frequency-domain characteristics of the univariate pth-order autoregressive process have been examined in this chapter. The starionary process was approximated by a process with fixed initial starting values and results in a Bayesian statistical approach for both time and frequency domains that do not require asymptotic approximations.

Although one may use asymptotic theory to examine the approach discussed, asymptotic results were not required to derive the techniques.

Some of the estimates produced were shown to be optimal and some were based on convenience. Convenience was chosen to provide techniques that are computationally feasible for microcomputers.

When dealing with the spectrum or squared gain of a univariate autoregressive process, one might produce individual or simultaneous interval estimates with an infinite upper limit due to the convenience and lack of optimality. Such an unfortunate result is possible with the usual techniques based on autoregressive processes. However, the Bayesian approach described seems less sensitive to producing infinite upper limits than some other approaches. Further, when dealing with univariate processes, this problem can be avoided altogether by working directly with the spectrum or squared gain of the filter of the autoregressive process. Rather than encountering the possibility of infinite upper limits, one encounters the possibility of negative or zero lower limits. Interpretation of the squared gain or spectrum of the filter is straightforward. Peaks of the spectrum of the process correspond to troughs of the spectrum of the filter.

Clearly, there is additional work that needs to be completed. Work is currently under way to produce optimal HPD intervals for the spectrum, squared gain, and phase at any particular frequency. The approach to frequency-domain estimation described in this chapter is currently being adapted to moving average processes. Further, work or multivariate processes is under way. However, the results presented in this chapter require the use of reciprocals. This approach cannot be applied to the multivariate case. Other approaches are currently being examined.

In summary, this chapter has described an approach to frequency-domain estimation for univariate autoregressive processes which does not use asymptotic approximations, is computationally feasible for small computers, and is accessible to a wide audience.

ACKNOWLEDGMENT

The author would like to thank the editor and referees for suggestions which have significantly improved this chapter.

REFERENCES

Akaike, H. (1978). Time Series Analysis and Control Through Parametric Models, in *Applied Time Series Analysis* (D. Findley, ed.) Academic Press, New York, pp. 1–23.

Blackman, R. B., and J. W. Tukey (1959). *The Measurement of Power Spectra*, Dover, New York.

Bloomfield, P. (1976). *Fourier Analysis of Time Series: An Introduction*, Wiley, New York.

Box, G. E. P. and G. M. Jenkins (1976). *Time Series Analysis: Forecasting and Control*, Rev. ed, Holden-Day, San Francisco.

Box, G. E. P., and G. C. Tiao (1973). *Bayesian Inference in Statistical Analysis*, Addison-Wesley, Reading, Mass.

Brigham, E. O. (1974). *The Fast Fourier Transform*, Prentice-Hall, Englewood Cliffs, N. J.

Brillinger, D. R. (1975). *Time Series: Data Analysis and Theory*, Holt, Rinehart and Winston, New York.

Broemeling, L. D. (1985). *Bayesian Analysis of Linear Models*, Marcel Dekker, New York.

Cook, P. J. (1983). A Bayesian analysis of autoregressive processes: time and frequency domain, Ph. D. dissertation, Oklahoma State University.

Cook, P. J. (1985). Bayesian autoregressive spectral analysis, *Commun. Stat. Theory Methods* 14(5), 1001–1018.

Cook, P. J. (1987). Bayesian autoregressive filter and spectrum estimation, *Comm. in Stat. Theory Methods* 16(9), 2697–2715.

DeGroot, M. (1970). *Optimal Statistical Decisions*, McGraw-Hill, New York.

Dzhaparidze, K. (1986). *Parameter Estimation and Hypothesis Testing in Spectral Analysis of Stationary Time Series*, Springer-Verlag, New York.

Fuller, W. A. (1976). *Introduction to Statistical Time Series*, Wiley, New York.

Graybill, F. A. (1976). *Theory and Application of the Linear Model*, Duxbury Press, Belmont, Calif.

Jeffreys, H. (1961). *Theory of Probability*, 3rd ed., Clarendon Press, Oxford.

Koslov, J. W., and R. H. Jones (1985). A unified approach to confidence bounds for the autoregressive spectral estimator, *J. Time Ser. Anal.* 6(3), 141–151.

Miller, R. G. (1981). *Simultaneous Statistical Inference*, 2nd Ed., Springer-Verlag, New York.

Morris, J. (1977). Forecasting the sunspot cycle, *J.R. Stat. Soc. Ser. A* 140, 437–447.

Newton, H. J., and M. Pagano (1982). *Simultaneous Confidence Bands for Autoregressive Spectra*, Technical Report N-32, Texas A & M University, College Station, Tex.

Pandit, S. M., and S. M. Wu (1983). *Time Series and Systems Analysis with Applications*, Wiley, New York.

Parzen, E. (1979). Time Series Modeling, Spectral Analysis and Forecasting, in *Directions in Time Series, Proceedings of the IMS Special Topics Meeting on Time Series Analysis* (D. R. Brillinger and G. C. Tiao, eds.), Iowa State University, Ames, Iowa, pp. 80–111.

Press, S. J. (1982). *Applied Multivariate Analysis: Using Bayesian and Frequentist Methods of Inference, 2nd ed.*, R. E. Krieger, Melbourne, Fla.

Priestley, M. B. (1981). *Spectral Analysis and Time Series*, Academic Press, New York.

Robinson, E. A. (1983). *Multichannel Time Series Analysis with Digital Computer Programs*, 2nd ed., Goose Pond Press, Houston, Tex.

Robinson, E. A., and S. Treitel (1980). *Geophysical Signal Analysis*, Prentice-Hall, Englewood Cliffs, N. J.

Shore, R. W. (1980). A Bayesian Approach to the Spectral Analysis of Stationary Time Series, in *Bayesian Analysis in Econometrics and Time Series Analysis* (A. Zellner and J. Kadane, eds.), North-Holland, Amsterdam, pp. 83–180.

Shove, D. J. (1983). *Sunspot Cycles*, Hutchinson Ross, Stroudsburg, Pa.

Shuster, A. (1898). On the investigation of hidden periodicities, *Terr. Mag.* 3(13), 13–46.

Slutzky, E. (1937). The summation of random causes as the source of cyclic processes, *Econometrica* 5, 105–146.

Zellner, A. (1971). *An Introduction to Bayesian Inference in Econometrics*, Wiley, New York.

6

Recursive Estimation for Nonlinear Dynamic Systems

HAROLD W. SORENSON United States Air Force, Headquarters, USAF/CCN, Washington, District of Columbia

The recursive estimation of parameters and states of linear and nonlinear stochastic dynamic systems can be attacked directly using a Bayesian approach. The solution for linear, Gaussian systems has been known for more than two decades. It provides the basis for the extended Kalman filter, which has become the de facto standard approach for solving nonlinear problems. The closed-form solution of the Bayesian recursion equations for nonlinear problems is restricted to a few simple cases primarily of academic interest. Numerical algorithms have been proposed and developed, but they have seen little application because of the inherent computational burden. The recent "explosion" in digital technology suggests that some practical and important problems can be addressed now using the Bayesian approach. As a result, the problem of developing numerical algorithms for the solution of the Bayesian recursion relations is being readdressed in the context of emerging digital technologies.

The discussion contained in this chapter is intended to provide a perspective on the Bayesian approach to developing recursive estimation algorithms for nonlinear stochastic problems. After a brief review of the literature, two specific algorithms for solving nonlinear problems are stated and discussed. Then the algorithms are applied to two nonlinear problems and numerical results are presented. The problems of bearings only, passive tracking, and combined system identification and state estimation for linear constant-coefficient systems are discussed. The numerical results are meant to be illustrative rather than exhaustive.

1. INTRODUCTION AND STATEMENT OF THE PROBLEM

Consider the following general problem. Measurements $\mathbf{z}_k$ of the output of a system are available at a discrete set of times t_k. Based on assumptions regarding the mathematical *structure of the system*, the measurements are used to *estimate variables of the model that are best* in some sense. The rationale for the estimator generally is based on some *general performance measure* and the behavior of the estimator is judged by *asymptotic measures* or by *observed properties*. For this discussion the output of the system is defined to be the signal $\mathbf{y}_k$. The measurements are related to the signal according to

$$\mathbf{z}_k = \mathbf{y}_k + \mathbf{v}_k, \qquad k = 1, 2, \ldots, \tag{1.1}$$

where $\mathbf{v}_k$ represents measurement noise.

In (1.1), a structure has been defined. Other assumptions are possible, but the additive character displayed in (1.1) is widely used and serves as a basis for most analyses. It suggests the intuitively appealing idea that the instrument errors $\mathbf{v}_k$ occur separately from the signal process $\mathbf{y}_k$. Examples in which the error magnitude is influenced by the magnitude of the signal are not difficult to imagine. Nonetheless, the assumption of additive noise is sufficiently realistic that attention is directed to (1.1) throughout most of this discussion.

Equation (1.1) provides a natural starting point for the discussion of *parameter* and *state* estimators, *linear* and *nonlinear* estimation problems, *deterministic* and *probabilistic* approaches to the estimation problem, and the *performance analysis of estimators*. All of these considerations arise as a consequence of the structural and probabilistic assumptions that are defined for the signal and for the noise. In the succeeding paragraphs, the discussion focuses on the structural assumptions that distinguish the problems and emphasizes the commonality of concept and results wherever possible. A more detailed discussion of parameter estimation is provided by Sorenson (1980) and can be used as a source of more detailed discussion of many of the topics discussed herein.

1.1 Nonlinear Estimation Problem

Given a probabilistic formulation of the problem, a general solution of the problem is provided by taking a *Bayesian approach* that provides the *posterior distribution* of the variables given the measurement data. A disadvantage of the Bayesian approach is that it produces a function (i.e., the posterior distribution) that must be translated into a finite-dimensional description of the variables that are being estimated. In linear, Gaussian problems, the desired finite-dimensional characterization is produced in a natural manner by the statistics of the posterior distribution. Because of the difficulty in obtaining an analytic form for the posterior, most approaches to the estimation problem focus on the development of *point estimators* of the unknown variables. Because there are an unlimited number of possible point estimates, it is necessary to introduce an ordering by defining the concept of *best estimators.* This discussion focuses on the computation of the posterior with secondary interest in the estimators that can be derived therefrom.

Given (1.1), it is common to assume structural properties for the signal that permits discrimination between it and the noise. In particular, the following structural properties are introduced.

1. The signal is assumed to have a structure that provides a common relationship among measurement sampling times.
2. The noise is assumed to have a different character than the signal. By assuming that it varies substantially, the effect of the noise may be reduced in a nontrivial fashion. The greatest contrast is provided by assuming that noise samples are statistically independent.

From knowledge of the system, one can hypothesize a mathematical model of the signal that provides a relationship between the signal values at different sampling times. Generally, this modeling effort produces a description of the signal that reduces the uncertainty regarding the signal to a few parameters. For this discussion it is assumed that variables can be defined that provide a structural relationship between signal samples at different times. Suppose that the signal is represented as

$$\mathbf{y}_k = \mathbf{h}_k(\mathbf{x}_k), \tag{1.2}$$

where $\mathbf{h}_k$ is known and the variables $\mathbf{x}_k$ are described through a stochastic difference equation,

$$\mathbf{x}_{k+1} = \mathbf{f}_k(\mathbf{x}_k) + \mathbf{w}_k. \tag{1.3}$$

The variables $\mathbf{x}_k$ are referred to as *state* variables, and (1.2) and (1.3) provide the basis for the *nonlinear state* estimation problems.

The noise $\mathbf{v}_k$ must have characteristics that permit discrimination, using the measurements, between the signal and noise. The structure that was given the signal in the preceding discussion implies considerable dependence between samples. As a contrast, it would be desirable if the noise samples exhibited no predictable relationship. As the noise is unknown, it is reasonable to describe the noise, and to an extent discussed below, the signal in a probabilistic context. More specifically, the noise is assumed to be an independent sequence. That is, the noise samples manifest no probabilistic dependence. For convenience, the noise sequences are assumed to be

Gaussian with known means and covariances and will be referred to as *white* noise. The general discussion of Bayesian estimators does not require the Gaussian assumption, but the more specific algorithmic discussions rely on this assumption.

The general model used throughout this presentation is summarized below as Equation I. Because of the importance of linear systems, notation for this special case is introduced in Equation I. The tabulation that is presented will be referred to throughout the discussion and is intended to provide a convenient reference for assumptions and notation. Note that it is assumed throughout that the initial state $\mathbf{x}_1$ is Gaussian with mean value $\boldsymbol{\mu}_1$ and covariance matrix M_1. Similarly, the noise sequences are assumed to be Gaussian with the statistics stated in Equation I. Throughout this discussion, a Gaussian variable $\mathbf{x}$ with mean value $\boldsymbol{\mu}$ and covariance matrix Σ is denoted as $N_X(\boldsymbol{\mu}, \Sigma)$.

Based on the Equation I, the following state estimation problem can be formulated and is considered in the subsequent sections. For state estimation, the variables change with time and introduces the need to define the *time* at which the state is to be estimated. When $\mathbf{x}_k$ is to be estimated from $\mathbf{z}(k) = \{\mathbf{z}_1, \mathbf{z}_2, \ldots, \mathbf{z}_k\}$, the *filtering* problem is obtained. *Prediction* and *smoothing* problems are defined when $\mathbf{x}_{k+j}$ is to be estimated from $\mathbf{z}(k)$ for $j > 0$ and $j < 0$, respectively. For this discussion, attention is restricted to the filtering problem.

THE GENERAL MODEL: EQUATION I

Measurements:

$$\mathbf{z}_k = \mathbf{h}_k(\mathbf{x}_k) + \mathbf{v}_k, \qquad k = 1, 2, \ldots$$

or

$$\mathbf{z}_k = H_k \mathbf{x}_k + \mathbf{v}_k \qquad \text{(linear model)}$$

The noise sequence is white with the Gaussian density $N_{vk}(0, R_k)$.

Dynamics:

$$\mathbf{x}_{k+1} = \mathbf{f}_k(\mathbf{x}_k) + \mathbf{w}_k, \qquad k = 1, 2, \ldots$$

or

$$\mathbf{x}_{k+1} = \phi_k \mathbf{x}_k + \mathbf{w}_k \qquad \text{(linear model)}$$

The initial state $\mathbf{x}_1$ is Gaussian $N_{x1}(\boldsymbol{\mu}_1, M_1)$ and is independent of $\mathbf{v}(k)$, $\mathbf{w}(k)$ for all k.

The system noise is white with the Gaussian density $N_{wk}(0, Q_k)$.

When an estimate is required for each k as k increases, one is led naturally to the following definition.

RECURSIVE FILTERING PROBLEM Given an estimate of $\mathbf{x}_{k-1}$ based on $\mathbf{z}(k-1)$, say $\hat{\mathbf{x}}_{k-1|k-1}$, estimate $\mathbf{x}_k$ from $\hat{\mathbf{x}}_{k-1|k-1}$ and $\mathbf{z}_k$.

1.2 Bayesian Approach to the Solution

The recursive filtering problem is solved in general by using Bayes' rule. The general solution involves Bayes' rule to incorporate the effects of each new measurement and is supplemented by a convolution equation that describes the effect of the time variation of the state variables. To obtain the recursive solution, it is necessary to assume that $\mathbf{v}_k$ describes an independent noise sequence.

Assuming that $\mathbf{v}_k$ and $\mathbf{w}_k$ are independent, noise sequences, the posterior density of $\mathbf{x}_k$ given $\mathbf{z}(k)$ can be determined recursively. The recursion is presented below as Equation II. As with Equation I, this tabulation is intended to serve as a convenient reference. Note that the filtering density $p(\mathbf{x}_k|\mathbf{z}(k))$ is produced from Bayes' rule with $p(\mathbf{x}_k|\mathbf{z}(k-1))$ serving as the prior for the state $\mathbf{x}_k$ given the measurements $\mathbf{z}(k-1)$. The density $p(\mathbf{z}_k|\mathbf{x}_k)$ is

BAYESIAN RECURSION RELATIONS: EQUATION II

Filtering Density:

$$p(\mathbf{x}_k|\mathbf{z}(k)) = c_k p(\mathbf{z}_k|\mathbf{x}_k)p(\mathbf{x}_k|\mathbf{z}(k-1)), \qquad k = 1, 2, \ldots$$

where

$$\frac{1}{c_k} = \int p(\mathbf{z}_k|\mathbf{x}_k)\mathrm{p}(\mathbf{x}_k|\mathbf{z}(k-1))\,d\mathbf{x}_k$$

Initial Conditions:

$$p(\mathbf{x}_1|\mathbf{z}(-1)) \equiv p(\mathbf{x}_1)$$

Prediction Density:

$$p(\mathbf{x}_{k+1}|\mathbf{z}(k)) = \int p(\mathbf{x}_{k+1}|\mathbf{x}_k)p(\mathbf{x}_k|\mathbf{z}(k))\,d\mathbf{x}_k$$

Gaussian Noise Priors:

$$p(\mathbf{z}_k|\mathbf{x}_k) = N_{vk}(\mathbf{z}_k - \mathbf{h}_k(\mathbf{x}_k), R_k)$$
$$p(\mathbf{x}_{k+1}|\mathbf{x}_k) = N_{wk}(\mathbf{x}_{k+1} - \mathbf{f}_k(\mathbf{x}_k), Q_k)$$

---

determined from the noise density $p(\mathbf{v}_k)$. The prediction density $p(\mathbf{x}_{k+1}|\mathbf{z}(k))$ requires the *convolution* of $p(\mathbf{x}_{k+1}|\mathbf{x}_k)$ and $p(\mathbf{x}_k|\mathbf{z}(k))$. The recursion is initiated using the prior for $\mathbf{x}_1$. The state estimation problem requires consideration of the time at which a state estimate is to be determined and incorporates the propagation effects through the prediction density. The nonlinear convolution complicates, greatly, the computational burden and arises because of the presence of the system noise $\mathbf{w}_k$. This requirement represents the complexity introduced by the Markov character of the state.

2. SOLVING THE NONLINEAR ESTIMATION PROBLEM

In this section the general, nonlinear system I is considered and can be regarded as a very complete description of realistic estimation/identification problems. As stated, it also constitutes a problem about which it is very difficult to obtain any meaningful insights in a closed form. Consequently, attention is directed toward using the Bayesian approach as the basis for numerical studies of specific types of systems. Many methods for the numerical solution of the Bayesian recursion relations have been proposed and are reviewed briefly below. To illustrate the insights possible using numerical solutions, the p-vector approach is defined. The algorithm is applied to the bearings-only, passive tracking problem. Then the combined problem of system estimation/system identification is considered and another algorithm using the Bayesian approach is developed and illustrated through its application to a simple numerical example.

A landmark in the development of stochastic control theory in a Bayesian context for which the nonlinear estimation problem is fundamental was provided by the germinal work of Fel'dbaum (1965). His work was extended and expanded by Aoki (1967). These two books provide the basis for much of the work that has been done since 1965. The growth of various aspects of the field has been charted through several books, of which Aström (1970), Anderson and Moore (1979), Jazwinski (1970), Meditch (1969), and Sworder (1966) are prominent. The control problem is addressed briefly herein in order to emphasize further the power of the Bayesian approach. However, most of the discussion concerns the problems of state estimation and system identification.

2.1 Important Special Case: The Linear Filtering Problem

It is not generally possible to obtain analytic solutions for the Bayesian recursion relations. The principal exception occurs for systems that are linear and Gaussian. The solution for this general class provides the basis for most of the work that has been done for nonlinear systems, so it is desirable to consider these systems in some detail.

Consider the following system. The system is described by the linear difference equation

$$\mathbf{x}_k = \phi_{k-1}\mathbf{x}_{k-1} + \Gamma_{k-1}\mathbf{u}_{k-1} + \mathbf{w}_{k-1}. \tag{2.1}$$

At each sampling time the state is observed through noisy measurement data described by

$$\mathbf{z}_k = H_k\mathbf{x}_k + \mathbf{v}_k. \tag{2.2}$$

As above, the noise sequences are zero mean, white, and Gaussian. The inputs $\mathbf{u}(k)$ are assumed to be known at t_k. They are included to emphasize that they do not affect the problem significantly and will play a more prominent role in the discussion of Section 4.2.

Consider the recursive filtering problem and the solution of the Bayesian recursion relations. Since system (2.1) is linear and the initial state and input noise are Gaussian, the state at any time is Gaussian. This, coupled with the Gaussian character of the noise and the assumed linearity in (2.2), implies that the measurements are Gaussian for all k. Although possibly not obvious without some reflection, these observations imply that the posterior density $p(\mathbf{x}_k|\mathbf{z}(k))$ is Gaussian for all k. The posterior density $p(\mathbf{x}_k|\mathbf{z}(k))$ can be described entirely in terms of the mean $\hat{\mathbf{x}}_{k|k}$ and covariance matrix $P_{k|k}$. It can be shown (e.g., Sorenson, 1970a), in a straightforward but tedious manner, that these statistics are given by the Kalman filter equations, as given in Equation III. Equation III describes the mean, covariance, and therefore the Gaussian posterior density function for the system (2.1)–(2.2). It is interesting to note that the covariance is independent of the measurements and,

KALMAN FILTER (LINEAR, GAUSSIAN SYSTEM): EQUATION III

Moments of Filtering Density:

$$\hat{\mathbf{x}}_{k|k} = \hat{\mathbf{x}}_{k|k-1} + K_k[\mathbf{z}_k - H_k\hat{\mathbf{x}}_{k|k-1}]$$

$$K_k = P_{k|k-1}H_k^T[H_kP_{k|k-1}H_k^T + R_k]^{-1}$$

$$P_{k|k} = P_{k|k-1} - K_kH_kP_{k|k-1}$$

where

$$\hat{\mathbf{x}}_{1|0} \equiv \boldsymbol{\mu}_1 \qquad P_{1|0} \equiv M_1$$

Moments of Prediction Density:

$$\hat{\mathbf{x}}_{k+1|k} = \phi_k\hat{\mathbf{x}}_{k|k} + \Gamma_k\mathbf{u}_k$$

$$P_{k+1|k} = \phi_kP_{k|k}\phi_k^T + Q_k$$

---

consequently, the control inputs. Therefore, the control policy cannot affect the shape of the posterior density.

2.2 Computational Approach to the Nonlinear Filtering Problem

Even for the linear system, the solution of the filtering problem is given in the form of a computational algorithm. For nonlinear and/or non-Gaussian systems, one is forced also to seek computational algorithms. Unfortunately, it is not possible generally to reduce the problem to simple algebraic equations (e.g., as with the Kalman filter). Instead, various approximations must be introduced to obtain any kind of numerical solution. A sizable literature describing various approaches to the numerical solution on the nonlinear filtering problem has developed. It is not feasible to attempt a comprehensive discussion of all these methods. In this section some of the approaches that have been proposed are reviewed, and then the discussion concentrates on an approach that is relatively simple and provides useful results for many problems. References to many other papers are included so that the reader can seek greater depth on specific approaches from the original sources.

The general solution of the nonlinear filtering problem is given by the posterior density function $p(\mathbf{x}_k|\mathbf{z}(k))$ as described by the Bayesian recursion relations II. The numerical solution is discussed now for systems having the form I. While the noise sequences and initial state are assumed to be Gaussian, most of the discussion in this section applies to arbitrary noise distributions. A more detailed discussion of the problem of solving the Bayesian recursion relations is given by Sorenson (1974).

A large number of methods for the evaluation of the Bayesian recursion relations have been proposed and studied. These methods have the common characteristic that the calculations are performed after defining a "grid." The grid points provide a finite collection on which approximations can be based. Obviously, these points are contained in a finite region of state space even though the integrations generally are carried out over infinite intervals. Thus the functions must be such that there is negligible probability mass outside the region containing the grid points. The manner in which the grid is defined is an important consideration in the development of an algorithm.

Consider a possible approach to the evaluation of the nonlinear convolution required to determine the prediction density as presented in Equation II. Suppose that a specific value is prescribed for $\mathbf{x}_{k+1}$, so the integration will yield a well-defined number. The numerical integration essentially requires that the integral be replaced by a summation involving a discretization of the integration variable $\mathbf{x}_k$. The manner in which the discretization or grid points are defined may be accomplished arbitrarily or as an integral part of the quadrature method. For example, in an nth-order Gauss-Hermite quadrature, the grid points are chosen as the zeros of the nth Hermite polynomial. Let $\boldsymbol{\eta}_{kj}$ $\{j = 1, 2, \ldots, n_k\}$ denote the n_k grid points for the variable $\mathbf{x}_k$. Furthermore, suppose that $\mathbf{x}_{(k+1)i}$ is regarded as the ith grid point for the discretization of the variable $\mathbf{x}_{k+1}$ into n_{k+1} points. The convolution is replaced by

$$p(\mathbf{x}_{(k+1)i}) = \sum_{j=1}^{n_k} \alpha_j p(\boldsymbol{\eta}_{kj}) p[\mathbf{x}_{(k+1)i} - f(\boldsymbol{\eta}_{kj})], \qquad i = 1, 2, \ldots, n_{k+1}.$$

The coefficients α_j represent the weighting coefficients of the numerical scheme. Clearly, if there are a large number of grid points, the storage and computational burden can be sizable.

Because of the storage and computational burden implied by solving the Bayesian recursion relations, it is natural to seek ways in which these requirements can be reduced. The earliest, and by far the most extensively applied approach, was motivated by the existence of the general solution for linear Gaussian systems III. In this case a single grid point is defined at each sampling time. Then the system equations **f** and **h** are linearized relative to the grid point. This approximation of the system implies that the state and measurement perturbations are Gaussian and that the Kalman filter equations III can be applied directly. When the grid point is chosen as the "best" estimate, the resulting estimator is called the *extended Kalman filter* (EKF) (Sorenson, 1966; Anderson and Moore, 1979).

This is apparently the simplest possible approach since it involves a single grid point and linear equations at each sampling time. In addition, the grid point at the kth time is obtained directly from the previous grid point and the approximate system equations. It is also the most crude approximation and its validity depends heavily on the quality of the linear approximation. The EKF is stated as Equation IV. Note that the system matrices are defined from

EXTENDED KALMAN FILTER (EKF): EQUATION IV

$$\hat{\mathbf{x}}_{k|k} = \hat{\mathbf{x}}_{k|k-1} + K_k[\mathbf{z}_k - \mathbf{h}_k(\hat{\mathbf{x}}_{k|k-1})]$$

$$\hat{\mathbf{x}}_{k+1|k} = \mathbf{f}_k(\hat{\mathbf{x}}_{k|k})$$

where

$$H_k \equiv \frac{\partial \mathbf{h}_k(\hat{\mathbf{x}}_{k|k-1})}{\partial \mathbf{x}_k}$$

$$\phi_k \equiv \frac{\partial \mathbf{f}_k(\hat{\mathbf{x}}_{k|k})}{\partial \mathbf{x}_k}$$

the perturbation equations. The gain matrix K_k and the covariance matrices, $P_{k|k}$ and $P_{k|k-1}$, are determined using these gradient matrices.

Practical experience has demonstrated that the assumptions inherent in the extended Kalman filter are often valid and satisfactory results are obtained. There are also well-known disadvantages and difficulties associated with the application of the extended Kalman filter. The manifestation of these difficulties is commonly referred to as the "divergence problem" (Sorenson, 1985, p. 127). Divergence is said to occur when the actual error in the estimate becomes inconsistent with the error covariance matrix approximation provided by the filter equations. This situation arises because of errors in the filter model, either as a result of errors in the basic model or as a result of the linearization errors.

In the Bayesian context, the use of the extended Kalman filter implies that the posterior density is Gaussian. This can be an extremely poor approximation of the actual density function if all possible values of the state $\mathbf{x}$ are considered. A number of generalizations to include higher-order perturbations have been proposed. Since the approximations can be regarded as being most accurate in some neighborhood of the single grid (or reference) point, we refer to them as *local methods.* The obvious extension (Sorenson and Stubberud, 1968a; Athans et al., 1968; Jazwinski, 1970) is to consider retaining the second-order terms in the expansion of the system functions $\mathbf{f}$ and $\mathbf{h}$. Commonly, the assumption is made that posterior density is still Gaussian even with the presence of the second-order terms. This assumption is made to make possible the computation of the higher-order moments and thereby to overcome the "moment closure problem."

The first serious attempt to eliminate the Gaussian assumption involved the use of Gram-Charlier or Edgeworth expansions (Sorenson and Stubberud, 1968b). The expansion is a series of Hermite polynomials which are orthogonal with respect to a Gaussian distribution and can be used to represent a wide class of density functions. The initial use of this non-Gaussian approximation was based on a perturbative approximation. As a consequence, it suffered from the disadvantage that a large number of terms were required to obtain a reasonable approximation of a distinctly non-Gaussian density. The behavior of the estimator obtained from this density approximation was found to be very sensitive to the quality of the approximation. When the infinite series is truncated, as it must be for practical application, the resulting series can become negative over portions of the state space. Consequently, the density approximation is not itself a density. This can introduce unexpected influences into the behavior of the estimator, particularly if the integral over the region in which the function is nonpositive has a nontrivial value. Subsequently, other density approximations using the Edgeworth expansion have been proposed (Kizner, 1969; Srinivasan, 1970).

This local method seems to be most useful when the posterior density is unimodal even though it is not Gaussian.

The obvious disadvantage of the local methods stems from the use of a single grid point on which to base the approximation. Several methods have been proposed that attempt to improve the approximation by considering the density at many points selected throughout the region containing nonnegligible probability mass. These methods can be regarded as representing specific examples of ways in which the numerical integrations just discussed can be accomplished. Some of these global approximations are reviewed in the following paragraphs.

Quite possibly, the first step toward the development of a global method was taken by Magill (1965), with a subsequent generalization by Hilborn and Lainiotis (1969). They considered linear systems with unknown parameters. To deal with this nonlinear problem, a grid was established by discretizing the unknown parameters and by considering the resulting collection of linear filtering problems. A global method for the general nonlinear filtering problem was proposed by Bucy (1969) when he introduced the point-mass method. This approach was elaborated on by Bucy and Senne (1970) at the First Symposium on Nonlinear Estimation in 1970. At this same meeting Alspach and Sorenson (1970) proposed the Gaussian sum approximation as an alternative approach. Subsequent Symposia on Nonlinear Estimation included many extensions and the introduction of other techniques. Center (1972) provided a unifying theoretical framework by considering the problem in the context of generalized least squares. His approach permits, conceptually at least, the development of a countless number of approximations. In the Second Symposium on Nonlinear Estimation, Center (1971) discussed as specific examples the point-mass, Gaussian sum, and Edgeworth expansion approximants. Later (1972) he also discussed the spline approximation method proposed by de Figueiredo and Jan (1971).

All specific global methods must provide solutions of the following general problems:

1. An initial grid must be defined. It is important that the region encompassed by the grid includes the true value of the state. In addition, the number and manner in which the grid points are distributed within the approximation region must be defined.
2. A procedure must be given for defining the grid at each subsequent sampling time. While the grid could be the same throughout, the dynamic nature of the problem and the desire for computational efficiency indicate the advisability of redefining the grid at each sampling time.
3. Given the grid, a method must be selected for approximating the functions and/or for carrying out Bayes' rule calculations. The approxi-

mation method and the grid selection method are not unrelated and the implementation of a particular method may require interaction between the two considerations.

3. TWO APPROACHES TO THE COMPUTATIONAL SOLUTION

In this section two computational algorithms are presented and discussed. First, the solution of the general Bayesian recursion relations is considered and the p-vector approach is developed. In Section 3.2 a more structured problem is considered and an algorithm that is tailored to the structure is presented. The two approaches provide a perspective on the range of algorithms that is made possible by taking the Bayesian point of view.

3.1 The p-Vector Approach

A simple computational algorithm will be developed in this section that emphasizes the fundamental difficulties faced by an algorithm developer. As described, there are three basic computational problems: the measurement update, the propagation through the dynamic model, and the convolution implied by the additive system noise. Each problem is addressed specifically in the following paragraphs. A more sophisticated approach is proposed by Kramer (1985).

Measurement update: determining $p(\mathbf{x}_k|\mathbf{z}(k))$.

Suppose that the one-step prediction density is known. At the time of the first measurement, it is assumed that

$$p(\mathbf{x}_1|\mathbf{z}(0)) \equiv p(\mathbf{x}_1),$$

where $p(\mathbf{x}_1)$ is the prior density assumed for the initial state $\mathbf{x}_1$. At subsequent times, the prediction density is known as the output of the computational algorithm.

To begin computations, a set of grid points must be defined. Denote these n_1 grid points as $\boldsymbol{\mu}_1(n_1) = \{\boldsymbol{\mu}_{1i}; i = 1, 2, \ldots, n_1\}$. The number n_1 of grid points must be chosen by the user, as is the location of the grid points. Certainly, the extent of the grid must cover the region of state space for which the prior density is nonnegligible.

Consider the effect of the first measurement, $\mathbf{z}_1$. Recall that the filtering density satisfies

$$p(\mathbf{x}_1|\mathbf{z}_1) = c_1 p(\mathbf{x}_1)p(\mathbf{z}_1|\mathbf{x}_1),$$

where

$$\frac{1}{c_1} = \int p(\mathbf{x}_1)p(\mathbf{z}_1|\mathbf{x}_1)\,d\mathbf{x}_1.$$

Assuming that the measurement noise is Gaussian with zero mean and covariance R_1, it follows that

$$p(\mathbf{z}_1|\mathbf{x}_1) = N_V(\mathbf{z}_1 - \mathbf{h}(\mathbf{x}_1), R_1).$$

The density function can be evaluated at each of the n_1 grid points. Then it is straightforward to evaluate

$$\frac{p(\boldsymbol{\mu}_{1i}|\mathbf{z}_1)}{c_1} = p(\boldsymbol{\mu}_{1i})N_V(\mathbf{z}_1 - \mathbf{h}(\boldsymbol{\mu}_{1i}), R_1); \qquad i = 1, 2, \ldots, n_1. \tag{3.1}$$

The normalization constant c_1 needs to be computed to ensure that the density values remain reasonable (e.g., do not tend to zero). Note that c_1 does *not* change the shape of the posterior density; it scales every point in an identical manner.

OBSERVATION 1 Incorporation of the effects of a measurement does *not* affect the grid points; it serves only to modify the probability density at each grid point.

OBSERVATION 2 Given the prior density and the noise density functions, there is no approximation involved in determining the posterior density, except for the calculation of the normalization constant c_1.

OBSERVATION 3 The calculations described for the first measurement are identical for all subsequent measurement times.

OBSERVATION 4 Implicit in this computation is the realization that the Bayesian solution of the parameter estimation problem (i.e., $\mathbf{x}_1 = \mathbf{x}_k = \boldsymbol{\theta}$ for all k) can be accomplished *without* approximation and in a particularly straightforward manner.

Updates for system dynamics

Suppose that the filtering density $p(\mathbf{x}_k|\mathbf{z}(k))$ is known at the grid points $\boldsymbol{\mu}_k(n_k)$ and that the state noise is additive. Let $\mathbf{s}_k$ be defined as

$$\mathbf{s}_k \equiv \mathbf{f}_k(\mathbf{x}_k)$$

and consider the determination of the posterior density $p(\mathbf{s}_k|\mathbf{z}(k))$. Clearly, this is obtained as a nonlinear transformation of the state $\mathbf{x}_k$ given the measurements $\mathbf{z}(k)$ and is determined in a reasonably straightforward manner using well-known results from basic probability theory. For this discussion, assume that $\mathbf{f}_k$ is a one-to-one invertible mapping of $\mathbf{x}_k$ to $\mathbf{s}_k$. Extension to many-to-one mappings is possible using known results but will not be treated here.

To determine the posterior density, first let

$$\boldsymbol{\eta}_{ki} \equiv \mathbf{f}_k(\boldsymbol{\mu}_{ki}), \qquad i = 1, 2, \ldots, n_k.$$

Next, let $\mathbf{f}_k^{-1}$ denote the inverse of $\mathbf{f}_k$ and define $|J|$ as the absolute value of the Jacobian of the derivatives $(\partial \mathbf{f}_k^{-1} | \partial \mathbf{s}_k)$ of the inverse function. Then it is well known that $p(\mathbf{s}_k | \mathbf{z}(k))$ can be computed according to

$$p(\boldsymbol{\eta}_{ki} | \mathbf{z}(k)) = p(\boldsymbol{\mu}_{ki} | \mathbf{z}(k)) |J(\boldsymbol{\eta}_{ki})|. \tag{3.2}$$

Thus the density value at the new grid point $\boldsymbol{\eta}_{(k+1)i}$ is the value of the density at the previous grid point $\boldsymbol{\mu}_{ki}$ scaled by $|J|$ evaluated at the new grid point.

It may not be possible to determine the inverse function $\mathbf{f}_k^{-1}$ and, therefore, its derivatives. Then it is reasonable to use the inverse function theorem to represent the local representation of the inverse of $\mathbf{f}_k$ at $\boldsymbol{\mu}_{ki}$. In other words, consider the linear approximation

$$\mathbf{s}_k = \mathbf{f}_k(\boldsymbol{\mu}_{ki}) + \frac{\partial \mathbf{f}_K(\boldsymbol{\mu}_{ki})}{\partial \mathbf{x}_k} (\mathbf{x}_k - \boldsymbol{\mu}_{ki}).$$

The inverse of this function is found to be

$$\mathbf{x}_k = \left[\frac{\partial \mathbf{f}_k(\boldsymbol{\mu}_{ki})}{\partial \mathbf{x}_k} \right]^{-1} \left[\mathbf{s}_k - \mathbf{f}_k(\boldsymbol{\mu}_{ki}) + \frac{\partial \mathbf{f}_k(\boldsymbol{\mu}_{ki})}{\partial \mathbf{x}_k} \boldsymbol{\mu}_{ki} \right],$$

where, clearly,

$$\frac{\partial \mathbf{x}_k}{\partial \mathbf{s}_k} = \left[\frac{\partial \mathbf{f}_k(\boldsymbol{\mu}_{ki})}{\partial \mathbf{x}_k} \right]^{-1}.$$

A natural approximation follows by introducing

$$J(\boldsymbol{\eta}_{ki}) \equiv \left[\frac{\partial \mathbf{f}_k(\boldsymbol{\mu}_{ki})}{\partial \mathbf{x}_k} \right]^{-1}. \tag{3.3}$$

Using this value for J, the posterior density at the grid points $\boldsymbol{\mu}_{(k+1)i}$ can be determined.

OBSERVATION 1 With linear dynamics, no approximation occurs in accomplishing the dynamic update of the posterior density.

OBSERVATION 2 If the inverse of the dynamics can be determined explicitly, no approximation should be necessary in the dynamic update. Otherwise, the approximation described above can be used.

Determining the prediction density

If there is no system noise (i.e., $\mathbf{w}_k \equiv 0$ for all k), then

$$p(\mathbf{x}_{k+1}|\mathbf{z}(k)) = p(\mathbf{s}_k|\mathbf{z}(k)).$$

Otherwise, the prediction density is determined by convolving the update density and the noise density since

$$\mathbf{x}_{k+1} = \mathbf{s}_k + \mathbf{w}_k.$$

Assuming that $\mathbf{w}_k$ is Gaussian, the prediction density is given by

$$p(\mathbf{x}_{k+1}|\mathbf{z}(k)) = \int p(\mathbf{s}_k|\mathbf{z}(k))N_W(\mathbf{x}_{k+1} - \mathbf{s}_k, Q_k)\, d\mathbf{s}_k.$$

The update density is known at the grid points $\boldsymbol{\eta}_{ki}$ but not at intermediate points. Thus the convolution must be approximated using the available information. The evaluation of a discrete convolution is a reasonable approach. If the grid points are equally spaced, a fast Fourier transform (FFT) can be utilized. Unfortunately, the dynamic update may destroy a constant spacing. Then it may not be possible to utilize the computational efficiency of the FFT. Nonetheless, the computation of a discrete convolution to determine the prediction density at a prescribed set of grid points $\boldsymbol{\mu}_{(k+1)i}$ is straightforward to accomplish.

OBSERVATION 1 Since N_W is known for all $\mathbf{w}$, it is apparent that this density can be evaluated at any values of the argument $(\mathbf{x}_{k+1} - \mathbf{s}_k)$. Thus the discrete convolution

$$p(\boldsymbol{\mu}_{(k+1)j}|\mathbf{z}(k)) = \sum_{i=1}^{n_k} p(\boldsymbol{\eta}_{ki}|\mathbf{z}(k))N_W(\boldsymbol{\mu}_{(k+1)j} - \boldsymbol{\eta}_{ki}, Q_k) \tag{3.4}$$

can be evaluated for each grid point $\boldsymbol{\mu}_{(k+1)j}$.

OBSERVATION 2 An algorithm has been defined which leaves the choice of the grid points entirely to the discretion of the user. In fact, the selection of the grid is an important part of the algorithm. Recall that there is no approximation in the determination of the filtering density no matter how many or where the grid points are defined. Similarly, there should be little approximation error in the dynamics update, none if the system is linear. Thus the grid point selection is important in controlling approximation errors, primarily in accomplishing the convolution necessitated by the additive system noise.

OBSERVATION 3 If there is no state noise, the grid evolves in a natural manner. The number of grid points does not change unless the user chooses to eliminate or introduce points. Grid points change in accordance with the dynamical model and do not require respecification as the measurement samples are processed. However, the presence of state noise causes the respecification of the grid at every dynamics update time. Of course, the same number of grid points can be used at every sampling time. Because of the spreading of the densities caused by the convolution, one can consider scaling the existing grid points by a constant factor whose value can be based on the standard deviation of the state noise.

OBSERVATION 4 The grid points for the prediction density, following the convolution caused by adding the state noise, can be defined to accommodate the evolving region containing nonnegligible probability mass. The number of grid points can be selected to equal the number at the preceding sampling time. Alternatively, the number can be reduced or increased as the user chooses.

A detailed discussion has been given of the problem of determining the posterior density of the state given measurements. Knowledge of the function is essential for any analysis of the nonlinear filtering problem. The representation developed for the algorithm can be restated to provide a definition of the *p-vector* (i.e., the probabilistic state) of the system.

DEFINITION The p-vector of the system is defined to be the variables $\{\boldsymbol{\mu}_{ki}, \alpha_{ki}; i = 1, 2, \ldots, n_k\}$, where the $\boldsymbol{\mu}_{ki}$ represent grid points. The α_{ki} represent the value of the prediction density $p(\boldsymbol{\mu}_{ki}|\mathbf{z}(k-1))$. The number of grid points n_k can be defined to be a constant as discussed above. If n represents the dimension of the state, the dimension of the p-vector is $(n+1)n_k$ for all k.

The dynamics of the p-vector are defined above. Basically, the grid points $\boldsymbol{\mu}_{ki}$ evolve according to the model of the system dynamics. State noise can introduce a scaling of the grid point to account for the convolutional spreading. The update of the density values involves as many as three steps. First, the measurement processing produces the filtering density. Then the dynamics transform the density. Finally, the additive noise requires the evaluation of a convolution for each grid point. The composition of each of these operations defines the dynamic model of the density values α_{ki}.

To conclude this discussion, let us summarize some important features regarding the algorithm that has been defined.

1. In (3.2), measurements $\mathbf{z}_k$ serve to modify the weighting coefficients. Because of the exponential character of N_V, the terms exhibiting the smallest residuals will be given substantially greater weight. Consequently, measurements cause the density function to become more concentrated about the mean value.

2. Measurements do *not* affect the location of the grid points.
3. Measurements do *not* affect the number of terms directly. If the coefficients are reduced below a prescribed threshold, one may choose to eliminate the corresponding term from the sum. Thus the number of terms in the representation can be reduced in forming $p(\mathbf{x}_k|\mathbf{z}(k))$, thereby lessening the computational burden.
4. If the dynamic model includes control inputs $\mathbf{u}_k$ with the model

$$\mathbf{x}_{k+1} = \mathbf{f}_k(\mathbf{x}_k, \mathbf{u}_k) + \mathbf{w}_k,$$

the control variables $\mathbf{u}_k$ explicitly influence the location of the grid points. The grid points are used to form the residuals in (3.2). Thus the controls $\mathbf{u}_k$, implicitly, influence the weighting coefficients in (3.2). These two effects have come to be referred to as the "dual" control aspect of stochastic control theory. The control sequence $\mathbf{u}(k)$ is used not only to achieve a control objective (e.g., drive the grid points to a terminal state), but also to learn (e.g., by influencing the weighting coefficients) about the state of the system.
5. The maximum a posteriori estimator of the state can be obtained by identifying the grid point associated with the *largest* weighting coefficient.
6. The minimum mean square estimator of the state is obtained from the application of numerical integration methods to the p-vector information. Higher-order moments are determined in similar ways.

3.2 System Identification/State Estimation Problem

The Bayesian approach can be used as the basis for considering the combined problems of state estimation and system identification for linear dynamic systems. Although the p-vector algorithm can be used, the structure of the system can be exploited to simplify the structure of the computational algorithm. Unlike the p-vector, which can be extended immediately to arbitrary noise distributions, the algorithm developed in this section requires the Gaussian assumption in order to use the Kalman filter. The general problem will be considered first and the general structure of an algorithm will be stated. Then attention will be directed to a simple, scalar system and numerical results will be presented.

Consider the following linear system:

$$\mathbf{x}_{k+1} = F_k(\boldsymbol{\theta})\mathbf{x}_k + G_k(\boldsymbol{\theta})\mathbf{u}_k + Q_k(\boldsymbol{\theta})\mathbf{w}_k \tag{3.5}$$

$$\mathbf{z}_k = H_k(\boldsymbol{\theta})\mathbf{x}_k + R_k(\boldsymbol{\theta})\mathbf{v}_k, \tag{3.6}$$

where the input signal $\mathbf{u}_k$ can be regarded as either a known function of time or as a feedback signal generated from the measurement sequence $\mathbf{z}(k)$. The initial state $\mathbf{x}_1$ is assumed to be Gaussian with mean value $\hat{\mathbf{x}}_{1|0}$ and covariance $P_{1|0}$. The sequences $\mathbf{w}(k)$ and $\mathbf{v}(k)$ are stationary Gaussian white noise with zero means and covariance matrix I. Note that uncertainty regarding the noise covariance is modeled through the matrix factors $Q_k(\boldsymbol{\theta})$ and $R_k(\boldsymbol{\theta})$. The noise sequences are mutually independent and independent of the initial state $\mathbf{x}_1$. The unknown parameters $\boldsymbol{\theta}$ are assumed to be Gaussian with mean value $\hat{\boldsymbol{\theta}}_{1|0}$ and covariance Σ_1. Furthermore, $\boldsymbol{\theta}$ and $\mathbf{x}_1$ are assumed to be independent.

The combined *system identification/state estimation problem* for the system (3.5)–(3.6) is easily stated. From the measurements $\{\mathbf{z}_1, \mathbf{z}_2, \ldots, \mathbf{z}_k\} \equiv \mathbf{z}(k)$ and knowledge of the inputs $\{\mathbf{u}_1, \mathbf{u}_2, \ldots, \mathbf{u}_{k-1}\} \equiv \mathbf{u}(k-1)$, estimate the state $\mathbf{x}_k$, and identify the model parameters $\boldsymbol{\theta}$.

Basically, this is a nonlinear estimation problem since $\boldsymbol{\theta}$ and $\mathbf{x}_k$ appear multiplicatively in the model. As a result, there is no closed-form solution for the estimator of $[\boldsymbol{\theta}, \mathbf{x}_k]$. Instead, estimators have been proposed which are motivated by reasonable arguments and ad hoc approximations. For example, the extended Kalman filter has been used for this problem with varying success. Ljung (1979) proposed a modification of the EKF for which he claims convergence. Using Bayes' rule, the posterior density for $[\boldsymbol{\theta}, \mathbf{x}_k]$, given the input $\mathbf{u}(k-1)$ and output $\mathbf{z}(k)$, can be determined. In particular, it can be shown that the general form of Bayes' rule for this problem has the following recursive description.

Filtering density:

$$p(\boldsymbol{\theta}, \mathbf{x}_k|\mathbf{z}(k), \mathbf{u}(k-1)) = \frac{p_V(\mathbf{z}_k|\mathbf{x}_k, \boldsymbol{\theta})p(\boldsymbol{\theta}, \mathbf{x}_k|\mathbf{z}(k-1), \mathbf{u}(k-1))}{p(\mathbf{z}_k|\mathbf{z}(k-1), \mathbf{u}(k-1))}, \tag{3.7}$$

where

$$p(\mathbf{z}_k|\mathbf{z}(k-1), \mathbf{u}(k-1)) = \int p_V(\mathbf{z}_k|\mathbf{x}_k, \boldsymbol{\theta})p(\boldsymbol{\theta}, \mathbf{x}_k|\mathbf{z}(k-1), \mathbf{u}(k-1))\, d\mathbf{x}_k.$$

At $k = 1$, initial conditions are defined as

$$p(\boldsymbol{\theta}, \mathbf{x}_1|\mathbf{z}(0), \mathbf{u}(0)) \equiv p(\boldsymbol{\theta}, \mathbf{x}_1).$$

Prediction density:

$$p(\boldsymbol{\theta}, \mathbf{x}_{k+1}|\mathbf{z}(k), \mathbf{u}(k)) = \int p_W(\mathbf{x}_{k+1}|\boldsymbol{\theta}, \mathbf{x}_k, \mathbf{z}(k), \mathbf{u}(k))p(\boldsymbol{\theta}, \mathbf{x}_k|\mathbf{z}(k), \mathbf{u}(k))\, d\mathbf{x}_k. \tag{3.8}$$

These densities assume a more specific form when the Gaussian assumptions regarding the initial state $\mathbf{x}_1$, the transition parameter $\boldsymbol{\theta}$, and the noise sequences $\mathbf{w}(k)$, $\mathbf{v}(k)$ are introduced. Then the filtering and prediction densities reduce to a product of the Gaussian density described by the Kalman filter, given $\boldsymbol{\theta}$, and the posterior density for $\boldsymbol{\theta}$. These results are stated and discussed in the following paragraphs.

Filtering density/Gaussian:

$$p(\boldsymbol{\theta}, \mathbf{x}_k|\mathbf{z}(k), \mathbf{u}(k-1)) = p(\boldsymbol{\theta}|\mathbf{z}(k), \mathbf{u}(k-1))N_{k|k}(\hat{\mathbf{x}}_{k|k}, P_{k|k}), \tag{3.9}$$

where $N_{k|k}$ represents a Gaussian density with moments described by the Kalman filter equations. In writing the equations for $\hat{\mathbf{x}}_{k|k}$ and $P_{k|k}$ using III, remember that they are functions of the unknown parameters $\boldsymbol{\theta}$. The posterior density for $\boldsymbol{\theta}$ can be written as

$$\begin{aligned} p(\boldsymbol{\theta}|\mathbf{z}(k), \mathbf{u}(k)) &= p(\boldsymbol{\theta}|\mathbf{z}(k), \mathbf{u}(k-1)) \\ &= c_k N_I(0, \Pi_{k|k-1})p(\boldsymbol{\theta}|\mathbf{z}(k-1), \mathbf{u}(k-1)), \end{aligned} \tag{3.10}$$

where

$$\begin{aligned} \mathbf{i}_k &= \mathbf{z}_k - H_k\hat{\mathbf{x}}_{k|k-1}, \\ \Pi_{k|k-1} &= H_k P_{k|k-1} H_k^T + R_k R_k^T, \\ \frac{1}{c_k} &= \int N_I(0, \Pi_{k|k-1})p(\boldsymbol{\theta}|\mathbf{z}(k-1), \mathbf{u}(k-1)\, d\boldsymbol{\theta}. \end{aligned}$$

Initial conditions:

$$\begin{aligned} p(\boldsymbol{\theta}|\mathbf{z}(0), \mathbf{u}(0)) &= p(\boldsymbol{\theta}) = N_\theta(\hat{\boldsymbol{\theta}}_{1|0}, \Sigma_1), \\ p(\boldsymbol{\theta}, \mathbf{x}_1) &= p(\boldsymbol{\theta})p(\mathbf{x}_1), \\ p(\mathbf{x}_1) &= N_1(\hat{\mathbf{x}}_{1|0}, P_{1|0}) \end{aligned}$$

Prediction density/Gaussian prior:

$$p(\boldsymbol{\theta}, \mathbf{x}_{k+1}|\mathbf{z}(k), \mathbf{u}(k)) = p(\boldsymbol{\theta}|\mathbf{z}(k), \mathbf{u}(k))N_{k+1|k}(\hat{\mathbf{x}}_{k+1|k}, P_{k+1|k}), \tag{3.11}$$

where $\hat{\mathbf{x}}_{k+1|k}$ and $P_{k+1|k}$ are given by the Kalman filter Equation III.

Equations (3.9), (3.10), and (3.11) provide a general basis for the study of the state estimation/state identification problem. Consider some important observations regarding these equations.

OBSERVATION 1 There have been *no* assumptions regarding the form of the estimation and identification algorithms. As discussed below, the development does suggest a natural and intuitively appealing structure for an estimation algorithm.

OBSERVATION 2 No assumptions regarding the source of the input sequence have been made other than to assume $\mathbf{u}(k)$ is a known function of time (i.e., an open-loop policy) or a known function of the output sequence $\mathbf{z}(k)$ (i.e., a closed-loop policy).

OBSERVATION 3 The Kalman filter provides a natural basis for updating the parameters of the conditional densities $N_{k+1|k}$ and $N_{k|k}$ in (3.9) and (3.11).

Note that *no* modification of the state estimator is suggested by these relations in distinction with Ljung's approach in modifying the EKF.

To expand on this point, consider the maximum a posteriori estimator of $\mathbf{x}_k$ and $\boldsymbol{\theta}$ as obtained by considering (3.9)–(3.11). These estimators must satisfy the condition

$$\frac{\partial p(\boldsymbol{\theta}, \mathbf{x}_k | \mathbf{z}(k), \mathbf{u}(k))}{\partial \mathbf{x}_k} = 0, \tag{3.12}$$

$$\frac{\partial p(\boldsymbol{\theta}, \mathbf{x}_k | \mathbf{z}(k), \mathbf{u}(k))}{\partial \boldsymbol{\theta}} = 0. \tag{3.13}$$

The first condition is seen to be satisfied by choosing

$$\hat{\mathbf{x}}_{\mathrm{MAP}}(k|k) = \hat{\mathbf{x}}_{k|k}(\hat{\boldsymbol{\theta}}_{\mathrm{MAP}}(k|k))$$
$$\hat{\mathbf{x}}_{\mathrm{MAP}}(k|k-1) = \hat{\mathbf{x}}_{k|k-1}(\hat{\boldsymbol{\theta}}_{\mathrm{MAP}}(k|k)).$$

Thus the Kalman filter provides the MAP estimator of the state. Note, however, that it must be evaluated with the MAP estimator of $\boldsymbol{\theta}$. This estimator is not readily obtained in a closed-form from (3.12)–(3.13). Consider this problem below.

OBSERVATION 4 The determination of the MAP estimator of $\boldsymbol{\theta}$ using (3.12)–(3.13) leads to complicated algebraic equations that must be solved since $\boldsymbol{\theta}$ enters $p(\boldsymbol{\theta}, \mathbf{x}_k|\mathbf{z}(k), \mathbf{u}(k))$ through $p(\boldsymbol{\theta})|\mathbf{z}(k), \mathbf{u}(k))$, $\hat{\mathbf{x}}_{k|k}$, and $P_{k|k}$. For the purposes of this analysis, it is important to recognize that it is more useful to defer the question of obtaining a suitable estimator. Instead, examine the characteristics of the density function itself. Thus consider the numerical evaluation of (3.9)–(3.11) for a range of parameter values. Some representative results are presented below that illustrate the insights that can be gained.

OBSERVATION 5 The input signal enters the densities *only* through the updating relation for $\hat{\mathbf{x}}_{k+1|k}$ in (3.11). Thus the input only shifts $N_{k+1|k}$ but does not otherwise alter the density function as it involves the state. But this is a manifestation of the well-known separation theorem for linear, Gaussian systems and quadratic performance indices.

OBSERVATION 6 The input signal *can* have a profound effect on the nature of the density function relative to the transition parameter $\boldsymbol{\theta}$. However, it enters in a subtle fashion. The input affects the propagation of the state, which is then reflected through the measurement and, more significantly, through the innovation sequence $\mathbf{i}_k$.

To describe the behavior of the density function and to illustrate the influence of the input signal, it is useful to consider numerical examples. The example in Section 4.2 is representative of more extensive studies and demonstrate the utility of the general approach. Other discussion and results are provided by Kramer (1985).

4. SOME NUMERICAL ILLUSTRATIONS OF THE BAYESIAN APPROACH

Two problems are considered from a Bayesian point of view in this section. The bearings-only, passive tracking problem has proven to be a difficult challenge for the extended Kalman filter (EKF) (Chou, 1977; Aidala and Hammel in Sorenson, 1985). In Section 4.1 the *p*-vector is used to address this problem and results are compared with an EKF. The simple example of the combined state estimation/state identification problem is introduced in Section 4.2 to illustrate the approach developed in Section 3.2.

Because of the system model, the *p*-vector solution involves *no* approximations, assuming that the model is a valid representation of the physical system. The numerical results demonstrate insights into the estimation problem that are not possible when the extended Kalman filter is used. Similarly, the algorithm used for the state estimation/system identification problem involves no approximations and provides a perspective on the estimation problem that does not emerge from the EKF. Thus a complete numerical solution of the nonlinear estimation problem is provided for both systems.

The Bayesian approach produces the numerical representation of the posterior density function. Having this density function, the analyst has considerable amount of information about the estimator that is presented most effectively using graphical displays. Several types of displays are used in the examples to illustrate the potential richness of the solution. The use of graphical outputs in conjunction with the Bayesian solution seems to the author to be an important union that these examples are intended to emphasize.

4.1 Bearings-Only Target Tracking

The bearings-only target tracking problem arises in engineering applications in which the position and velocity of an object must be estimated using a sensor that provides a measurement of the angle defining the line of sight from the observer to the object. No information about the distance (or range) from the observer to the target (or object) is available. The requirement to estimate the range using only angular measurements has been recognized for a long time to be a difficult problem. This problem arises, for example, in determining planetary orbits from telescopic measurements.

For many applications, and for the purpose of this discussion, both the observer and the target are moving and the relative position and velocity are estimated. Motion is assumed to occur in a plane so that it is sufficient to estimate two position and two velocity variables. The state vector $\mathbf{x}$ is four-dimensional and is defined explicitly in the following paragraph.

Suppose that a target is being tracked using a device that produces measurements of bearing from the tracking platform to the target. Both the platform and the target can move in some well-behaved manner that will be represented as involving piecewise constant velocity.

Let $\mathbf{x}$ represent the four-dimensional state vector for the target with east and north components denoted as the variables x_1 and x_3. The corresponding velocity components are denoted as the variables x_2 and x_4, respectively. Because the velocity is constant, the dynamic model of the system is described in the following manner:

$$\begin{aligned}
x_{1T}(k+1) &= x_{1T}(k) + x_{2T}(k)\,\Delta t \\
x_{2T}(k+1) &= x_{2T}(k) \\
x_{3T}(k+1) &= x_{3T}(k) + x_{4T}(k)\,\Delta t \\
x_{4T}(k+1) &= x_{4T}(k)
\end{aligned}$$

or

$$\mathbf{x}_T(k+1) \equiv \begin{bmatrix} x_{1T}(k+1) \\ x_{2T}(k+1) \\ x_{3T}(k+1) \\ x_{4T}(k+1) \end{bmatrix} = \begin{bmatrix} 1 & \Delta t & 0 & 0 \\ 0 & 1 & 0 & 0 \\ 0 & 0 & 1 & \Delta t \\ 0 & 0 & 0 & 1 \end{bmatrix} \begin{bmatrix} x_{1T}(k) \\ x_{2T}(k) \\ x_{3T}(k) \\ x_{4T}(k) \end{bmatrix}.$$

Suppose that the platform has the identical, constant-velocity model

$$\mathbf{x}_p(k+1) = \phi_k \mathbf{x}_p(k).$$

The state of the target relative to the tracker is defined as

$$\mathbf{x}_R \equiv \mathbf{x}_T - \mathbf{x}_P,$$

which, since ϕ_k is common to both $\mathbf{x}_T$ and $\mathbf{x}_P$, produces a simple model

$$\mathbf{x}_R(\mathbf{k}+1) = \phi_k \mathbf{x}_R(k).$$

For simplicity, random effects (i.e., $\mathbf{w}_k$) are regarded as negligible. Also, to reflect a maneuver for the tracker, the velocity can be changed instantaneously to a new constant value. The change is assumed to be known. Measurements of the bearing β_k from the platform to the target are available at times t_k, $k = 1, 2, \ldots,$ where

$$\beta_k = \tan^{-1}\left[\frac{x_{2T}(k) - x_{2p}(k)}{x_{1T}(k) - x_{1p}(k)}\right] + v_k.$$

The measurement noise v_k is assumed to be zero mean, Gaussian, and white with variance σ_v^2.

For this example the dynamics are linear but the measurement is nonlinear. This problem is known to be difficult for the EKF (e.g., Chou, 1977) and to result sometimes in biased estimates and/or filter divergence. A few numerical results are presented below to illustrate the utility of the p-vector approach. Recall that the linearity of the dynamics and the absence of system noise means that the Bayesian recursion relations can be evaluated without approximation errors (i.e., no convolution is required). Results have been selected for an instance in which the EKF performance is unsatisfactory. The p-vector performance illustrates the type of information that is gained in a relatively straightforward manner.

Five grid points are defined for each component of the state vector. Thus a total of 625 grid points are used in all the results that are shown. As there is no system noise and there was no attempt to eliminate unlikely trajectories, the number of grid points is constant at 625. Effectively, 625 trajectories and the associated weights are evaluated at each sampling time. Because of the simple dynamic model, the calculations are simple and readily performed. Results of four types are presented below to summarize the results. First, an x–y plot is shown as Figure 1(A) of the behavior of the platforms, the EKF, and the four most significant elements of the p-vector. Then estimates of the relative velocity are shown in Figure 1(B). Next, the behavior of the magnitude of the four largest density values on the grid is shown in Figure 1(C). Finally, the number of grid points for which the density value exceeds a threshold value is shown in Figure 1(D) as the sampling time increases. This provides a measure of the concentration of the density about the peak. Moments of the posterior density are not displayed, nor is the density itself presented. These types of presentation format are reserved for the second example discussed in Section 4.2.

The results presented in Figure 1 were chosen for a situation and initial conditions that caused the EKF to diverge. The initial grid for the p-vector was selected so that the actual state at the initial time was enclosed within the area defined by the grid. This ensures that the state at subsequent sampling times is enclosed within the grid area.

(A)

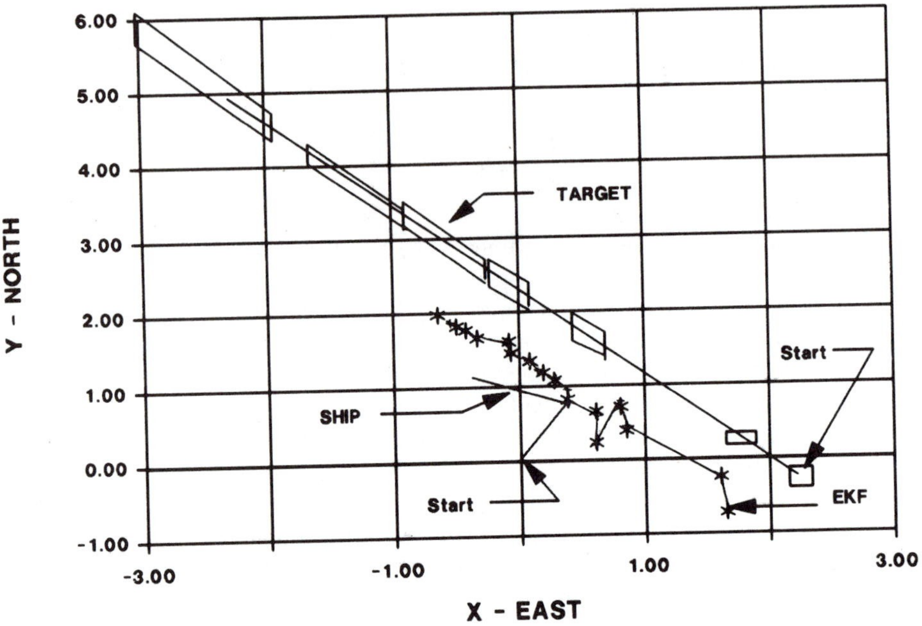
6.00
5.00
4.00
3.00
2.00
1.00
0.00
-1.00
Y - NORTH
TARGET
Start
SHIP
Start
EKF
-3.00
-1.00
1.00
3.00
X - EAST

(B)

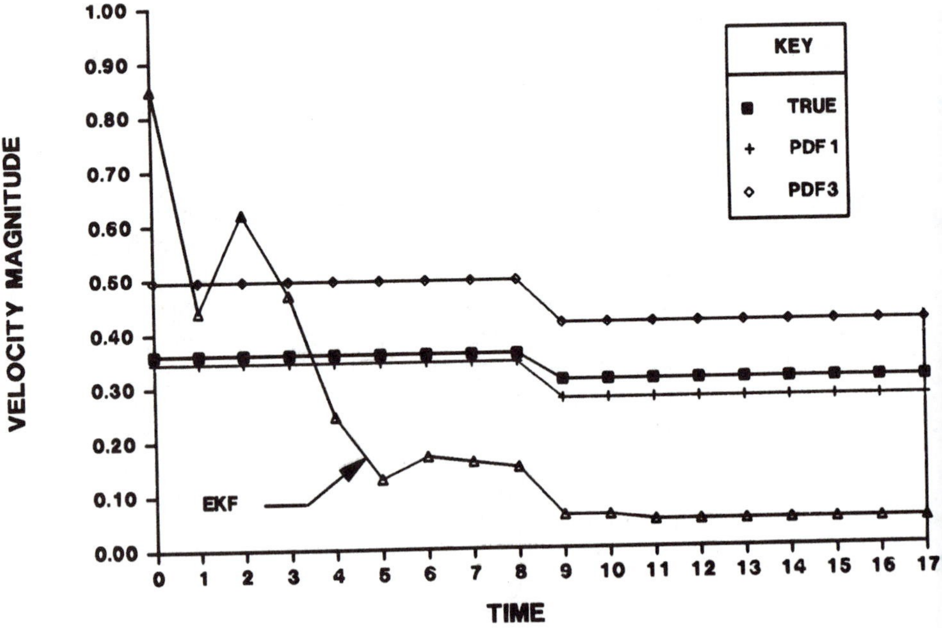
1.00
0.90
0.80
0.70
0.60
0.50
0.40
0.30
0.20
0.10
0.00
VELOCITY MAGNITUDE
KEY
TRUE
PDF 1
PDF3
EKF
0 1 2 3 4 5 6 7 8 9 10 11 12 13 14 15 16 17
TIME

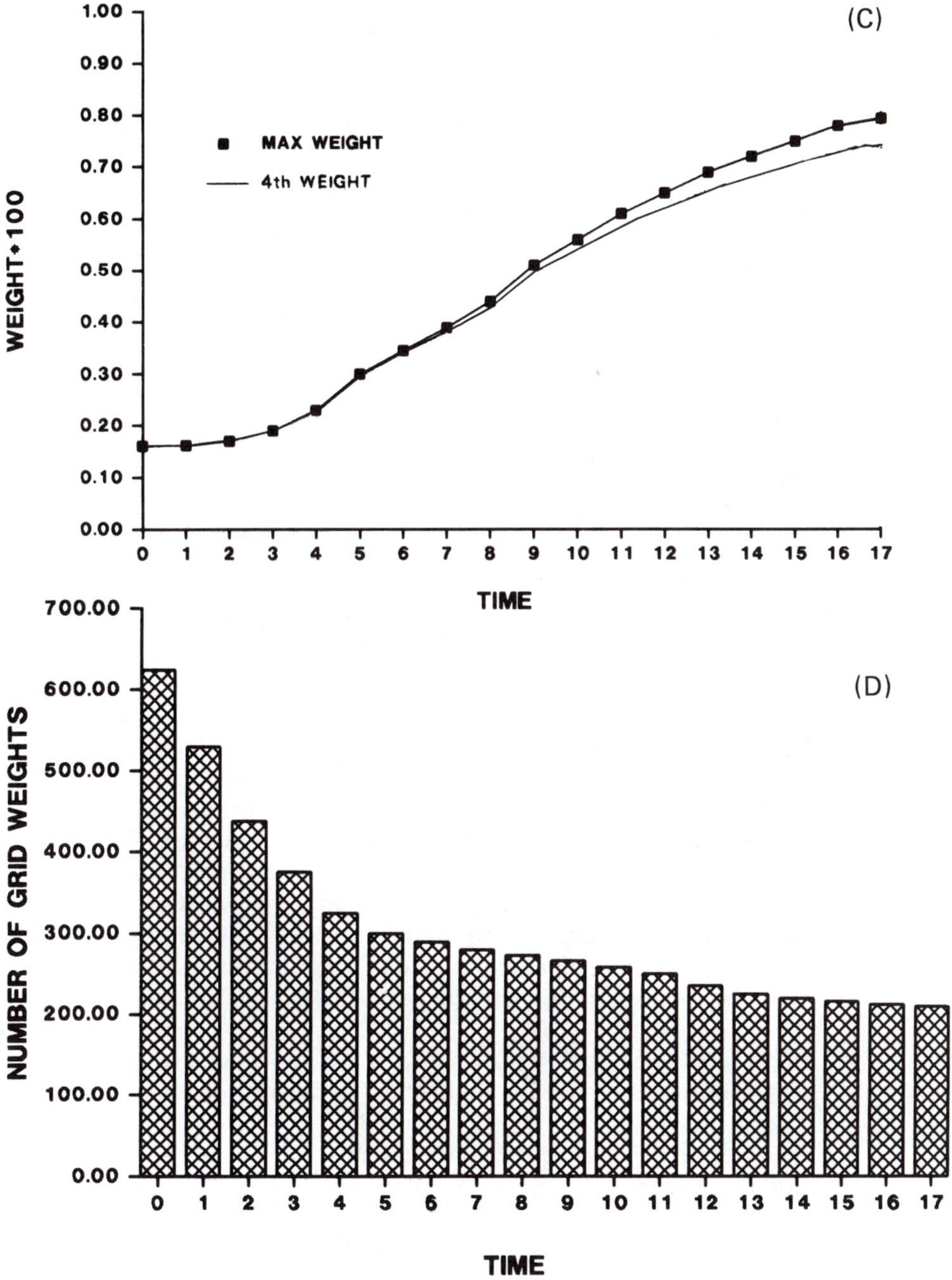

Figure 1. (A) Platform and target tracks; (B) velocity behavior; (C) behavior of maximal weight; (D) frequency above 0.015.

Case 1: initial conditions

True relative state: $\mathbf{x}^T = (2.25, -0.25, -0.25, 0.25)$
EKF initial state: $\hat{\mathbf{x}}_{0|-1}^T = (1.7, -0.8, -0.7, -0.3)$
Bearing error standard deviation $= \sigma_v = 0.5°$
Initial grid: ($5^4 = 625$ points) – uniform weights:

x_1	2.1	(2.2)	2.3	2.4	2.5
x_2	−0.6	−0.5	(−0.3)	−0.2	−0.1
x_3	−0.5	−0.4	−0.3	(−0.2)	−0.1
x_4	0.1	(0.2)	0.3	0.4	0.5

Notes:
1. Grid points nearest the true values have been circled.
2. The grid encloses the true state value.
3. The EKF was initialized to cause divergence to illustrate one of the observed behavior patterns.

Results: Results for 17 measurements are shown in Figure 1. The tracking platform executes a 90° turn at the ninth sampling time in an effort to enhance observability.

1. As is apparent in Figure 1(A), the EKF exhibits divergence with a collapse of the range estimate.
2. The bearing measurements provide limited information regarding velocity, although the system is theoretically observable. Figure 1(B) shows the y-component of velocity, the EKF estimate, and grid point values.
3. The p-vector exhibits the difficulty in estimating velocity. Only the four grid tracks having maximum weighting are shown in Figure 1(B). Near the end of the observation interval, the bearing changes so little that the track uncertainly becomes very elongated along the line of sight, as seen in Figure 1(A).
4. The p-vector grid encloses the actual track throughout the sampling interval. This must occur as a result of the initial grid conditions and the state model.
5. The p-vector weights become concentrated around the target track. As shown in Figure 1(C), weights increase by a factor of 5, but additional increases appear doubtful.
6. The concentration is emphasized by the frequency plot [Figure 1(D)]. The number of grid points with weights greater than 0.0015 (i.e., the initial uniform weight) diminishes from 625 to 210, but the number is changing slowly by the end of the observation interval. This indicates

that some convergence to a more centralized density has occurred, but the convergence rate becomes very small.

4.2 Scalar State Estimation/System Identification Problem

Consider the following stochastic system with scalar state x_k and scalar measurement z_k described by

$$x_{k+1} = ax_k + u_k + w_k$$
$$z_k = x_k + v_k, \qquad k = 1, 2, \ldots .$$

The input signal u_k can be regarded as a function of time or as a feedback signal generated from the measurements $\{z_1, z_2, \ldots, z_k\}$. The initial state x_1 is Gaussian with mean $(\hat{x}_1)$ and variance P_1. The sequences $\mathbf{w}(k)$ and $\mathbf{v}(k)$ are stationary Gaussian white noise with zero mean and variances Q and R, respectively. The noise sequences are assumed to be mutually independent and independent of the initial state x_1. Suppose that the transition variable a is a Gaussian random variable with mean $\hat{a}_1$ and variance σ_1^2. Further, a and x_1 are assumed to be independent.

For this discussion, assume that the state and measurement noise variances, Q and R, are equal to 4. The prior mean and variance for the transition parameter a will be taken to be 0.4 and 0.16, respectively. The initial state has mean zero and variance 100. This completes the definition of the parameters of the problem.

To begin, consider the system identification/state estimation problem when there are no inputs [i.e., $\mathbf{u}(k) = 0$ for all k]. By computing the posterior density $f(a, x_k|\mathbf{z}(k), \mathbf{u}(k-1))$, one can consider several questions. For this illustration, results are presented regarding the following questions:

1. How is the posterior density affected by the true value of the transition parameter a?
2. How does the extended Kalman filter perform as an estimator of a and x_k?
3. Does the posterior density converge to a Gaussian-like distribution?
4. Do different noise realizations affect the maximum of the distribution or the spread of the distribution?
5. How accurate an estimator of a and x_k is provided by the maxima of the posterior density?

These questions merely scratch the surface regarding the types of investigation that are made possible through the evaluation of the posterior density relation. No other approach permits investigation of these questions in the general and insightful manner that these results provide. Consider results that speak to the preceding questions.

Case 1

$$a_{\text{TRUE}} = 0.4$$
$$x_{\text{TRUE}}(1) = -5$$

In Figure 2(A), a two-dimensional tabulation of $f(a, x_1|z(1))$ is provided for specific values of a and x_1. The transition parameter is discretized (horizontal axis) in units of 0.1, whereas the state (vertical axis) is discretized in units of 1. The number in each cell represents the value of the density function at the grid point except for the appropriate normalization constant (approximately 1000 in this instance). For this plot, a single measurement z_1 has been processed. As is seen by inspecting Equation (3.9), this measurement provides information only regarding x_1 and not a. Thus the marginal density $f(x_1|z_1)$ displayed in the right-hand column is no longer the prior Gaussian, whereas the marginal density displayed in the lower row is given as $f(a|z(1)) = f(a)$.

This type of plot is repeated for 500, 1000, and 1500 samples in Figure 2(B), (C), and (D) and these plots can be used to discuss some of the earlier questions.

1. Observe that the densities retain their unimodality. In fact, they remain reasonably symmetric and Gaussian-like. Further analysis indicates that

A = 0.4

	(B) K = 500 X(500) = 0.9							(C) K = 1000 X(1000) = 1.6						(D) K = 1500 X(1500) = -2.2			
	PARAMETER A						F(X/Z)	PARAMETER A					F(X/Z)	PARAMETER A			F(X/Z)
STATE X	0	0.1	0.2	0.3	0.4	0.5		0.2	0.3	0.4	0.5	0.6		0.3	0.4	0.5	
-8	0	0	0	0	0	0	0	0	0	0	0	0	0	0	0	0	0
-7	0	0	0	0	0	0	0	0	0	0	0	0	0	0	2	0	2
-6	0	0	0	0	0	0	0	0	0	0	0	0	0	0	11	2	13
-5	0	0	1	3	1	0	5	0	0	0	0	0	0	0	49	9	58
-4	0	0	6	17	4	0	27	0	0	3	1	0	4	0	129	22	151
-3	0	1	21	59	13	0	94	0	0	17	8	0	25	1	210	35	246
-2	0	2	44	129	30	0	205	0	1	59	28	0	88	1	211	35	247
-1	0	3	59	174	48	0	276	0	2	131	60	0	193	(0)	132	22	154
0	0	(2)	48	145	35	0	230	(0)	2	178	81	0	261	0	51	8	59
1	0	1	24	75	18	0	118	0	2	149	68	0	219	0	12	2	14
2	0	0	7	24	6	0	37	0	1	78	36	0	115	0	2	0	2
3	0	0	1	5	1	0	7	0	0	25	12	0	37	0	0	0	0
4	0	0	0	1	0	0	1	0	0	5	2	0	7	0	0	0	0
F(A/Z) =	0	9	211	632	148	0		0	8	646	296	0		2	809	135	

○ —DENOTES THE EKF ESTIMATE

K = 1 A = 0.4 X(1) = -3.4

STATE X		-0.5					0.0					0.5					1.0					F(X/Z)
-12	0	0	0	0	0	0	0	0	0	0	0	0	0	0	0	0	0	0	0	0	0	0
-11	0	0	0	0	0	0	0	0	0	0	0	0	0	0	0	0	0	0	0	0	0	0
-10	0	0	0	0	0	0	0	0	0	1	1	1	0	0	0	0	0	0	0	0	0	3
-9	0	0	0	0	1	1	1	1	2	2	2	2	2	1	1	1	1	0	0	0	0	18
-8	0	0	1	1	2	2	3	4	4	5	5	5	4	4	3	2	2	1	1	0	0	49
-7	0	1	1	2	3	5	6	8	9	10	10	10	9	8	6	5	3	2	1	1	0	100
-6	1	1	2	4	5	7	10	12	14	16	16	16	14	12	10	7	5	4	2	1	1	160
-5	1	2	3	4	7	9	12	15	18	19	20	19	18	15	12	9	7	4	3	2	1	200
-4	1	2	3	4	6	9	12	14	17	18	19	18	17	14	12	9	6	4	3	2	1	191
-3	1	1	2	3	5	6	8	10	12	13	14	13	12	10	8	6	5	3	2	1	1	136
-2	0	1	1	2	3	4	5	6	7	8	(8)	8	7	6	5	4	3	2	1	1	0	82
-1	0	0	0	1	1	2	2	3	3	3	3	3	3	3	2	2	1	1	0	0	0	33
0	0	0	0	0	0	1	1	1	1	1	1	1	1	1	1	1	0	0	0	0	0	11
1	0	0	0	0	0	0	0	0	0	0	0	0	0	0	0	0	0	0	0	0	0	0
F(A/Z) =	4	8	13	21	33	46	60	74	87	96	99	96	87	74	60	46	33	21	13	8	4	983

○—DENOTES THE EKF ESTIMATE

Figure 2. (A) Posterior $f(x(1), a|\mathbf{z}(1))$; (B)–(D) posterior $f(x(k), a|\mathbf{z}(k))$.

a Gaussian distribution is a good approximation for the marginal densities except for small (i.e., less than 50 samples) sample sizes.

2. The true value of the transition parameter a is 0.4. The true value of the state changes at each sample time and is indicated on the plots. Note that the peak of the density function provides a reasonable estimate of a and $x(k)$ but is certainly not error-free. The distribution retains considerable "spread" (i.e., nontrivial variance) even after processing 1500 samples. The error in $[\hat{a}_{MAP}, \hat{x}_{MAP}(T)]$ is consistent with the variance of the density function.
3. The observation that the density function exhibits a substantial variance after 1500 samples implies slow convergence for any identification algorithm. It is useful to consider the behavior of the marginal density $f(a|\mathbf{z}(k))$. This density is plotted in Figure 3. These densities change very slowly and the results suggest that convergence in any practical situation (finite sample sizes) is not to be expected. The standard deviation is inversely proportional to the maximum value of the density and is, therefore, decreasing. The peak of the density, which serves as the MAP estimator of a, remains close to 0.4.

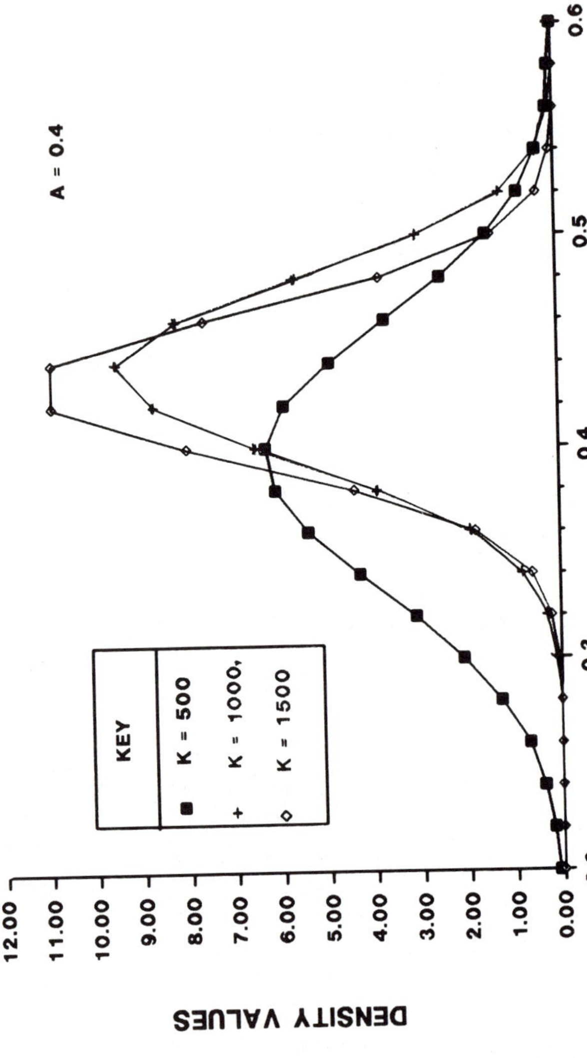

Figure 3. Posterior $f(a|\mathbf{z}(k))$.

Case 2.

$$a_{\text{TRUE}} = 0.9$$

$$x_{\text{TRUE}}(1) = -5$$

Consider now the effect of a larger value for the transition parameter a. Instead of the overdamped system in Case 1, the system is assumed to be underdamped with a transition parameter value that is near to the stability boundary. In investigating the convergence of many identification algorithms, considerable attention is given to cases in which a is near 1. The analysis indicates that one should expect better performance from a suitable algorithm as the stability boundary is neared. This observation is illustrated in the two-dimensional plots of $f(a, x(k)|\mathbf{z}(k))$ provided in Figure 4 and confirmed in Figure 5 with plots of the marginal density $f(a|\mathbf{z}(k))$. Note that the marginal density appears to be converging to a Gaussian distribution with a diminishing variance. The maximum of the density function provides an accurate estimate of a. Also, the EKF estimate is well behaved and reasonable.

A = 0.9

	(A) K = 500 X(500) = 2.32					(B) K = 1000 X(1000) = 0.62					(C) K = 1500 X(1500) = -2.32				
	PARAMETER A				F(X/Z)	PARAMETER A				F(X/Z)	PARAMETER A				F(X/Z)
STATE X	0.7	0.8	0.9	1.0		0.7	0.8	0.9	1.0		0.7	0.8	0.9	1.0	
-7	0	0	0	0	0	0	0	0	0	0	0	0	19	0	19
-6	0	0	0	0	0	0	0	1	0	1	0	0	90	0	90
-5	0	0	0	0	0	0	0	10	0	10	0	0	285	0	285
-4	0	0	0	0	0	0	0	44	0	44	0	0	590	0	590
-3	0	0	2	0	2	0	0	131	0	131	0	0	805	0	805
-2	0	0	13	0	13	0	0	255	0	255	0	0	723	0	723
-1	0	1	50	0	51	0	0	327	0	327	0	0	427	0	427
0	0	1	129	0	130	0	0	276	0	276	0	0	166	0	166
1	0	2	220	0	222	0	0	153	0	153	0	0	43	0	43
2	0	2	247	0	249	0	0	56	0	56	0	0	7	0	7
3	0	2	183	0	185	0	0	13	0	13	0	0	1	0	1
4	0	1	89	0	90	0	0	2	0	2	0	0	0	0	0
5	0	0	29	0	29	0	0	0	0	0	0	0	0	0	0
6	0	0	6	0	6	0	0	0	0	0	0	0	0	0	0
F(A/Z) =	0	9	969	0		0	0	1268	0		0	0	3159	0	

◯ —DENOTES THE EKF ESTIMATE (circled: 247 in (A), 327 in (B), 723 in (C))

Figure 4. Posterior $f(x(k), a|\mathbf{z}(k))$.

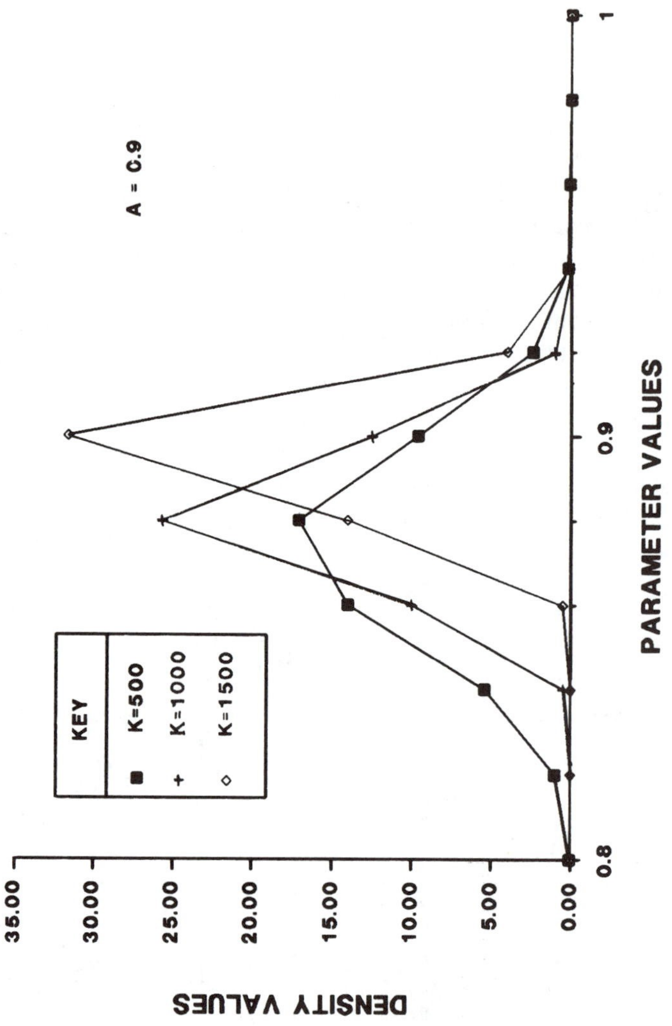

Figure 5. Posterior densities for $a = 0.4$.

Case 3

$$a_{\mathrm{TRUE}} = 0.1$$
$$x_{\mathrm{TRUE}}(1) = -5$$

The results presented above relate to a single realization of the system and measurement noise sequences. Examine the influence that different noise realizations can have on the densities. As noted above, the behavior becomes more erratic as $|a|$ approaches zero. To emphasize the possible influence of the noise realization, assume that the true value of the transition parameter is 0.1. The marginal density after 500 samples is presented in Figure 6 for three different noise realizations. Except for the noise realization, the parameters of the runs are identical.

The noise realization is seen to have a substantial influence on the location of the peak of the density. The variance is *not* affected to as great an extent and the Gaussian-like character of the density is *not* changed, at least not through any cursory examination. The behavior of the EKF is very sensitive to the realization.

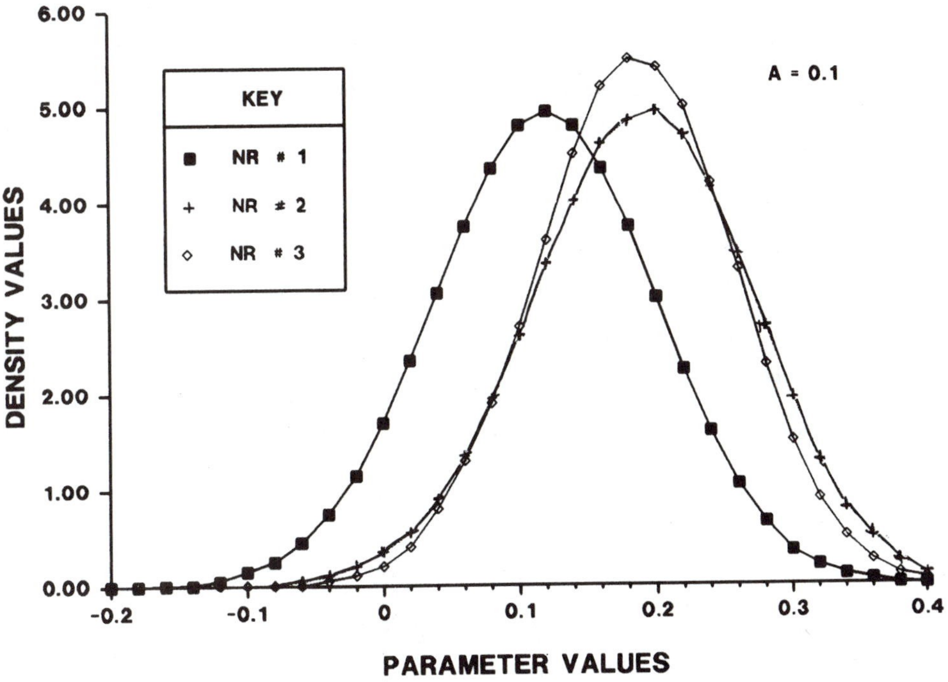

Figure 6. Noise realizations.

This completes the illustration of the system identification/state estimation problem in the absence of any input signal. It is well known (e.g., see Aström, 1970) that persistently exciting inputs are necessary for system identification. The remaining discussion of this example considers the effect of different input signals on the ability to identify the transition parameter.

The system identification problem becomes more tractable when the system is driven by known inputs. This point is illustrated by considering the example with a variety of input signals having the same energy. The noise realizations are the same in each case, as are all parameters of the problem. It was seen earlier that $a_{\mathrm{TRUE}} = 0.4$ led to a difficult identification problem, and this case will be reconsidered.

The effect of known input signal is compared with the zero-input case. Four types of input signals are considered.

1. Impulse every 10 samples:

$$u(t) = \begin{cases} 4\sqrt{10}, & t = 10i + 1, \quad i = 0, 1, 2, \ldots \\ 0, & \text{all other } t. \end{cases}$$

2. Pseudorandom sequence:

$$u(t) = 4 \operatorname{sgn} n(t),$$

where $n(t)$ is a sample chosen from a uniformly distributed unit variance white noise generator.

3. White noise:

$$u(t) = v(t),$$

where $v(t)$ is a sample from a Gaussian white noise sequence with zero mean and variance 4.

4. Sinusoidal input:

$$u(t) = \frac{160}{\pi} \sin \frac{\pi t}{5}.$$

The energy in the input signal for a 10-sample interval is the same,

$$E = \sum_{i=1}^{10} E[u^2(i)] = 160,$$

and is equal to four times the average energy in the system noise and in the measurement noise.

The joint and marginal densities for the zero input and for the input signals are shown in Figure 7 after processing 500 measurement samples. Results for the pulse, pseudorandom, and white noise inputs are very similar, so only the results for white noise are shown. Some interesting conclusions are apparent.

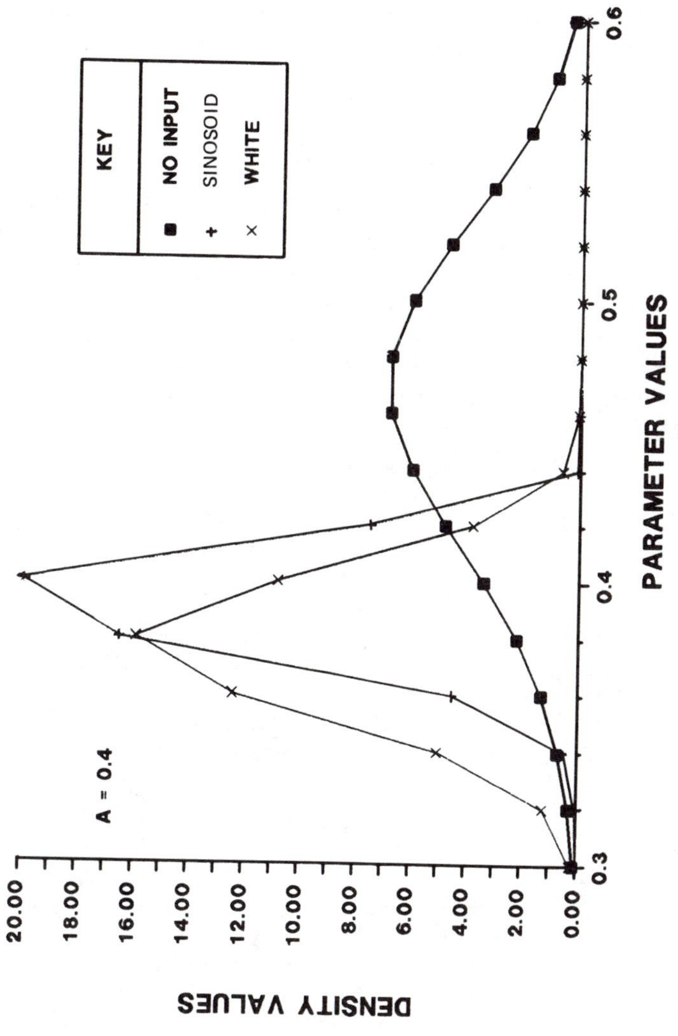

Figure 7. Input signal effect.

1. The input signals force the state to assume values that are generally larger than the noise signals (i.e., the output signal-to-noise ratio is increased). Consequently, the marginal density $f(a|\mathbf{z}(k))$ has a much smaller variance. This is apparent by considering the variances displayed on Figure 8(B) and (C). The variance is reduced approximately by one-third through the introduction of input signals.
2. The greatest reduction in variance comes from the use of a sinusoidal signal. This conforms with analytical results published in earlier studies. However, the results are not very sensitive to the form of the input signal. The primary influence is the input signal energy.
3. The extended Kalman filter performs satisfactorily. In fact, it serves as a reasonable approximation of a maximum a posteriori estimator. This improvement in behavior reflects our earlier comments regarding the apparent coupling between the identification algorithm and the input signal.

K = 500 A = 0.4

	(A) NO INPUT X(500) = -1.4					(B) SINUSOIDAL INPUT X(500) = -5.05					(C) WHITE NOISE INPUT X(500) = -1.44				
	PARAMETER A				F(X/Z)	PARAMETER A				F(X/Z)	PARAMETER A				F(X/Z)
STATE X	0.3	0.4	0.5	0.6		0.2	0.3	0.4	0.5		0.2	0.3	0.4	0.5	
-11	0	0	0	0	0	0	0	5	0	5	0	0	0	0	0
-10	0	0	0	0	0	0	0	31	0	31	0	0	0	0	0
-9	0	0	0	0	0	0	0	130	0	130	0	0	0	0	0
-8	0	0	0	0	0	0	0	333	0	333	0	0	0	0	0
-7	0	1	1	0	2	0	0	527	0	527	0	0	2	0	2
-6	0	4	8	1	13	0	0	516	0	516	0	0	15	0	15
-5	1	17	34	3	55	0	0	312	0	312	0	0	67	0	67
-4	2	49	91	7	149	0	0	117	0	117	0	1	190	0	191
-3	4	85	152	11	252	0	0	27	0	27	0	2	330	0	332
-2	4	91	159	11	265	0	0	4	0	4	0	2	356	0	358
-1	3	61	103	7	174	0	0	0	0	0	0	1	237	0	238
0	1	25	42	3	71	0	0	0	0	0	0	1	97	0	98
1	0	6	11	1	18	0	0	0	0	0	0	0	25	0	25
2	0	1	2	0	3	0	0	0	0	0	0	0	0	0	0
F(A/Z) =	15	340	603	44		0	0	2002	0		0	7	1323	0	

○ – DENOTES THE EKF ESTIMATE

Figure 8. Posterior $f(x(500), a|\mathbf{z}(500))$.

5. SUMMARY AND CONCLUDING REMARKS

By taking a Bayesian point of view, a numerical approach to nonlinear filtering and system parameter identification can be developed that provides insights into the problem that can be obtained only with great difficulty, if at all, by conventional methods of analysis. With this approach, questions of algorithm form and their convergence properties can be ignored. One can examine observability and identifiability by graphical means that emphasizes the important qualitative aspects of these concepts.

The Bayesian approach implies a subjective view of probability that seems appropriate in many important problems. One seldom has a frequency-related basis for assuming probability distributions. But it is frequently possible to define subjectively measures of one's understanding and belief about parameter values. These values can be introduced easily into the algorithms that have been described. In fact, it is simple to examine the effect of different assumptions and to compare results. The Bayesian viewpoint can be used to examine problems extending beyond estimation theory. Its utility for input signal synthesis is suggested in Example 4.2, and it is possible to carry this analysis much further.

To conclude, the assertion pursued throughout this chapter is that the Bayesian approach provides a framework within which systematic numerical studies of specific nonlinear systems can be pursued. The validity of the assertion is strengthened by recent and continuing growth in digital and computer graphics capability. Clearly, the evaluation of the Bayesian recursion relation implies a considerable computational burden. The understanding of numerical results is enhanced by appropriate graphical displays. The examples presented in Section 4 were chosen to illustrate some of the displays that are possible.

REFERENCES

Alspach, D. L., and H. W. Sorenson (1970). Approximation of Density Functions by a Sum of Gaussians for Nonlinear Estimation, in *Proceedings of the First Symposium on Nonlinear Estimation*, San Diego, Calif.

Anderson, B. D. O., and J. B. Moore (1979). *Optimal Filtering*, Prentice-Hall, Englewood Cliffs, N. J.

Aoki, M. (1967). *Optimization of Stochastic Systems*, Academic Press, New York.

Aström, K. J. (1970). *Introduction to Stochastic Control Theory*, Academic Press, New York.

Athans, M., R. P. Wishner and A. Bertolini (1968). Suboptimal state estimations for continuous time nonlnear systems from discrete noisy measurements, *IEEE Trans. Autom. Control* AC-13, 504–514.

Bucy, R. S. (1969). Bayes' theorem and digital realization for nonlinear filters, *J. Astronaut. Sci.* 17, 80–94.

Bucy, R. S., and K. D. Senne (1970). Realization of Optimum Discrete-Time Nonlinear Estimators, in *Proceedings of the First Symposium on Nonlinear Estimation*, San Diego, Calif.

Center, J. L. (1971). Practical Nonlinear Filtering of Discrete Observations by a Generalized Least-Squares Approximation of the Conditional Probability, in *Proceedings of the 2nd Symposium on Nonlinear Estimation*, San Diego, Calif.

Center, J. L. (1972). Practical nonlinear filtering based on generalized least-squares approximation of the conditional probability distribution, Ph.D. dissertation, Washington University, St. Louis.

Chou, S. I. (1977). Some Drawbacks of Extended Kalman Filters in ASW Passive Angle Tracking, in *Proceedings of the Symposium on Passive Target Tracking*, Naval Postgraduate School, Monterey, Calif.

de Figueiredo, R. J. P., and J. G. Jan (1971). Spline Filters, in *Proceedings of the 2nd Symposium on Nonlinear Estimation*, San Diego, Calif.

Fel'dbaum, A. A. (1965). Optimal Control Systems, Academic Press, New York.

Hilborn, C. G., and D. G. Lainiotis (1969). Optimal estimation in the presence of unknown parameters, *IEEE Trans. Syst. Man Cybern.* SMC-5, 38–43.

Ho, Y. C., and R. C. K. Lee (1964). A Bayesian approach to problems in stochastic estimation and control, *IEEE Trans. Autom. Control* AC-43, 333—339.

Jazwinski, A. H. (1970). *Stochastic Processes and Filtering Theory*, Academic Press, New York.

Kizner, W. (1969). *Optimal Nonlinear Estimation Based on Orthogonal Expansions*, Technical Report 32-1366, Jet Propulsion Laboratory, Pasadena, Calif.

Kramer, S. (1985). The Bayesian approach to recursive state estimation: implementation and application, Ph.D. dissertation, University of California, San Diego.

Ljung, L. (1979). Asymptotic behaviour of the extended Kalman filter as a parameter estimator for linear systems, *IEEE Trans. Autom. Control* AC-24, 36–50.

Magill, D. T. (1965). Optimal adaptive estimation of sampled stochastic processes, *IEEE Trans. Autom. Control* AC-10, 434–439.

Meditch, J. S. (1969). *Stochastic Optimal Linear Estimation and Control*, McGraw-Hill, New York.

Sorenson, H. W. (1966). Kalman Filtering Techniques, in *Advances in Control Systems*, Vol. 3 (C, T, Leondes, ed.), Academic Press, New York.

Sorenson, H. W. (1974). On the development of practical nonlinear filters, *Inf. Sci.* 7, 253–270.

Sorenson, H. W. (1980). *Parameter Estimation: Principles and Problems*, Marcel Dekker, New York.

Sorenson, H. W., ed. (1985). *Kalman Filtering: Theory and Application*, IEEE Press, New York.

Sorenson, H. W., and A. R. Stubberud (1968a). Recursive filtering for systems with small but nonnegligible nonlinearities, *Int. J. Control* 7, 271–280.

Sorenson, H. W., and A. R. Stubberud (1968b). Nonlinear filtering by approximation of the a posterori density, *Int. J. Control* 8, 33–51.

Srinivasan, K. (1970). State estimation by orthogonal expansion of probability distribution, *IEEE Trans. Autom. Control* AC-15, 3–10.

Sworder, D. (1966). *Optimal Adaptive Control Systems*, Academic Press, New York.

7

Nonnormal and Nonlinear Dynamic Bayesian Modeling

ANDY POLE, MIKE WEST, and P. JEFF HARRISON Department of Statistics, University of Warwick, Coventry, England

1. INTRODUCTION

The purpose of this chapter is to review a number of recent developments in statistical forecasting with nonlinear and nonnormal Bayesian dynamic models. The basic normal dynamic linear model (NDLM) was introduced in the early 1960s (Kalman, 1960; Kalman and Bucy, 1961), and Bayesian interpretations were given by a number of authors, including Ho and Lee (1964) and Jazwinski (1970). The first comprehensive Bayesian treatment of these models in the statistical literature was Harrison and Stevens (1976), and since then, motivated both by academic interest and the perceived requirements of practitioners, much work has been undertaken to extend the theoretical basis and range of application of these models.

We begin the discussion by recalling the basic form of the NDLM, and for simplicity, we shall confine attention to the univariate observation case.

Observation equation: $y_t = \mathbf{F}_t^T \boldsymbol{\theta}_t + v_t, \quad v_t \sim N(0, V_t)$ (1.1)

System equation: $\boldsymbol{\theta}_t = \mathbf{G}_t \boldsymbol{\theta}_{t-1} + \mathbf{w}_t, \quad \mathbf{w}_t \sim N(0, \mathbf{W}_t)$ (1.2)

where $\mathbf{F}_t$, $\mathbf{G}_t$, and $\mathbf{W}_t$ are all assumed known at time t with

y_t a univariate observation;

$\mathbf{F}_t$ a $p \times 1$ vector of regression variables;

$\boldsymbol{\theta}_t$ a $p \times 1$ vector of unknown parameters;

$\mathbf{G}_t$ a $p \times p$ state transition matrix;

V_t the observation variance;

$\mathbf{W}_t$ a $p \times p$ system variance matrix;

and we denote such a model by NDLM $\{\mathbf{F}, \mathbf{G}, V, \mathbf{W}\}_t$.

Dynamic state-space models of this form provide an excellent basis for constructing forecasting models for a number of reasons—simplicity, structuring, and insights into model construction prominent among these. Given a particular form of model characterized by $\{\mathbf{F}_i, \mathbf{G}_i\}$ in the DLM framework, a modeler may be quite prepared to postulate that this pair is locally adequate for much of the time in describing a given system. This is conceptually similar to the way in which a smooth function can at any point be locally approximated by a straight line. Thus, in general, a quantified statistical model is an approximate selective local description of a process whose validity decreases as it is used for prediction farther away from the environment (including time) on which its construction was based. Conditions change and hence the requirement for a dynamic specification.

For a simple system a deterministic model is often an appropriate representation, however, an inappropriate deterministic model can be disastrous since it lacks any facility for adapting to model departures. Moreover, it is seldom the case that real-world relationships are known in sufficient detail and values accurately measurable enough to permit such a deterministic model. These kinds of arguments provide a sound and powerful underpinning for the stochastic dynamic formulation of forecasting models.

The choice of the state-space form of the DLM as an appropriate framework within which to model many forecasting systems rests on the twin foundations of simplicity and interpretability. Its simplicity derives from having just two basic constituents: an observation equation relating the values actually recorded and of primary interest to the forecaster to underlying conditions; and a system equation that defines the evolutionary dynamics of the individual components describing these conditions. The componentwise structure of the system states (see below for details) facilitates understanding of the building blocks of the model and the formal structuring itself shows clearly how "everything hangs together."

In addition to this classical NDLM machinery, the Bayesian formulation includes the specification of beliefs about the state vector $\boldsymbol{\theta}_t$ in the form of prior and posterior probability distributions. The posterior distribution of the state vector at time $(t-1)$ will be denoted by $\boldsymbol{\theta}_{t-1}|D_{t-1} \sim N(\mathbf{m}_{t-1}, \mathbf{C}_{t-1})$, where D_{t-1} denotes all the relevant information up to and including time $(t-1)$, "$\sim$" means "distributed as" and $N(\mathbf{m}_{t-1}, \mathbf{C}_{t-1})$ is conventional notation for the normal (Gaussian) distribution with mean $\mathbf{m}_{t-1}$ and covariance $\mathbf{C}_{t-1}$. It is perhaps worthwhile emphasising at this point that the posterior distribution of $\boldsymbol{\theta}_t$ at any time will retain the normal form in such a model only if the prior distribution is also specified as normal. If we start an analysis at some time t_0 with initial beliefs represented by $N(\mathbf{m}_0, \mathbf{C}_0)$ and move the model mechanically through time the posterior distribution at any time is always normal (as are the implied priors) and there are convenient recurrence formulas for the parameters $\mathbf{m}_t$ and $\mathbf{C}_t$ which were originally defined by Kalman (1960).

$$\begin{aligned}\mathbf{m}_t &= \mathbf{a}_t + \mathbf{A}_t e_t \\ \mathbf{C}_t &= \mathbf{R}_t - \mathbf{A}_t \mathbf{A}_t^T Q_t^{-1}\end{aligned} \tag{1.3}$$

where

$$\begin{aligned}\mathbf{a}_t &= \mathbf{G}_t \mathbf{m}_{t-1} \\ \mathbf{R}_t &= \mathbf{G}_t \mathbf{C}_{t-1} \mathbf{G}_t^T + \mathbf{W}_t \\ \mathbf{A}_t &= \mathbf{R}_t \mathbf{F}_t Q_t^{-1} \\ Q_t &= \mathbf{F}_t^T \mathbf{R}_t \mathbf{F}_t - V_t \\ e_t &= y_t - \mathbf{F}_t^T \mathbf{a}_t.\end{aligned}$$

Furthermore, these recurrences remain valid even if we break out of the mechanistic mold mentioned above and include additional information in the prior specification at any time—provided that this takes the form of adjustments to the normal distribution parameters only and not of the prior distributional form. This is important since much of the flexibility of the Bayesian approach derives from the ability to formally incorporate new information in an analysis whenever appropriate in a unified manner.

Powerful and useful as the NDLMs have proved in practice, they are still quite restricted by the assumptions of normality and the imposition of linear structure throughout (the two are of course related). In many real situations either or both of these structures will be inappropriate even as a first approximation. The direct approach to transforming data may enable a solution to these problems in certain circumstances but in general two major disadvantages of this technique are apparent. First, the interpretability of the model parameters may become obscured, if not completely destroyed, and second, the existence of a suitable transformation may be precluded by the often conflicting requirements of linearity and normality.

In the remainder of this chapter we describe some developments of dynamic Bayesian models which have been designed to facilitate inclusion of nonlinear and nonnormal components in the analysis of time series and regression problems. In Section 3 we introduce the notion of the guide relationship as a means of modeling certain kinds of nonlinear structure while remaining within the normal distributional framework. These ideas are fundamental to the nonlinear, nonnormal theme of the chapter. In Section 4 we take this procedure a stage further, showing how the class of permissible distributions may be widened to include members of the one parameter exponential family. In Section 5 we describe the widest class of models included in our discussion, models with general observational and transitional distributions not restricted to any parametric family or considerations of conjugacy. In Section 6 we turn attention away from model formulation and analysis and review work on an equally important aspect of applied statistical modeling and forecasting, namely model monitoring and the assessment of model adequacy.

Before we begin our tour through the nonlinear, nonnormal world of dynamic Bayesian models, we first describe an extremely useful modification of the NDLM analysis designed to overcome a major problem with the practical implementation of these models. These ideas, presented in Section 2, carry over to the more general models in a direct manner, as will be shown.

2. DISCOUNT CONCEPT

Currently, applications of dynamic Bayesian models are dominated by the NDLM (although the situation is changing and we will see examples of more complex models in the later sections of this chapter). But even in this simplest of cases, a degree of statistical sophistication on the part of forecasters is a necessary prerequisite for successful implementation. At the very least, familiarity with the normal probability distribution and its properties and, ideally, with parametric modeling and state-space stochastic difference equations is required. Even in this desirable situation the specification of the observation and system variance matrices has been the source of much difficulty. We have found in practice that the observation variance is often confused with the one-step-ahead forecast error variance. Moreover, even experienced forecasters have little natural quantitative feel for the elements of the system variance matrix, $\mathbf{W}$, especially since it is not invariant to the measurement scale of the independent variables. The problem is exacerbated in time series models since $\mathbf{W}$ is then ambiguous in the sense that there can exist an uncountable set of such matrices which yield identical forecast functions.

To obviate these problems, a rather wide class of models which eliminates the need to specify the system variance matrix (directly) was introduced by Ameen and Harrison (1985a). In particular, a major parsimonious subset of these models based on the use of discounting techniques has proved extremely useful in practice. Many practitioners have a natural feel for the discounting concept, and furthermore, once the discount factors have been specified, standard techniques may be utilized for learning on-line about the observation variance. In this chapter, for the most part, we confine our attention to these discount models. For details of the more general normal weighted Bayesian models (NWBM), see Ameen and Harrison (1985a, b). Alternative procedures, which have been employed in the context of systems control, are discussed in a number of papers in the control literature (see, e.g., Sage and Melsa, 1971).

2.1 A Solution

In the system equation of the model (1.2), the stochastic term $\mathbf{w}_t$ leads to an increase in uncertainty or, equivalently, a loss in information about the state vector between times $(t-1)$ and t. More precisely, at time $(t-1)$ the state vector has posterior variance $\mathbf{C}_{t-1}$, which leads, via the evolution in (1.2), to the prior variance for $\boldsymbol{\theta}_t$ given by $\mathbf{G}_t\mathbf{C}_{t-1}\mathbf{G}_t^T + \mathbf{W}_t$. Letting $\mathbf{P}_t = \mathbf{G}_t\mathbf{C}_{t-1}\mathbf{G}_t^T$ denote the first term here, $\mathbf{P}_t$ may be regarded as the prior variance of a stable vector with no stochastic changes. Adding the evolution error term $\mathbf{w}_t$ to $\mathbf{G}_t\boldsymbol{\theta}_{t-1}$ to give the state vector $\boldsymbol{\theta}_t$ has the effect of increasing the uncertainty from the ideal $\mathbf{P}_t$ to the actual $\mathbf{R}_t = \mathbf{P}_t + \mathbf{W}_t$. This suggests two possible (not necessarily alternative) interpretations of the system equation: first (and conventionally), that the state evolution mechanism is inherently noisy and $\mathbf{w}_t$ is the formal representation of the noise component; and second, that our knowledge of the evolution is imperfect, with $\mathbf{w}_t$ describing our uncertainty. The latter interpretation leads us in a natural way to the discount concept.

The evolution error term models an additive increase in uncertainty, or loss of information, about the state vector between observations. To achieve a suitable degree of information decay over time, however, it is clear that the relative magnitudes of $\mathbf{W}_t$ and $\mathbf{P}_t$ are important. This leads to thinking in terms of a natural rate of decay of information that suggests a multiplicative rather than additive increase in uncertainty.

In the most basic scalar case we are lead to consider defining the prior variance for $\boldsymbol{\theta}_t$ as

$$R_t = \frac{P_t}{\delta}, \tag{2.1}$$

where δ is a discount factor, $0 < \delta \leqslant 1$. Thus the information loss is quantified as a percentage of the existing information as measured by covariance. The simplest extension to the vector case leads to a parsimonious form for the evolution variance matrix precisely of the same form as (2.1) ($\mathbf{R}_t$

and $\mathbf{P}_t$ are covariance "matrices"). This implies that

$$\mathbf{W}_t = \frac{\mathbf{P}_t(1-\delta)}{\delta}. \tag{2.2}$$

Thus $\mathbf{W}_t$ has precisely the same internal structure, in terms of covariance, as $\mathbf{P}_t$. The magnitude of variances and covariances is controlled by the discount factor, δ, with the implication that information decays at the same rate for each of the elements of the state vector. This procedure is applied for an individual model component. In practice, an NDLM $\{\mathbf{F}, \mathbf{G}, V, \mathbf{W}\}_t$ comprises an observation component $\{\mathbf{0}, \mathbf{0}, V, \mathbf{0}\}_t$ and several component NDLMs $\{\mathbf{F}_i, \mathbf{G}_i, 0, \mathbf{W}_i\}_t$ with $i = 1, \ldots, r$ and full rank $\mathbf{G}_i$s. The recommended construction is then $\mathbf{F}_t^T = \{\mathbf{F}_1^T, \ldots, \mathbf{F}_r^T\}_t$, $\mathbf{G}_t = \mathrm{diag}\{\mathbf{G}_1, \ldots, \mathbf{G}_r\}_t$, and $\mathbf{W}_t = \mathrm{diag}\{\mathbf{W}_1, \ldots, \mathbf{W}_r\}_t$—see the examples in the following section. It is then important to consider different discount factors $\delta_i (i = 1, \ldots, r)$ for each component and generally to recognize that the system evolutions of components are independent. This gives a structural form $\mathbf{P}_t = \mathbf{G}_t\mathbf{C}_{t-1}\mathbf{G}_t^T = [\mathbf{P}_{ij}]_t$ and corresponding form $\mathbf{R}_t = [\mathbf{R}_{ij}]_t$, $\mathbf{R}_{ii} = \mathbf{P}_{ii}/\delta_i$, $i = 1, \ldots, r$, and $\mathbf{R}_{ij} = \mathbf{P}_{ij}$ for $i \neq j$. This implies that $\mathbf{W}_i = \mathbf{P}_{ii}(1-\delta_i)/\delta_i$, $i = 1, \ldots, r$.

Note that we do not recommend that the off-diagonal blocks of $\mathbf{P}_t$ corresponding to different model components be discounted since, as we said above, it is often useful for the system evolution for the different blocks $\mathbf{G}_i$ to be considered as independent of one another. The information from the observation series is distributed among the elements of $\boldsymbol{\theta}_t$ and naturally produces parameter covariances. Now consider what happens in response to a disturbance in one or more of the parameters of a given block. This may be signaled by a suitably small discount factor for that block. However, in general it may not be desirable to apply a similar discount to the covariances with other blocks since this is likely to result in introducing the disturbance to those blocks.

Further generalizations and discount approaches are discussed in Ameen and Harrison (1985a,b) and West and Harrison (1986). The following examples illustrate the component-wise model construction process and provide practical guidance on the selection of suitable discount factors.

2.2 Illustration

Of particular interest for practical time series analysis and forecasting are dynamic models including polynomial trend, seasonal, and regression components. To illustrate the construction of a model from individual components and the discounting procedures outlined above, consider a scalar time series Y_t, $t = 1, 2, \ldots,$ such that

$$Y_t = \mu_{Tt} + \mu_{St} + v_t, \tag{2.3}$$

where μ_{Tt} is a trend term and μ_{St} a seasonal term. A widely used linear growth component NDLM for μ_{Tt} is defined as

$$\{\mathbf{F}_t, \mathbf{G}_t, 0, \mathbf{W}_t\} = \left\{\begin{pmatrix}1\\0\end{pmatrix}, \begin{pmatrix}1 & 1\\0 & 1\end{pmatrix}, 0, \mathbf{W}_{Tt}\right\}, \tag{2.4}$$

which can be written

$$\begin{aligned}\mu_{Tt} &= \mathbf{F}_{Tt}^T\boldsymbol{\theta}_{Tt}\\ \boldsymbol{\theta}_{Tt} &= \mathbf{G}_t\boldsymbol{\theta}_{T,t-1} + \mathbf{w}_{Tt},\end{aligned} \tag{2.5}$$

where $\boldsymbol{\theta}_{T,t}(1)$ represents the trend level at time t and $\boldsymbol{\theta}_{T,t}(2)$ the incremental growth between times $(t-1)$ and t. For the seasonal component, operational and modeling considerations mean that it is desirable in most cases to model seasonal effects in terms of the combination of a number of harmonic (cosine) waves. Such a representation is concise, parametrically parsimonious, and allows easy handling of the seasonal constraint. However, for communication it is important to acknowledge that conversion to seasonal effect parameters (factors) will be necessary and this is straightforward (see, e.g., Harrison and West, 1986). The NDLM $\{\mathbf{F}_S, \mathbf{G}_S, 0, \mathbf{W}_S\}$ representation for a single such harmonic is

$$\begin{aligned}\mathbf{F}_S &= (1, 0)^T\\ \mathbf{G}_S &= \begin{pmatrix}\cos j\omega & \sin j\omega\\ -\sin j\omega & \cos j\omega\end{pmatrix},\end{aligned} \tag{2.6}$$

where $\omega = \frac{1}{2}\pi/p$, $1 \leqslant j \leqslant \frac{1}{2}p$, and p is the period of the seasonality. For example, let $j = 1$ and consider the deterministic evolution of the state vector from initial values $\theta_{1,0}$ and $\theta_{2,0}$, then applying (2.6) gives the following periodic sequence:

$$\begin{aligned}\boldsymbol{\theta}_1 &= (\theta_{1,0}\cos\omega + \theta_{2,0}\sin\omega; -\theta_{1,0}\sin\omega + \theta_{2,0}\cos\omega)^T\\ \boldsymbol{\theta}_2 &= (\theta_{1,0}\cos 2\omega + \theta_{2,0}\sin 2\omega; -\theta_{1,0}\sin 2\omega + \theta_{2,0}\cos 2\omega)^T\\ \boldsymbol{\theta}_3 &= (\theta_{1,0}\cos 3\omega + \theta_{2,0}\sin 3\omega; -\theta_{1,0}\sin 3\omega + \theta_{2,0}\cos 3\omega)^T.\end{aligned}$$

A more complex seasonal pattern with multiple harmonics is constructed using components of this type combined as detailed above. A simple growth/seasonal NDLM may thus be written as

$$\left\{\begin{bmatrix}\mathbf{F}_T\\ \mathbf{F}_S\end{bmatrix}, \begin{bmatrix}\mathbf{G}_T & 0\\ 0 & \mathbf{G}_S\end{bmatrix}, V_t, \begin{bmatrix}\mathbf{W}_T & 0\\ 0 & \mathbf{W}_S\end{bmatrix}\right\}, \tag{2.7}$$

with observational noise V_t, a trend component NDLM $\{\mathbf{F}_T, \mathbf{G}_T, 0, \mathbf{W}_T\}$, and a seasonal component NDLM $\{\mathbf{F}_S, \mathbf{G}_S, 0, \mathbf{W}_S\}$ (which in general will include more than one harmonic, such subcomponents being brought together as a single seasonal model component).

Now consider the selection of discount factor in such a component model. Often, a quantified trend representation is only locally adequate in the very short term, whereas a quantified seasonal pattern is adequate over a longer period. Hence an appropriate trend discount factor δ_T will generally be much smaller than an appropriate seasonal discount factor δ_S. Suppose that $\boldsymbol{\theta}_T$ and $\boldsymbol{\theta}_S$ represent the component parameter vectors and that at time $(t-1)$,

$$\operatorname{var}\left[\begin{matrix}\boldsymbol{\theta}_T \\ \boldsymbol{\theta}_S\end{matrix}\middle| D_{t-1}\right] = \mathbf{C}_{t-1} = \begin{bmatrix}\mathbf{C}_T & \mathbf{C}_{TS} \\ \mathbf{C}_{ST} & \mathbf{C}_S\end{bmatrix}_{t-1}. \tag{2.8}$$

Then following the above,

$$\mathbf{R}_t = \mathbf{G}_t \begin{bmatrix}\dfrac{\mathbf{C}_T}{\delta_T} & \mathbf{C}_{TS} \\ \mathbf{C}_{ST} & \dfrac{\mathbf{C}_S}{\delta_S}\end{bmatrix}_{t-1} \mathbf{G}_t^T = \begin{bmatrix}\mathbf{P}_T & \mathbf{P}_{ST} \\ \mathbf{P}_{ST} & \mathbf{P}_S\end{bmatrix}_t \tag{2.9}$$

with

$$\mathbf{W}_t = \begin{bmatrix}(\delta_T^{-1}-1)\mathbf{P}_T & 0 \\ 0 & (\delta_S^{-1}-1)\mathbf{P}_S\end{bmatrix}_t. \tag{2.10}$$

In this way, although the information decay over the transition is allowed to be different for the two components individually, the covariance structure between the blocks is maintained. (Similar ideas are used in a multistate model discussed in Section 6.)

In practical applications it has been found that dynamic models are relatively robust to the values of discount factors within a suitable range (Ameen and Harrison, 1985a; West et al., 1985). For routine work linear trend discount factors typically lie between 0.8 and 0.95, seasonal discounts between 0.9 and 0.99, and those for regression parameters of independent variables between 0.95 and 0.99. In fact, a full likelihood is available for the discount factors, but in practice it is sufficient to experiment with a few values within the aforementioned ranges.

A few final remarks are in order concerning the discount procedure discussed above. First, it is important to recognize that the method of determining $\mathbf{W}_t$ does not alter the underlying structure of the model. Thus, updating and forecasting proceed in exactly the same manner as in the original Harrison and Stevens formulation, based on the Kalman filter recursions. Second, $\mathbf{W}_t$ can no longer be constant in general. Instead of this absolute constancy there is, in routine running of the model, an alternative invariance property, and that is invariance of $\mathbf{W}_t$ with respect to $\mathbf{C}_{t-1}$. Harrison and West (1986) discuss the use of discount factors in practice and give extensive examples.

The analysis thus far has assumed that the observational variance V_t is known at time t, and for k-step-ahead prediction, V_{t+k} is also known at time t. For situations in which this is not the case, a variety of approaches to learning on line about this component of the models have been suggested (Smith and West, 1980; Harrison and Stevens, 1976; Canterellis and Johnston, 1983; Ameen and Harrison, 1985a). Where the observational variance is constant, a tractable, fully Bayesian learning procedure is available. Using the standard inverse gamma conjugate prior for V we can derive recurrence formulae for the parameters of this distribution. For further details and examples of these methods, see Migon (1983).

3. NORMAL NONLINEAR DYNAMIC MODELS

The properties of the multivariate normal distribution mean that the linear structure of the basic NDLM leads to a very simple, tractable analysis of this class of models. In this section we show how, by an appropriate approximation procedure, this tractability can be maintained when we introduce nonlinear relationships. The simplest normal dynamic nonlinear model retains most of the structure and assumptions of the NDLM, with the exception that the linear link function between the observation series and the state variables is replaced by a nonlinear one. We use the term "guide relationship" to describe the latter for reasons that will manifest themselves as the analysis unfolds. The model has the following structure.

$$\begin{aligned}
&\text{Observation equation:} \quad && y_t = \Psi_t + v_t, && v_t \sim N(0, V_t) \\
& && \Psi_t \approx g(\boldsymbol{\theta}_t), && \\
&\text{System equation:} && \boldsymbol{\theta}_t = \mathbf{G}_t\boldsymbol{\theta}_{t-1} + \mathbf{w}_t, && \mathbf{w}_t \sim N(0, \mathbf{W}_t),
\end{aligned} \tag{3.1}$$

where $g(\cdot)$ is a nonlinear function, and the notation $\approx$ indicates a "guide relationship" [see the discussion following (3.2) below]. For our present purposes we take $g(\cdot)$ to be deterministic, but more generally, stochastic versions may be introduced (see Section 5 for a model of this kind). We assume that the joint prior distribution of Ψ_t and the state vector $\boldsymbol{\theta}_t$ are jointly normal, as in

$$\left.\begin{matrix}\Psi_t \\ \boldsymbol{\theta}_t\end{matrix}\right| D_{t-1} \sim N\left[\begin{pmatrix}\hat{\Psi}_t \\ \hat{\boldsymbol{\theta}}_t\end{pmatrix}, \begin{pmatrix} q_t & \mathbf{s}_t^T \\ \mathbf{s}_t & \mathbf{R}_t\end{pmatrix}\right], \tag{3.2}$$

where $\hat{\Psi}_t = E[g(\boldsymbol{\theta}_t)|D_{t-1}]$, $q_t = V[g(\boldsymbol{\theta}_t)|D_{t-1}]$, and $\mathbf{s}_t = \operatorname{cov}[\boldsymbol{\theta}_t, g(\boldsymbol{\theta}_t)|D_{t-1}]$. Furthermore, the marginal distributional assumptions on $\boldsymbol{\theta}_t$ from the NDLM are retained, in particular the discounting approach is maintained for specifying the prior variance matrix $\mathbf{R}_t$, and hence if, as specified, Ψ_t is a nonlinear function of $\boldsymbol{\theta}_t$, the joint distribution in (3.2) cannot be normal. Here the nature of the guide relationship is crucial. We do not regard $g(\cdot)$ as a completely specified component of the underlying mathematical model but merely as an aid to determining the first two joint moments of (3.2). In this way we can retain the joint normality assumption in this nonlinear case. An alternative justification for (3.2) is to regard it as a first-order approximation to the true distributional form. However, we prefer the former argument since it more nearly equates with how we envisage the model-building process. We are not sufficiently well informed to know the true nature of real-world relationships, but we can construct models that provide us with guidance in uncertain situations. The acid test of any such procedure is, of course, the performance of the models in practice (and this should be borne in mind particularly in later sections, where further approximations are introduced).

With the specification of (3.2) we once more have a model fully characterized by appropriate normal distributions. Hence the tractability of the standard NDLM follows and indeed the usual Kalman filter recursions are valid. From the structure of (3.1) and (3.2) it follows from standard theory that the posterior distribution of $\Psi_t|D_t$ is normal, whence the posterior distribution of the state vector is also normal, as can be seen from the following decomposition.

$$\begin{aligned} p(\boldsymbol{\theta}_t, \Psi_t|D_t) &= p(\boldsymbol{\theta}_t, \Psi_t|D_{t-1}, y_t) \\ &\propto p(y_t|\Psi_t)p(\boldsymbol{\theta}_t|\Psi_t, D_{t-1})p(\Psi_t|D_{t-1}) \\ &\propto p(\Psi_t|D_t)p(\boldsymbol{\theta}_t|\Psi_t, D_{t-1}). \end{aligned} \tag{3.3}$$

Thus information from y_t feeds through to $\boldsymbol{\theta}_t$ only via its influence on Ψ_t, and the posterior moments for $\boldsymbol{\theta}_t$ are then simply obtained from

$$\begin{aligned} E(\boldsymbol{\theta}_t|D_t) &= E[E(\boldsymbol{\theta}_t|\Psi_t, D_{t-1})|D_t] \\ V(\boldsymbol{\theta}_t|D_t) &= V[E(\boldsymbol{\theta}_t|\Psi_t, D_{t-1})|D_t] + E[V(\boldsymbol{\theta}_t|\Psi_t, D_{t-1})|D_t], \end{aligned} \tag{3.4}$$

and the evolution to time $t - 1$ priors follows from the system equation (3.1) and the succeeding discussion. Hence all the usual forms of recurrences are available for this model. As an illustration of this kind of analysis, recall the simple growth/seasonal model of Section 2.2. It is often noticed in commercial time series that data exhibit pronounced cyclical behavior in which the change from peak to trough increases with the average level. An excellent example of this phenomenon is the airline ticket sales data (see, e.g., Brown, 1963, p. 429); indeed, sales data for many goods demonstrate similar behavior.

A simple model which has been used successfully in these kinds of situations is given by

$$y_t = (\mu_{Tt} + \mu_{Rt})\mu_{St} + v_t, \tag{3.5}$$

but for the purposes of this section, we may ignore the regression component μ_{Rt}. The seasonal component of such a model can no longer have zero mean, so it is adjusted to have mean 1, giving the seasonal component model

$$\mu_{St} = 1 + \mathbf{F}_{St}^T \boldsymbol{\theta}_{St}, \tag{3.6}$$

where $\mathbf{F}_{St}$ and $\boldsymbol{\theta}_{St}$ are defined as in Section 2.2 using a Fourier representation of the seasonal form, and this implies that the seasonal factors sum to unity, rather than zero, as before. Computation of joint moments of $\boldsymbol{\theta}_t$ and $g(\boldsymbol{\theta}_t) = \boldsymbol{\theta}_{Tt}\boldsymbol{\theta}_{St}$ is straightforward in this bilinear case. For a practical application of such a model, see Harrison and West (1986).

4. NONLINEAR, NONNORMAL DYNAMIC MODELS

In the preceding section we saw how the standard NDLM could be generalized in a tractable way to nonlinear models using the notion of the guide relationship. However, the models therein are still somewhat hamstrung by the assumption of (joint) normality throughout, and while this may be an acceptable approximation in many situations, it will often be the case that something different or more general is desired. Perhaps the most evident reasons for uneasiness over the use of normal distributional assumptions are skewness, nonnegativity, discreteness, and outlier proneness of many data sets in forecasting. Clearly, we would like a more appropriate model to describe departures from normality.

The kind of situation depicted above is similar in many respects to that which existed pre-1970 in classical (i.e., non-Bayesian) modeling. In response to this, Nelder and Wedderburn (1972) introduced and developed a very wide class of models, called generalized linear models (GLMs), to facilitate nonnormal and/or nonlinear classical statistical analyses. Bayesian inference and data analysis for these kinds of models in a static setting were given in West (1985).

It seems appropriate then to build on the GLM approach in seeking a solution to the problems posed in the dynamic Bayesian context, and this was accomplished in West, Harrison, and Migon (1985)—hereafter referred to as WHM—on which we shall base our discussion. Thus the models we shall be looking at in this section, dynamic generalized linear models (DGLMs), are, as the name suggests, essentially dynamic versions of the GLM. However, in addition to the adaptive capability of dynamic models, the Bayesian approach used here provides a number of further advantages over GLMs. In particular, the use of conjugate prior distributions leads to closed-form updating and predictive distributions and computationally straightforward algorithms which are of great practical value.

4.1 The Model

The natural dynamic extension of the GLM is obtained by indexing throughout the unknown regression parameter, $\boldsymbol{\theta}$, by time, t, and introducing an evolutionary mechanism for the same. We may thus describe the following components of the DGLM.

Adhering to the notation of WHM, the observational or sampling density is the general exponential family, which may be written in the form

$$p(y_t|\eta_t, \phi) = \exp[\phi\{y_t\eta_t - a(\eta_t)\}]b(y_t, \phi), \tag{4.1}$$

where η_t is the natural parameter of the distribution such that $E[y_t|\eta_t, \phi] = \mu_t = \dot{a}(\eta_t)$, and ϕ is a scale parameter with $V[y_t|\eta_t, \phi] = \ddot{a}(\eta_t)\phi^{-1}$. To begin, we shall assume ϕ known (and hence drop the conditioning on this parameter, which should be taken as implicit), leaving discussion of the more complex unknown ϕ case until Section 4.2.

The natural conjugate prior distribution for η_t which provides the closed-form Bayesian analysis may be written as

$$p(\eta_t|D_{t-1}) = c(\alpha_t, \beta_t)\exp[\alpha_t\eta_t - \beta_t a(\eta_t)], \tag{4.2}$$

which is denoted $p(\eta_t|D_{t-1}) \sim \mathrm{CP}[\alpha_t, \beta_t]$ for some parameters α_t and β_t. From (4.1) and (4.2) we can immediately obtain the marginal predictive distribution for y_t using standard probability calculus as

$$p(y_t|D_{t-1}) = \frac{c(\alpha_t, \beta_t)}{c(\alpha_t + \phi y_t, \beta_t + \phi)} b(y_t, \phi) \tag{4.3}$$

and the posterior distribution for η_t is the updated conjugate form $\mathrm{CP}[\alpha_t + \phi y_t, \beta_t + \phi]$.

It remains to specify the nature of the relationship between the natural parameter of the sampling distribution η_t and the state vector $\boldsymbol{\theta}_t$, and the evolutionary mechanism for the latter. The first of these components follows the standard GLM formulation, that is,

$$g(\eta_t) = \lambda_t = \mathbf{F}_t^T\boldsymbol{\theta}_t, \tag{4.4}$$

where λ_t is the linear predictor in the GLM context, and $g(\eta_t)$ is a known nonlinear function. In the GLM setting the link function $g(\cdot)$ is used to determine estimates of the regression parameters (static analogs of our state parameters) and asymptotic theory invoked to justify the procedure. In the present Bayesian context, however, the link between $g(\eta_t)$ and λ_t is used only as a guide in determining the parameters α_t and β_t of the prior distribution on η_t. The arguments underpinning this use of the guide relation are precisely those discussed in the preceding section, "the [linear] model in $\boldsymbol{\theta}_t$ is a fiction, a modeler's guide to reality," WHM.

For the second component above we employ the usual discounting approach to the evolution of the state vector, but in order to allow for greater flexibility in the model, we work only in terms of the first two moments. Thus dropping the normal distributional assumption, if we have the posterior specification at time $(t-1)$,

$$\boldsymbol{\theta}_{t-1}|D_{t-1} \sim [\mathbf{m}_{t-1}, \mathbf{C}_{t-1}], \tag{4.5}$$

the evolution at time t is

$$\boldsymbol{\theta}_t|D_{t-1} \sim [\mathbf{a}_t, \mathbf{R}_t], \tag{4.6}$$

where $\mathbf{a}_t = \mathbf{G}_t\mathbf{m}_{t-1}$, $\mathbf{R}_t = \mathbf{G}_t\mathbf{C}_{t-1}\mathbf{G}_t^T + \mathbf{W}_t$, and the state transition and variance matrices, $\mathbf{G}_t$ and $\mathbf{W}_t$, respectively, are given, the latter defined using discount factors as in Section 2. Working with first and second moments only in this way provides a degree of model robustness against departures from the normality assumption (or any other we might entertain). The restrictive structures required by such assumptions are not imposed on the present model, thus allowing a wider variety of behavior compatible with the model. We do, however, have to pay a price for this greater generality and this is the introduction of a further approximation in the analysis, discussion of which is given below.

Equations (4.5) and (4.6) determine the prior mean and variance of λ_t and the prior covariance of λ_t and $\boldsymbol{\theta}_t$. For routine running of the model the prior specification is completed using these values and the guide relationship to assign appropriate values to the parameters α_t and β_t of the prior for η_t. That is, we have

$$\begin{aligned} E(\lambda_t|D_{-1}) &= \mathbf{F}_t^T\mathbf{a}_t = f_t \\ V(\lambda_t|D_{t-1}) &= \mathbf{F}_t^T\mathbf{R}_t\mathbf{F}_t = q_t \\ \operatorname{cov}(\boldsymbol{\theta}_t, \lambda_t|D_{t-1}) &= \mathbf{R}_t\mathbf{F}_t = \mathbf{s}_t. \end{aligned}$$

The first two of these are matched with the moments for λ_t derived from (4.2), yielding two equations for α_t and β_t, which may then be solved. In practice, we may at any time receive further subjective prior information which we may incorporate by overwriting the values of α_t and β_t with suitable alternatives. The flexibility of the sequential Bayesian analysis facilitates such intervention and accounts for much of the power of the models. Whatever values are judged appropriate, the forecast distribution for $y_t|D_{t-1}$ has the standard form in (4.3) and the posterior distribution for η_t has the conjugate form $\mathrm{CP}[\alpha_t + \phi y_t, \beta_t + \phi]$.

The only outstanding quantity necessary to complete the Bayesian analysis is the posterior distribution of the state vector $\boldsymbol{\theta}_t$. This is impossible to calculate precisely for the current model since we have only a partially specified prior, and furthermore the model does not provide a likelihood on $\boldsymbol{\theta}_t$. The likelihood or sampling distribution in (4.1) is a function of η_t, and while we do have the functional relationship between η_t and $\boldsymbol{\theta}_t$ given in (4.4), this is a guide relation. Again we emphasize the nature of this construct: It is a modeler's guide to reality which is not meant to be a formal means of deriving complete distributional forms of η_t from $\boldsymbol{\theta}_t$, or vice versa. However, to proceed to time $(t-1)$ we require only the first two moments of the posterior distribution [see (4.5) and the subsequent discussion].

The joint posterior distribution of $\boldsymbol{\theta}_t$ and η_t permits the following useful decomposition, independent of the choice of sampling or prior distributions, which is derived as in equation (3.3),

$$p(\boldsymbol{\theta}_t, \eta_t | D_t) \propto p(\eta_t | D_t) p(\boldsymbol{\theta}_t | \eta_t, D_{t-1}), \tag{4.7}$$

whence information in y_t feeds through to $\boldsymbol{\theta}_t$ only via its influence on η_t. Posterior moments for $\boldsymbol{\theta}_t$ may now theoretically be obtained from an application of familiar relationships as in

$$\begin{aligned} E(\boldsymbol{\theta}_t | D_t) &= E[E(\boldsymbol{\theta}_t | \eta_t, D_{t-1}) | D_t] \\ V(\boldsymbol{\theta}_t | D_t) &= V[E(\boldsymbol{\theta}_t | \eta_t, D_{t-1}) | D_t] + E[V(\boldsymbol{\theta}_t | \eta_t, D_{t-1}) | D_t]. \end{aligned} \tag{4.8}$$

Unfortunately, however, the conditional moments in (4.8) will be unknown complicated functions of η_t by virtue of the nonlinear guide relationship. Instead of nice completely specified joint normal distributions, we have only incomplete information on η_t and $\boldsymbol{\theta}_t$ given by the joint prior moments

$$\left. \begin{matrix} g(\eta_t) \\ \boldsymbol{\theta}_t \end{matrix} \right| D_{t-1} \sim \left[\begin{pmatrix} f_t \\ \mathbf{a}_t \end{pmatrix}, \begin{pmatrix} q_t & \mathbf{s}_t^T \\ \mathbf{s}_t & \mathbf{R}_t \end{pmatrix} \right] \tag{4.9}$$

together with the marginal posterior moments $g(\eta_t)|D_t \sim [g_t, p_t]$, where the moments g_t and p_t may be calculated from the fully conjugate posterior on η_t. With only this partial information available, an alternative method for determining the posterior moments of $\boldsymbol{\theta}_t$ is required.

The approach adopted by WHM uses the technique of linear Bayes estimation (Hartigan, 1969; Goldstein, 1976): In view of the linear relation $\lambda_t = \mathbf{F}_t^T \boldsymbol{\theta}_t$, it is natural to adopt a linear function of $g(\eta_t)$ as a predictor of $\boldsymbol{\theta}_t$. Specifically, suppose that we choose $\mathbf{d}$ such that $\mathbf{d} = \mathbf{d}_0 + \mathbf{d}_1 g(\eta_t)$, where $\mathbf{d}_0$ and $\mathbf{d}_1$ are chosen to minimize the overall quadratic risk (see WHM for justification)

$$r_t(\mathbf{d}) = \operatorname{tr} E(\mathbf{A}_t(\mathbf{d}) | D_{t-1}), \tag{4.10}$$

where $\mathbf{A}_t(\mathbf{d}) = E[(\boldsymbol{\theta}_t - \mathbf{d})(\boldsymbol{\theta}_t - \mathbf{d})^T | \eta_t, D_{t-1}]$. Now the joint moments of (4.7) serve to determine the required predictor, and it is easily seen that the optimal setting is

$$\mathbf{d} = \hat{\mathbf{a}}_t = \mathbf{a}_t + \frac{\mathbf{s}_t(g(\eta_t) - f_t)}{q_t}, \tag{4.11}$$

which gives the minimum value of $E[\mathbf{A}_t(\mathbf{d})|D_{t-1}]$, the estimated posterior variance of $\boldsymbol{\theta}_t$ as

$$\hat{\mathbf{R}}_t = \mathbf{R}_t - \frac{\mathbf{s}_t\mathbf{s}_t^T}{q_t}. \tag{4.12}$$

Finallly, we achieve the feedback of information from y_t to $\boldsymbol{\theta}_t$ by substituting the values $\hat{\mathbf{a}}_t$ and $\hat{\mathbf{R}}_t$ for the conditional mean and variance in (4.8) to obtain the updated posterior moments

$$E(\boldsymbol{\theta}_t|D_t) = \mathbf{m}_t \approx \mathbf{a}_t + \frac{\mathbf{s}_t(g_t - f_t)}{q_t}$$

$$V(\boldsymbol{\theta}_t|D_t) = \mathbf{C}_t \approx \mathbf{R}_t - \frac{\mathbf{s}_t\mathbf{s}_t^T(1 - p_t/q_t)}{q_t}. \tag{4.13}$$

This completes the updating equations for the DGLM; we may note that they have the same form as the standard Kalman filter equations with the relevant q_t, p_t substituted. [If $g(\eta_t)$ and $\boldsymbol{\theta}_t$ were jointly normally distributed, then this is necessarily true since linear Bayes estimates are identical for all joint distributions with the specified first two joint moments.]

At this point it is useful to note the similarities and contrasts of the DGLM and the normal nonlinear models of Section 3. First, it is important to recognize that the latter are not a special case of the DGLM—the underlying modeling assumptions have some crucial differences. In the DGLM there is no *direct* linkage of the state vector $\boldsymbol{\theta}_t$ with the observation y_t. The connection exists purely through the guide relationship between η_t (directly related to y_t) and λ_t (directly related to $\boldsymbol{\theta}_t$). For the model of Section 3, however, a direct link was constructed by the assumption of joint normality between $\boldsymbol{\Psi}_t$ and $\boldsymbol{\theta}_t$. The role of the guide relationship is similar in both models. It serves merely as a means of associating distributional moments from one side ($\boldsymbol{\Psi}_t$ or η_t) to the other ($\boldsymbol{\theta}_t$ or λ_t) and not as a mechanism through which full distributional forms are transformed.

Finally, the role of the linear state transition is rather more tenuously used in the DGLM. WHM emphasize the form of the time evolution in terms of the state moments [equations (4.5) and (4.6)], rather than explicitly stating the system equation. However, it is pointed out that this usual physical interpretation may be used (without the restriction to normality on $\mathbf{w}_t$).

4.2 Scale Parameters in the DGLM

The continuous members of the exponential family have natural scale parameters, and when these are unknown, a full Bayesian learning procedure is available as described for the normal models in Section 2. In discrete cases, however, ϕ is fixed, usually at unity. For such models West (1985) has developed a procedure based on familiar ideas for assessing model adequacy by embedding a proposed model within a larger, more general class indexed by some parameter ϕ. Inference on this parameter provides the means whereby the assumption of the original model (or distribution) can be examined. In this analysis a gamma prior for $\phi|D_{t-1}$ in the nonnormal models leads to an approximate gamma posterior $\phi|D_t \sim G(v_t/2, \delta_t/2)$, where $v_t = v_{t-1} + 1$ and $\delta_t = \delta_{t-1} + D(y_t|D_{t-1})$, the latter term being the residual deviance (see West, 1985).

The practical implementation procedure for these models suggested by WHM is: (1) Perform the analysis of Section 4.1 conditional on $\phi = 1$; and (2) calculate v_t and δ_t after the update in step 1. Then the posterior for ϕ provides a method of monitoring the fit of the standard model: $\phi \approx 1$ means that the model is acceptable while $\phi < 1$ ($\phi > 1$) indicates that there is less (more) variation in the data than allowed for by the standard specification and simplification (elaboration) of the model should be considered.

More general procedures for model monitoring and assessment are discussed in Section 6.

4.3 Example: Analysis of Survival Data

The concepts underlying the DGLM have recently been applied in developing models for survival data (Gamerman, 1986), which provide for nonproportional hazards via time-varying parameters. In the proportional hazards models of Cox (1972), we have a survival function $S(t)$ for an individual (or component) which defines the probability of surviving (or continued functioning) beyond a given time $t > 0$.

$$S(t) = \Pr(T > t) = \exp\left\{-\int_0^t \lambda(u)\,du\right\}, \tag{4.14}$$

where

$$\lambda(t) = \lambda_0(t)\exp(\mathbf{z}^T\boldsymbol{\beta}) \tag{4.15}$$

is the hazard function, with λ_0 the baseline hazard—a lower bound common to all members of a group, $\mathbf{z}$ a vector of regressor variables or covariates particular to an individual member of the group, $\boldsymbol{\beta}$ a vector of parameters modeling the effects of the covariates, and T the random variable denoting survival time. The basic model was extended to a piecewise model by Breslow (1972) wherein the baseline hazard is supposed constant within intervals $I_i = (t_{i-1}, t_i]$, but having different values in different intervals:

$$\ln[\lambda_0(t)] = \alpha_i, \qquad t \in I_i \tag{4.16}$$

for some given partition of the time axis with $t_0 < t_1 < \cdots$.

Beginning with (4.14) and (4.15), Gamerman constructs a dynamic Bayesian model by modeling time variation in the covariate and parameter vectors in addition to the baseline hazard. He argues that the unrestricted change in α_i between intervals is unrealistic and that a more appropriate formulation is obtained by positing an evolutionary change through time, the simplest such mechanism being the random walk. Furthermore, as we have argued repeatedly throughout this chapter, the influence of variates on an observable time series is likely to be changing rather than fixed; hence some time-varying relationship for $\boldsymbol{\beta}$ is appropriate. Finally, in many applications, particularly to human systems but also in mechanical systems, especially where feedback mechanisms may be operating, the values of variates themselves may change over time (these, of course, are known). With these new structures we now have the dynamic model defined by (4.14) with

$$\begin{aligned} \ln[\lambda(t)] &= [1, \mathbf{z}_i^T]\boldsymbol{\theta}_i, \qquad i \in I_i \\ \boldsymbol{\theta}_i &= \boldsymbol{\theta}_{i-1} + \mathbf{w}_i, \end{aligned} \tag{4.17}$$

where $\boldsymbol{\theta}_i^T = (\alpha_i, \boldsymbol{\beta}_i^T)$ and the stochastic term $\mathbf{w}_i$ has an appropriately specified set of covariances, typically set according to the discounting techniques of Section 2, to determine the extent of the variation in $\boldsymbol{\theta}_i$.

The sequential analysis of this model is based on a generalization of that in the DGLM. We have a piecewise exponential distribution for the survival time T and a logarithmic guide relation, which we can write as

$$\ln[\lambda_i^{(j)}] = \mathbf{z}_j^T\boldsymbol{\theta}_i, \qquad j = 1, \ldots, r_i,\ i = 1, \ldots, N, \tag{4.18}$$

where i indexes the interval, j indexes the individuals known to be alive (functioning) at the beginning of the interval I_i, and $\mathbf{z}_j$ is redefined to include the constant term. An important feature of this application is that *for observations within an interval there is no evolution*: $\mathbf{z}$ and $\boldsymbol{\theta}$ are constant, but this does not affect the way in which the analysis is performed. The likelihood for individual j (which has either ceased to function within or survived the interval—a third possibility of censorship is also present in some cases) combines with the prior on $\lambda_i^{(j)}|D_{i-1,j-1}$ (in an obvious notation) to yield the posterior $\lambda_i^{(j)}|D_{i-1,j}$. Then linear Bayes methods provide the moments of the updated posterior distribution for $\boldsymbol{\theta}_i|D_{i-1,j}$. With no evolution the implied priors for the next observation within the interval are just these posteriors.

This procedure is continued until all r_i observations have been processed, whence the evolution to interval I_{i+1} follows as usual; for example, the posterior state vector $\boldsymbol{\theta}_i|D_{i-1,r_i} \sim [\mathbf{m}_i, \mathbf{C}_i]$ leads to the updated prior $\boldsymbol{\theta}_{i+1}|D_i \sim [\mathbf{a}_{i+1}, \mathbf{P}_{i+1}]$, where $D_{i-1,r_i} = D_i$, $\mathbf{a}_{i+1} = \mathbf{G}_{i+1}(b_{i+1})\mathbf{m}_i$, $\mathbf{P}_{i+1} = \mathbf{G}_{i+1}(b_{i+1})\mathbf{C}_i\mathbf{G}_{i+1}^T(b_{i+1}) + \mathbf{W}_{i+1}$, and $\mathbf{G}_i$ is the system evolution matrix over interval I_i, generally dependent on b_i = length (I_i). Technical details are given in Gamerman (1986).

An application of the dynamic survival model to unemployment studies is reported in Gamerman and West (1986) which dramatically illustrates the benefits that can be derived in comparison with the static proportional hazards models. A model is constructed for unemployment durations (the survival times) in the United Kingdom recorded by the Department of Health and Social Security (DHSS) cohort study of the unemployed. The objective is to assess and predict the effects of various socioeconomic factors (the covariates of our model) on the lengths of time that people continue to claim benefit after initially registering as unemployed.

The data set contains information on a total of 2332 men initially registering as unemployed in autumn of 1978 and covering the next 52 weeks. Failures (i.e., removals from the unemployment register) are recorded on a daily basis and the data are uncensored apart from truncation at the end of the study period. Information on about 800 socioeconomic factors was recorded and updated for each person on three occasions during the year. For this analysis a small subset of eight of these was chosen from about 20 identified as of primary interest in previous studies (Narendranathan et al., 1985):

1. Age of individual in years
2. Marriage indicator (1 if married, 0 if single)
3. Education indicator (1 if minimum secondary school leaving certificate, 0 otherwise)
4. Staying power indicator (1 if less than 12 months in last full-time job, 0 otherwise)
5. Previous unemployment indicator (1 if registered in last 12 months, 0 otherwise)
6. Local unemployment rate (%)
7. Log (estimated income at work)
8. Log (unemployment benefit)

For further details of the data set and these regressors in particular, see the references sited above.

The conjugate prior for the natural parameter (exponential likelihood) is the gamma distribution, that is,

$$\lambda_i^{(j)}|D_{i-1,j-1} \sim \text{gamma}\,(\alpha_{ij}, \gamma_{ij}), \tag{4.19}$$

where

$$\alpha_{ij} = q_{ij}^{-1}$$
$$\gamma_{ij} = q_{ij}^{-1}\exp(-f_{ij})$$

[recall (4.9) and the discussion following (4.18): i here indexes the current interval and j indexes individuals within the interval]. The corresponding posterior distribution in this case is gamma $(\alpha_{ij} + \delta_{ij}, \gamma_{ij} + t_{ij} - t_{i-1})$, where

$$\delta_{ij} = \begin{cases} 0, & \text{if individual } j \text{ remained on the register: } t_{ij} = t_i \\ 1, & \text{if individual } j \text{ left the register at time } t_{ij} < t_i. \end{cases}$$

This posterior becomes the prior for individual $j + 1$—recall that there is no evolution within an interval—and the process is repeated until all individuals on the register at the beginning of the interval have been processed. The "final" posterior of this procedure then goes through the random walk evolution, (4.17), to become the initial prior for interval I_{i+1}.

The initial prior specification used here was relatively vague, with $\mathbf{m}_0 = \mathbf{0}$ and $\mathbf{C}_0 = 1000\mathbf{I}_9$. A discount factor of 0.9 was selected, reflecting a prior belief in a relatively smooth, steady evolution of the state vector which is consistent with slowly varying effects of the regressor variables.

The results of this analysis are extremely interesting. The baseline hazard is quite stable—although there is a slight suggestion of an increase in the first six months, followed by a corresponding decline thereafter. The three most important covariates, 8, 7, and 1, respectively, all exhibit a similar pattern of development in their explanatory power over the year. This feature is most noticeable in the coefficient of unemployment benefit. Initially, such income has a marked negative effect; during the first five months of unemployment, a high unemployment income markedly depresses the likelihood of a return to work. As time passes, however, this effect decays, with the estimated coefficient being rather stable near zero over the latter half of the year, so this variate has little explanatory power here. Similar trajectories are apparent for the coefficients of income at work and age. This tendency toward zero is consistent with a view that strong and common social effects eventually tend to dominate the income factors that are so important in the early stages.

The education and staying power indicators have fairly constant coefficients, while the (absolute) effect of the local unemployment rate is increasing over time, although the uncertainty associated with this is quite high and prevents this factor from dominating as an explanatory variable. Qualitatively, the major divergence with previous studies, apart from the clearly evident time variation in effects, concerns the unemployment indicator 4. In Narendranathan et al. (1985) it was suggested that the coefficient was positive; this study clearly indicates a negative value. Gamerman and West

(1986) offer the explanation that people with staying power can be considered more attractive from an employer's point of view.

After fitting the model, unemployment spells were forecast for two hypothetical persons whose major characteristics (relevant to the model) contrasted as follows. Individual 1 is younger than individual 2, receives lower unemployment benefit, does not rate on the educational indicator, and has an "unfavorable" employment background. Notice that in this example the ability of the model to allow changing values of the covariates and corresponding parameters over time is likely to be crucial given the great instability of many economies at the present time. In particular, changing levels of unemployment benefit and the educational index (through retraining for example) are very likely to occur.

Now, unlike the proportional hazards, models whose survival functions are related through power transformations and therefore cannot cross, the flexibility of the dynamic models allows for a variety of relationships, and in this illustration the survival curves intersect after a period of $4\frac{1}{2}$ months. The explanation offered for this behavior is that the age and high unemployment benefit of individual 2 lead to a relatively lower likelihood of an early return to work (recall the discussion above). After a few months, however, his prospects increase relative to individual 1 because of the decay of the age and income effects, while education and employment history factors remain in his favor. Clearly, these features are a consequence of the particular model employed, but the important point is that many different kinds of behavior can be straightforwardly modeled in the dynamic framework presented, permitting a much richer and potentially vastly superior analysis to be undertaken.

5. GENERAL DYNAMIC NONLINEAR MODELS

5.1 Model and Analysis

In Section 3 we saw how the standard NDLM could be extended to models with nonlinear responses in terms of the state parameter by introducing approximate analyses based on guide relationships. In Section 4 we described a methodology for extending the range of noise distributions to a parametric family which includes the normal as a special case. In the present section these generalizations are taken a step further by considering models with general observation and transitional distributions not restricted to the exponential class or considerations of conjugacy.

An early attempt at a computationally feasible method for estimating nonlinear dynamic models is Sorenson and Alspach (1971). The basic idea here is to replace the normal distributional framework of the models with one composed of convex linear combinations of normals. In this way nonnormal error sequences can be directly approximated, but in addition, models with nonlinear structures can be accommodated since "many of the properties and concomitant difficulties of the [nonlinear] approximation are exhibited by considering linear systems which are influenced by non-Gaussian noise."

The initial prior density of the state vector and all observation and system noise terms are represented in the form of discrete normal mixtures, as, for example,

$$p(\boldsymbol{\theta}_0|D_0) = \sum_{i=1}^{l} \alpha_{i0} N(\mathbf{m}_{i0}, \mathbf{C}_{i0}), \tag{5.1}$$

where $\alpha_{i0} \geqslant 0$ and $\sum_{i=1}^{l} \alpha_{i0} = 1$, and in fact this restriction on the mixture weights is imposed throughout, so that the resulting densities are proper. With the linear state and observation equations such a setup permits direct use of the Kalman filter updating equations applied to each element of the prior density combined in turn with each element of the observation noise density in order to obtain the corresponding components of the posterior density. A similar process is applied to this posterior to obtain the prior density for the following period. Thus, in essence, the single NDLM is replaced by a whole panoply of NDLMs whose combination approximates the envisioned nonnormal/nonlinear situation. There is, however, a computational difficulty with this procedure, and this is the exponential increase in the number of terms required to define the prior/posterior distributions as each observation is processed. Two arguments are advanced for reducing this number at any stage in order to obviate the problem. First, as the weights of some elements become small, their contribution to the density will be negligible and hence can be ignored. Second, some elements will have similar locations, and these may be combined into a single representative element with an appropriately adjusted weight. [This kind of mixture of normals model with mixture collapsing for computational simplicity is used in the different context of model monitoring in Harrison and Stevens (1976)—see Section 6 for a brief discussion of the ideas involved.]

Masreliez (1975) describes methods for estimating the state vector of a DLM which has normal and nonnormal noise components. Specifically, an approximation is proposed for the sequence of on-line estimates of the state in the two complementary cases of nonnormal observation noise with normal system noise, and normal observation noise with nonnormal system noise. In the first case the state prior distribution is assumed to be normal at each stage, while the observation forecast distribution is constrained only to be twice differentiable. It is then straightforward to compute an exact expression for the posterior moments of $\boldsymbol{\theta}_t$ which may be expressed in the form

$$\mathbf{m}_t = \mathbf{a}_t + \mathbf{P}_t\mathbf{F}_t^T b_t(y_t)$$
$$\mathbf{C}_t = \mathbf{P}_t - \mathbf{P}_t\mathbf{F}_t^T\mathbf{B}_t(y_t)\mathbf{F}_t\mathbf{P}_t, \tag{5.2}$$

where $b_t(y_t)$ is the score function of the one-step forecast density $p(y_t|D_{t-1})$ with derivative $\mathbf{B}_t(y_t)$ and all other quantities are as previously defined. These posterior moments determine the prior moments for the next period by applying the system equation in the usual manner, but the form of the prior is assumed to be normal—an approximation unless the observation noise is also normal. It will be noted that this use of exact moments together with an imposed (approximate) distributional form is precisely the kind of procedure used in previous sections under the aegis of the guide relation. However, the Masreliez formulation is much more difficult to implement than any of the other procedures discussed in this chapter since the convolution of the observation noise density with the state prior density requires, except in the special case of normal mixtures already described, numerical integration techniques.

More recently, a very general class of nonnormal, nonlinear models has been used successfully in forecasting industrial time series. Once more the guide relationship is crucial in obtaining a tractable analysis. Using elements of the analysis of Sections 3 and 4, construct a model having the following probabilistic structure. The state vector at time $(t-1)$ is assumed to have a posterior distribution known only partially in terms of the first two central moments as

$$\boldsymbol{\theta}_{t-1}|D_{t-1} \sim [\mathbf{m}_{t-1}, \mathbf{C}_{t-1}]. \tag{5.3}$$

We apply to this an evolutionary process (in general nonlinear) as before such that the resulting prior for the state vector at time t has mean and covariance

$$\boldsymbol{\theta}_t|D_{t-1} \sim [\mathbf{a}_t, \mathbf{R}_t], \tag{5.4}$$

where $\mathbf{a}_t = \mathbf{G}_t\mathbf{m}_{t-1}$ (which is a linear approximation if the evolution is nonlinear) and $\mathbf{R}_t$ is obtained from $\mathbf{C}_{t-1}$ using the now familiar discount technique.

Now, recalling the structure of the model in Section 4, in particular equation (4.9), we let ψ be a parameter related to the state vector, our knowledge of the linkage being represented through the guide relationship $g(\cdot)$ such that we have the partial prior specification,

$$\left.\begin{matrix}\psi_t \\ \boldsymbol{\theta}_t\end{matrix}\right| D_{t-1} \sim \left[\begin{pmatrix} f_t \\ \mathbf{a}_t \end{pmatrix}, \begin{pmatrix} q_t & \mathbf{s}_t^T \\ \mathbf{s}_t & \mathbf{R}_t \end{pmatrix}\right]. \tag{5.5}$$

Here the full joint distribution is left unspecified; we assume only that the marginal prior for $\psi_t|D_{t-1}$ is given in any particular application. [Thus we could, if occasion demanded, still have a normal prior for ψ_t while retaining the more general specification for the marginal prior on $\boldsymbol{\theta}_t$ and the joint prior in (5.5).] In many cases we will take this, or a one-to-one transformation, as conjugate to the sampling density, although this is by no means necessary.

With the model defined in this way, the updating procedures for $\boldsymbol{\theta}_t$ and ψ_t follow along the same lines as for the DGLM in Section 4. The posterior distribution for $\psi_t|D_t$ is obtained directly from the sampling density for y_t and the prior for $\psi_t|D_{t-1}$ (in the conjugate case no numerical integrations are necessary). Then, for the state parameter $\boldsymbol{\theta}_t$, we update the moments using the linear Bayes methodology. The resulting recurrence relations for these quantities are precisely those given in Section 4, since the structure of the models is identical on replacing λ_t there with ψ_t here. Alternatively, recall the results of that section in the normal case. The properties of the multivariate normal distribution mean that the linear Bayes estimates, which are distributionally invariant, coincide with the actual moments in that case. Another derivation of these recurrences based on minimizing quadratic forms is given in Migon and Harrison (1983) and Migon (1983).

5.2 Application: Modeling the Impact of Advertising

Migon and Harrison (1983) gave an application of the general dynamic nonlinear model to advertising data. The objective of this work (undertaken jointly with a market research company) was to build a descriptive model relating television advertising to consumer awareness, reasons for which were listed as including the following:

1. The ability of reviewing previous advertising campaigns
2. A capability for correcting the tendency for many currently used pretesting methods to give a systematically misleading feedback to the creative department
3. For assisting in more rational advertising decisions
4. For developing improved relationships between product sales and advertising

Following previous studies in this area and after further discussions with market researchers, a set of properties required for a satisfactory modeling of the relationship was defined to include, minimally:

1. Diminishing return rate from increased advertising (a saturation effect)
2. Exponential decay in awareness (consumer memory loss)
3. Threshold awareness in the absence of TV advertising (general knowledge including the effects of advertising in other media)
4. Maximum level of awareness (the limit of the effect in property 1)

5. Sampling variation dependent on awareness level and sample size
6. General slow change in the quantities describing properties 1 to 5
7. Likelihood of a sudden marked change in awareness response resulting from a major change in advertisement

Furthermore, the market research company had a shopping list of additional facilities which they desired in the model including:

1. Missing observations capability
2. Means of including subjective information
3. Capacity for rapid response to signaled changes in advertising and for assessing such changes
4. "What if" facility for assessing model adequacy and enabling reason 3
5. Filtering facility for estimating underlying awareness based on data up to 5 weeks on

To begin the model construction process, a set of "sophisticatedly simple" mathematical models is postulated which encapsulate the features described in properties 1 to 5. For example, the hypothesis of diminishing returns in property 1 is modeled as

$$E = d[1 - \exp(-ka)], \tag{5.6}$$

where E is the effect of advertising; a the adstock, a measure of total advertising effort; and k and d parameters to calibrate the relationship suitably. These components are then combined to yield the complete dynamic specification as follows. With N_t people sampled in week t, the number of people aware, U_t, is modeled as a conditional binomial,

$$U_t|N_t, \psi_t \sim B(N_t, \psi_t), \tag{5.7}$$

where ψ_t is the stochastically perturbed weekly awareness (property 5). The deterministic component of ψ_t is modeled with a guiding relationship derived from the mathematical models as

$$p_t = \mu_t + d_t - [d_t - \lambda_t E_{t-1}] \exp(-k_t x_t), \tag{5.8}$$

where μ_t is the threshold awareness (property 3), $\mu_t + d_t$ the asymptotic awareness (property 4), λ_t the memory decay rate (property 5), x_t a measure of the addition to adstock in week t, p_t the proportional awareness in the population, and k_t the advertising response factor. The stochastic element in the guide relationship for ψ_t is necessary to allow for regional variation about the population awareness because the weekly sample is taken from a selected subset of TV regions.

The model is completed by a simple random walk for the parameter evolution chosen in accordance with the sixth property, and finally, property

7 is an inherent feature of the Bayesian sequential analysis. All the additional requirements of the market research company are standard features of the dynamic Bayesian models.

In Figure 1 the data on a fast-moving consumer product are recorded. The plot includes the observed weekly proportional consumer awareness and the one-step-ahead point predictions, the weekly TVRs (additions to adstock), and periods of missing observations, which correspond to no-awareness samples. The two most striking features of these data are the decay in awareness during periods of no advertising evident on two occasions, and the sharp nonlinear response to an advertising burst. Also clear is the speed with which the model forecasts respond to the advertising campaigns. The first advertisement was successful in generating consumer awareness; however, in

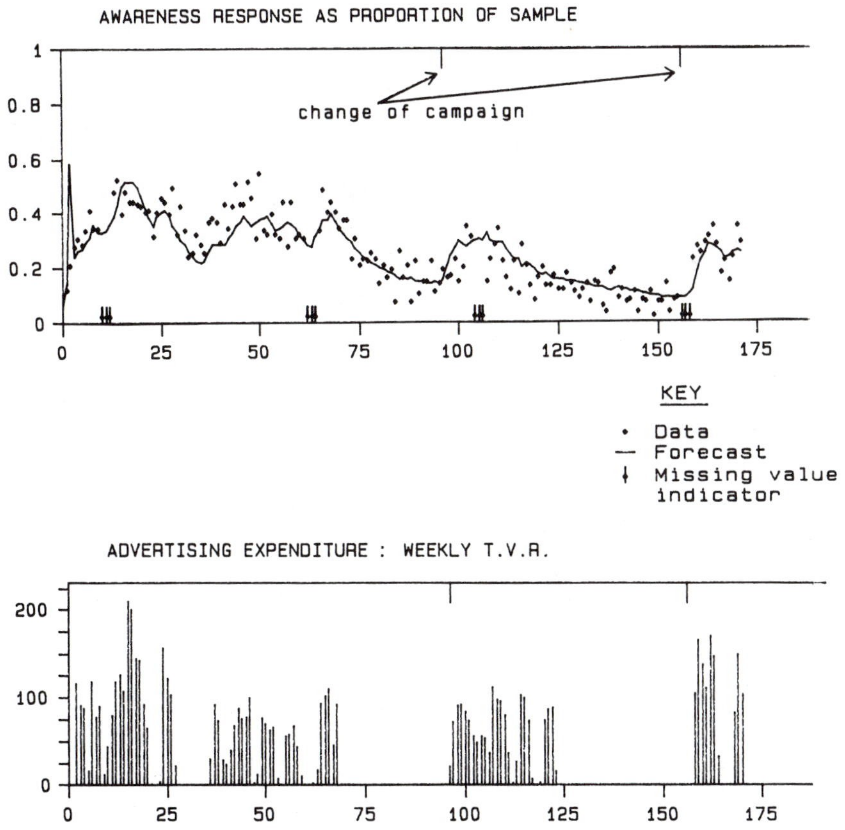

Figure 1. Advertising example: inputs, data, and one-step ahead forecasts.

week 96 it was replaced by a qualitatively different advertisement which unfortunately was soon assessed by the model to be relatively poor. In fact, the latter advertisement was estimated to require about five times as much exposure as the original in order to achieve the same increase in awareness response. In turn, this advertisement was replaced in week 158 by another qualitatively different advertisement that was estimated to improve the awareness response over its immediate predecessor by a factor of 3.

6. FORECAST MONITORING AND MODEL ADAPTATION

The basic underlying motivation for the construction and analysis of the models presented in this chapter is forecasting. The objective is to facilitate wise decision making by predicting future values of the observation series accurately and with as little uncertainty as possible. These requirements are usually subject to certain caveats, resource constraints being among the most obvious and commonly occurring. Under such circumstances it is important to be able to determine how well a given model is performing and to be able to react appropriately to this information. This is the subject of the remainder of the chapter.

The questions that immediately arise in response to these seemingly innocuous but practically formidable requirements are: How do we check on the performance of our model (and of course we are specifically concerned with forecast performance)? How can we allow the model to adapt in the light of changing circumstances (other than those specifically modeled by the dynamic features of the model)? More specifically, what we desire is an ability to recognize, and a facility for adapting appropriately to, situations in some way at odds with the standard model. The kinds of features of interest that occur in practice are outliers, abrupt or gradual parametric (i.e., structural) change, increased variation or uncertainty, and so on. To be more concrete, we consider the situation where we have developed a standard model from our arsenal of dynamic Bayesian models from which we require to produce a sequence of one-step-ahead forecasts through time. How, then, do we monitor the adequacy of these forecasts and adapt the model to perceived shortcomings?

In Harrison and Stevens (1976), the proposed solution for NDLMs was to set up a multistate model, that is, a collection of models in addition to the standard model, each allowing for the possibility of some specific kind of structural departure from the standard model. Forecasts are then in the form of discrete mixtures (of the individual model forecasts), where the mixture weights reflect the relative plausibility of the alternative models as measured by their posterior probabilities. This technique is essentially a dynamic application of the model elaboration technique familiar in robustness studies, and in this sense the mixture forecasts may be regarded as robust. Multistate procedures are also discussed in Willsky (1976). The references cited therein describe a number of different approximation techniques which may be used to overcome the problem of exponential increase in the number of submodels—recall the discussion of this issue in Section 5.

West (1986b) proposes a similar procedure for use in the nonnormal, nonlinear models of primary interest in this paper. Specifically, he considers the class of DGLMs (Section 4) and defines the class of alternative models for the nonstandard observations by an appropriate setting of the parameters of the prior distribution of the natural parameter, η_t, of the sampling density. This procedure has an interesting by-product in the normal case since the sampling density, conditional on η_t, does not depend on the state (outlier, standard, etc.) at any time, whereas in the original Harrison-Stevens formulation the sampling density for the outlier model is in the form of a contaminated normal. Apart from this, the scheme for the DGLM is conceptually identical with the Harrison-Stevens multistate scheme.

Multistate models have proved extremely effective in identifying and adapting to changes with a minimum loss in predictive performance. However, the approach does suffer from a number of important drawbacks which have hampered its implementation in some cases. Perhaps the most fundamental is the inherent increase in complexity occasioned by the need to specify a whole set of submodels in an attempt to protect forecasts from a range of possible exceptional and unforeseeable circumstances. The standard model is likely to be composed of a number of elements (e.g. the familiar trend/seasonal type models), and to allow for possible structural changes in each of these, a number of alternative models must be elaborated. This increased structural complexity proceeds in parallel with greatly increased parametrization, both of which make such models difficult to think about (and possibly interpret) and can lead to marked reluctance on the part of practitioners to accept them. Moreover, the implications of additional subjective information have to be considered carefully in respect to each submodel, and the latter have all to be fully updated as each new observation becomes available. The resulting heavy demand on computational resources raises yet more obstacles, especially in oft-occurring situations where many series have to be forecast simultaneously.

Ameen and Harrison (1985) developed a modified multistate procedure which has major advantages in circumstances where computational problems are to the fore. The scheme provides a monitor which explicitly signals periods when the standard model is failing to perform satisfactorily, and improved predictive performance is achieved as a consequence of the procedure adopted for constructing the alternative models. A similar scheme for nonnormal, nonlinear models has recently been developed in West (1986b).

In the Ameen-Harrison scheme the monitor is a cumulative sum (CUSUM) procedure originally developed for use in forecasting by Harrison and Davies (1964), which utilizes the one-step-ahead forecast errors. In essence, the idea is that for forecasts that are "close" to the realized values of the series, contributions to the CUSUM will be small. Thus poor forecast performance is flagged if the value of the CUSUM exceeds certain limits—which depend on the error structure of the model. It is applied in the following way. The standard model is run as usual until the monitor signals some kind of departure from the model as evidenced by accumulated poor predictive performance in the CUSUM. Thereupon the multistate system is invoked [with appropriately selected prior (sub) model probabilities] and this system continues in operation until the posterior probability of the standard model achieves some prespecified level. On achieving this threshold the CUSUM is reset; the standard model is reverted to and continues to be used until the next breakdown is signaled. Parameter distributions will, of course, have been updated during the multistate phase, and if marked persistent structural change has occurred, may be quite different from their values when the monitor was triggered. In addition, during the single-state phases the alternative models are kept in preparation for future multistate running. This does not, however, involve the full updating procedure used in the latter phases (which would imply only minimal computational advantages over the full-blown multistate system associated with not having to compute the forecast densities of the alternative models). The general principle is to derive for each alternative model the marginal prior for the block characterizing the change from the corresponding posterior but to take all other information from the posterior of the standard model. This aims at keeping the information on the other components as stable as possible in order to prepare good estimates of the change. Otherwise, the covariance structure between the components might produce violent fluctuations in the estimate of the presumed stable components. In addition, a set of model probabilities is calculated using Bayes' theorem, $P_t^{(i)} \propto L(y_t|M^{(i)})P_{t-1}^{(i)}$, in an obvious notation.

Such CUSUM schemes are generally inappropriate diagnostic tools for nonnormal models because of the intractability of the resulting statistics. However, Spall (1985) describes such a procedure that relies on asymptotic theory developed for nonnormal state-space models in Spall and Wall (1984). In this scheme the model is assumed to have the basic linear structure in (1.1) and (1.2), but some (or possibly all) of the noise distributions can be of arbitrary—even unknown—form. The procedure is a non-Bayesian standard hypothesis test applied to a statistic—whose precise form depends on information about possible model departures—composed of a sum of normalized Kalman filter estimates of the state.

Willsky (1976) reviews a variety of failure-detection procedures for dynamic systems. Only linear systems are considered, although it is stated that "the basic concepts, if not the detailed analyses, carry over to other classes of systems," and this will be seen in a particular case below. These approaches center on testing the appropriateness of the state-space representation of the system under consideration, and as such are not directly tailored to the monitoring of forecast performance. However, the two aspects are inextricably linked and procedures of both types are complementary in a complete forecasting and modeling system.

In two recent papers (West, 1986a; West and Harrison, 1986), an approach to forecast monitoring has been developed which is designed to address the model complexity issue raised by the multistate models. Essentially, forecasts from the standard model are assessed in comparison with those of a single alternative model and appropriate actions initiated in response. The concept of comparing just two models representing "fail" and "no-fail" which underlies this approach can be seen in an early paper on system failure detection (Newbold and Ho, 1968), which is also discussed in the aforementioned review in Willsky (1976). Comparison is made between the likelihood of the observation from the forecast distribution under the standard and alternative models; that is, we compute the Bayes factor (assuming equal prior probabilities at each stage). Thus small values of $p_S(y_{t+1}|D_t)$ are judged as potentially troublesome (requiring direct intervention) only if the corresponding value of $p_A(y_{t+1}|D_t)$ is suitably larger (where $p_S(y_{t+1}|D_t)$ and $p_A(y_{t+1}|D_t)$ are the observed values of the predictive densities under the standard and alternative models, respectively). Hence, by monitoring the value of the Bayes factor at each time, we can assess the adequacy of the predictive performance of the standard model with respect to that of the alternative. This alternative model is constructed sequentially such that a range of possible departures from the standard model can be represented while remaining impartial in respect of which of these kinds may actually have occurred. In this way we have an automatic procedure for monitoring forecast performance and signaling possible breakdowns, but needless to say, additional subjective information which may become available at any time is incorporated directly into the standard model in the usual manner (by, for example, appropriate adjustments to the prior moments of the state vector).

In the discussion of the WH monitoring scheme above, we have mentioned only the procedure for identifying possible structural change at any time on the basis of the current observation. More general methods for identifying changes that may not be apparent by these single observation comparisons, for example, a gradual structural shift greater than the evolutionary mechanism allows for but small enough not to trigger the monitor as described, have also been developed. West and Harrison (1986) provide operational details of these schemes and give illustrative examples. Further applications are examined in more detail in Harrison and West (1986), where these simple monitoring schemes are used to provide an effective and robust approach to time series decomposition and smoothing suitable for routine application in time series analysis.

Given a suitable monitoring procedure, the next requirement is to specify the actions to be taken when model failure is detected. The Bayesian formulation provides for the incorporation of any external information which may become available, but equally important is the action to be taken when no such information is available. Such circumstances call for an automatic model adaptation procedure, and one possible scheme is described below.

A simple and convenient procedure is to increase the uncertainty in the prior distribution as measured by the prior covariance matrix $\mathbf{R}_t$. Notice that a procedure of this sort can easily be implemented in any of the models described in this chapter; in particular, there are no complications arising from the use of guide relationships or partially specified prior distributions. In the absence of information about the precise type of change expected, a simple intervention to increase uncertainty about the entire state vector—by adjusting downward the component discount factors—will allow the model to adapt more rapidly to new data and hence identify the change. (It may be useful to recall at this point the discussion of model component stability and discount factors in Section 2.) Typically, these intervention discount factors are used only at the time of the monitor signal. Elaborations and operational aspects of this adaptation procedure are described with examples in West and Harrison (1986).

REFERENCES

Ameen, J. M., and P. J. Harrison (1985a). Normal Discount Bayesian Models, in *Bayesian Statistics 2* (J. M. Bernardo et al., eds.), North-Holland, Amsterdam, pp. 271–294.

Ameen, J. M., and P. J. Harrison (1985b). Discount Bayesian Multi-process Modeling with CUSUMS, in *Time Series Analysis: Theory and Practice 5* (O. D. Anderson, ed.), North-Holland, Amsterdam, pp. 117–134.

Breslow. A. N. (1972). “Comment” on Cox’s (1972) paper, *J. R. Stat. Soc. Ser, B* 34, 216–217.

Breslow, A. N. (1974). Covariance analysis of censored survival data, *Biometrics* 30, 89–99.

Brown, R. G. (1963). *Smoothing, Forecasting and Prediction of Discrete Time Series*, Prentice-Hall, Englewood Cliffs, N. J.

Canterellis, N., and F. R. Johnston (1983). On line variance estimation for the steady state Bayesian forecasting model, *J. Time Ser. Anal.* 3, 225–234.

Cox, D. R. (1972). Regression models and life tables (with discussion), *J. R. Stat. Soc. Ser. B* 34, 187–220.

Gamerman, G. (1986). *Dynamic Bayesian Models for Survival Data*, Warwick Research Report 75, University of Warwick, Coventry, England.

Gamerman, D., and M. West (1986). An Application of Dynamic Survival Models in Unemployment Studies, in *Proceedings of the 2nd. Conference on Practical Bayesian Statistics.*

Goldstein, M. (1976). Bayesian analysis of regression problems, *Biometrika* 63, 51–58.

Harrison, P. J., and O. L. Davies (1964). The use of cumulative sum (CUSUM) techniques for the control of routine forecasts of product demand, *J. Oper. Res. Soc. Am.* 12, 325–333.

Harrison, P. J., and C. F. Stevens (1976). Bayesian forecasting (with discussion), *J. R. Stat. Soc. Ser. B* 38, 205–247.

Harrison, P. J., and M. West (1986). Practical Bayesian Forecasting, in *Proceedings of the 2nd Conference on Practical Bayesian Statistics.*

Hartigan, J. A. (1969). Linear Bayesian methods, *J. R. Stat. Soc.* 31, 446–454.

Ho, Y. C., and R. C. K. Lee (1964). A Bayesian approach to problems in stochastic estimation and control, *IEEE Trans. Autom. Control* 9, 333–339.

Jazwinski, A. H. (1970). *Stochastic Processes and Filtering Theory*, Academic Press, New York.

Kalman, R. E. (1960). An approach to linear filtering and prediction theory, *Trans. ASME*, 82D, 35–45.

Kalman, R. E., and R. S. Bucy (1961). New results in linear filtering and prediction theory, *Trans. ASME*, 83D, 15–108.

Masreliez, C. J. (1975). Approximate non-Gaussian filtering with linear state and observation relations, *IEEE Trans. Autom. Control* 20, 107–110.

McCullagh, P., and J. A. Nelder (1983). *Generalized Linear Models*, Chapman & Hall, London.

Migon, H. S. (1983). An approach to non-linear Bayesian forecasting problems with applications, *Ph.D. thesis*, Warwick University.

Migon, H. S., and P. J. Harrison (1983). An Application of Non-linear Bayesian Forecasting to Television Advertising, in *Bayesian Statistics 2* (J. M. Bernardo et al. eds.), North-Holland, Amsterdam, pp. 681–696.

Narendranathan, W., S. Nickell, and J. Stern (1985). Unemployment benefits revisited, *Econ. J.* 95, 307–329.

Nelder, J. A., and R. W. M. Wedderburn (1972). Generalized linear models, *J. R. Stat. Soc. Ser. A* 135, 370–384.

Newbold, P. M., and Y. C. Ho (1968). Detection of changes in the characteristics of a Gauss-Markov process, *IEEE Trans. Autom. Control* 4, 707–718.

Sage, A. P., and J. L. Melsa (1971). *Systems Identification*, Academic Press, New York.

Smith, A. F. M., and M. West (1980). Monitoring renal transplants: an application of the multi-process Kalman filter, *Biometrics* 39, 867–878.

Sorenson, H. W., and D. L. Alspach (1971). Recursive Bayesian estimation using Gaussian sums, *Automatica* 7, 465–479.

Spall, J. C. (1985). Validation of state space models from a single realization of non-Gaussian measurements, *IEEE Trans. Autom. Control* 30, 1212–1214.

Spall, J. C., and K. D. Wall (1984). Asymptotic distribution theory for the Kalman filter state estimator, *Commun. Stat. Theory Methods* 13, 1981–2003.

West, M. (1985). Generalized Linear Models: Scale Parameters, Outlier Accomodation and Prior Distributions, in *Bayesian Statistics 2* (J. M. Bernardo et al., eds.), North-Holland, Amsterdam, pp. 531–558.

West, M. (1986a). Bayesian model monitoring, *J. R. Stat. Soc. Ser. B*. 48, 70–78.

West, M. (1986b). *Non-normal Multi-process Models*, Warwick Research Report 81, University of Warwick, Coventry, England.

West, M., and P. J. Harrison (1986). Monitoring and adaptation in Bayesian forecasting models, *J. Am. Stat. Assoc.* 81, 741–750.

West, M., P. J. Harrison, and H. S. Migon (1985). Dynamic generalized linear models and Bayesian forecasting (with discussion), *J. Am. Stat. Assoc.* 80, 73–97.

Willsky, A. S. (1976). A survey of design methods for failure detection in dynamic systems, *Automatica* 12, 601–611.

8

New Exact Nonlinear Filters

FREDERICK E. DAUM Raytheon Company, Wayland, Massachusetts

1. INTRODUCTION

In this chapter we survey the progress in nonlinear filtering theory over the last 10 years, including several new exact recursive filters for nonlinear and/or non-Gaussian problems. These new filters are generalizations of the Kalman filter. Kalman solved the linear filtering problem with Gaussian noise in 1960. The Kalman filter theory has been extremely useful in a wide variety of practical applications (see Sorenson, 1985, or Hutchinson, 1984). On the other hand, when the approximations of linearity and/or Gaussian noise are inadequate, the Kalman filter performs poorly. This has motivated the intensive search for nonlinear filters over the last quarter century (see Sorenson, 1974).

Unfortunately, the quest for nonlinear filters has proven to be very difficult. Most published results focus on approximate solutions to nonlinear or non-Gaussian problems (e.g., Wishner et al., 1969; Schwartz and Stear, 1968; Alspach and Sorenson, 1972; Willman, 1981; Fisher, 1967). These approximations work well in some applications but are unsatisfactory for others. No general theory provides error bounds on the performance of these approximate filters. Researchers have little theoretical insight into why certain approximations work well sometimes but not at other times.

Rather than attempting to approximate the solution to nonlinear or non-Gaussian problems, one might try to find exact recursive filters for certain special problems. This approach has been pursued vigorously over the last 10 years. In a seminal paper, Beneš (1981) derived an exact recursive filter for nonlinear problems that satisfy certain special conditions. The Beneš filter has stimulated considerable research with the goal of discovering other exact recursive nonlinear filters (e.g., see Mitter, 1983; Baras, 1980; Ocone et al., 1982; Wong, 1983; Beneš, 1985). This work has used Lie algebraic methods and has assumed that the measurements are made continuously in time. In contrast, no Lie algebraic theory is used in this chapter. Instead, the results are obtained by direct manipulation of the Fokker-Planck equation and Bayes' rule. Thus the theory is accessible to a wider audience, considering that Lie algebras are not the stock in trade of many statisticians, econometricians, or engineers. Some interesting connections between the methods in this paper and Lie algebras are discussed in Section 7.

Figure 1 shows the relationship between the Kalman filter, the Beneš (1981) filter, and the new filters discussed in this paper. The most general filter shown in Figure 1 (Daum, 1986a) is based on the exponential family of probability densities. The exponential family was identified by Fisher (1934) as a distinguished class of densities. In particular, for smooth nowhere-vanishing densities, the exponential family is the only class that has a sufficient statistic with fixed finite dimension. This basic result, known as the Fisher-Darmois-Koopman-Pitman theorem, has been generalized and proved with increasing rigor by many researchers (e.g., see Barankin and Maitra, 1963; Daum, 1986d). The original F-D-K-P theorem was developed in the context of classical statistics. In fact, the remarks of Koopman (1936) are explicitly anti-Bayesian. It is ironic, therefore, that this fundamental result from classical statistics should be so useful in the Bayesian context of this paper. Sawitzki (1979) should be given credit for connecting the exponential family of densities and the Bayesian filtering problem.

The practical significance of a fixed finite-dimensional sufficient statistic is that the storage requirements and computational complexity do not grow as more and more measurements are accumulated. Roughly speaking, if a filtering problem does not have a sufficient statistic with fixed finite dimension, its computational complexity is equivalent to solving a parabolic partial differential equation over R^n in real time. This is generally not possible for practical applications, even with the most advanced parallel computers available in the foreseeable future.

As shown in Figure 1, the Kalman filter is a special case of the general theory in this paper. This is because the Kalman filter is based on the Gaussian probability density, and the Gaussian density is a special case of the exponential family. Moreover, most of the "named" probability densities are from exponential families, because these are the most useful densities for estimation and decision theory, owing to the existence of sufficient statistics with fixed finite dimension. The exponential family is a basic assumption in

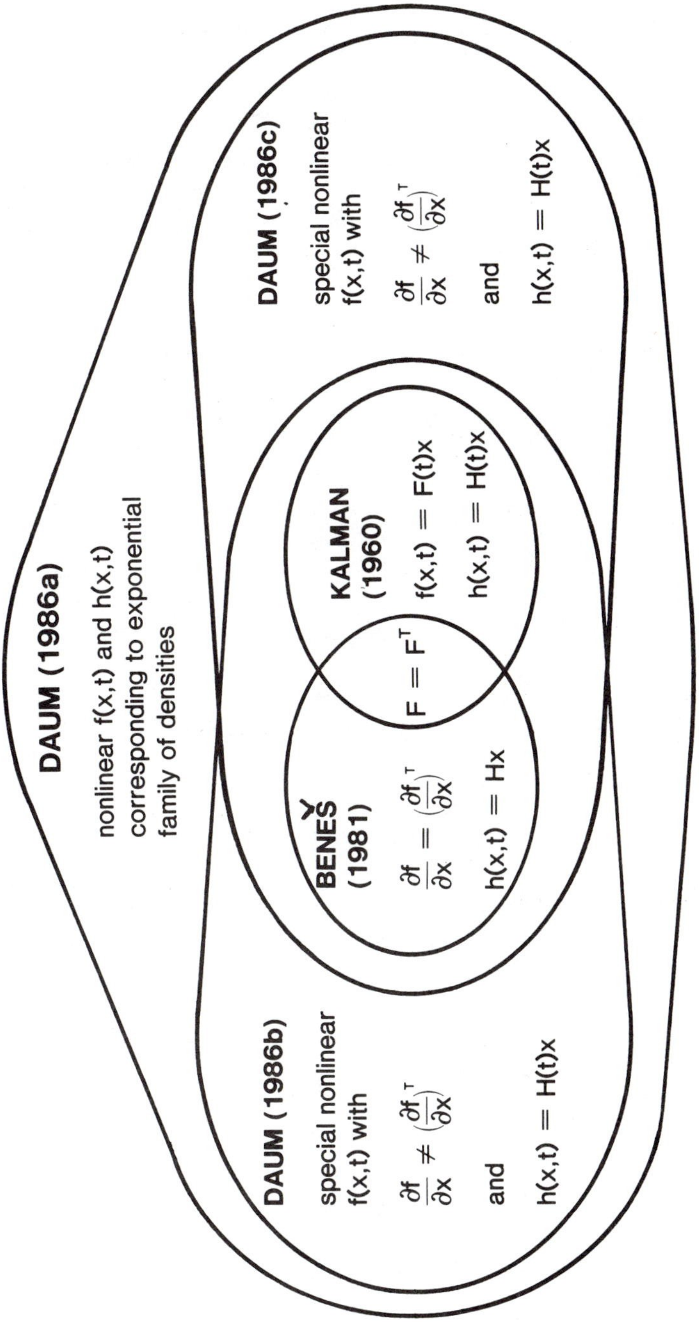

Dynamics: $dx = f(x,t)\,dt + G(t)\,dw$

Measurements: $z(t_k) = h(x(t_k), t_k) + v_k$

Figure 1. Comparison of exact nonlinear filters.

the estimation of generalized linear models of Nelder and Wedderburn (1972), as well as the more recent work of West et al. (1985).

The remainder of this chapter is organized as follows: The mathematical statement of the Bayesian filtering problem is given in Section 2. The general methodology for solving this problem is explained in Section 3, which includes a discussion of the heuristics that were used to discover these new filters. The main results are given in Section 4. Several new approximation methods are suggested in Section 5. An application of the new theory is given in Section 6, and a fundamental flaw in the extended Kalman filter is noted. This flaw is not shared by the new theory presented here. In Section 7 the Lie algebraic approach is compared to the methods in this chapter. The results in this chapter assume that the probability densities are smooth and nowhere vanishing. If this is not the case, it is still possible to find exact recursive filters within the family of "generalized exponential" densities. Results along these lines are noted in Section 8.

The mathematical tools used in this chapter are very elementary: Bayes' rule and the Fokker-Planck equation. The basic facts required are summarized on pages 163–165 in Jazwinski (1970). A brief tutorial on the F-P equation is given in the Appendix to this chapter. The Bayesian approach to filtering problems is explained in Ho and Lee (1964). Sorenson (1974) provides an excellent survey of the nonlinear filtering problem. Further background material on nonlinear filtering is given by Bucy (1970), Frost and Kailath (1971), Davis and Marcus (1981), and Marcus (1984).

2. BAYESIAN FILTERING PROBLEM

Consider the problem of estimating a R^n-valued random variable, $x(t)$, that evolves in continuous time according to the Itô stochastic differential equation

$$dx(t) = f(x(t), t)\, dt + G(t)\, dw \tag{2.1}$$

given a sequence of R^m-valued discrete-time measurements

$$z_k = g(x(t_k), t_k, v_k) \tag{2.2}$$

in which v_k is an R^m-valued random variable independent of $\{w(t)\}$ and $x(t_0)$. The process noise, $w(t)$, is a R^n-valued Brownian motion independent of $x(t_0)$ with $E(dw\, dw^T) = I\, dt$. It is assumed that the measurement noise, v_k, has statistically independent values at distinct points in time. The probability density of z_k conditioned on $x_k = x(t_k)$ is assumed to be given, and it is denoted by $p(z_k|x_k)$ for $k = 1, 2, \ldots, N$. For example, if $z_k = h(x(t_k), t_k) + v_k$, and v_k is a Gaussian zero-mean random variable, then

$$p(z_k|x_k) = c_k \exp\left[-\tfrac{1}{2}(z_k - h(x_k, t_k))^T R_k^{-1}(z_k - h(x_k, t_k))\right], \tag{2.3}$$

where c_k is a positive real number and $R_k = E(v_k v_k^T)$ is assumed to be a given positive definite $m \times m$ matrix. It should be emphasized, however, that $p(z_k|x_k)$ is not required to be Gaussian in general.

The Itô equation (2.1) corresponds to the following formal differential equation:

$$\frac{dx}{dt} = f(x(t), t) + G(t)\frac{dw}{dt},$$

in which dw/dt is often interpreted as Gaussian zero-mean white noise. The Itô equation (2.1) is shorthand notation for an integral equation which avoids the embarrassment that dw/dt does not exist. Tutorial information on Itô equations can be found in Jazwinski (1970).

ASSUMPTION 1 The random variable, $x(t)$, has a nowhere-vanishing unconditional probability density, denoted by $p(x, t)$, that is twice continuously differentiable in x and continuously differentiable in t. Furthermore, $p(x, t)$ approaches zero sufficiently fast as $\|x\| \to \infty$, so that $p(x, t)$ satisfies the Fokker-Planck equation corresponding to (2.1).

ASSUMPTION 2 The density $p(z_k|x_k)$ is nowhere vanishing, and it is twice continuously differentiable in x_k and z_k.

ASSUMPTION 3 For a given initial condition, $p(x, t_k)$, the Fokker-Planck equation corresponding to (2.1) has a unique bounded solution for all x and $t_k \leqslant t \leqslant t_{k+1}$.

DEFINITION The sigma algebra generated by the set of all measurements up to and including time t_k is denoted by $Z_k = \{z_1, z_2, \ldots, z_k\}$.

DEFINITION Let $p(x, t|Z_k)$ denote the probability density of $x(t)$ conditioned on Z_k.

DEFINITION The conditional density is said to belong to an "exponential family" if and only if it can be represented in the form

$$p(x, t|Z_k) = a(x, t)b(Z_k, t) \exp\left[\theta^T(x, t)\phi(Z_k, t)\right],$$

in which $\theta(x, t)$ is a function of x and t, but it is not a function of Z_k; moreover, $\phi(Z_k, t)$ is a function of Z_k and t, but it is not a function of x. Similarly, $a(\cdot)$ and $b(\cdot)$ are nonnegative scalar-valued functions of the indicated arguments.

The goal of nonlinear filtering is to compute the conditional density $p(x, t|Z_k)$ for $t = t_k$. Given this conditional density, any estimate of $x(t)$ can be computed, at least in principle. For example, one might compute the conditional mean of $x(t)$:

$$\hat{x}(t) = E(x(t)|Z_k) = \int xp(x, t|Z_k)\,dx. \tag{2.4}$$

In certain special cases, this n-fold integral for $x(t)$ can be evaluated in closed form (e.g., see Daum, 1986c). However, there is no general theory that predicts when this is possible. In a given practical application, a symbol manipulation language (e.g., MACSYMA or REDUCE) could be used to explore this possibility. Numerical evaluation of this n-fold integral in real time is infeasible for most practical problems, unless $p(x, t|Z_k)$ has a very special structure (see Schwartz, 1985; Davis and Rabinowitz, 1984). The computational complexity of evaluating general n-fold integrals numerically in real time is well beyond the state of the art, even using the most advanced VLSI parallel processing for the foreseeable future.

As an alternative, one could compute the maximum a posteriori estimate of $x(t)$ from $p(x, t|Z_k)$. In certain special cases, this might be done in closed form, but in general numerical techniques would be required. For example, the conjugate gradient algorithm is a numerical method that is feasible for almost all practical problems, and it is ideally suited to modern VLSI parallel processing technology (e.g., see McBryan, 1986). Good references on numerical methods for unconstrained nonlinear optimization are Sorenson (1980), Press et al. (1986), and Powell (1986). It turns out that the exponential family has a number of nice convergence and stability properties using modified Levenberg algorithms, as reported by Osborne (1987).

In this chapter the random process, $x(t)$, evolves in continuous time, but the measurements are made in discrete time. This "hybrid" setup contrasts with most filtering theory, which deals with continuous-time random processes and continuous-time measurements, or else discrete-time random processes and discrete-time measurements. A notable exception to the norm is the book by Jazwinski (1970). The reasons for this hybrid approach are threefold. First, most physical processes are conveniently modeled in terms of differential equations (object motion, electrical circuits, chemical processes, nuclear reactions, biological phenomena, etc.). Second, the continuous-time model of $x(t)$ allows the use of the elegant formalism of the Fokker-Planck equation. Third, in practical applications, filtering algorithms are implemented with digital computers, owing to their superior flexibility, accuracy, and dynamic range as compared to currently available analog devices. The use of a digital computer requires that the measurements be made in discrete time. These points are reinforced in Hutchinson's (1984) survey of Kalman filter applications.

Many researchers prefer to work with continuous-time measurement models ($dz = h(x, t)\,dt + dv$). For example, the entire Lie algebraic theory of nonlinear filtering is based on this model (e.g., Mitter, 1983; Brockett, 1981; Beneš, 1985; Davis and Marcus, 1981). New exact nonlinear filters with continuous-time observations are reported in Daum (1985b, 1987a).

3. METHODOLOGY AND HEURISTICS

The filtering problem defined in Section 2 is hopeless; it is much too general to solve. An acceptable solution for a practitioner is some expression for $p(x, t|Z_k)$ in terms of a fixed finite number of parameters. In the terminology of classical statistics, this means that $p(x, t|Z_k)$ admits a "sufficient statistic" with a fixed finite dimension. The dimension is "fixed" in the following sense: As more and more measurements, z_k, are accumulated, the number of parameters required to characterize $p(x, t|Z_k)$ remains fixed. More precisely, $\psi = \psi(Z_k)$ is said to be a "sufficient statistic" if and only if $p(x, t|Z_k) = p(x, t|\psi_k)$ for all x, t, and Z_k. In other words, knowing ψ_k is just as good as knowing Z_k. If ψ_k has a fixed finite dimension as the dimension of Z_k grows, $p(x, t|Z_k)$ must be a very degenerate function! Yet this is exactly what happens in the Kalman filter problem. The sufficient statistic in this special case is the n-dimensional conditional mean vector of $x(t)$ and the $n \times n$ error covariance matrix; the dimension is therefore fixed at $n + n(n + 1)/2$ for all values of k. In summary, filtering problems with fixed finite-dimensional sufficient statistics are, in some vague intuitive sense, extremely rare, but if such problems can be identified and solved, the reward is very great.

The basic lesson from this discussion is that the central task of nonlinear filter theory is to *discover* what class of problems have recursive finite-dimensional solutions. Moreover, the process of "discovery" must be done in some systematic fashion; otherwise, the chance of success is rather small. With this idea in mind, the following strategy seems sensible: Start with a function $p(x, t|Z_k)$ that is known to have a sufficient statistic with fixed finite dimension, and characterize the class of filtering problems for which this is the solution. This sounds like cheating, and of course it is, but it works, and therefore it will be adopted without hesitation. The critical reader will realize that this method is standard in applied mathematics, and thus no claim for originality is made here.

Table 1 summarizes the results of the program outlined above. The column labeled $p(x, t|Z_k)$ gives the assumed form for the unnormalized conditional density. This is the starting point for the process of discovery. The column labeled "filter" lists the exact nonlinear recursive filter that results from each $p(x, t|Z_k)$. The first entry in Table 1 is the Kalman filter. The

Table 1 Conditional Densities for Exact Recursive Filters

Conditional density, $p(x, t \mid Z_k)$	Filter
1. η = Gaussian	Kalman (1960)
2. $\eta \exp\left[\int^{x} f(x)\, dx\right]$	Beneš (1981)
3. $\eta P_{ss}^{\alpha}(x)$	Daum (1986c)
4. $\eta q^{\alpha}(x, t)$	Daum (1986c)
5. $\eta Q(x, t)$	Daum (1986b)
6. $p(x, t) \exp[\theta^{T}(x, t)\psi(Z_k, t)]$	Daum (1986a)

unnormalized Gaussian density has the form

$$\eta = \exp\left[-\tfrac{1}{2}(x - m)^{T} P^{-1}(x - m)\right], \tag{3.1}$$

in which m is the conditional mean of x and P is the $n \times n$ error covariance matrix of x. The next entry is the Beneš (1981) filter. Here η has the form defined by (3.1), but m and P are *not* necessarily the mean and covariance matrix of x. Nevertheless, the sufficient statistic for the Beneš filter is still (m, P), which has fixed finite dimension $n + n(n + 1)/2$, exactly as in the case of the Kalman filter. The symbol η has this more general meaning for all the remaining entries in Table 1. For the symbol $\exp[\int^{x} f(x)\, dx]$ to make sense, the R^{n}-valued function $f = f(x)$ must be the gradient of a scalar-valued function, and this is one of the requirements for the Beneš theory.

The next filter in Table 1, Daum (1986c), was discovered using the following heuristics. Focus attention on the form of $p(x, t \mid Z_k)$ for the Beneš filter, and ask the question: What is the significance of the factor $\exp[\int_{0}^{x} f(x)\, dx]$? After some reflection, it is apparent that $\exp[\int_{0}^{x} f(x)\, dx]$ is the square root of the unnormalized steady-state density of x, denoted by $P_{ss}(x)$, in the special case that $f = f(x)$ is the gradient of a scalar-valued function and $dx = f(x)\, dt + dw$. Next, ask whether this interesting factor could be replaced by $\sqrt{P_{ss}(x)}$ even if $f(x)$ is not the gradient of a scalar-valued function. It turns out that this is indeed possible. Next, consider the possibility that $\sqrt{P_{ss}(x)}$ could be replaced by $P_{ss}^{\alpha}(x)$ in which $0 < \alpha < 1$. This series of heuristics leads to the third filter in Table 1.

The fourth filter was discovered using a similar series of heuristics. Ask what is special about the steady-state density of x as compared to the transient density, $p(x, t)$. This results in $p(x, t \mid Z_k) = \eta p^{\alpha}(x, t)$. Next, ask what is special about the initial condition $p = p(x, t_0)$ for the Fokker-Planck equation compared to a more general initial condition. This results in $p(x, t|Z_k) = \eta q^{\alpha}(x, t)$, in which $q(x, t)$ is the solution of the Fokker-Planck equation corresponding to (2.1), but without requiring that $q(x, t_0) = p(x, t_0)$.

The fifth filter in Table 1, Daum (1986b), was discovered by asking the following question. What is special about $q^{\alpha}(x, t)$ compared to some more general scalar-valued function $Q = Q(x, t)$?

So far, all the filters in Table 1 have a sufficient statistic with fixed finite dimension of precisely $n + n(n + 1)/2$, exactly as in the Kalman filter. The Gaussian factor, η, is the only part of $p(x, t|Z_k)$ that depends on Z_k. The other factor was allowed only to depend on x and t. All these filters require that the measurement vector has the special form: $z_k = H_k x_k + v_k$, in which v_k is a Gaussian zero-mean random variable independent of x_k. This is precisely the same assumption as in the Kalman filter theory.

The final filter in Table 1, Daum (1986a), does *not* require that z_k is linear in $x(t_k)$ and that v_k is Gaussian. In fact, the only specific restriction on z_k is that $p(z_k|x_k)$ is a member of an exponential family of densities that is compatible, in some sense, with the dynamics (2.1). It turns out that this $p(x, t|Z_k)$ is the most general form for a conditional density from an exponential family. The sufficient statistic is $\psi = \psi(Z_k, t)$, and its dimension is fixed and finite, but it is not necessarily $n + n(n + 1)/2$. The motivation for considering this exponential family is twofold: First, all the examples in Table 1 so far have been from exponential families. Second, the Fisher-Darmois-Koopman-Pitman theorem, suitably generalized for random processes, says that this is the most general class of smooth nowhere-vanishing conditional densities that have fixed finite-dimensional sufficient statistics (Daum, 1986d).

Table 1 lists the conditional densities that correspond to the Kalman filter and five new nonlinear filters. These densities are all guaranteed to admit some kind of sufficient statistic with fixed finite dimension. This is true by construction. For the first five filters in Table 1, the sufficient statistic is (m, P), whereas for the sixth filter, the sufficient statistic is ψ. The only remaining question is: What class of nonlinear problems are solved by these densities? In other words, what class of nonlinear dynamics, $f(x, t)$, and measurement densities, $p(z_k|x_k)$, correspond to the conditional densities in Table 1? This question can be answered by substituting a given density into the Fokker-Planck equation and finding under what conditions this provides a solution. The Fokker-Planck (F-P) equation is a linear second-order parabolic partial differential equation (PDE) that governs the evolution of *both* the conditional density, $p = p(x, t|Z_k)$, and the unconditional density, $p = p(x, t)$. The F-P equation corresponding to the dynamics (2.1) is

$$\frac{\partial p}{\partial t} = -\frac{\partial p}{\partial x} f - p \operatorname{tr}\left(\frac{\partial f}{\partial x}\right) + \frac{1}{2} \operatorname{tr}\left(\mathrm{GG}^T \frac{\partial^2 p}{\partial x^2}\right) \tag{3.2}$$

Tutorial material on the F-P equation is given in the Appendix. Starting with the trial solutions in Table 1, this PDE can be solved using the classical idea of "separation of variables" (e.g., see Koornwinder, 1980). According to Slater and Frank (1933): "It is a method for reducing the partial differential equation to a set of ordinary [differential] equations, and we shall find it very useful. In fact, it is so valuable that practically the only partial differential equations which can be solved at all are those for which this method can be used."

Separation of variables to solve the F-P equation is a very natural approach here. In particular, the exponential family of densities is in precisely the right form for separation of variables! It is surprising that the connection between separation of variables and the exponential family of densities has not been exploited previously. There is an interesting relationship between separation of variables and the Lie algebraic approach to nonlinear filtering discussed in Section 7. Separation of variables is a standard technique in modern control theory. For example, it is used to derive the matrix Riccati equation from the Hamilton-Jacobi-Bellman PDE for the optimal control of a linear plant with quadratic performance criterion.

4. MAIN RESULTS

Table 1 lists the Kalman filter, along with five new exact nonlinear recursive filters. However, Table 1 does not describe the class of nonlinearities, $f = f(x, t)$, that result in exact solutions. For the first five filters, the class of nonlinearities is defined in Table 2. The sixth filter in Table 1 is much more general than the others, and its description is given later in this section. As explained in Section 3, all the filters listed in Table 2 require that $z_k = H(t_k)x(t_k) + v_k$, in which v_k is Gaussian zero-mean measurement noise independent of x_k, whereas the sixth filter in Table 1 allows a more general model of the measurements.

In Table 2 the various matrices and vectors (A, b, c, D, E) are allowed to be functions of time, but they cannot be functions of x. The condition given for the Kalman filter, $\partial f/\partial x = A(t)$, is merely another way of expressing the requirement that $f(x, t)$ must be a linear function of x: $f(x, t) = A(t)x + b(t)$. The condition on $f(x, t)$ is written in this form to emphasize its similarity to the other conditions listed in Table 2.

The first condition on $f(x)$ for the Beneš filter is another way of requiring that $f = f(x)$ is the gradient of some scalar-valued function of x. The second condition on $f(x)$ for the Beneš filter is obviously satisfied for any linear $f(x)$, but it is also satisfied by many nonlinear functions. For example, $f(x) = \tanh x$ satisfies these conditions. Other specific examples are given in Beneš (1981) and Daum (1984). If the first condition on $f(x)$ for the Beneš

Table 2 Class of Dynamics for Exact Recursive Filters

Filter	Class of dynamics
1. Kalman (1960)	$\dfrac{\partial f}{\partial x} = A(t)$
2. Beneš (1981)	$\dfrac{\partial f}{\partial x} = \left(\dfrac{\partial f}{\partial x}\right)^T$ and $\|f(x)\|^2 + \operatorname{tr}\left(\dfrac{\partial f}{\partial x}\right) = x^T A x + b^T x + c$
3. Daum (1986c)	$f - \alpha Q r^T = Dx + E$ and $\operatorname{tr}\left(\dfrac{\partial f}{\partial x}\right) + \dfrac{\alpha}{2} r Q r^T = x^T A x + b^T x + c$ where $r = \dfrac{\partial}{\partial x} \log P_{ss}(x)$
4. Daum (1986c)	Same as filter 3, but with $r = \dfrac{\partial}{\partial x} \log q(x, t)$
5. Daum (1986b)	$\dfrac{\partial f}{\partial x} - \left(\dfrac{\partial f}{\partial x}\right)^T = D^T - D$ and $\dfrac{\partial f}{\partial t} + \dot{D}x + \dot{E} = -\dfrac{\partial f^T}{\partial x} f - \dfrac{1}{2}\left[\dfrac{\partial}{\partial x} \operatorname{tr}\left(\dfrac{\partial f}{\partial x}\right)^T + (2A + D^T D)x + D^T E + b\right.$
6. Daum (1986a)	See Theorems 4.1 and 4.2

filter is satisfied, and if $f(x)$ is linear in x, $f(x) = Ax$, this implies that $A = A^T$. Therefore, the eigenvalues of A must be real. This is a rather severe restriction on the class of linear dynamics that the Beneš filter can handle. In summary, the Beneš filter can solve certain nonlinear problems that the Kalman filter cannot handle, but the Kalman filter can solve many linear problems that the Beneš filter cannot. This situation is depicted in the Venn diagram (Figure 1).

Intuitively, one expects that there should be a single filter that solves all linear problems as well as the nonlinear Beneš problems. This is indeed the case, as shown in Figure 1. It is easy to verify that filters 3, 4, and 5 in Tables 1 and 2 each solve all the Kalman and Beneš problems. In addition, these filters solve a wide class of nonlinear problems that neither the Kalman nor Beneš filter can solve. For specific examples, see Daum (1986b,c). The relationship between all of these filters is summarized in the Venn diagram (Figure 1).

So far, the conditional densities for these new filters have been described (Table 1), and the specific dynamics that admit these densities are listed in Table 2, but the filters themselves have not been displayed. In the present context, a "filter" is an algorithm for recursively propagating a fixed finite-dimensional sufficient statistic. These algorithms are summarized in Table 3. In particular, Table 3 shows the differential equations that govern the evolution of m and P between measurement times for filters 1 to 5 in Tables 1 and 2. Recall that m is R^n-valued and P is a $n \times n$ matrix for all these filters. The sufficient statistic for the sixth filter is the M-dimensional vector ψ, and it propagates between measurements as shown in Table 3.

When a measurement is made at time t_k, the values of m and P, for the first five filters, are all updated with the same equations:

$$m_k = \bar{m}_k + \bar{P}_k H_k^T R_k^{-1}(z_k - H_k \bar{m}_k) \tag{4.1}$$

$$P_k = \bar{P}_k - \bar{P}_k H_k^T (H_k \bar{P}_k H_k^T + R_k)^{-1} H_k \bar{P}_k, \tag{4.2}$$

in which (m_k, P_k) is the value of the sufficient statistic immediately after the kth measurement, and $(\bar{m}_k, \bar{P}_k)$ is the value immediately before the measurement at time t_k. The values of $(\bar{m}_k, \bar{P}_k)$ are obtained by integrating the ordinary differential equations given in Table 3 over the interval $t_{k-1} \leqslant t \leqslant t_k$. The initial conditions for these ordinary differential equations

Table 3 Propagation of Sufficient Statistics Between Measurement Times

Filter	Propagation equations
1. Kalman (1960)	$\dot{m} = Am$ $\dot{P} = AP + PA^T + GG^T$
2. Beneš (1981)	$\dot{m} = -PAm - \frac{1}{2}Pb$ $\dot{P} = I - PAP$
3. Daum (1986c)	$\dot{m} = 2(\alpha - 1)PAm + Dm + (\alpha - 1)Pb + E$ $\dot{P} = 2(\alpha - 1)PAP + DP + PD^T + Q$
4. Daum (1986c)	Same as filter 3
5. Daum (1986b)	$\dot{m} = -(2PA + D)m - E - Pb$ $\dot{P} = -2PAP - PD^T - DP + I$
6. Daum (1986a)	$\dot{\psi} = A^T\psi + \Gamma$ where $\Gamma = (\Gamma_1, \Gamma_2, \ldots, \Gamma_M)^T$ with $\Gamma_j = \psi^T B_j \psi$

(ODEs) are obtained from (4.1) and (4.2) at the time of the previous measurement.

Inspection of Table 3 shows that filters 3 and 4 degenerate to filter 2 in the special case of $D = 0$, $E = 0$, $Q = I$, and $\alpha = \frac{1}{2}$. This is exactly what one should expect from the dynamics defined in Table 2 (see Daum, 1986c). Also, note that for the Kalman filter, m is the conditional mean of x, and P is the error covariance matrix, but (m, P) does *not* have this significance for the other filters.

The remainder of this section will be devoted to the last filter listed in Tables 1 to 3 (Daum, 1986a). As shown in the Venn diagram (Figure 1), this filter includes all the others as special cases.

THEOREM 4.1 Consider the Bayesian filtering problem stated in Section 2. Suppose that Assumptions 1 to 3 are satisfied. Also, suppose that there exists a sufficient statistic of fixed finite dimension M for this problem, and that $p(x, t|Z_k)$ is from an exponential family of densities. Under these conditions, the exact unnormalized probability density of $x(t)$ conditioned on Z_k can be represented as

$$p(x, t|Z_k) = p(x, t) \exp [\theta^T(x, t)\psi(t)], \tag{4.3}$$

in which $\psi(t)$ is an R^M-valued sufficient statistic that can be computed recursively as follows. There exists a $M \times M$ matrix-valued function of time, $A = A(t)$, and a set of $M \times M$ matrix-valued functions of time, $B_j = B_j(t)$ for $j = 1, 2, \ldots, M$, such that $\psi(t)$ satisfies the following ODE between measurements $(t_k \leqslant t \leqslant t_{1+1})$:

$$\frac{d\psi(t)}{dt} = A^T(t)\psi(t) + \Gamma(t), \tag{4.4}$$

in which $\Gamma = (\Gamma_1, \ldots, \Gamma_M)^T$ where $\Gamma_j = \psi^T B_j \psi$ for $j = 1, 2, \ldots, M$. The initial condition for (4.4) at time t_k is

$$\psi(t_k) = \bar{\psi}(t_k) + \left[\frac{\partial}{\partial \theta} \log p(z_k|x)\right]^T, \tag{4.5}$$

in which $\bar{\psi}(t_k)$ is the value of ψ immediately before a measurement at time t_k, and $\psi(t_k)$ is the value of ψ immediately after the measurement z_k at time t_k. The value of $\bar{\psi}(t_k)$ is the solution of (4.4) at time t_k as a result of integrating (4.4) over the interval $t_{k-1} \leqslant t \leqslant t_k$. If there were no previous measurements, the initial condition for (4.5) is $\bar{\psi}(t_1) = 0$.

Furthermore, under the conditions assumed above, the following set of PDEs has a solution for $\theta = \theta(x, t)$ such that condition (4.8) is satisfied for some R^M-valued function $c = c(z_k, t_k)$ for all z_k. In condition (4.8), x_0 is a fixed but arbitrary point in R^n.

$$\frac{\partial \theta}{\partial t} = \frac{\partial \theta}{\partial x}(Qr^T - f) + \tfrac{1}{2}\xi - A\theta \tag{4.6}$$

$$\frac{1}{2}\frac{\partial \theta}{\partial x} Q \left(\frac{\partial \theta}{\partial x}\right)^T = \sum_{j=1}^{M} \theta_j B_j \tag{4.7}$$

$$\log \frac{p(z_k|x)}{p(z_k|x_0)} = c^T(z_k, t_k)\theta(x, t_k), \tag{4.8}$$

in which

$$r = r(x, t) = \frac{\partial p(x, t)}{\partial x}\bigg/ p(x, t), \qquad \xi_j = \text{tr}\left(Q \frac{\partial^2 \theta_j}{\partial x^2}\right), \qquad \xi = (\xi_1, \ldots, \xi_M)^T,$$

$$\theta = (\theta_1(x, t), \ldots, \theta_M(x, t))^T \qquad Q = GG^T.$$

Proof. The basic idea for this proof is to substitute the exponential density, $p(x, t|Z_k)$, into the Fokker-Planck equation, and solve the resulting PDE by the method of "separation of variables." Further details are given in Daum (1985a, 1988).

THEOREM 4.2 Consider the Bayesian filtering problem stated in Section 2, and suppose that assumptions 1 to 3 are satisfied. Assume that there exists a $M \times M$ matrix-valued function of time, $A = A(t)$, and a set of $M \times M$ matrix-valued functions of time, $B_j = B_j(t)$ for $j = 1, 2, \ldots, M$, such that the PDEs (4.6)–(4.7) have a solution, $\theta = \theta(x, t)$ that satisfies (4.8) for some R^M-valued function $c = c(z_k, t_k)$ for all z_k. Furthermore, suppose that there exists a R^M-valued solution, $\psi(t)$, to (4.4), such that $p(x, t) \exp[\theta^T(x, t)\psi(t)]$ is finite for all x and t. Under these conditions, the estimation problem has a R^M-valued sufficient statistic, $\psi(t)$, and the unnormalized conditional density is given by (4.3).

Proof. This is a direct verification using the Fokker-Planck equation and Bayes' rule. Mathematical induction is used to prove that $p(x, t|Z_k)$ maintains the form (4.3) from one measurement time to the next, similar to the proof in Daum (1986c).

REMARK 1 The practical significance of these equations is that the estimation problem is factored into two parts: (1) the off-line calculation of $\theta(x, t)$ and $p(x, t)$ which do not depend on the measurements Z_k, and (2) the on-line calculation of $\psi(t)$ via (4.4) and (4.5), which depends on the measurements. The equations for $\theta(x, t)$ and $p(x, t)$ are PDEs which can be solved by numerical methods if required, whereas $\psi(t)$ is governed by an ODE, which could also be solved by numerical methods if necessary. The basic idea is that engineers might be willing to solve ODEs on-line in real time (as is commonplace nowadays with extended Kalman filters), but they are generally unable to solve PDEs on-line in real time.

REMARK 2 Notice that (4.4) is a vector Riccati equation for the sufficient statistic $\psi(t)$. This is analogous to the matrix Riccati equation in the Kalman-Bucy filter. Also, (4.5) is analogous to the difference equation to update the conditional mean in the Kalman filter.

REMARK 3 It is easy to handle the more general Itô equation:

$$dx = f(x, t)\,dt + G(x, t)\,dw \tag{4.9}$$

for smooth nowhere-singular $n \times n$ matrix-valued functions $G(x, t)$. This would be accomplished by a simple transformation of coordinates.

REMARK 4 Inspection of (4.5) might suggest that $\psi(t_k)$ depends on x; nevertheless, $\psi(t_k)$ is independent of x. It is obvious from (4.8) that

$$\frac{\partial}{\partial \theta} \log p(z_k|x) = c^T(z_k, t_k),$$

which is independent of x. It follows that $\psi(t_k) = \bar{\psi}(t_k) + c(z_k, t_k)$ is an equivalent form for (4.5).

REMARK 5 The corresponding filtering problem with continuous-time observations is solved in Daum (1987a).

REMARK 6 Mitter (1979) has noted the relevance of "gauge transformations" to nonlinear filtering theory. Inspection of (4.3) shows that it is what physicists call a "local" gauge transformation from $p(x, t)$ to $p(x, t|Z_k)$. That is, the Fokker-Planck equation is "invariant" under this transformation, in the sense that both $p(x, t)$ and $p(x, t|Z_k)$ must satisfy the same Fokker-Planck equation.

REMARK 7 One can think of $\theta(x, t)$ as defining a "natural" coordinate system for a given estimation problem (see Amari, 1982). It is a preferred coordinate system because the Fokker-Planck equation is "separable" in these coordinates. In this sense $\theta(x, t)$ is analogous to the "special functions" of mathematical physics (e.g., Bessel functions, parabolic cylinder functions, etc.), which are of interest because certain PDEs are "separable" in these special coordinates. Intuitively, the reason for solving the PDEs (4.6)–(4.7) is to find this preferred coordinate system.

REMARK 8 Note that the PDEs (4.6)–(4.7) and the boundary condition (4.8) cannot determine $\theta = \theta(x, t)$ uniquely. This is because θ enters the original problem through (4.3), and it is obvious that any other pair (θ', ψ') will work as well if $\theta' = \Lambda\theta$ and $\psi' = \Lambda\psi$, in which Λ is any orthogonal matrix (i.e., $\Lambda^T\Lambda = I$). Therefore, if (4.6)–(4.7) are to be solved numerically, some other condition on $\theta = \theta(x, t)$ must be appended to obtain a unique solution. Asking a computer to solve a problem that does not have a unique solution is asking for trouble. Moreover, asking for the minimum norm solution will not determine a unique solution either.

REMARK 9 The Kalman filter is a special case of the filter derived here. This is because a Gaussian density is a member of an exponential family. To be more specific, consider the problem of estimating a R^1-valued random variable x, with the usual Kalman filtering setup. In this special case, $\theta(x) = (x, x^2)^T$ and ψ is R^2 valued. Condition (4.8) corresponds to the fact that $p(z_k|x)$ is Gaussian in this special case. Moreover, (4.6)–(4.7) are obviously satisfied in this case. If x is R^n valued, θ and ψ are R^M valued with $M = n + n(n + 1)/2$, and $\theta(x) = (x_1, x_2, \ldots, x_n, x_1x_2, x_1x_3, \ldots, x_1^2, \ldots, x_n^2)^T$. The sufficient statistic, ψ, in this case merely encodes the same information as the conditional mean and covariance matrix in the standard Kalman filter equations.

5. GLOBAL AND LOCAL APPROXIMATIONS

In practical applications it is rather unlikely that a given problem will satisfy the theoretical conditions required in Table 2 exactly. Therefore, some kind of approximation is inevitable. This is the same situation that obtains in the case of the Kalman filter; problems with exactly linear dynamics and linear measurements are virtually nonexistent. The widespread use of the "extended" Kalman filter derives from the ease with which a given nonlinear problem can be approximated by a linear one. Fortunately, a number of similar approximation methods can be used here as well.

First, a simple local approximation for filters 3 and 4 in Table 2 will be suggested. In particular, define the two quantities

$$V(x) = \mathrm{tr}\left(\frac{\partial f}{\partial x}\right) + \frac{\alpha}{2} rQr^T \tag{5.1}$$

$$W(x) = f - \alpha Qr^T. \tag{5.2}$$

If there are no values of A, b, D, and E that satisfy the theoretical conditions in Table 2 exactly, then approximate values of A, b, D, and E could be computed by a second-order Taylor series expansion of $V(x)$ and a first-order Taylor series expansion of $W(x)$ about the conditional mean $\hat{x} = E(x|Z_k)$. More specifically, the desired approximations are

$$V(x) \sim x^T A x + b^T x + c \tag{5.3}$$

$$W(x) \sim Dx + E. \tag{5.4}$$

An elementary calculation shows that the Taylor series approximation suggested above results in

$$A = \frac{1}{2}\frac{\partial^2 V(\hat{x})}{\partial x^2} \tag{5.5}$$

$$b^T = \frac{\partial V(\hat{x})}{\partial x} - 2\hat{x}^T A \tag{5.6}$$

$$D = \frac{\partial W(\hat{x})}{\partial x} \tag{5.7}$$

$$E = W(\hat{x}) - D\hat{x}. \tag{5.8}$$

These approximate values of A, b, D, and E can be used in the filter equations in Table 3, analogous to the extended Kalman filter. A similar local approximation for filter 5 in Table 2 is suggested in Daum (1986b).

An alternative approximation for filters 3 and 4 is to find values of A, b, c, D, E that result in a least squares approximation for (5.3)–(5.4). Intuitively, a weighted least squares approximation, using $p(x|Z_k)$ as a weighting factor, should perform well.

Next, several approximation techniques for filter 6 are considered. The first method has a "global" character, and it is based on the Ritz-Galerkin technique for the numerical solution of PDEs (Kantorovich and Krylov, 1958). In particular, a convenient parameterization of $\theta(x, t)$ could be selected, and the values of $A(t)$, $B_j(t)$, and $c(z_k, t_k)$ can be computed to approximate (4.6)–(4.8) in a least squares sense, using the Ritz-Galerkin formalism. For the special case of Gaussian measurement noise (2.3), it is always possible to satisfy (4.8) exactly by including $h(x, t)$ and $h^T(x, t)R^{-1}h(x, t)$ in the linear span of $\theta(x, t)$. More generally, if $p(z_k|x_k)$ is from a non-Gaussian exponential family, then (4.8) is satisfied by definition. If this were the case, then $c(z_k, t_k)$ would be known exactly, and the Ritz-Galerkin method would be used to approximate $A(t)$, $B_j(t)$ and the parameters defining $\theta(x, t)$.

The second approximation method is "local," and it is similar in concept to the extended Kalman filter. In particular, one can choose a convenient form for $\theta(x, t)$ and compute A, B_j, and $c(z_k, t_k)$ to satisfy (4.6)–(4.8) exactly at a given value of $\theta(x, t)$. An obvious choice is the point $\theta = \theta(\hat{x}, t)$, in which $\hat{x}$ is the conditional mean of $x(t)$. There are many possible variations on this theme. For example, if the measurement noise is Gaussian (2.3), or if $p(z_k|x_k)$ is from an exponential family, then (4.8) can be satisfied exactly, and only A and B_j must be approximated locally. To be more specific, define V and W as

$$V = \frac{\partial\theta}{\partial x}(Qr^T - f) + \frac{1}{2}\xi - \frac{\partial\theta}{\partial t} \tag{5.9}$$

$$W = \frac{1}{2}\frac{\partial\theta}{\partial x}Q\left(\frac{\partial\theta}{\partial x}\right)^T. \tag{5.10}$$

Comparison with (4.6)–(4.7) shows that the desired local approximations are

$$A \sim \frac{\partial V}{\partial\theta} \tag{5.11}$$

$$B_j \sim \frac{\partial W}{\partial\theta_j} \tag{5.12}$$

evaluated at some point, such as $\hat{x}$.

As noted in Section 4, the Kalman filter is a special case of the new filters in this paper. Intuitively, one would expect that these new approximations should be superior to the extended Kalman filter in many practical applications. Monte Carlo simulations are required, however, to test this intuitive notion for a number of specific examples. Such Monte Carlo simulations will be reported in the next section.

6. APPLICATION OF NEW THEORY

In this section the new theory is applied to a common radar tracking problem. The problem is to estimate the position and velocity of an object that is observed with a radar. The radar provides measurements of range, azimuth, and elevation at discrete points in time. For many radar applications, the standard engineering approach to this problem is to use an extended Kalman filter (EKF) (see Jazwinski, 1970; Wishner, 1969; Daum, 1983). This requires a linear approximation of either the measurement equations or the differential equations of motion or both. In contrast, the new exact filter in Theorem 4.2 does not require such an approximation. For this application it turns out that the EKF suffers from a fundamental flaw, called "false cross-range observability" (see pp. 281–282 in Daum, 1983). In particular, the EKF is very optimistic about its ability to estimate angle rates from a sequence of range measurements. This optimism is due to the linear approximation inherent in the EKF. It has become standard engineering practice to mitigate this nonlinear effect by using one or more ad hoc modifications to the EKF (see Daum, 1983). Although these ad hoc measures can improve performance somewhat, they do not address the basic flaw in the EKF itself.

Suppose that the state vector, $x(t)$, is modeled as position and velocity in three Cartesian coordinates: $x(t) = (x, y, z, \dot{x}, \dot{y}, \dot{z})^T$. If object acceleration ($\ddot{x}$, $\ddot{y}$, $\ddot{z}$) is modeled as Gaussian zero-mean white noise, and $p(x, t_0)$ is Gaussian, then $p(x, t)$ is simply a Gaussian density. For this special case, one solution of (4.6)–(4.7) is $\theta = (x, y, z, \dot{x}, \dot{y}, \dot{z}, x^2, y^2, z^2, \dot{x}^2, \dot{y}^2, \dot{z}^2, xy, xz, yz, x\dot{x}, x\dot{y}, x\dot{z}, y\dot{x}, y\dot{y}, y\dot{z}, z\dot{x}, y\dot{z}, z\dot{y}, z\dot{z}, \dot{x}\dot{y}, \dot{x}\dot{z}, \dot{y}\dot{z})^T$ for a suitable choice of A and B_j.

Consider what happens to ψ_k when the radar makes a measurement of range, $R = \sqrt{x^2 + y^2 + z^2}$. Assume that R^2 is measured by the radar with a Gaussian zero-mean error with known variance. (This is an unorthodox model of radar range measurement error, which is very convenient for the new theory, and it is just as good a physical model as the standard one.) Inspection of (4.5) shows that the only components of ψ_k that are changed are the seventh, eighth, and ninth, corresponding to the terms (x^2, y^2, z^2) in θ. The first three components of ψ_k, corresponding to (x, y, z) in θ, are not affected by a range measurement. Intuitively, the new nonlinear filter thinks that it can deduce something about (x^2, y^2, z^2) from a range measurement, but nothing is learned directly about (x, y, z).

In contrast, the EKF will not change the coefficients of (x^2, y^2, z^2) as a function of measured range, because these are elements of the covariance matrix. But rather, the EKF will change the coefficients of (x, y, z), which are elements of the conditional mean vector. Intuitively, the EKF thinks that it can learn something about (x, y, z) from a range measurement, but nothing is learned about (x^2, y^2, z^2) directly. For certain parameter values, the EKF thinks that it can deduce the sign of (x, y, z) from a single range measurement. This is clearly wrong, and it is one manifestation of "false cross-range observability." This basic flaw in the EKF is due to linearization of the measurement equation. The EKF cannot be told about the nonlinear relationship between (x, y, z) and R.

Monte Carlo simulations for the new nonlinear filter show that the estimation accuracy is very close to the generalized Cramér-Rao bound, whereas several EKFs produce estimation error variances in velocity that are up to four times this theoretical lower bound.

7. RELATIONSHIP TO LIE ALGEBRAIC APPROACH

It is extremely rare for a Bayesian filtering problem to have an exact recursive solution in terms of a sufficient statistic with a fixed finite dimension. For this to happen, the conditional density, $p(x, t|Z_k)$, must be a very special type of function (see Section 3). Considering the apparent difficulty of finding finite-dimensional filters, Brockett decided that it would be useful to develop a theory that could predict when a given estimation problem has such a solution (see Brockett, 1981). This theory has been developed for continuous-time observations using Lie algebraic techniques in Hazewinkel and Marcus (1982), Mitter (1983), Ocone (1980), Chaleyat-Maurel and Michel (1984), Marcus (1984), Baras (1980), and others.

A Lie algebra (LA) is a collection of elements $\{A, B, C, \ldots\}$ that has two special properties: (1) it is a vector space, and (2) a bracket operation is defined on any two elements of LA, such that if A and B are in LA, then $[A, B]$ is also in LA, and the bracket satisfies the following two identities for all elements A, B, C:

(a) $[A, A] = 0$

(b) $[A, [B, C]] + [B, [C, A]] + [C, [A, B]] = 0.$

Property (a) is called skew symmetry, and property (b) is the "Jacobi identity." For example, the collection of all $n \times n$ real matrices is a Lie algebra with the bracket operation defined as $[A, B] = AB - BA$. In this example the bracket measures the extent to which A and B commute; if $AB = BA$, then $[A, B] = 0$. Lie algebras in which $[A, B] = 0$ for all elements A, B are called "Abelian." It turns out that all finite-dimensional Lie algebras can be represented by collections of matrices. Many Lie algebras are not Abelian, but they might have other special properties. For example, if the Lie algebra can be represented by strictly lower triangular matrices, it is called "nilpotent"; if it can be represented by lower triangular matrices, it is called "solvable." These special properties are useful in characterizing nonlinear filtering problems with finite-dimensional sufficient statistics (e.g., see Marcus, 1984). More generally, Lie algebras have been studied intensively as a basic tool to understand Lie groups. The solutions to certain ordinary and partial differential equations form Lie groups. Intuitively, the Lie algebra is the vector space that corresponds to the tangent space on the solution manifold of these differential equations. The Lie group is often too difficult to analyze directly, and the Lie algebra is generally a simpler object to study, owing to its vector space structure. There is an enormous amount known about linear algebra, and thus much can be said about the representations of finite-dimensional Lie algebras as matrices. This is the main reason that mathematicians "like" Lie algebras: they are a powerful tool to study nonlinear ordinary differential equations and partial differential equations using linear algebra.

In the context of nonlinear filtering, Brockett (1981) formulated a criterion for deciding whether a given filtering problem has a finite-dimensional sufficient statistic. In particular, this will happen when the so-called "estimation Lie algebra" is finite dimensional. The estimation Lie algebra is generated by the operators in the Zakai stochastic partial differential equation (see Daum, 1987a). Although this sounds complicated, it is a beautifully simple idea with great intuitive appeal. Perhaps the most accessible introduction to Lie algebras and nonlinear systems is the book by Casti (1985). The books by Gilmore (1974) and Olver (1986) are both lucid and full of concrete examples using differential equations. The book by Amari (1985) shows the connection between classical statistics and classical differential geometry. For example, the Fisher information matrix is the metric tensor for a certain Riemann manifold.

In this paper, an analogous theory is derived for discrete-time measurements without using Lie algebraic methods. Theorem 4.1 is based on two main ideas: (1) the classical notion of separation of variables in partial differential equations (PDEs), and (2) the generalized Fisher-Darmois-Koopman-Pitman theorem for densities of exponential type. The conditions in Theorem 1 take the form of a set of PDEs. In a given coordinate system, Brockett's Lie algebraic criterion for the existence of finite-dimensional filters can also be expressed as a system of PDEs. Moreover, Miller (1977) has developed a Lie algebraic theory for the separation of variables in second-order linear PDEs. Figure 2 illustrates the relationship between these ideas.

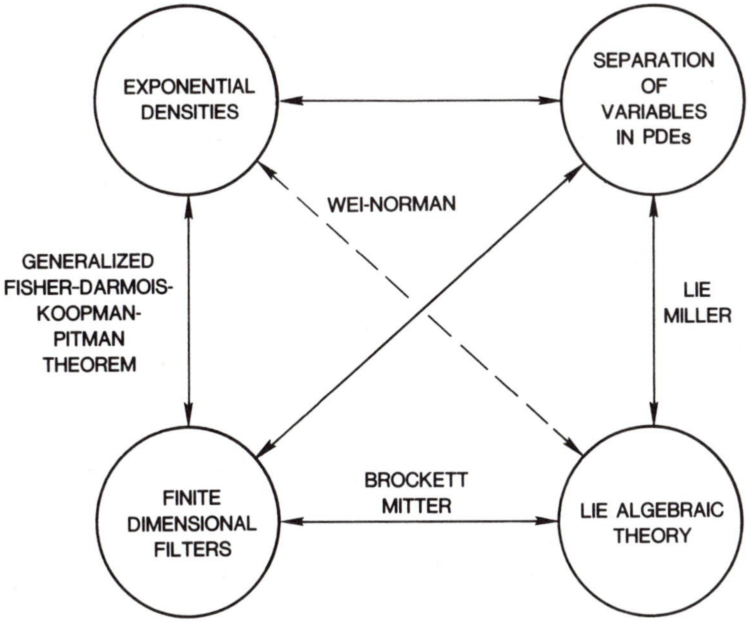

Figure 2. Circle of ideas relating to finite-dimensional filters.

There are two main differences between these results and the Lie algebraic theory. First, this chapter provides explicit filtering equations, whereas there is no analogous result from the Lie algebraic theory. Second, the Lie algebraic theory is "coordinate free," whereas the results given here are not. These two facts are intimately related. In particular, explicit filter equations can be obtained here because a special coordinate system was chosen in which the Fokker-Planck equation "separates." One can think of $\theta(x, t)$ as defining a set of preferred coordinates (Amari, 1982). Intuitively, the purpose of solving the PDEs (4.6)–(4.7) is to find this preferred coordinate system.

On the other hand, there is a deep relationship between Lie algebras and "separation of variables." In particular, much of Lie's work was motivated by the desire to systematize the classical methods of solving PDEs (see van der Waerden, 1985, pp. 146–147; Gilmore, 1974, p. 100). Lie's first theorem is a statement about separation of variables (Gilmore, 1974, p. 96). Lie's second theorem can be proved with the standard separation of variables argument that was used in this paper (Gilmore, 1974, p. 102). In a certain sense, this explains why Lie algebras are relevant to nonlinear filtering theory.

Separation of variables (SOV) is an extremely powerful but vague technique to solve PDEs. Miller's excellent book (1977) develops a Lie algebraic theory of separation of variables, but it fails to provide any mathematical definition of SOV! This lacuna was noted by Koornwinder (1980). Moreover, according to Olver (1986), "except for special classes of partial differential equations, such as Hamilton-Jacobi and Helmholtz equations, the precise connections between symmetries [i.e., Lie algebras] and separation of variables is not well understood at present." Even for the Hamilton-Jacobi equation, the theory of SOV is not adequate. The classical results of Stäckel apply only for orthogonal coordinates under rather restrictive conditions (e.g., see Goldstein, 1980). The theory of SOV for the Hamilton-Jacobi equation is still an active area of research (see Kalnins and Miller, 1986, 1985).

8. GENERALIZED EXPONENTIAL FAMILIES

The theory in this chapter has been developed assuming that all the probability densities are smooth and nowhere vanishing. These assumptions are not merely technical details, but rather, they have important practical implications. In particular, exact recursive filters with fixed finite dimensions can be constructed for problems that do not satisfy assumptions 1 and 2. For example, Servi and Ho (1981) have derived such filters for linear systems with uniformly distributed measurement noise and process noise. Moreover, the conditional density is not from an exponential family, but rather, it is uniform.

The uniform density is a member of a "generalized exponential family" (GEF). The GEF does not require that the densities are smooth and nowhere vanishing, but rather only an algebraic condition is required (see Lauritzen, 1975; Johansen, 1977). This algebraic condition is the existence of a certain homomorphism, which is analogous to Brockett's Lie algebraic homomorphism condition (see Brockett, 1981).

ACKNOWLEDGMENT

I would like to thank the several anonymous reviewers of this chapter. It was a pleasure to read their many thoughtful and helpful comments.

APPENDIX: THE FOKKER-PLANCK EQUATION

A fundamental tool for analyzing nonlinear stochastic differential equations is the Fokker-Planck (F-P) equation. It is a linear second-order parabolic partial differential equation that governs the evolution of the unconditional density, $p(x, t)$, for all time, as well as the conditional density, $p(x, t|Z_k)$, between measurement times $t_k \leqslant t < t_{k+1}$. Suppose that the random process $x(t)$ satisfies the Itô stochastic differential equation

$$dx = f(x, t)\, dt + G(t)\, dw(t). \tag{A.1}$$

The corresponding F-P equation is

$$\frac{\partial p}{\partial t} = -\frac{\partial p}{\partial x} f - p \operatorname{tr}\left(\frac{\partial f}{\partial x}\right) + \frac{1}{2} \operatorname{tr}\left(GG^T \frac{\partial^2 p}{\partial x^2}\right). \tag{A.2}$$

There are two main reasons that we like to study this PDE rather than the Itô equation: (1) the F-P equation is linear, in contrast to the nonlinear Itô equation; and (2) the F-P equation is deterministic, whereas the Itô equation is stochastic. In particular, several hundred years of mathematical work has supplied us with powerful methods to solve deterministic linear PDEs (e.g., separation of variables).

A short but formal calculation shows how the F-P equation is derived from (A.1). First, define the complex-valued function of x and λ,

$$\phi = \exp(i\lambda^T x), \tag{A.3}$$

in which $i = \sqrt{-1}$ and λ is R^n valued. Next, expand ϕ in a Taylor series to second order in x,

$$d\phi = \frac{\partial \phi}{\partial x} dx + \frac{1}{2} dx^T \frac{\partial^2 \phi}{\partial x^2} dx. \tag{A.4}$$

Substituting (A.1) into (A.4) results in

$$d\phi = \frac{\partial \phi}{\partial x} f\, dt + \frac{\partial \phi}{\partial x} G\, dw + \frac{1}{2} f^T \frac{\partial^2 \phi}{\partial x^2} f(dt)^2 + dw^T G^T \frac{\partial^2 \phi}{\partial x^2} f\, dt$$
$$+ \frac{1}{2} dw^T G^T \frac{\partial^2 \phi}{\partial x^2} G\, dw. \tag{A.5}$$

Taking expected values of both sides of (A.5) with respect to the density p yields

$$d\bar{\phi} = \overline{\frac{\partial \phi}{\partial x} f}\, dt + \frac{1}{2} \overline{f^T \frac{\partial^2 \phi}{\partial x^2} f}\, (dt)^2 + \frac{1}{2} \operatorname{tr}\left(GG^T \overline{\frac{\partial^2 \phi}{\partial x^2}} \right) dt, \tag{A.6}$$

in which $\bar{\phi} = E(\phi)$ denotes the expected value of ϕ, which is also the Fourier transform of $p(x, t)$. To obtain (A.6), the following two properties of the Brownian motion, $w(t)$, were used: $E(dw) = 0$ and $E(dw\, dw^T) = I\, dt$. Next, divide (A.6) by dt and take the limit as $dt \to 0$:

$$\frac{\partial \bar{\phi}}{\partial t} = \overline{\frac{\partial \phi}{\partial x} f} + \frac{1}{2} \operatorname{tr}\left(GG^T \overline{\frac{\partial^2 \phi}{\partial x^2}} \right). \tag{A.7}$$

From the definition of $\phi = \exp(i\lambda^T x)$, it follows that $\partial\phi/\partial x = i\lambda^T \phi$ and $\partial^2\phi/\partial x^2 = -\lambda\lambda^T \phi$, which results in

$$\frac{\partial \bar{\phi}}{\partial t} = i\lambda^T \overline{f\phi} - \frac{1}{2} \operatorname{tr}(GG^T \lambda\lambda^T)\bar{\phi}. \tag{A.8}$$

Finally, taking the inverse Fourier transform of (A.8) produces the F-P equation (A.2), after recognizing that $\overline{f\phi}$ is the Fourier transform of fp, and using the fact that $\bar{\phi}$ is the Fourier transform of p:

$$\bar{\phi} = E(\phi) = \int p(x, t) \exp(i\lambda^T x)\, dx. \tag{A.9}$$

A superb lucid survey of the F-P equation, including its derivation, solutions, and utility, is given by Fuller (1969), without using Itô calculus. Jazwinski (1970) derives the F-P equation using Itô calculus but with a different method than the one sketched above. The recent book by Risken (1984) is encyclopedic and practical.

REFERENCES

Alspach, D. L., and H. W. Sorenson (1972). Nonlinear Bayesian estimation using Gaussian sum approximations, *IEEE Trans. Autom. Control* AC-17, 439–448.

Amari, S. (1982). Differential geometry of curved exponential families—curvature and information loss, *Ann. Stat.* 10, 357–385.

Amari, S. (1985). *Differential-Geometrical Methods in Statistics*, Springer-Verlag, Berlin.

Barankin, E. W., and A. P. Maitra (1963). Generalization of the Fisher-Darmois-Koopman-Pitman theorem on sufficient statistics, *Sankhyā Ser. A* 25, 217–244.

Baras, J. S. (1980). Group Invariance Methods in Nonlinear Filtering of Diffusion Processes, in *Proceedings of the 1980 IEEE Decision and Control Conference, pp. 72–79.*

Beneš, V. E. (1981). Exact finite-dimensional filters for certain diffusions with nonlinear drift, *Stochastics* 5, 65–92.

Beneš, V. E. (1985). New exact nonlinear filters with large Lie algebras, *Syst. Control Lett.* 5, 217–221.

Brockett, R. W. (1979). Classification and Equivalence in Estimation Theory, in *Proceedings of the 1979 IEEE Conference on Decision and Control*, pp 172–175.

Brockett, R. W. (1981). Nonlinear Systems and Nonlinear Estimation Theory, in *Stochastic Systems: The Mathematics of Filtering and Identification and Applications* (M. Hazewinkel and J. C. Willems, eds.), D. Reidel, Dordrecht, The Netherlands.

Bucy, R. S. (1970). Linear and nonlinear filtering, *Proc. IEEE* 58, 854–864.

Casti, J. L. (1985). *Nonlinear System Theory*, Academic Press, Orlando, Fla.

Chaleyat-Maurel, M., and D. Michel (1984). Des Résultats de Non Existence de Filtre de Dimension Finie, *Stochastics* 13, 83–102.

Daum, F. E. (1984). Exact Finite Dimensional Nonlinear Filters for Continuous Time Processes with Discrete Time Measurements, in *Proceedings of the 1984 IEEE Conference on Decision and Control*, pp. 16–22.

Daum, F. E. (1985a). Exact Nonlinear Filters, Exponential Densities and Separation of Variables, in *Proceedings of the 23rd Allerton Conference*, University of Illinois, Urbana, Ill.

Daum, F. E. (1985b). Exact Solution to the Zakai Equation for Certain Diffusions, in *Proceedings of the 1985 IEEE Control and Decision Conference*, pp. 1964–1965.

Daum, F. E. (1986a). Exact Nonlinear Recursive Filters, in *Proceedings of the 20th Conference on Information Sciences and Systems*, Princeton University, Princeton, N.J., pp. 516–519.

Daum, F. E. (1986b). New Nonlinear Filters and Exact Solutions of the Fokker-Planck Equation, in *Proceedings of the American Control Conference*, pp. 884–888.

Daum, F. E. (1986c). Exact finite dimensional nonlinear filters, *IEEE Trans. Autom. Control* AC-31(7), 616–622.

Daum, F. E. (1986d). The Fisher-Darmois-Koopman-Pitman Theorem for Random Processes, in *Proceedings of the 1986 IEEE Conference on Decision and Control*, pp. 1043–1044.

Daum, F. E. (1987a). Solution of the Zakai equation by separation of variables, *IEEE Trans. Autom. Control* AC-32(10).

Daum, F. E. (1988). Exact nonlinear recursive filters with non-Gaussian measurements, to be published in *IEEE Trans. Inf. Theory*.

Daum, F. E., and R. J. Fitzgerald (1983). Decoupled Kalman filters for phased array radar tracking, *IEEE Trans. Autom. Control* AC-28, 269–283.

Davis, M. H. A., and S. I. Marcus (1981). An Introduction to Nonlinear Filtering, in *Stochastic Systems: The Mathematics of Filtering and Identification and Applications* (M. Hazewinkel and J. C. Willems, eds.), D. Reidel, Dordrecht, The Netherlands.

Davis, P. J., and P. Rabinowitz (1984). *Methods of Numerical Integration*, 2nd ed. Academic Press, Orlando, Fla. Chap. 5.

Fisher, J. R. (1967). Optimal Nonlinear Filtering, in *Advances in Control Systems* (C. T. Leondes, ed.), Academic Press, New York, pp. 197–300.

Fisher, R. A. (1934). Two new properties of mathematical likelihood, *Proc. R. Soc. London Ser. A* 144, 285–307.

Frost, P. A., and T. Kailath (1971). An innovations approach to least-squares estimation:—III. Nonlinear estimation in white Gaussian noise, *IEEE Trans. Autom. Control* AC-16(3), 217–226.

Fuller, A. T. (1969). Analysis of nonlinear stochastic systems by means of the Fokker-Planck equation, *Int. J. Control* 9, 603–655.

Gilmore, R. (1974). *Lie Groups, Lie Algebras, and Some of Their Applications*, Wiley, New York.

Goldstein, H. (1980). *Classical Mechanics*, 2nd ed., Addison-Wesley, Reading, Mass., pp. 449–457.

Hazewinkel, M., and S. I. Marcus (1982). On Lie algebras and finite dimensional filtering, *Stochastics* 7, 29–62.

Ho, Y. C., and R. C. K. Lee (1964). A Bayesian approach to problems in stochastic estimation and control, *IEEE Trans. Autom. Control* AC-9, 333–339.

Hutchinson, C. E. (1984). The Kalman filter applied to aerospace and electronic systems, *IEEE Trans. Aerosp. Electron. Syst.* 500–504.

Jazwinski, A. H. (1970). *Stochastic Processes and Filtering Theory*, Academic Press, New York.

Johansen, S. (1977). Homomorphisms and General Exponential Families, *Recent Developments in Statistics* (J. R. Barra et al., eds.), North-Holland, Amsterdam, pp. 489–499.

Kalman, R. E. (1960). A new approach to linear filtering and prediction problems, *Trans. ASME J. Basic Eng.* 82D, 35–45.

Kalnins, E. G., and W. Miller (1985). Generalized Stäckel matrices, *J. Math. Phys.* 26(9), 2168–2173.

Kalnins, E. G., and W. Miller (1986). Separation of variables on *n*-dimensional Riemann manifolds, *J. Math. Phys.* 27(7), 1721–1736.

Kantorovich, L. V., and V. I. Krylov (1958). *Approximate Methods of Higher Analysis* (translated by C. D. Benster), Interscience, New York, pp. 258–283.

Koopman, B. O. (1936). On distributions admitting a sufficient statistic, *Trans. Am. Math. Soc.* 39, 399–409.

Koornwinder, T. H. (1980). A Precise Definition of Separation of Variables, in *Geometrical Approaches to Differential Equations* (R. Martini, ed.), Springer-Verlag, Berlin, pp. 240–263.

Lauritzen, S. L. (1975). General exponential models for discrete observations, *Scand. J. Stat.* 2, 23–33.

Marcus, S. I. (1984). Algebraic and geometric methods in nonlinear filtering, *SIAM J. Control Optimization* 22,(6), 817–844.

McBryan, O. A., and E. F. Van de Velde (1986). Parallel Algorithms for Elliptic Equations, in *New Computing Environments: Parallel, Vector and Systolic* (A. Wouk, ed.), SIAM, Philadelphia, pp. 236–270.

Miller, W. (1977). *Symmetry and Separation of Variables*, Addison-Wesley, Reading, Mass.

Mitter, S. K. (1979). On the analogy between mathematical problems of nonlinear filtering and quantum physics, *Ricerche Autom.* 10(2), 163–216.

Mitter, S. K. (1983). Lectures on Nonlinear Filtering and Stochastic Control, in *Non-linear Filtering and Stochastic Control, Proceedings of the 3rd 1981 Session of CIME, Cortona, Italy,* (S. K. Mitter and A. Moro, eds.), Springer-Verlag, New York, pp. 170–207.

Nelder, J. A., and R. W. M. Wedderburn (1972). Generalized linear models, *J. R. Stat. Soc. Ser. A* 135, 370–384.

Ocone, D. (1980). Nonlinear Filtering Problems with Finite Dimensional Estimation Algebras, in *Proceedings of the Joint Automatic Control Conference.*

Ocone, D., J. S. Baras, and S. I. Marcus (1982). Explicit filters for diffusions with certain nonlinear drifts, *Stochastics* 8, 1–16.

Olver, P. J. (1986). *Applications of Lie Groups to Differential Equations*, Springer-Verlag, New York.

Osborne, M. R. (1987). Estimating nonlinear models by maximum likelihood for the exponential family, *SIAM J. Sci. Stat. Comput.* 8(3), 446–456.

Powell, M. J. D. (1986). Convergence properties of algorithms for nonlinear optimization, *SIAM Rev.* 28(4), 487–500.

Press, W. H., B. P. Flannery, S. A. Teukolsky and W. T. Vetterling (1986). *Numerical Recipes*, Cambridge University Press, Cambridge: Chaps. 10 and 14.

Risken, H. (1984). *The Fokker-Planck Equation*, Springer-Verlag, Berlin.

Sawitzki, G. (1979). Exact Filtering in Exponential Families: Discrete Time, in *Stochastic Control Theory and Stochastic Differential Systems* (M. Kohlmann and W. Vogel, eds.), Springer-Verlag, New York, pp. 554–558.

Schwartz, C. (1985). Numerical integration in many dimensions: I and II, *J. Math. Phys.* 26(5), 951–957.

Schwartz, L., and E. B. Stear (1969). A computational comparison of several nonlinear filters, *IEEE Trans. Autom. Control* AC-13, 83–86.

Servi, L. D., and Y. C. Ho (1981). Recursive estimation in the presence of uniformly distributed measurement noise, *IEEE Trans. Autom. Control* AC-26, 563–565.

Slater, J. C., and N. H. Frank (1933). *Introduction to Theoretical Physics*, McGraw-Hill, New York, pp. 163–164.

Sorenson, H. W. (1974). On the development of practical nonlinear filters, *Inf. Sci.* 7, 253–270.

Sorenson, H. W. (1980). *Parameter Estimation*, Marcel Dekker, New York: Chap. 6.

Sorenson, H. W., ed. (1985). *Kalman Filtering: Theory and Applications*, IEEE Press, New York.

van der Waerden, B. L. (1985). *A History of Algebra*, Springer-Verlag, Berlin.

West, M., P. J. Harrison, and H. S. Migon (1985). Dynamic generalized linear models and Bayesian forecasting (with discussion), *J. Am. Stat. Assoc.* 80, 73–97.

Willman, W. W. (1981). Edgeworth expansions in state perturbations estimation, *IEEE Trans. Autom. Control* AC-26(2), 493–498.

Wishner, R. P., J. A. Tabaczynski, and M. Athans (1969). A comparison of three non-linear filters, *Auromatica* 5, 487–496.

Wong, W. S. (1983). New classes of finite-dimensional nonlinear filters, *Syst. Control Lett.* 3, 155–164.

9

Bayesian Approach to Robustifying the Kalman Filter

DANIEL PEÑA Universidad Politecnica de Madrid, Madrid, Spain

IRWIN GUTTMAN Statistics Department, University of Toronto, Toronto, Ontario, Canada

1. INTRODUCTION

An important problem in the engineering literature is to estimate the state of a system that changes over time, using a set of measurements made on the system. In mathematical terms the problem can be expressed as

$$\begin{aligned} \mathbf{x}_t &= f(\mathbf{x}_{t-1}) + \boldsymbol{\varepsilon}_t \\ \mathbf{y}_t &= g(\mathbf{x}_t) + \mathbf{v}_t, \end{aligned} \tag{1.1}$$

where the first equation relates the present and future state variables of the system, and the second relates the observed measurement ($\mathbf{y}_t$) with the state, $\mathbf{x}_t$, of the system. The first equation, called the state equation, assumes that the future state cannot be exactly determined by the present because of the effect of many unknown factors summarized in the random variable $\boldsymbol{\varepsilon}_t$. Similarly, the random term $\mathbf{v}_t$ reflects some measurement error. The objective of the analysis is to estimate the state $\mathbf{x}_{t+k}$ given the set of observations $\mathbf{y}_1, \ldots, \mathbf{y}_t$. When $k > 0$, we have a forecasting problem, when $k = 0$, a filtering problem, and when $k < 0$, a smoothing problem. To build a control system to guide the state of the system, a recursive relation is needed, which is called the filtering of the state, and carried out at each point of time. This has motivated the search for recursive solutions and has become a key element of research since the late 1950s. An obvious first step to resolve the problem is to make a first-order Taylor expansion of the f and g function in (1.1) that linearizes the problem. Then, using the properties of orthogonal projection on linear spaces, Kalman (1960) derived a recursive algorithm to estimate the state of the system. It turns out that Kalman estimates are efficient; that is, they are unbiased and have minimum variance when and only when all the variables involved are normal and, consequently, only if the functions f and g are linear. Therefore, in the last 25 years there has been a continuous and steady line of research to generalize the Kalman solution (see Sorenson and Alspach, 1971; Bar-Shalom, 1978; Spall and Wall, 1984; Makowski, 1986; Daum, 1986).

An important concern has been to increase the robustness of the filter to handle the case of heavy-tailed state and measurement noise distributions, and considerable research has been carried out that applies ideas from robust estimation of this problem (see Masreliez, 1975; Masreliez and Martin, 1977; Tsai and Kurz, 1983).

A situation in which discordant measurements often arise is target tracking problems. Nahi (1969) assumed a nonzero probability that any observation consists of noise alone (and therefore has no information about the state of the target) and developed a recursive modified Kalman filter algorithm to deal with that problem. Singer and Sea (1973) analyzed the problem of several targets and false alarms, and Athans et al. (1977) extended the work of Nahi (1969) presenting three useful algorithms. Bar-Shalom (1978) surveys these papers and related work of this area.

In statistics, the idea of recursive estimation goes back to Gauss, who derived the recursive relations to update the parameters of a regression model when new data are observed (see Young, 1984). Gauss' results were mainly unnoticed. Indeed, many years later, starting from scratch, Plackett (1950) presented recursive relations in matrix form to help resolve the updating problem in linear regression. The idea of using the state-space engineering representation to formulate statistical models does not appear, by and large, in the statistical and econometric literature until the early to mid-1970s. Since then, the Kalman filter has found many applications in statistics. Some of these will be reviewed in this paper.

The state-space formulation and the idea of recursive updating of information are ideally suited for the Bayesian approach and in fact, most of the development of Kalman filtering ideas in statistics have a clear Bayesian flavor (see, e.g., Ho and Lee, 1964; Aoki, 1967; Harrison and Stevens, 1976; West et al., 1985).

In the statistical literature the concern about the heavy-tailed distributions arises only in the measurement equation that represents the real data observed. The research on robust procedures for recursive estimating using the Kalman filter has been directed primarily to applying M-estimators and influence function ideas to robustifying the filtering process (see Kleiner et al.,

1979; West, 1981; Campbell, 1982; Martin, 1984). West (1984) considered as error distribution function, the family of densities generated by

$$p(\varepsilon) = \int_0^{\infty} \lambda^{1/2}\phi(\lambda^{1/2}\varepsilon)f(\lambda)\, d\lambda,$$

where ϕ is the standard normal probability density function (p.d.f.), and f is taken to be a positive function on $(0, \infty)$, and uses it to derive Bayesian estimators in regression models.

In this chapter we allow for the presence of outliers in the data assuming a scale-contaminated normal distribution and apply a Bayesian approach to obtain a robust filter for the problem. We remark that our motivation (and subsequent mathematical development) arises from statistical needs, as opposed to needs of the type typically found in the engineering literature. The paper is organized as follows. In Section 2 we briefly present the Kalman filter and summarize its Bayesian derivation. In Section 3 we discuss the sampling model suggested and obtain the new filter. In Section 4 we sketch the application of our filter to Bayesian forecasting, linear regression, autoregressive time series estimation, and multivariate estimation. In Section 5 we present an example of the performance of the new filter when outliers are present. Finally, in Section 6 we provide some concluding remarks.

2. STANDARD KALMAN FILTER

Suppose that we observe at time t a $(p \times 1)$ response vector that is related to a nonobservable $(r \times 1)$ vector of parameters $\boldsymbol{\theta}_t$;

$$\mathbf{y}_t = A_t\boldsymbol{\theta}_t + \boldsymbol{\varepsilon}_y; \qquad \boldsymbol{\varepsilon}_y \sim N(\mathbf{0}, C_t), \tag{2.1}$$

where A_t is a $p \times r$ matrix of known coefficients and the measurement noise $\boldsymbol{\varepsilon}_y$ follows a normal distribution with zero mean and known positive definite covariance matrix C_t. The vector of state parameters $\boldsymbol{\theta}_t$ evolves over time following the linear structure:

$$\boldsymbol{\theta}_t = \boldsymbol{\Omega}_t\boldsymbol{\theta}_{t-1} + \boldsymbol{\varepsilon}_\theta, \qquad \boldsymbol{\varepsilon}_\theta \sim N(\mathbf{0}, R_t), \tag{2.2}$$

where Ω is a $r \times r$ known matrix and R_t is a $r \times r$ positive semidefinite matrix. Equation (2.1) is usually called the measurement equation and (2.2) the state equation. In statistical applications, the vector $\boldsymbol{\theta}_t$ usually represents the parameters of a given model, whereas in engineering applications, $\boldsymbol{\theta}_t$ describes the state of the physical system under study. The Kalman filter is a system of recursions to estimate the parameter vector before a new observation arrives, to forecast this observation, and to update the state vector once the new observation is known.

The structure of the filter can be derived in a Bayesian framework as follows. Suppose that we are at the first stage (i.e., $t = 1$), so that we have not observed any data, and that we have a prior distribution for the parameter vector $\boldsymbol{\theta}_0$ that is normal with mean $\boldsymbol{\mu}_0$ and covariance matrix V_0. Then, using (2.2), the distribution for the state vector in this period, $\boldsymbol{\theta}_t (t = 1)$, will be normal with parameters

$$E[\boldsymbol{\theta}_t | \mathbf{y}_{t-1}] = \boldsymbol{\mu}_{t|t-1} = \Omega_t \boldsymbol{\mu}_{t-1} \tag{2.3}$$

$$\operatorname{var}[\boldsymbol{\theta}_t | \mathbf{y}_{t-1}] = V_{t|t-1} = \Omega_t V_{t-1} \Omega_t^T + R_t, \tag{2.4}$$

where we will use the notation $\boldsymbol{\mu}_{t|t-1}$ to mean the expected value of $\boldsymbol{\theta}_t$ when $\mathbf{y}_{t-1}$ has been observed. [In this first period ($t = 1$) no data have yet been observed.] The next stage is to forecast the expected new observation $\mathbf{y}_t$ and according to (2.1), $\mathbf{y}_t$ will have a normal distribution with parameters

$$E[\mathbf{y}_t | \mathbf{y}_{t-1}] = \hat{\mathbf{y}}_t = A_t \boldsymbol{\mu}_{t|t-1} \tag{2.5}$$

$$\operatorname{var}[\mathbf{y}_t | \mathbf{y}_{t-1}] = M_t = A_t V_{t|t-1} A_t^T + C_t. \tag{2.6}$$

We have used the generic index t, since it is clear that equations (2.3)–(2.6) hold for any t, not only $t = 1$.

When the new observation $\mathbf{y}_t$ is made available, the parameter vector is updated according to Bayes' rule

$$p(\boldsymbol{\theta}_t | \mathbf{y}_t) \propto p(\mathbf{y}_t | \boldsymbol{\theta}_t) p(\boldsymbol{\theta}_t | \mathbf{y}_{t-1}),$$

and using standard Bayesian theory (see Guttman and Peña, 1984) the posterior will be normal with parameters:

$$\boldsymbol{\mu}_{t|t} = \boldsymbol{\mu}_{t|t-1} + V_{t|t-1} A_t^T M_t^{-1} (\mathbf{y}_t - A_t \boldsymbol{\mu}_{t|t-1}) \tag{2.7}$$

and

$$V_{t|t}^{-1} = A_t^T C_t^{-1} A_t + V_{t|t-1}^{-1}, \tag{2.8}$$

so that, as is easily verified,

$$V_{t|t} = V_{t|t-1} - V_{t|t-1} A_t^T M_t^{-1} A_t V_{t|t-1}, \tag{2.9}$$

where we recall that M_t is given at (2.6).

Notice the updating pattern from $\boldsymbol{\mu}_{t|t-1}$, the prior mean, to $\boldsymbol{\mu}_{t|t}$, contained in (2.7). Indeed, this equation implies that the current information $\boldsymbol{\mu}_{t|t}$ about the process parameters $\boldsymbol{\theta}_t$, given $\mathbf{y}_t$, is the prior information of $\boldsymbol{\theta}_t$ given $\mathbf{y}_{t-1}$, $\boldsymbol{\mu}_{t|t-1}$, plus an updating term, obtained by "filtering" the deviation of $\mathbf{y}_t$ from its predictive expectation [see (2.5)] by use of the matrix

$$\mathrm{KF}_t = V_{t|t-1} A^T M_t^{-1}. \tag{2.10}$$

Indeed, the matrix KF_t of (2.10) is referred to as the Kalman gain matrix, and

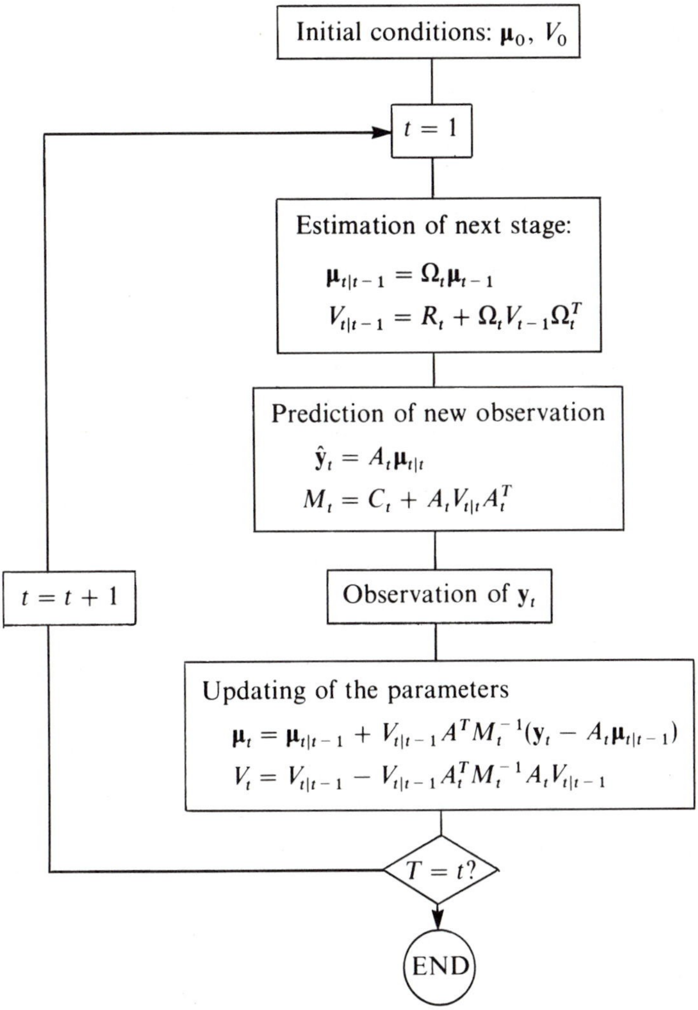
Initial conditions: $\boldsymbol{\mu}_0$, V_0
$t = 1$
Estimation of next stage:
$\boldsymbol{\mu}_{t|t-1} = \Omega_t \boldsymbol{\mu}_{t-1}$
$V_{t|t-1} = R_t + \Omega_t V_{t-1} \Omega_t^T$
Prediction of new observation
$\hat{\mathbf{y}}_t = A_t \boldsymbol{\mu}_{t|t}$
$M_t = C_t + A_t V_{t|t} A_t^T$
Observation of $\mathbf{y}_t$
Updating of the parameters
$\boldsymbol{\mu}_t = \boldsymbol{\mu}_{t|t-1} + V_{t|t-1} A^T M_t^{-1}(\mathbf{y}_t - A_t \boldsymbol{\mu}_{t|t-1})$
$V_t = V_{t|t-1} - V_{t|t-1} A_t^T M_t^{-1} A_t V_{t|t-1}$
$T = t?$
END
$t = t + 1$

Figure 1

we note that an alternative form is

$$\mathrm{KF}_t = V_{t|t} A^T C_t^{-1}.$$

Note, too, the update of $V_{t|t}$ contained in (2.12b); for example, we update $V_{t|t-1}^{-1}$, the precision of the prior of $\boldsymbol{\theta}_t$, given $\mathbf{y}_{t-1}$, by adding the precision $A_t^T C_t^{-1} A_t$ of the regression process parameter $\boldsymbol{\theta}_t$ of (2.1) to obtain $V_{t|t}^{-1}$, and so on. To enter the next stage ($t = 2$), set $\boldsymbol{\mu}_{t|t} = \boldsymbol{\mu}_t$ and $V_{t|t} = V_t$, and repeat the previous analysis. The loop continues in the same way for $t = 3, 4, \ldots$. This is illustrated in Figure 1 for the general case of having reached state $t - 1$ and observed $\mathbf{y}_{t-1}$, and so on.

There are several comments to be made about the foregoing Kalman filter. Essentially, the whole process is a least squares procedure, as the reader will no doubt have known or guessed. Least squares estimates are well known to be nonrobust to outlying observations (see Andrews et al., 1972), and in the Kalman filter these observations may have devastating consequences in some situations, such as line-process control of mass-produced items. Much evidence exists that data almost always contain a small proportion of spuriously generated observations (i.e., not in the manner intended) [see the general discussion in the paper by Box and Tiao (1968) and Guttman (1973)]. For these reasons we replace the assumption (2.1) by a more realistic sampling model, and in the following section investigate what form the ensuing "Kalman filtering" process will take.

3. ROBUST ALGORITHM

3.1 Different Sampling Model

As indicated in the preceding section, we wish to establish procedures that are robust to outliers in that they accommodate the appearance of aberrant observations appropriately—roughly speaking, giving small weight to observations that seem spuriously generated, and large weight to seemingly "good" observations.

As spuriousness that gives rise to outliers often means that the error distributions involved have tails heavier than those of the normal distribution, we generalize below the method of accommodating outliers used by Box and Tiao (1968) and replace the assumption (2.1) by the so-called scaled-contaminated model (SCM). This model was introduced into statistical practice by Jeffreys (1961) and has been used by Box and Tiao (1968) to "robustify" estimation in the standard linear model, by Abraham and Box (1979) to accommodate outliers in time series, and so on. Indeed, Cheng and Box (1980) have shown that the SCM model represents a sensible model in many situations where spuriousness is feared.

The SCM model, simply stated, is that

$$\mathbf{y}_t = A_t\boldsymbol{\theta}_t + \boldsymbol{\varepsilon}_y \tag{3.1}$$

where

$$\boldsymbol{\varepsilon}_y \sim \alpha_1 N(\mathbf{0}, C_1) + \alpha_2 N(\mathbf{0}, C_2). \tag{3.2}$$

In (3.2) we assume that the known constants α_1, α_2 are such that $\alpha_2 = 1 - \alpha_1$, and that, as is common in most applications (see Guttman, 1973), $\alpha_2 \in (0, 0.15)$. Further, we also assume that C_1 and C_2 are known, and such that by any measure, C_2 is larger than C_1. For example, it could be that (for σ_y^2 a known scalar)

$$C_1 = \sigma_y^2 I, \qquad C_2 = k^2\sigma_y^2 I, \qquad k^2 > 1.$$

The prescription (3.2) says that with small probability α_2, $\mathbf{y}_t$ is generated spuriously from $N(A_t\boldsymbol{\theta}_t, C_2)$, and so on.

In addition to (3.1) and (3.2), we also, in this section, make the assumption (2.2) of the preceding section and inquire into the question of how the change in the distribution of the noise $\boldsymbol{\varepsilon}_y$ affects the updating procedure discussed in the introductory section. For convenience we list the assumptions for this section at this point:

(i) (3.1) and (3.2) hold.

(ii) $\boldsymbol{\theta}_t = \Omega_t\boldsymbol{\theta}_{t-1} + \boldsymbol{\varepsilon}_\theta$, $\boldsymbol{\varepsilon}_\theta \sim N(\mathbf{0}, R_t)$.

(iii) $\boldsymbol{\theta}_{t-1} = \boldsymbol{\mu}_{t-1} + \boldsymbol{\varepsilon}_{\theta_{t-1}}$, $\boldsymbol{\varepsilon}_{\theta_{t-1}} \sim N(\mathbf{0}, V_{t-1})$ for given $\mathbf{y}_{t-1}$. (3.3)

We assume as before that we are at $t = 1$ and before observing any data we have a prior distribution for $\boldsymbol{\theta}_0$ that is normal with mean $\boldsymbol{\mu}_0$ and covariance V_0. Then the distribution of $\boldsymbol{\theta}_t (t = 1)$ given $\mathbf{y}_{t-1}$ will be

$$N_{\theta_t}(\boldsymbol{\mu}_{t|t-1}, V_{t|t-1}), \tag{3.4}$$

where

$$\boldsymbol{\mu}_{t|t-1} = \Omega_t\boldsymbol{\mu}_{t-1}, \qquad V_{t|t-1} = R_t + \Omega_t V_{t-1}\Omega_t^T.$$

We can now determine the predictive distribution $h(\cdot|\mathbf{y}_{t-1})$, say of $\mathbf{y}_t$, given $\mathbf{y}_{t-1}$. Formally, this is defined as

$$h(\mathbf{y}_t|\mathbf{y}_{t-1}) = \int_{\theta_t} f(\mathbf{y}_t|\boldsymbol{\theta}_t)p(\boldsymbol{\theta}_t|\mathbf{y}_{t-1})\,d\boldsymbol{\theta}_t, \tag{3.5}$$

and here f is dictated by (3.1), while p is obtained from (3.4). The result of doing the integration (3.5), as proved in the Appendix to this chapter, is as follows:

$$h(\mathbf{y}_t|\mathbf{y}_{t-1}) \sim \alpha_1 N_{\mathbf{y}_t}(A_t\boldsymbol{\mu}_{t|t-1}, M_{t,1}) + \alpha_2 N_{\mathbf{y}_t}(A_t\boldsymbol{\mu}_{t|t-1}, M_{t,2}) \tag{3.6a}$$

with

$$M_{t,i} = C_i + A_t V_{t|t-1} A_t^T, \qquad i = 1, 2, \tag{3.6b}$$

and will have moments

$$E(\mathbf{y}_t|\mathbf{y}_{t-1}) = \hat{\mathbf{y}}_t = A_t \boldsymbol{\mu}_{t|t-1} \tag{3.6c}$$

$$V(\mathbf{y}_t|\mathbf{y}_{t-1}) = \alpha_1 M_{t,1} + \alpha_2 M_{t,2}. \tag{3.6d}$$

3.2 Resulting Recursive Algorithm

The results (3.4) and (3.6) are statements that can be made before seeing $\mathbf{y}_t$. Once $\mathbf{y}_t$ is observed, we are in the position of being able to find the (posterior) distribution of $\boldsymbol{\theta}_t$, given $\mathbf{y}_t$. It is proved in the Appendix that the posterior distribution of $\boldsymbol{\theta}_t$, given $\mathbf{y}_t$, is

$$p(\boldsymbol{\theta}_t|\mathbf{y}_t) = \sum_{i=1}^{2} \alpha_{t,i} N_{\boldsymbol{\theta}_t}(\boldsymbol{\mu}_{t|t}^{(i)}, V_{t|t}^{(i)}), \tag{3.7a}$$

where

$$\boldsymbol{\mu}_{t|t}^{(i)} = \boldsymbol{\mu}_{t|t-1} + V_{t|t-1} A_t^T M_{t,i}^{-1} (\mathbf{y}_t - A_t \boldsymbol{\mu}_{t|t-1}) \tag{3.7b}$$

$$V_{t|t}^{(i)} = V_{t|t-1} - V_{t|t-1} A_t^T M_{t,i}^{-1} A_t V_{t|t-1} \tag{3.7c}$$

and where

$$\alpha_{t,i} = \frac{\alpha_i f(\mathbf{y}_t|A_t\boldsymbol{\mu}_{t|t-1}, M_{t,i})}{\sum_{j=1}^{2} \alpha_j f(\mathbf{y}_t|A_t\boldsymbol{\mu}_{t|t-1}, M_{t,j})}, \tag{3.7d}$$

with f denoting the density of the normal multivariate distribution, where, in general,

$$f(\mathbf{y}|\boldsymbol{\eta}, M) = (2\pi)^{-p/2}|M|^{-1/2} \exp\left[-\tfrac{1}{2}(\mathbf{y} - \boldsymbol{\eta})^T M^{-1}(\mathbf{y} - \boldsymbol{\eta})\right]. \tag{3.7e}$$

The reader will recognize that the denominator of (3.7d) is, on using (3.7e), the predictive density h of $\mathbf{y}_t$, given $\mathbf{y}_{t-1}$, stated in (3.6). Indeed, using (3.7e) in (3.7d) yields

$$\alpha_{t,1} = \left(1 + \frac{\alpha_2}{\alpha_1}\left(\frac{|M_{t,1}|}{|M_{t,2}|}\right)^{1/2} \times \exp\left\{\tfrac{1}{2}(\mathbf{y}_t - A_t\boldsymbol{\mu}_{t|t-1})^T(M_{t,1}^{-1} - M_{t,2}^{-1})(\mathbf{y}_t - A_t\boldsymbol{\mu}_{t|t-1})\right\}\right)^{-1},$$

$$\alpha_{t,2} = 1 - \alpha_{t,1}. \tag{3.7f}$$

From (3.7a)–(3.7f) we easily find (see the Appendix for proofs)

$$\begin{aligned} E(\boldsymbol{\theta}_t|\mathbf{y}_t) &= \boldsymbol{\mu}_{t|t-1} + V_{t|t-1}A_t^T\{\alpha_{t,1}M_{t,1}^{-1} + \alpha_{t,2}M_{t,2}^{-1}\}(\mathbf{y}_t - A_t\boldsymbol{\mu}_{t|t-1}) \\ &= \boldsymbol{\mu}_{t|t} \\ V(\boldsymbol{\theta}_t|\mathbf{y}_t) &= V_{t|t-1} - V_{t|t-1}A_t^T B_t A_t V_{t|t-1} = V_{t|t}, \end{aligned} \tag{3.8}$$

where

$$\begin{aligned} B_t = \alpha_{t,1}M_{t,1}^{-1} + \alpha_{t,2}M_{t,2}^{-1} - \alpha_{t,1}\alpha_{t,2}(M_{t,1}^{-1} - M_{t,2}^{-1})(\mathbf{y}_t - A_t\boldsymbol{\mu}_{t|t-1}) \\ \times (\mathbf{y}_t - A_t\boldsymbol{\mu}_{t|t-1})^T(M_{t,1}^{-1} - M_{t,2}^{-1}). \end{aligned} \tag{3.8a}$$

The quantities $\alpha_{t,1}$ and $\alpha_{t,2} = (1 - \alpha_{t,1})$ are the posterior probabilities that $\mathbf{y}_t$ has come from the intended source [i.e., $N(A_t\boldsymbol{\theta}_t, C_1)$] and the spurious source [i.e., $N(A_t\boldsymbol{\theta}_t, C_2)$], respectively. The posterior expectation is made up of two parts: the posterior expectation $\boldsymbol{\mu}_{t|t-1}$ of $\boldsymbol{\theta}_t$, given $\mathbf{y}_{t-1}$ [see (3.4)] and a deviation of $\mathbf{y}_t$ from its (marginal) predictive expectation $A_t\boldsymbol{\mu}_{t|t-1}$, but this time the filtering (gain) matrix is the weighted sum of two Kalman gain matrices, where the weights are the estimates $\alpha_{t,i}$, commented on just above. Put another way, the gain matrix involves the weighted sum of the predictive precisions that would be involved if sampling was from either $N(A_t\boldsymbol{\theta}_t, C_1)$ or $N(A_t\boldsymbol{\theta}_t, C_2)$, with weights the probabilities that y_t was so sampled. For more introspection about the updated variance-covariance $V_{t|t}$, we first invite the reader to inspect the updated $V_{t,t}^{(i)}$'s, and to acquaint themselves with the details of how these are used by reading the proof of (3.8a) in the Appendix.

To continue this procedure, we now use the following scheme:

$$\text{Set } \boldsymbol{\mu}_t = \boldsymbol{\mu}_{t|t} \quad \text{and} \quad V_t = V_{t|t} \qquad \text{[see (3.8) and (3.8a)].} \tag{3.9}$$

Following Heyde and Johnstone (1979), we let

$$p(\boldsymbol{\theta}_t|\mathbf{y}_t) \simeq N(\boldsymbol{\mu}_t, V_t). \tag{3.10}$$

[We remark that Heyde and Johnstone (1979) show that the limiting distribution of the posterior distribution as "t goes to infinity" is normal, under very general assumptions about the joint distribution of the y_t.] We may now enter the next stage—we replace (iii) of (3.3) with:

$$\text{Given } \mathbf{y}_t, \quad \boldsymbol{\theta}_t = \boldsymbol{\mu}_t + \boldsymbol{\varepsilon}_{\boldsymbol{\theta}_t}, \quad \boldsymbol{\varepsilon}_{\boldsymbol{\theta}_t} \sim N(\mathbf{0}, V_t) \tag{3.11}$$

and making the obvious modifications in (3.1) and (ii) of (3.3)—t is replaced by $(t + 1)$—we repeat the process above. This is illustrated in Figure 2 for the general case.

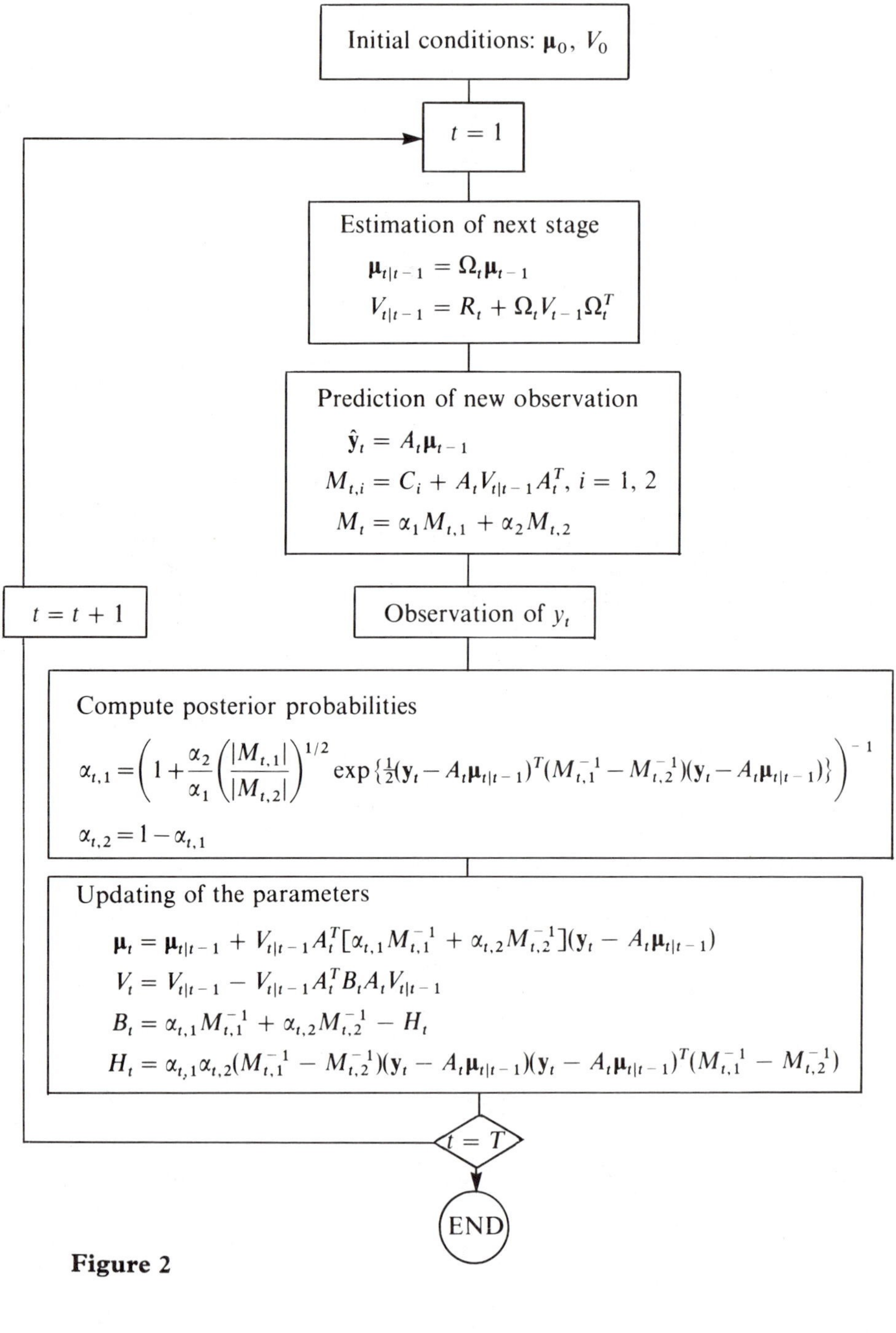
Initial conditions: $\boldsymbol{\mu}_0$, V_0
$t = 1$
Estimation of next stage
$\boldsymbol{\mu}_{t|t-1} = \Omega_t \boldsymbol{\mu}_{t-1}$
$V_{t|t-1} = R_t + \Omega_t V_{t-1} \Omega_t^T$
Prediction of new observation
$\hat{\mathbf{y}}_t = A_t \boldsymbol{\mu}_{t-1}$
$M_{t,i} = C_i + A_t V_{t|t-1} A_t^T$, $i = 1, 2$
$M_t = \alpha_1 M_{t,1} + \alpha_2 M_{t,2}$
$t = t + 1$
Observation of y_t
Compute posterior probabilities
$\alpha_{t,1} = \left(1 + \frac{\alpha_2}{\alpha_1} \left(\frac{|M_{t,1}|}{|M_{t,2}|} \right)^{1/2} \exp\{\tfrac{1}{2}(\mathbf{y}_t - A_t \boldsymbol{\mu}_{t|t-1})^T (M_{t,1}^{-1} - M_{t,2}^{-1})(\mathbf{y}_t - A_t \boldsymbol{\mu}_{t|t-1})\} \right)^{-1}$
$\alpha_{t,2} = 1 - \alpha_{t,1}$
Updating of the parameters
$\boldsymbol{\mu}_t = \boldsymbol{\mu}_{t|t-1} + V_{t|t-1} A_t^T [\alpha_{t,1} M_{t,1}^{-1} + \alpha_{t,2} M_{t,2}^{-1}](\mathbf{y}_t - A_t \boldsymbol{\mu}_{t|t-1})$
$V_t = V_{t|t-1} - V_{t|t-1} A_t^T B_t A_t V_{t|t-1}$
$B_t = \alpha_{t,1} M_{t,1}^{-1} + \alpha_{t,2} M_{t,2}^{-1} - H_t$
$H_t = \alpha_{t,1} \alpha_{t,2} (M_{t,1}^{-1} - M_{t,2}^{-1})(\mathbf{y}_t - A_t \boldsymbol{\mu}_{t|t-1})(\mathbf{y}_t - A_t \boldsymbol{\mu}_{t|t-1})^T (M_{t,1}^{-1} - M_{t,2}^{-1})$
$t = T$
END

Figure 2

4. SCOPE OF THE NEW FILTER

In the preceding section we developed a filter that may be used in various situations. To illustrate this, we discuss several instances where the new filter of Section 3 may be applied.

4.1 Bayesian Forecasting

One of the best-known applications of the Kalman filter in statistics is to that of Bayesian forecasting, as developed by Harrison and Stevens (1976). These authors used the standard formulation developed in the engineering literature for the linear control problem that we have presented in our Section 2, calling it the dynamic linear model. They showed that most statistical models can be considered under this general framework, which thus allows a unified approach, namely, by using the Kalman filter for recursive estimation of the parameters. Harrison and Stevens (1976) then applied these ideas to Bayesian forecasting.

As our robust filtering includes the standard algorithm as a particular case, the robust version of the algorithm we have developed can be applied to any of the forecasting models discussed by these authors.

To illustrate the behavior of the algorithm, we discuss its application to one of the most widely used model for forecasting, called the steady model (Harrison and Stevens, 1976). This model is a particular case of the general formulation (2.1) and (2.2) with $A_t = 1, \Omega_t = 1$ and with y_t and θ_t scalars. Then $\boldsymbol{\mu}_{t|t-1} = \mu_{t-1}$ and $V_{t|t-1} = R_t + V_{t-1}$. The updating equations for the standard recursive estimation of this model are, assuming that $R_t = R$ and $C_t = C$,

$$\mu_t = \mu_{t-1} + \frac{R + V_{t-1}}{C + R + V_{t-1}}(y_t - \mu_{t-1}) \tag{4.1}$$

$$V_t = \frac{C(R + V_{t-1})}{C + R + V_{t-1}}. \tag{4.1a}$$

After some iteration the system will reach a stable state in which $V_t = V_{t-1} = a$. Calling $\theta = C(C + R + a)^{-1}$, we see that (4.1) can be written as

$$\mu_t = \theta\mu_{t-1} + (1 - \theta)y_t, \tag{4.2}$$

and applying it recursively and assuming that $\mu_0 = 0$, we obtain

$$\mu_t = \sum_{i=1}^{t-1} \theta^i(1 - \theta)y_{t-i}. \tag{4.3}$$

We have, then, from (4.3) the result that the steady model is equivalent to an integrated first-order moving average process (see Box and Jenkins, 1970), which is equivalent to exponential smoothing.

To apply the robust Kalman filtering algorithm described in Section 3, we assume that $c_1 = \sigma^2$, $c_2 = k^2\sigma^2$; then

$$\mu_t = \mu_{t-1} + (R + V_{t-1})\left(\frac{\alpha_{t,1}}{\sigma^2 + R + V_{t-1}} + \frac{\alpha_{t,2}}{k^2\sigma^2 + R + V_{t-1}}\right)(y_t - \mu_{t-1}), \tag{4.4}$$

and assuming as before a stable state with $V_t = V_{t-1} = a$, then calling $\theta_1 = \sigma^2(\sigma^2 + R + a)^{-1}$ and $\theta_2 = (k^2\sigma^2)(k^2\sigma^2 + R + a)^{-1}$, we have

$$\mu_t = (\alpha_{t,1}\theta_1 + \alpha_{t,2}\theta_2)\mu_{t-1} + [(1 - \theta_1)\alpha_{t,1} + (1 - \theta_2)\alpha_{t,2}]y_t.$$

This expression shows that μ_t is obtained as a weighted combination of μ_{t-1} and y_t, as in the standard case, but now the weights are changing in every iteration and depend on the probabilities $\alpha_{t,1}$ and $\alpha_{t,2}$. Setting $\theta_{(t)} = \alpha_{t,1}\theta_1 + \alpha_{t,2}\theta_2$, we can write

$$\mu_t = \theta_{(t)}\mu_{t-1} + (1 - \theta_{(t)})y_t, \tag{4.5}$$

and now the expression for μ_t as a function of the observations is

$$\mu_t = \sum_{i=0}^{n} \frac{1 - \theta_{(t-i)}}{\theta_{(t-1)}} y_{t-i}\left(\prod_{j=0}^{i} \theta_{(t-j)}\right). \tag{4.6}$$

To interpret this equation, let us denote by $w(r; i, j, k, \ldots, h)$ the probability that r observations came from the "spurious" or "bad" population at times i, j, $k, \ldots, h$, where there are r symbols $i, j, \ldots, h$. Then

$$w(r; i, j, k, \ldots, h) = \left(\prod_{\substack{t=1 \\ t \neq i, j, \ldots, h}}^{n} \alpha_{t,1}\right) \alpha_{i2}\alpha_{j2}\alpha_{k2} \cdots \alpha_{h2} \tag{4.7}$$

and let us call $\hat{\mu}_t(r;\ i,\ j,\ k, \ldots, h)$ the smoothing associated with this combination of observations [see (4.3)]. This smoothing is the result of assuming that

$$\mu_t = \theta_{(1)}\mu_{t-1} + (1 - \theta_{(1)})y_t \tag{4.7a}$$

for observations y_t, $t = 1, \ldots, n$, but $t \neq i, j, k, \ldots, h$, and

$$\mu_t = \theta_{(2)}\mu_{t-1} + (1 - \theta_{(2)})y_t \tag{4.7b}$$

for $t = i, j, k, \ldots, h$. Then it is straightforward to show that

$$\mu_t = \sum w(r; i, j, \ldots, h)\hat{\mu}(r; i, j, \ldots, h), \tag{4.8}$$

where the summation is over all the 2^t possible combinations of observations chosen from $y_1, \ldots, y_t$. The main advantage of our algorithm (4.5) is that the 2^t exponential smoothing factors are the result of the recursive relationship (4.4) and are not computed separately but globally and here in a very direct way.

4.2 Robust Linear Regression

As mentioned previously, Box and Tiao (1968) assumed a scale-contaminated normal distribution for the noise of a regression model and used Bayesian estimation methods to obtain a "robust" estimation procedure that downweights suspicious observations in the linear model. Additionally, Chen and Box (1979) showed that given appropriate values to the parameters of the noise distributions, the Box-Tiao weights can reproduce functions for downweighting residuals using M-estimators that have been proposed on empirical grounds by Andrews et al. (1972). Although this approach provides a general way to deal with outliers in the linear model, the computations needed are cumbersome, because the posterior distribution of the parameter vector is a weighted average of 2^n posterior distributions. Box and Tiao (1968) suggested that one need compute only the first few leading terms, but with a large set of data the computational burden is still heavy, and there are no clear rules about how many terms we would need to obtain a proper approximation. Little (1983) has explored the relationship between the Bayesian weights and some influence measures proposed for the linear model and has used this relationship to suggest an algorithm to determine which of the weights matter, and their subsequent computation. We show in this section how we can compute the posterior mean in a simple way using the robust algorithm we have suggested.

The regression can be written as the state-space model

$$
\begin{aligned}
y_t &= \mathbf{x}_t' \boldsymbol{\beta}_t + \varepsilon_t \\
\boldsymbol{\beta}_t &= \boldsymbol{\beta}_{t-1} = \boldsymbol{\beta}
\end{aligned}
\tag{4.9}
$$

where y_t is now scalar in the observation equation and the regression parameter vector is constant over time. Assuming that the noise ε_t has a scale-contaminated normal distribution,

$$
\varepsilon_t \sim (1-\alpha)N(0, \sigma^2) + \alpha N(0, k^2\sigma^2), \qquad k^2 > 1, \tag{4.10}
$$

for fixed values of α, k^2, and σ^2, we then have a particular case of our previous formulation. In practice, however, none of the three parameters are known. Although one can assume α and k^2 known and make a sensitivity study afterward (see Box and Tiao, 1968), the problem of estimating σ^2 remains. The choice of an estimator for σ^2 is rather arbitrary. We start with a preliminary estimate of σ^2, say $\hat{\sigma}^2_{(1)}$, obtained from the residuals of the least squares fit that leads to the estimate $\hat{\boldsymbol{\beta}}_{(1)}$, and then compute an estimate $\hat{\boldsymbol{\beta}}_{(2)}$ using the robust Kalman filtering algorithm with the value $\hat{\sigma}^2_{(1)}$ for the variance. We then compute the new residuals and a new estimator $\hat{\sigma}^2_{(2)}$ for the variance. Using this estimator a new application of the algorithm is made which provides a new estimator $\hat{\boldsymbol{\beta}}_{(3)}$ and, consequently, a new set of residuals. From them a new variance estimator $\hat{\sigma}^2_{(3)}$ is built and the process is repeated until convergence.

In this particular case, writing $V_t = \sigma^2 \dot{V}_t$ and $m_{it} = \sigma^2 \dot{m}_{it}$, the computations needed are

$$\dot{m}_{1t} = 1 + \mathbf{x}_t' V_{t-1} \mathbf{x}_t = 1 + h_{(1)} \tag{4.11}$$

$$\dot{m}_{2t} = k^2 + h_{(1)} \tag{4.12}$$

$$\alpha_{t,1} = \left(1 + \frac{\alpha}{1-\alpha}\sqrt{\frac{\dot{m}_{1t}}{\dot{m}_{t2}}}\exp\left\{\frac{1}{2}\left[\left(\frac{y_t - \mathbf{x}_t'\hat{\boldsymbol{\beta}}_{t-1}}{\hat{\sigma}_{(t)}}\right)^2\left(\frac{1}{\dot{m}_{1t}} - \frac{1}{\dot{m}_{2t}}\right)\right]\right\}\right)^{-1} \tag{4.13}$$

$$\hat{\boldsymbol{\beta}}_{(t)} = \hat{\boldsymbol{\beta}}_{(t-1)} + \dot{V}_t \mathbf{x}_t \left(\frac{\alpha_{t,1}}{\dot{m}_{1t}} + \frac{\alpha_{t,2}}{\dot{m}_{2t}}\right) e_{(t)} \tag{4.14}$$

$$e_{(t)} = y_t - \mathbf{x}_t'\hat{\boldsymbol{\beta}}_{(t-1)} \tag{4.15}$$

$$\dot{V}_t = \dot{V}_{t-1} - \dot{V}_{t-1}\mathbf{x}_t b_t \mathbf{x}_t' \dot{V}_{t-1}, \tag{4.16}$$

where

$$b_t = \frac{\alpha_{t,1}}{\dot{m}_{1t}} + \frac{\alpha_{2,t}}{\dot{m}_{2t}} - \alpha_{t,1}\alpha_{t,2}\left(\frac{1}{\dot{m}_{1t}} - \frac{1}{\dot{m}_{2t}}\right)^2\left(\frac{e_{(t)}}{\hat{\sigma}_{(t)}}\right)^2. \tag{4.17}$$

A simple alternative for an estimator of scale is the median absolute deviation from the sample median $\tilde{y}$, given by

$$\text{MAD} = \underset{t}{\text{median}}\,\{|y_t - \tilde{y}|\}.$$

This estimator is known to be more efficient for the contaminated normal distribution than the sample variance, but is biased. A (robust) unbiased estimator of σ^2 has been provided by Andrews et al. (1972) and is given by

$$\hat{\sigma}_r^2 = (0.64)^{-1}\,\underset{i}{\text{median}}\,\{e_i\}, \tag{4.18}$$

where the e_i are the residuals from the least squares fit. The final covariance matrix for the parameters will be $V_t = \hat{\sigma}_r^2 \dot{V}_t$, where $\dot{V}_t$ is computed using (4.16).

Equation (4.14) shows that the estimate $\hat{\boldsymbol{\beta}}_{(t)}$ is a linear combination of estimators that would have been obtained assuming that ε_t is distributed as $N(0, \sigma^2)$ and $N(0, k^2\sigma^2)$, respectively, with weights that are the posterior probabilities $\alpha_{t,i}$'s, discussed before. As for the variance, instead of the least squares expression,

$$\dot{V}_t^{-1} = (V_{t-1}^{-1} + \mathbf{x}_t\mathbf{x}_t')^{-1}, \tag{4.19}$$

where $V_{t-1}^{-1} = X'X$, now equation (4.11) imposes an adaptive formulation:

$$\dot{V}_t^{-1} = (V_{t-1}^{-1} + a_t\mathbf{x}_t\mathbf{x}_t')^{-1}, \tag{4.20}$$

where $a_t = b_t/(1 - h_{(1)}b_t)$ depends on the posterior probabilities of the observation being an outlier. For example, it is straightforward to see that if $\alpha_{t,1} = 1$, $a_t = 1$, and if $\alpha_{t,2} = 1$, $a_t = k^{-2}$.

This algorithm provides a useful computational device to estimate $\boldsymbol{\beta}$ when the set of data is large. However, when the sample is small, there are two problems that make this approximation a crude one. First, the distribution for $\boldsymbol{\beta}$ at each state is not normal, but is a t-distribution when σ^2 is unknown. Second, with a small set of data, the ordering of the observations can influence the results. These two problems are not important for a large data set because (1) the t-distribution, when the number of degrees of freedom is large, is very well approximated by the normal, and (2) the order of the observations is not too important when the sample size is large, and a stable state has been reached.

The above is in contrast with the standard Kalman filter, in which σ^2 is not needed, while here the variances m_{it} are required, to compute the posterior probabilities that an observation has been spuriously generated, and for the variance of the estimation itself.

4.3 Autoregressive Time Series Estimation

Assuming in the previous case that $\mathbf{x}_t' = (y_{t-1}, \ldots, y_{t-p})$, the model reduces to the state-space estimation of an autoregressive process. Abraham and Box (1979) have studied inference in this type of model when there is a small probability that "bad" observations occur. The exact Bayesian solution is again difficult to compute because it involves, as in the regression case, the computation of 2^n distributions.

Although in this case the previous approach can be used to obtain a robust estimate of the parameters when all the data have been collected, the algorithm can also be used as a robust procedure for on-line estimation when the observations are received sequentially. In the latter case, however, a way to compute an estimate of σ^2 given the observed data is needed because at every instant t only observations $y_1, \ldots, y_t$, will be available.

The solution we suggest is to start with a robust estimate of the variance, and update this estimation when the new residual $y_t - \hat{y}_t$ is computed. The algorithm will be defined by the same equations but now (4.13) and (4.17) are

$$\alpha_{t,1} = \left(1 + \frac{\alpha}{1-\alpha}\sqrt{\frac{\dot{m}_{1t}}{\dot{m}_{2t}}}\exp\left\{\frac{1}{2}\left[\left(\frac{y_t - \mathbf{x}_t'\hat{\boldsymbol{\beta}}_{t-1}}{\hat{\sigma}}\right)^2\left(\frac{1}{\dot{m}_{1t}} - \frac{1}{\dot{m}_{2t}}\right)\right]\right\}\right)^{-1} \quad (4.21)$$

$$b_t = \frac{\alpha_{t,1}}{\dot{m}_{1t}} + \frac{\alpha_{t,2}}{\dot{m}_{2t}} - \alpha_{1t}\alpha_{2t}\left(\frac{1}{\dot{m}_{1t}} - \frac{1}{\dot{m}_{2t}}\right)^2\left(\frac{e_{(t)}}{\hat{\sigma}_t}\right)^2 \quad (4.22)$$

with

$$e_{(t)} = y_t - \hat{\beta}_{t1}y_{t-1} - \cdots - \dot{\beta}_{tp}y_{t-p} \quad (4.23)$$

$$\hat{\sigma}_t = \frac{\underset{t}{\text{median}}\,\{e_{(t)}\}}{0.64}. \quad (4.24)$$

4.4 Multivariate Estimation

In the previous examples the observed vector $\mathbf{y}_t$ was a scalar, but the algorithm is also very simple when $\mathbf{y}_t$ is a vector. As an example, we revise only the multivariate regression model. The standard formulation is

$$\begin{aligned} \mathbf{y}_t &= \mathbf{H}_t\boldsymbol{\mu}_t + \boldsymbol{\beta}_t\boldsymbol{\varepsilon} \\ \boldsymbol{\beta}_t &= \boldsymbol{\beta}_{t-1} = \boldsymbol{\beta}, \end{aligned} \quad (4.25)$$

with

$$H_t = \begin{bmatrix} \mathbf{x}_t' & \mathbf{0}' & \cdots & \mathbf{0}' \\ & \ddots & & \\ \mathbf{0}' & & & \mathbf{x}_t' \end{bmatrix}, \qquad \boldsymbol{\beta}_t = \begin{bmatrix} \boldsymbol{\beta}_{1t} \\ \vdots \\ \boldsymbol{\beta}_{kt} \end{bmatrix}, \quad (4.26)$$

where $\mathbf{x}_t'$ is the $1 \times p$ vector of explanatory variables and $\boldsymbol{\beta}_{it}$ is the $p \times 1$ vector of parameters linked to component y_{it} of $\mathbf{y}_t$. The noise $\boldsymbol{\varepsilon}_t$ is assumed to have a mixed distribution and we can impose different structures, depending on the particular problem on hand. For instance, if

$$C_1 = \begin{bmatrix} \sigma_{11} & \sigma_{12} & \cdots & \sigma_{1k} \\ \sigma_{k1} & \cdots & & \sigma_{kk} \end{bmatrix}, \quad (4.27)$$

the presence of a multivariate outlier can be modeled assuming that they come from a distribution with covariance matrix

$$C_2 = \begin{bmatrix} k_1^2\sigma_{11} & & \\ & \ddots & \\ & & k_k^2\sigma_{kk} \end{bmatrix}, \qquad k_j^2 > 1, \quad (4.28)$$

which implies that the components of an outlier are unrelated. This is referred to as external structure for spurious observations. Another example is $C_2 = h^2 C_1$, $h^2 > 1$, so that the components of outlying observations are related in the same way as observations from the good or intended source, but variances of outlying components are larger. This is referred to as internal multivariate structure for spurious observations.

5. AN EXAMPLE

To check the performance of the filter that we have developed, we made extensive computer simulations which have indeed verified the robustness of our filter. As an example, Table 1 and Figure 3 show 31 observations generated from the steady model of Section 4.1,

$$\begin{aligned} \mu_t &= \mu_{t-1} + \varepsilon_\mu, \qquad & \varepsilon_\mu &\sim N(0, 1) \\ y_t &= \mu_t + \varepsilon_y, \qquad & \varepsilon_y &\sim N(0, 1). \end{aligned} \tag{5.1}$$

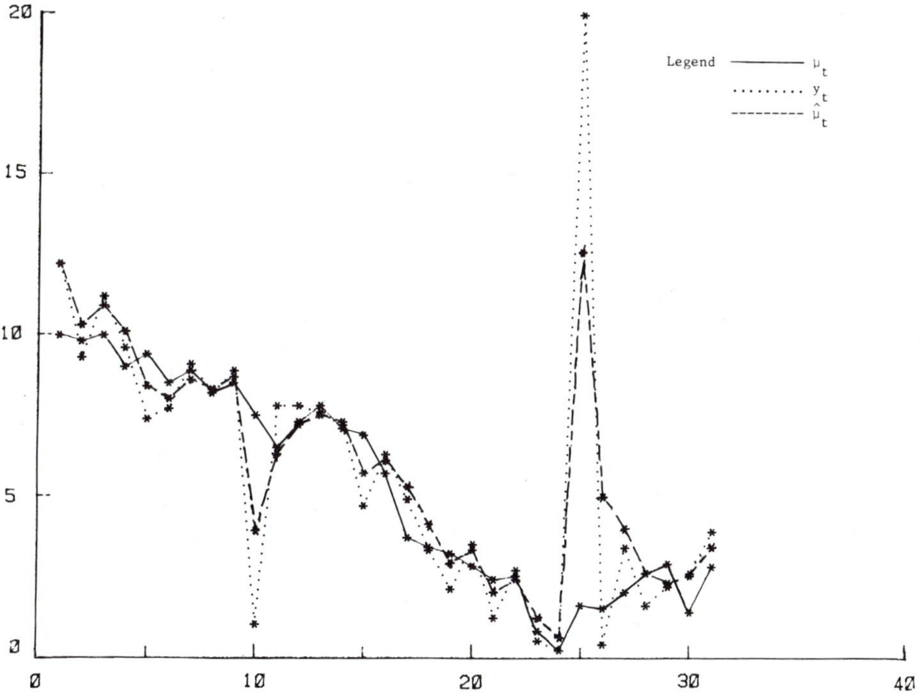

Figure 3. Plots of the true μ_t's, observed y_t's, and the estimates $\hat{\mu}_t$ of μ_t obtained from standard Kalman filtering.

Table 1. Data y_t and μ_t's Obtained Using (5.1), Along with Estimates $\hat{\mu}_t$ of μ_t Obtained Using the Standard Kalman Filter [see (4.1) and Figure 1] and Those Using the Robust Filter [see (4.4) and Figure 2[a]].

			Standard filtering		Robust filtering		
t	μ_t	y_t	$\hat{\mu}_t$ [see (4.1)]	V_t – (4.1a) ($V_0 = 10{,}000$)	$\hat{\mu}_t$ [see (4.4)]	$\alpha_{t,1}$ ($\alpha = 0.05$)	V_t ($V_0 = 10{,}000$)
1	10.00	12.18	12.18	1.0	12.17	0.95	2.2
2	9.83	9.32	10.28	0.7	10.18	0.90	1.3
3	9.98	11.20	10.85	0.6	10.87	0.98	0.7
4	8.99	9.59	10.07	0.6	10.07	0.97	0.7
5	9.36	7.41	8.42	0.6	8.64	0.85	1.1
6	8.50	7.69	7.97	0.6	8.01	0.98	0.7
7	8.90	9.06	8.64	0.6	8.66	0.98	0.7
8	8.20	8.17	8.35	0.6	8.36	0.98	0.6
9	8.47	8.86	8.67	0.6	8.66	0.98	0.6
10	7.46	1.00	3.93	0.6	8.19	0.00	1.5
11	6.49	7.79	6.32	0.6	7.91	0.98	0.7
12	7.34	7.79	7.23	0.6	7.84	0.98	0.7
13	7.82	7.62	7.47	0.6	7.71	0.98	0.6

14	7.06	7.19	7.30	0.6	7.39	0.98	0.6
15	6.85	4.71	5.70	0.6	6.00	0.84	1.1
16	5.67	6.28	6.06	0.6	6.17	0.98	0.7
17	3.69	4.88	5.33	0.6	5.38	0.97	0.7
18	3.37	3.34	4.10	0.6	4.18	0.94	0.8
19	3.25	2.08	2.85	0.6	2.92	0.93	0.8
20	2.81	3.53	3.27	0.6	3.31	0.98	0.7
21	2.36	1.25	2.02	0.6	2.09	0.93	0.8
22	2.46	2.70	2.44	0.6	2.47	0.98	0.7
23	0.82	0.48	1.23	0.6	1.29	0.94	0.7
24	0.24	0.19	0.59	0.6	0.61	0.98	0.7
25	1.62	20.00	12.59	0.6	1.82	0.00	1.6
26	1.46	0.35	5.03	0.6	0.79	0.97	0.8
27	1.96	3.42	4.03	0.6	2.28	0.87	1.0
28	2.62	1.64	2.55	0.6	1.86	0.98	0.7
29	2.95	2.17	2.31	0.6	2.05	0.98	0.6
30	1.40	2.64	2.52	0.6	2.41	0.98	0.6
31	2.84	3.87	3.35	0.6	3.29	0.97	0.7

[a]The observations y_{10} and y_{25} are outliers, replacing the original y_{10} and y_{25}. Values were taken from output computed to four decimals, and rounded off.

In the language of Section 4.1, $R = 1 = C$. Two outliers have been introduced in the measurement variable at points $t = 10$ and 25. We have assumed a vague prior for μ_0, characterized by a prior mean that is far from this initial state—the assumed mean for the prior is 15 and in the simulation, we have arbitrarily set the value of μ_1 as 10. Further, we have assumed a large prior standard deviation—$\sigma_0 = 100$. Figure 3 plots y_t, μ_t and also shows the performance of the standard Kalman filter [see Figure 1 and (4.1)–(4.1a)] by plotting the updated μ_t's, called $\hat{\mu}_t$, of (4.1) using the true values for the standard deviation of the errors of the measurement and state equation. Figure 3 shows the well-known result that the effects of initial conditions dies out rapidly not affecting the performance of the filter after a few steps. It is also seen that both outliers strongly affect the estimation of the state μ_t for their respective periods ($t = 10$ and 25) and distort for a while the performance of the filter. Also, Table 1 shows that the estimate of the variance converges quickly to the asymptotic value, a, where using (4.1a), a is the

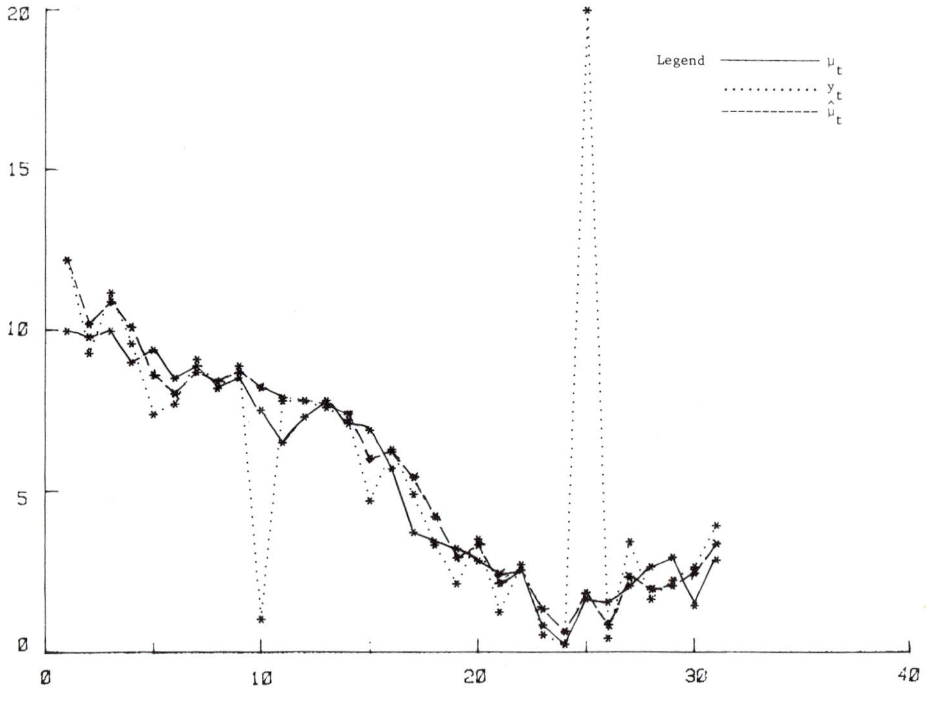

Figure 4. Plots of the true μ_t's, observed y_t's, and the estimates $\hat{\mu}_t$ of μ_t obtained from the robust Kalman filter.

solution of

$$a^2 + aR - CR = 0. \tag{5.2}$$

For $C = R = 1$, $a = 0.618 \simeq 0.6$.

Figure 4 shows the behavior with the same data using our filter [see (4.4)]. We have assumed the same initial conditions, with $\alpha_1 = 0.95$ and standard deviations for the mixture of $\sqrt{C_1} = 1$, $\sqrt{C_2} = 5$. The figure shows that for "good" observations the behavior of our filter is the same as the standard filter, but when an observation is detected as being an outlier (see the probabilities $\alpha_{t,1}$ of Table 1), it is practically disregarded from the computations. For example, Table 1 also shows that the posterior probability for the two "outlier" observations ($t = 10, 25$) coming from the aberrant-observation distribution is practically one, and although the standard deviation of the "bad" distribution is five times that of the true distribution, the outlier observations occurring at $t = 10$ and 25 are, to all practical effects, eliminated. [See Table 1, and in particular the column of $\hat{\mu}_t$'s, the updated μ_t's, for robust filtering given by (4.4).]

These results are fairly robust to the value of α, the prior probability of being an outlier, in the range [0.01, 0.1], and also robust to the values of the ratio C_2/C_1 when $C_2/C_1 > 3$. Figure 4 portrays graphically how well our filter works—the $\hat{\mu}_t$'s are much closer to μ_t's for all t, and especially at $t = 10$ and 25, then the μ_t's obtained using the standard Kalman filter, portrayed in Figure 3.

6. CONCLUSIONS

Real data sets almost always contain outlying (extreme) observations, and outliers are particularly damaging in on-line control situations in which the data are processed recursively. Thus an extremely bad value can distort the whole mechanism of control and make the process very unstable. In industrial practice, for example, all types of ad hoc procedures have been developed to cope with this situation (see Mah and Tamhane, 1982; Crowe et al., 1983), but a general methodology is needed.

In this chapter we offer a relatively simple model and obtain a procedure to deal with the problem. To represent the appearance of bad observations, a scale-contaminated normal distribution has been assumed for the measurement error. We have chosen this model because other authors have demonstrated that its use provides sensible solutions to other statistical problems, for example, in linear model estimation.

In fact, we have shown in this chapter how a Bayesian approach allows the development of a simple recursive estimation algorithm that has the desired property of "filtering" bad (i.e., extreme) observations. Indeed, extreme values are downweighted by their posterior probability of being spurious, and the estimates of the parameters are updated, recursively, accordingly.

Finally, we apply our model to the case of exponential smoothing with contaminated error, and show that the parameter estimates obtained from the resulting algorithm are a weighted combination of certain 2^n smoothing schemes. The application of this procedure to a broad range of statistical estimation problems is discussed briefly.

ACKNOWLEDGMENTS

This research was partially supported by NSERC (Canada) under Grant A8743 and by the Comision Asesora de Investigación Cientifica y Técnica, Spain. The authors are very grateful to Tom Leonard, University of Wisconsin–Madison, for valuable discussions and comments on this paper.

APPENDIX

This appendix is devoted to the derivation of results (3.6), (3.7), and (3.8).

We have from (3.5), with f given by (3.1) and p given by (2.3), that the predictive distribution of y_t, given y_{t-1}, is such that

$$h(\mathbf{y}_t|\mathbf{y}_{t-1}) = \sum_{i=1}^{2} \alpha_i \int_{\boldsymbol{\theta}_t} f(\mathbf{y}_t|A_t\boldsymbol{\theta}_t, C_i) f(\boldsymbol{\mu}_{t|t-1}, V_{t|t-1})\, d\boldsymbol{\theta}_t, \tag{A.1}$$

where the density f is defined in (3.7d). But the integral operation is clearly equivalent to the argument used in the standard formulation with C_i, replacing C so that

$$h(\mathbf{y}_t|\mathbf{y}_{t-1}) = \sum_{i=1}^{2} \alpha_i f(\mathbf{y}_t|A_t\boldsymbol{\mu}_{t|t-1}, M_{t,i}), \tag{A.2}$$

where $M_{t,i}$ is as defined in (3.6b), and the result (3.6) is now proved.

To derive the results (3.7), we note that (3.4) gives the prior for $\boldsymbol{\theta}_t$, while the distribution specified by (3.1) dictates the likelihood of $\mathbf{y}_t$, given $\boldsymbol{\theta}_t$. Hence the posterior of $\boldsymbol{\theta}_t$, given $\mathbf{y}_t$, is such that

$$p(\boldsymbol{\theta}_t|\mathbf{y}_t) \propto \sum_{i=1}^{2} \alpha_i f(\mathbf{y}_t|A_t\boldsymbol{\theta}_t, C_i) f(\boldsymbol{\theta}_t|\boldsymbol{\mu}_{t|t-1}, V_{t|t-1}). \tag{A.3}$$

The summand of (A.3) requires the "completion of a square" operation to form a quadratic form in $\boldsymbol{\theta}_t$. However, since the "constants" left over in the operation depend on i, they cannot be absorbed by the constant of proportionality. Now for the ith term, we have in the exponent, apart from the $-\frac{1}{2}$,

$$(\mathbf{y}_t - A_t\boldsymbol{\theta}_t)^T C_i^{-1}(\mathbf{y}_t - A_t\boldsymbol{\theta}_t) + (\boldsymbol{\theta}_t - \boldsymbol{\mu}_{t|t-1})^T V_{t|t-1}^{-1}(\boldsymbol{\theta}_t - \boldsymbol{\mu}_{t|t-1}), \tag{A.4}$$

and it is easy to see that (A.4) may be written as

$$[\boldsymbol{\theta}_t - V_{t|t}^{(i)}\mathbf{b}_t^{(i)}]^T V_{t|t}^{(i)-1}[\boldsymbol{\theta}_t - V_t^{(i)}\mathbf{b}_t^{(i)}] + \mathbf{y}_t^T C_i^{-1}\mathbf{y}_t + \boldsymbol{\mu}_{t|t-1}^T V_{t|t-1}^{-1}\boldsymbol{\mu}_{t|t-1} - \mathbf{b}_t^{(i)T} V_{t|t}^{(i)}\mathbf{b}_t^{(i)}, \tag{A.5}$$

where

$$V_{t|t}^{(i)} = \{V_{t|t-1}^{-1} + A_t^T C_i^{-1} A_t\}^{-1} = V_{t|t-1} - V_{t|t-1} A_t^T M_{t,i}^{-1} A_t V_{t|t-1} \tag{A.6}$$

and $M_{t,i} = C_i + A_t V_{t|t-1} A_t^T$ [see (A.4) and (A.13)], and where

$$\mathbf{b}_t^{(i)} = V_{t|t-1}^{-1}\boldsymbol{\mu}_{t|t-1} + A_t^T C_i^{-1}\mathbf{y}_t \tag{A.7}$$

[see (A.4)].

Now the last three terms in (A.5) may be written as

$$\mathbf{y}_t^T[C_i^{-1} - C_i^{-1}A_t V_{t|t}^{(i)} A_t^T C_i^{-1}]\mathbf{y}_t + \boldsymbol{\mu}_{t|t-1}^T[V_{t|t-1}^{-1} - V_{t|t-1}^{-1}V_{t|t}^{(i)}V_{t|t-1}] - 2y_t^T C_i^{-1} A_t V_{t|t}^{(i)} V_{t|t-1}^{-1}\boldsymbol{\mu}_{t|t-1}, \tag{A.8}$$

and it is straightforward to show that (A.8) may be written as

$$\mathbf{y}_t^T M_{t,i}^{-1}\mathbf{y}_t + \boldsymbol{\mu}_{t|t-1}A_t^T M_{t,i}^{-1}A_t\boldsymbol{\mu}_{t|t-1} - 2\mathbf{y}_t^T M_{t,i}^{-1}A_t\boldsymbol{\mu}_{t|t-1}, \tag{A.9}$$

which is, of course,

$$(\mathbf{y}_t - A_t\boldsymbol{\mu}_{t|t-1})^T M_{t,i}^{-1}(\mathbf{y}_t - A_t\boldsymbol{\mu}_{t|t-1}).$$

Hence $p(\boldsymbol{\theta}_t|\mathbf{y}_t)$ of (A.3) is such that

$$p(\boldsymbol{\theta}_t|\mathbf{y}_t) \propto \sum_{i=1}^{2} \alpha_i |C_i^{-1}|^{1/2}|V_{t|t-1}^{-1}|^{1/2} \times \exp[-\tfrac{1}{2}(\mathbf{y}_t - \boldsymbol{\mu}_{t|t-1})^T M_{t,i}^{-1}(\mathbf{y}_t - A_t\boldsymbol{\mu}_{t|t-1})] \times \exp[-\tfrac{1}{2}(\boldsymbol{\theta}_t - \boldsymbol{\mu}_{t|t}^{(i)})^T V_{t|t}^{(i)-1}(\boldsymbol{\theta}_t - \boldsymbol{\mu}_{t|t-1}^{(i)})], \tag{A.10}$$

where we have written $\boldsymbol{\mu}_{t|t}^{(i)} = V_{t|t}^{(i)}\mathbf{b}_t^{(i)}$. Using standard Kalman filter results in

$$\boldsymbol{\mu}_{t|t}^{(i)} = \boldsymbol{\mu}_{t|t-1} + V_{t.t}^{(i)} A_t^T C_i^{-1}(\mathbf{y}_t - A_t\boldsymbol{\mu}_{t|t-1}) \tag{A.11}$$

or

$$\boldsymbol{\mu}_{t|t}^{(i)} = \boldsymbol{\mu}_{t|t-1} + V_{t|t-1}A_t^T M_{t,i}^{-1}(\mathbf{y}_t - A_t\boldsymbol{\mu}_{t|t-1}). \tag{A.11a}$$

Using the definition of the multivariate normal density given in (3.7e), we now easily find that

$$p(\boldsymbol{\theta}_t|\mathbf{y}_t) = K \sum_{i=1}^{2} \alpha_i \frac{|C_i^{-1}|^{1/2}|V_{t|t-1}^{-1}|^{1/2}}{|M_{t,i}^{-1}|^{1/2}|V_{t|t}^{(i)-1}|^{1/2}} f(\mathbf{y}_t|A_t\boldsymbol{\mu}_{t|t-1}, M_{t,i}) \times f(\boldsymbol{\theta}_t|\boldsymbol{\mu}_t^{(i)}, V_{t|t}^{(i)}), \tag{A.12}$$

and it is easy to see that the determinants involved are such that the indicated products and ratios involved is 1. Finally, integrating with respect to $\boldsymbol{\theta}_t$ yields

$$K^{-1} = \sum_{i=1}^{2} \alpha_i f(\mathbf{y}_t|A_t\boldsymbol{\mu}_{t|t-1}, M_{t,i}) = h(\mathbf{y}_t|\mathbf{y}_{t-1}) \tag{A.13}$$

so that using this in (A.12), we now have

$$p(\boldsymbol{\theta}_t|\mathbf{y}_t) = \sum_{i=1}^{2} \alpha_{t,i} f(\boldsymbol{\theta}_t|\boldsymbol{\mu}_{t|t}^{(i)}, V_{t|t}^{(i)}), \tag{A.14}$$

with $\alpha_{t,i}$ as advertised in (3.7d).

Now using (A.14), we can calculate moments. We have

$$E(\boldsymbol{\theta}_t|\mathbf{y}_t) = \sum_{i=1}^{2} \alpha_{t,i} \int \boldsymbol{\theta}_t f(\boldsymbol{\theta}_t|\mu_{t|t}^{(i)}, V_{t,t}^{(i)})\, d\boldsymbol{\theta}_t$$

$$= \sum_{i=1}^{2} \alpha_{t,i} \boldsymbol{\theta}_{t|t}^{(i)} \tag{A.15}$$

and using (A.11a), which is the result (3.7a), we have

$$E(\boldsymbol{\theta}_t|\mathbf{y}_t) = \boldsymbol{\mu}_{t|t-1} + \sum_{1}^{2} \alpha_{t,i} V_{t|t-1} A_t^T M_{t,i}^{-1}(\mathbf{y}_t - A_t\boldsymbol{\mu}_{t|t-1}), \tag{A.16}$$

which may be expressed as in (3.8), or as above in (A.15).

To find the variance-covariance, we first determine $E(\boldsymbol{\theta}_t\boldsymbol{\theta}_t^T|\mathbf{y}_t)$ and then use the identity

$$V(\boldsymbol{\theta}_t|y_t) = E(\boldsymbol{\theta}_t\boldsymbol{\theta}_t^T|\mathbf{y}_t) - [E(\boldsymbol{\theta}_t|\mathbf{y}_t)][E(\boldsymbol{\theta}_t|\mathbf{y}_t)]^T. \tag{A.17}$$

Now from (A.14) we have

$$E(\boldsymbol{\theta}_t\boldsymbol{\theta}_t^T|\mathbf{y}_t) = \sum_{i=1}^{2} \alpha_{t,i} E(\boldsymbol{\theta}_t^T\boldsymbol{\theta}_t|\boldsymbol{\mu}_{t|t}^{(i)}, V_{t|t}^{(i)})$$

$$= \sum_{i=1}^{2} \alpha_{t,i}[V_{t|t}^{(i)} + \mu_{t|t}^{(i)}\boldsymbol{\mu}_{t|t}^{(i)T}]. \tag{A.18}$$

Now $V_{t,t}^{(i)}$ is given in (A.6), so that we have

$$E(\boldsymbol{\theta}_t\boldsymbol{\theta}_t^T|\mathbf{y}_t) = V_{t|t-1} - \sum_{i=1}^{2} \alpha_{t,i} V_{t|t-1} A_t^T M_{t,i}^{-1} A_t V_{t|t-1} + \sum_{i=1}^{2} \alpha_{t,i} \boldsymbol{\mu}_{t|t}^{(i)}\boldsymbol{\mu}_{t|t}^{(i)T}. \tag{A.19}$$

Substituting (A.16) in (A.19), and doing some straightforward but tedious algebra, and remembering that $\alpha_{t,2} = 1 - \alpha_{t,1}$, we find that

$$V(\theta_t|\mathbf{y}_t) = E(\theta_t\theta_t^T|\mathbf{y}_t) - \left\{\mu_{t|t-1} + \sum_1^2 \alpha_{t,i} V_{t|t-1} A_t^T M_{t,i}^{-1}(\mathbf{y}_t - A_t\boldsymbol{\mu}_{t|t-1})\right.$$

$$\left. \times \boldsymbol{\mu}_{t|t-1} + \sum \alpha_{t,i} V_{t,t-1} A_t^T M_{t,i}^{-1}(\mathbf{y}_t - A_t\boldsymbol{\mu}_{t|t-1})\right\}^T \quad \text{(A.20)}$$

takes the form advertised in (3.8a).

REFERENCES

Abraham, B., and G. E. P. Box (1979). Bayesian analysis of some outlier problems in times series, *Biometrika* 66, 229–236.

Andrews, D. F., P. J. Bickel, F. R. Hampel, P. J. Huber, W. H. Rogers, and J. W. Tukey (1972). *Robust Estimates of Location*, Princeton University Press, Princeton, N.J.

Aoki, M. (1967). *Optimization of Stochastic Systems*, Academic Press, New York.

Athans, M., R. H. Whiting, and M. Gruber (1977). A suboptimal estimation algorithm with probabilistic editing for false measurements with applications to target tracking with wake phenomena, *IEEE Trans. Autom. Control* 372–384, June.

Bar-Shalom, Y. (1978). Tracking methods in a multitarget environment, *IEEE Trans. Autom. Control* 1, 618–626, August.

Box, G. E. P., and G. M. Jenkins (1970). *Time Series Analysis Forecasting and Control*, Holden-Day, San Francisco.

Box, G. E. P., and G. C. Tiao (1968). A Bayesian approach to some outlier problems, *Biometrika* 55, 119–129.

Bryson, A. E., Jr., and Y. C. Ho (1975). *Applied Optimal Control*, Hemisphere Publishing, New York.

Campbell, K. (1982). Recursive computation of M-estimates for the parameters of a finite autoregressive process, *Ann. Stat.* 10(2), 442–453.

Chen, G., and G. E. P. Box (1979). *Further Study of Robustification via a Bayesian Approach*, Technical Report 1998, Mathematics Research Center, University of Wisconsin—Madison.

Cheng, G., and G. E. P. Box (1980). *Implied Assumptions for Some Proposed Robust Estimators*, Technical Summary Report 1997, Mathematics Research Center, University of Wisconsin—Madison.

Crowe, C. M., et al. (1983). Reconciliation of process flow rates by matrix projection, *AIChE J.* 29(6), 881–888.

Daum, F. E. (1986). Exact finite dimensional non-linear filters, *IEEE Trans. Autom. Control* AC-31, 616–662.

Guttman, I. (1973). Care and handling of univariate or multivariate outliers in detecting spuriosity—a Bayesian approach, *Technometrics* 15, 723–738.

Guttman, I., and D. Peña (1984). *Robust Kalman Filtering and its Applications*, Technical Report, Mathematics Research Center, University of Wisconsin—Madison.

Harrison, P. J., and C. F. Stevens (1976). Bayesian forecasting (with discussion), *J. R. Stat. Soc. Ser. B* 38, 205–247.

Heyde, C. C., and I. M. Johnstone (1979). On asymptotic posterior normality for stochastic processes, *J. R. Stat. Soc. Ser. B* 41, 184–189.

Ho, Y. C., and R. C. K. Lee (1964). A Bayesian Approach to Problems in Stochastic Estimation and Control, in *Proceedings of the Joint Automatic Control Conference*, Stanford University, Stanford, Calif., pp. 382–387.

Jeffreys, H. (1961). *Theory of Probability*, 3rd ed., Clarendon Press, Oxford.

Kalman, R. E. (1960). A new approach to linear filtering and prediction problems, *Trans. ASME* 82D, 35.

Kleiner, B., R. D. Martin, and D. J. Thomson (1979). Robust estimation of power spectra, *J. R. Stat. Soc. Ser. B* 41(3), 313–351.

Little, J. K. (1983). *Regression Diagnostics and the Bayesian Analysis of the Scale-Contaminated Normal Model*, Technical Report 713, Department of Statistics, University of Wisconsin—Madison.

Mah, R. S. H., and A. C. Tamhane (1982). Detection of gross errors in process data, *AIChE J.* 28(5), 828–830.

Makowski, A. M. (1986). Filtering formulae in non-Gaussian initial conditions, *Stochastics* 16, 1–24.

Martin, R. D. (1984). A Review of Some Aspects of Robust Inference for Times Series, in *Statistical Signal Processing* (E. J. Wegman and J. G. Smith, eds.), Marcel Dekker, New York.

Masreliez, C. J. (1975). Approximate non-Gaussian filtering with linear state and observation relations, *IEEE Trans. Autom. Control* AC-20, 107–110.

Masreliez, C. J., and R. D. Martin (1977). Robust Bayesian estimation for the linear model and robustifying the Kalman filter, *IEEE Trans. Autom. Control* AC-22(3), 361–371.

McWorter, A., W. A. Spivey, and W. J. Wrobleski (1976). Sensitivity analysis of varying parameter economic models, *Int. Stat. Rev.* 44(2), 265–282.

Nahi, N. E. (1969). Optimal recursive estimation with uncertain observation, *IEEE Trans. Inf. Theory* 457–462, July.

Plackett, R. L. (1950). Some theorems on least squares, *Biometrika* 37, 149–157.

Singer, R. A., and R. G. Sea (1973). New results in optimizing surveillance system tracking and data correlation performance in dense multitarget environment, *IEEE Trans. Autom. Control* 571, 582, December.

Sorenson, H. W., and D. L. Alspach (1971). Recursive Bayesian estimation using Gaussian sums, *Automatica* 6, 465–479.

Spall, J. C., and K. D. Wall (1984). Asymptotic distribution theory for the Kalman filter state estimator, *Commun. Stat. Theory Methods* 13, 1981–2003.

Stein, C. (1956). Inadmissibility of the usual estimator for the mean of multivariate normal distribution, *3rd Berk. Symp. Math. Stat. Prob.* 1, 197–206.

Tsai, C., and L. Kurz (1983). An adaptive robustizing approach to Kalman filtering, *Automatica* 19(3), 278–288.

West, M. (1981). Robust sequential approximate Bayesian estimation, *J. R. Stat. Soc. Ser. B* 43(2), 157–166.

West, M. (1984). Outlier models and prior distributions in Bayesian linear regression, *J. R. Stat. Soc. Ser. B* 46(3), 431–439.

West, M., P. J. Harrison, and H. S. Migon (1985). Dynamic generalized linear models and Bayesian forecasting, *J. Am. Stat. Assoc.* 80(389), 73–97.

Young, P. (1984). *Recursive Estimation and Time-Series Analysis*, Springer-Verlag, Berlin.

10

Modeling and Filtering for Discretely Valued Time Series

JAMES V. WHITE The Analytic Sciences Corporation, Reading, Massachusetts

1. INTRODUCTION

A discretely valued time series is a sequence of symbols from a finite set. Such time series are termed *categorical* in this chapter, because in applications the symbols usually represent the names of categories. An example of a categorical time series is the following sequence of meteorological data, which describe cloud conditions observed at 12-hour intervals: . . . , partly cloudy, scattered clouds, clear, cloudy, ...

The main goal of this chapter is to describe a family of discrete-state models and an associated recursive Bayesian filtering algorithm, which do for categorical series what the family of Gaussian state-space models and the Kalman filtering algorithm do for numerical time series.

1.1 Applications of Categorical Time Series

Weather forecasters are interested in computing the conditional probabilities of future cloud conditions, given observations of past and present weather. To do this coherently, stochastic models are first developed for the categorical weather series, and then the models are used to compute the conditional probabilities in accordance with Bayes' rule.

Some numerical time series are actually categorical. For example, the categorical sequence, . . . 30, 18, 24, . . . , is a list of weather conditions described by standard synoptic weather codes. The code number 30 represents "dust storm," 18 represents "squalls within sight of the station," and 24 represents "freezing drizzle." Although integers are used as code symbols, there is no physically meaningful arithmetic defined on the set of weather codes, and only a partial ordering of the symbols makes physical sense.

As discussed by Bucy et al. (1972), a quantized (i.e., discretely valued) time series can accurately approximate a numerical time series that is governed by *nonlinear* differential equations driven by noise. Two advantages of a discretely valued approximation are the following:

1. The approximation can be made to converge to the numerical process as the number of quantization levels (i.e., categories) increases.
2. The Bayesian filter for the discretely valued series can be implemented exactly using simple matrix algebra, whereas the Bayesian filtering equations for the numerical time series involve multiple integrals.

This discrete approach to the Bayesian nonlinear filtering of numerical data was introduced by Bucy (1969) as a point-mass approximation of the probability densities and was used successfully to analyze the performance of numerical nonlinear filters in Bucy et al. (1971, 1972, 1973, 1980, 1983) and Perrenod and Bucy (1985).

The problem of automatic speech recognition has been pursued by Levinson et al. (1983) and Rabiner and Juang (1985) using categorical time series. Individual words are probabilistically related to sequences of tokens that are computed by a signal processor from numerical acoustic time series. Algorithms for computing the probabilities of different words, given the series of tokens, are used to recognize specific words automatically. (The acoustic waveform data are typically sampled with a frequency of about 7 kHz, while the derived categorical series of tokens has the much smaller sampling frequency of about 70 Hz.)

The author has recently developed categorical time-series models for technology growth. In this application, the growth of software technologies and hardware technologies are modeled, together with their impacts on the effectiveness of specific systems based on these technologies. In the simplest models, each technology and each measure of effectiveness is categorized using a few categories. For example, a particular state variable in such a model may assume the values 1, 2, or 3, where 1 indicates that a particular technology has a relatively low probability of existing, 2 indicates that it has a moderate probability, and 3 indicates that it has a high probability, where the meaning of "low," "medium," and "high" probabilities depends on the particular model. More complicated models are constructed by increasing the number of categories.

1.2 Overview

The remainder of this chapter is organized as follows. In Section 2 we describe a versatile family of stochastic discrete-state models for categorical time series. The output of such a model is a random function of the state variable of a finite Markov chain. This type of model is called a "hidden Markov model" in the literature on speech processing (Rabiner and Juang, 1986). The conditional probability distributions of the observed output and the state variable of the chain satisfy linear difference equations. Therefore, a matrix notation for these state models is defined, which parallels the familiar conventions of state-space modeling in linear systems theory.

A recursive Bayesian filtering algorithm for discrete-state models is derived and discussed in Section 3. To emphasize the similarities and differences between this categorical filter and the Kalman filter for Gauss-Markov time series, a matrix notation is introduced that matches many of the conventions for Kalman filtering in Gelb (1974). The state vector in these nonlinear filtering equations is the conditional probability distribution of the chain's state variable, and the output of the filter at time t is the conditional distribution of the next symbol to be observed at time $t + 1$, given all the previously observed symbols. There are no covariances or conditional means being computed, just discrete probability distributions. In fact, for most categorical time series there is no physically meaningful arithmetic defined on the set of observed symbols; therefore, the means and covariances of observed symbols are either not defined or physically irrelevant.

In Section 4 we address the problems of measuring the inaccuracy of a stochastic discrete-state model and comparing the inaccuracies of two-candidate models. Conventional inaccuracy measures based on root-mean-square (rms) error do not usually apply, because most categorical time series have neither metrics nor arithmetics defined on their symbol sets. Therefore, conventional rms errors cannot even be computed. The *inaccuracy* of a stochastic model (as opposed to its *utility* for achieving a particular goal) depends only on the probability distributions of the model and the truth. It is now known from information theory and statistical theory that there is only one measure of model accuracy that satisfies reasonable axioms for an additive measure of information about model inaccuracy (Kannappan, 1972; Aczel and Daroczy, 1975; Shore and Johnson, 1980; Larimore and Mehra, 1985; Matsuoka and Ulrych, 1986). This inaccuracy measure was extensively applied by Kullback (1959), who called it discrimination information (DI). More recently, DI has also been called the Kullback-Liebler number, Kullback information, cross entropy, and directed divergence.

In Section 4 we address three main topics: (1) the practical application of DI to categorical modeling and its connection with log likelihoods, (2) the use of categorical filters for computing the exact log likelihood statistics of discrete-state models on finite time series, and (3) a Bayesian comparison of model inaccuracies, which includes the computation of the Akaike information criterion (Akaike, 1973, 1974, 1976).

2. STATE MODELS

A categorical time series, starting at time $t = 1$ and ending at time $t = T$, is represented by the sequence $(y_1, y_2, \ldots, y_T)$, which is also denoted $\{y\}_T$. The observed symbol at time t is represented as y_t. A *stochastic model* for the underlying random process of the time series is a mathematical description that determines the probability of every admissible sequence $\{y\}_T$.

In this section we describe a family of stochastic discrete-state models for categorical time series. The technical approach is to model the conditional probability distributions of future symbols as a function of the state of a finite Markov chain. The resulting models are nonlinear, but their probability distributions satisfy linear difference equations that are similar to the state equations of Gaussian random processes having values in linear spaces.

2.1 Notation

Vectors (column matrices) are denoted by boldface type (e.g., $\mathbf{x}$ or $\mathbf{X}$), and $\mathbf{X}^T$ denotes the transpose of $\mathbf{X}$. Random variables and the observed values of random variables are denoted by lowercase letters; when there is the possibility of confusion, a prime is used to denote observed values. Discrete probability distributions are represented as uppercase letters. The random variables in this discussion have only a finite number of integer values. Therefore, it is convenient to use the shorthand notation $[n]$ to represent the set $\{1, 2, \ldots, n\}$. For example, suppose that x is a random variable having four values. Then $x \in [4]$, and the vector $\mathbf{X}$ denotes the following probability distribution:

$$\mathbf{X} = \begin{bmatrix} \Pr(x = 1) \\ \Pr(x = 2) \\ \Pr(x = 3) \\ \Pr(x = 4) \end{bmatrix}.$$

The fact that x has the probability distribution $\mathbf{X}$ is implied by the use of uppercase and lowercase versions of "x." The notation $y \sim \mathbf{X}$ is used to indicate that either the random variable y has the same probability distribution as x, or the number y is a realization from the distribution $\mathbf{X}$. The context determines which meaning is intended. The notation $y|x \sim \mathbf{Y}$ means that $\mathbf{Y}$ is the conditional distribution of y given x.

The finite sequence of random variables $(x_1, x_2, \ldots, x_t)$ is denoted $\{x\}_t$ for $t = 1, 2, \ldots$. The empty sequence is $\{x\}_0$, and a sequence of unbounded length is denoted $\{x\}$. The symbols x_t and $\{x\}_t$ may represent either random variables or realized values of random variables; the context determines which of these interpretations is appropriate.

The conditional probability distribution of the random variable x, given the categorical time series $\{y\}_T$, is denoted by the vector $\mathbf{X}[\{y\}_T]$. Brackets and bold type are also used to indicate a specific column of a matrix: for example, $\mathbf{H}[k]$ denotes the kth column of the matrix H.

2.2 Markov Chains

A finite *Markov chain* (Kemeny and Snell, 1960; Meyer, 1975) is denoted by the triple $(n, \mathbf{X}_1, \Phi)$, where n is the number of states (an integer), $\mathbf{X}_1$ the initial state probability distribution (an n-vector), and Φ the $n \times n$ matrix of state transition probabilities. Every Markov chain can be represented by a sequence of random variables $\{x\}$ with $x_t \in [n]$. The value of x_t denotes the state of the chain at time t, and therefore the random variable x_t is referred to as the state variable at time t. The following notation is used for the initial state probability distribution:

$$X_1(k) = \Pr(x_1 = k) \qquad (k \in [n]) \tag{2.1}$$

$$\mathbf{X}_1 = [X_1(1) \quad X_1(2) \quad \cdots \quad X_1(n)]^T. \tag{2.2}$$

The elements of the transition matrix Φ are the conditional probabilities of the future state, given the present state. For example, the element in the jth row and kth column is defined as

$$\Phi(j, k) = \Pr(x_{t+1} = j | x_t = k) \qquad (j, k \in [n]). \tag{2.3}$$

To obtain consistency with the established matrix notation for linear systems and Kalman filtering, the transition matrix Φ defined by (2.3) is the *transpose* of the usual transition matrix in Markov chain theory. (The columns of Φ sum to unity.) This transposition is consistent with probability distributions being represented by column matrices rather than rows.

From (2.3) it follows that the probability distribution of the state variable, $\mathbf{X}_t$, satisfies the following linear difference equation, given the initial distribution $\mathbf{X}_1$:

$$\mathbf{X}_{t+1} = \Phi \mathbf{X}_t \qquad (t = 1, 2, \ldots). \tag{2.4}$$

For example, if there are two states $(n = 2)$, so that $x_t \in \{1, 2\}$ and

$\Pr(x_{t+1} = 1|x_t = 1) = 0.9$ and $\Pr(x_{t+1} = 2|x_t = 2) = 0.8$, then

$$\Phi = \begin{bmatrix} 0.9 & 0.2 \\ 0.1 & 0.8 \end{bmatrix}, \tag{2.5}$$

where the off-diagonal elements of Φ are selected so that each column sums to unity. Suppose that the chain starts in state 1 with probability 1. Then $\mathbf{X}_1 = [1 \quad 0]^T$, and from (2.4), $\mathbf{X}_2 = \Phi\mathbf{X}_1 = [0.9 \quad 0.1]^T$, $\mathbf{X}_3 = \Phi\mathbf{X}_2 = [0.83 \quad 0.17]^T$, and so on.

A realization (i.e., sample path) of the state sequence $\{x\}$ can be computed for the Markov chain $(n, \mathbf{X}_1, \Phi)$ as follows:

$$x_1 \sim \mathbf{X}_1 \tag{2.6}$$

$$x_{t+1} \sim \mathbf{\Phi}[x_t] \qquad (t = 1, 2, \ldots). \tag{2.7}$$

In (2.6) the initial state is drawn from the distribution $\mathbf{X}_1$. From then on, the state at time $t + 1$ is drawn from the distribution $\mathbf{\Phi}[x_t]$ (where $\mathbf{\Phi}[k]$ denotes the kth column of the matrix Φ).

As an example, consider the transition matrix defined by (2.5) and the initial state $x_1 = 2$. Then $x_2 \sim \mathbf{\Phi}[x_1] = \mathbf{\Phi}[2] = [0.2 \quad 0.8]^T$.

2.3 Outputs from a Markov Chain

A categorical time series $\{y\}_T$ is modeled as a sequence of outputs from a Markov chain (i.e., the observed symbol y_t is the output of the chain at time t). In developing a simple matrix notation, it is convenient to represent each category, or observed symbol, as an integer, so that if there are m categories, then $y_t \in [m]$. With this convention, an m-valued output from a Markov chain at time t is *defined* to be a random variable, $y_t \in [m]$, whose distribution depends only on the current state. In other words, the conditional probability distribution of y_t (denoted $\mathbf{Y}_t[\{x\}_t, \{y\}_{t-1}]$ and conditioned on the sequence of past and present states $\{x\}_t$ and the sequence of past outputs $\{y\}_{t-1}$) is by definition a function of nothing but the chain's current state x_t. Therefore, the conditional distribution of y_t can be expressed as an m-vector probability distribution, denoted $\mathbf{H}[x_t]$, which depends only on the current state:

$$\mathbf{Y}_t[\{x\}_t, \{y\}_{t-1}] = \mathbf{H}[x_t] \qquad (t = 1, 2, \ldots). \tag{2.8}$$

For the purpose of making probability statements about the present output, y_t, (2.8) states that knowledge of the current state x_t is equivalent to knowing all states, past and present, and all past output symbols.

Given a realization of states, $\{x\}$, a corresponding realization of outputs is computed by drawing a sequence of numbers, $y_t \in [m]$, such that $y_t \sim \mathbf{H}[x_t]$ for $t = 1, 2, \ldots$.

For each state of the chain, there is a conditional probability distribution for the output y_t. If $x_t = k$, then this distribution is represented by the following m-vector:

$$\mathbf{H}[k] = \begin{bmatrix} \Pr(y_t = 1 | x_t = k) \\ \Pr(y_t = 2 | x_t = k) \\ \vdots \\ \Pr(y_t = m | x_t = k) \end{bmatrix} \qquad (k \in [n]). \tag{2.9}$$

By arranging these distributions in order, the $m \times n$ output probability matrix H is defined as

$$H = [\mathbf{H}[1] \quad \mathbf{H}[2] \quad \cdots \quad \mathbf{H}[n]]. \tag{2.10}$$

The H matrix is used to express the distribution of the output y_t in terms of the distribution of the state x_t:

$$\mathbf{Y}_t = H\mathbf{X}_t \qquad (t = 1, 2, \ldots). \tag{2.11}$$

For example, if there are two states ($n = 2$, $x_t \in \{1, 2\}$), two possible values for the output ($m = 2$, $y_t \in \{1, 2\}$), and if $Pr(y_t = 1 | x_t = 1) = 0.7$ and $Pr(y_t = 2 | x_t = 2) = 0.4$, then

$$H = \begin{bmatrix} 0.7 & 0.6 \\ 0.3 & 0.4 \end{bmatrix}, \tag{2.12}$$

where the off-diagonal elements are selected to make the columns of H sum to unity. Further, if the state distribution is $\mathbf{X}_t = [0.3 \quad 0.7]^T$, then from (2.11) the conditional distribution of the output is $\mathbf{Y}_t = H\mathbf{X}_t = [0.63 \quad 0.37]^T$.

For future use in developing a recursive Bayesian filtering algorithm for state models, (2.11) is used to express the conditional distribution of the future output y_{t+1} in terms of the conditional distribution of the future state x_{t+1}, given the sequence of past and present outputs $\{y\}_t$:

$$\mathbf{Y}_{t+1}[\{y\}_t] = H\mathbf{X}_{t+1}[\{y\}_t]. \tag{2.13}$$

In (2.13) the conditional state distribution $\mathbf{X}_{t+1}[\{y\}_t]$ is computed by using (2.4) to write

$$\mathbf{X}_{t+1}[\{y\}_t] = \Phi\mathbf{X}_t[\{y\}_t]. \tag{2.14}$$

2.4 Definition of State Models

By combining the results of Sections 2.2 and 2.3, state models are now defined for categorical time series. In words, a state model for a categorical time series is a Markov chain with a finite number of outputs. More precisely, a state model for a time series $\{y\}_T$, with $y_t \in [m]$, is denoted by the quintuple $(m, n, \mathbf{X}_1, \Phi, H)$, where m is the number of observable categories, n the number of states, $\mathbf{X}_1$ the initial state probability distribution (an n-vector), Φ the $n \times n$ state transition probability matrix, and H the $m \times n$ output probability matrix. Later, in Section 4.2, it will be shown how the probability of observing any admissible time series can be computed for a specific state model by using a Bayesian recursive filtering algorithm.

The columns of Φ and H are conditional probability distributions. The kth column of Φ, denoted $\mathbf{\Phi}[k]$, is the conditional probability distribution of the state variable x_{t+1}, given that $x_t = k$; and likewise, $\mathbf{H}[k]$ is the conditional distribution of the output y_t, given that $x_t = k$.

The distributions of the state and output variables satisfy the following linear difference equations for $t = 1, 2, \ldots$, with $\mathbf{X}_1$ a given initial distribution:

$$\mathbf{X}_{t+1} = \Phi \mathbf{X}_t \tag{2.15}$$

$$\mathbf{Y}_t = H \mathbf{X}_t. \tag{2.16}$$

Equations (2.15) and (2.16) are equivalent to the linear deterministic system depicted in Figure 1. In this figure each block represents matrix multiplication by the indicated matrix, and the paths are labeled by the probability distributions of the state and output variables.

The following algorithm can be used to compute sample paths of the state and output sequences. (In this algorithm, the notation $x \sim \mathbf{X}$ means that x is a realization of the random variable having the probability distribution $\mathbf{X}$.)

Initialize:

$$x_1 \sim \mathbf{X}_1. \tag{2.17}$$

For $t = 1$ to T:

$$x_{t+1} \sim \mathbf{\Phi}[x_t] \tag{2.18}$$

$$y_t \sim \mathbf{H}[x_t]. \tag{2.19}$$

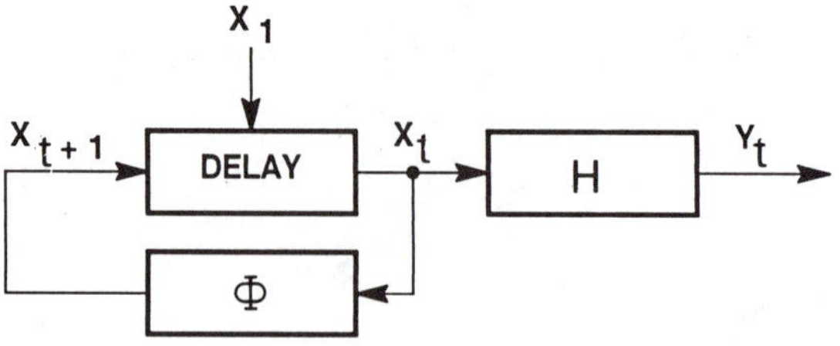

Figure 1. Deterministic linear system governing the probability distributions of a state model. (From White and Janota, 1985.)

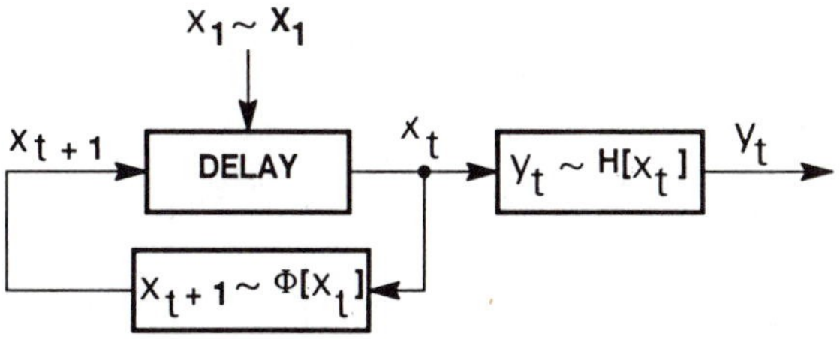

Figure 2. Stochastic nonlinear system governing the sample paths of the state and output variables of a state model. The symbol "∼" means "has the distribution." (From White and Janota, 1985.)

The sample-path algorithm described by (2.17) to (2.19) corresponds to the stochastic nonlinear system depicted in Figure 2. Both Figures 1 and 2 have similar topologies. Essential differences between Figures 1 and 2 are that Figure 1 is linear and deterministic, while Figure 2 is nonlinear and stochastic.

2.5 Examples of State Models

Three examples of state models are presented in this section. These models were developed from time series of meteorological cloud data sampled every 12 hours over a 3-month period. The section concludes with a discussion of multivariate and nonstationary models.

Cloud-data models

The first model, which is based on data from Las Vegas, Nevada, is depicted as a directed graph in Figure 3. There are two observables, termed C and N/C. The symbol C indicates that there is a cloud ceiling, while N/C indicates

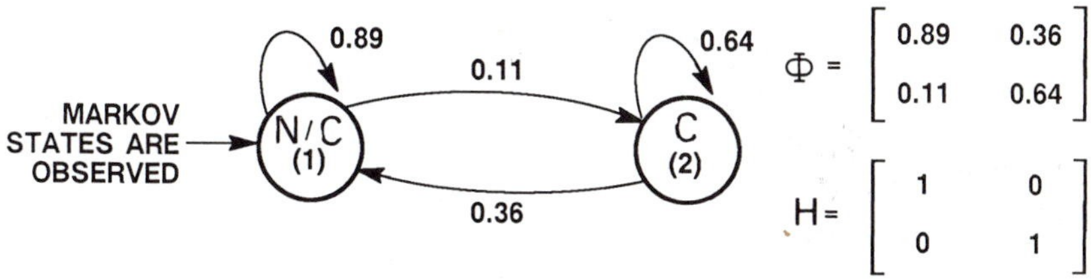

Figure 3. State model for cloud-ceiling data from Las Vegas, Nevada.

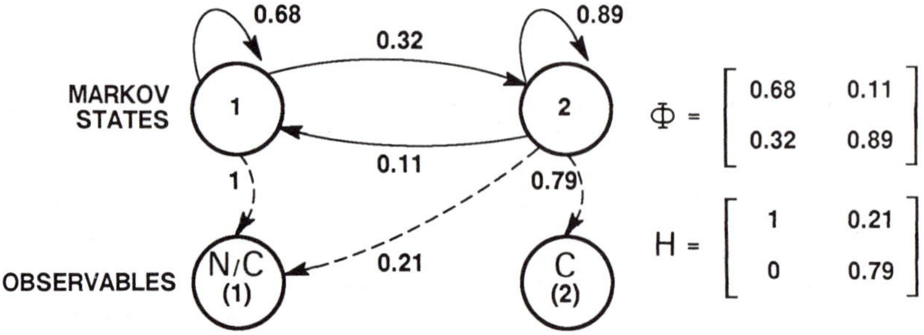

Figure 4. State model for cloud-ceiling data from Hyannis, Massachussetts.

there is no cloud ceiling. (In practice, more categories of cloud data are recorded.) In this model the observable categories happen to correspond exactly to the states of the Markov chain. In other words, the states of the chain are directly observed, and therefore, H is an identity matrix. The transition probabilities of the chain are the branch weights. A comparison of the graph with the transition matrix shows that the first column, $\mathbf{\Phi}[1]$, contains the probabilities of transitions leaving the first state (N/C), while the second column, $\mathbf{\Phi}[2]$, corresponds to the transitions leaving the second state (C).

The second model is based on data from Hyannis, Massachussetts, and is depicted in Figure 4. In this model the observables are distinct from the states of the Markov chain. Each dashed branch indicates the probability of observing a particular output while the chain is in a particular state. For example, the dashed branch from state 1 to output N/C (output 1) indicates

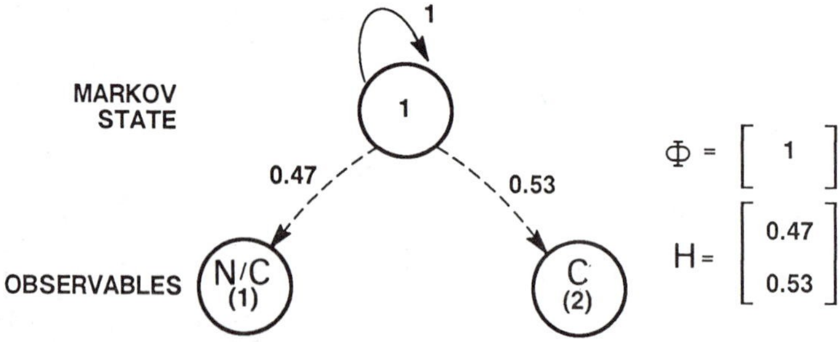

Figure 5. State model for cloud-ceiling data from Bedford, Massachussetts.

that output N/C occurs with probability 1.00 while the chain is in state 1. An examination of the output matrix shows that the first column, **H**[1], contains the probabilities of all dashed lines emanating from state 1. For example, the zero in **H**[1] corresponds to there being zero probability of observing C (output 2) while the chain is in state 1. In this model the number of states equals the number of outputs, but this is merely a coincidence.

The third model, depicted in Figure 5, is based on data from Bedford, Massachusetts. This is a degenerate one-state model. There is no memory in this model. Therefore, successive outputs are statistically independent of each other, which is the categorical equivalent of white noise. The fact that the output probabilities are nearly equal to 0.5 indicates that the cloud cover changes erratically over 12-hour intervals.

Multivariate and nonstationary models

Suppose that both cloud-ceiling and temperature data are measured simultaneously, that is, there are two channels of data, and the time series are multivariate. For this example the temperature categories are termed Above Normal, Normal, or Below Normal. (These categories are typically defined by requiring that Above and Below each occur 30% of the time.) The set of output symbols for the multivariate model is defined to be in one-to-one correspondence with the product of the two sets {C, N/C} and {Above, Normal, Below). Since there are six pairs of symbols in this product set, $y_t \in [6]$. One possible assignment of the model outputs is the following list:

$y_t = 1$ means (C, Above)
$y_t = 2$ means (C, Normal)
$y_t = 3$ means (C, Below)
$y_t = 4$ means (N/C, Above)
$y_t = 5$ means (N/C, Normal)
$y_t = 6$ means (N/C, Below).

This use of product sets to model multivariate time series can be extended to any number of simultaneously measured data channels.

At time t every discrete-state model is in exactly one state (represented by the value of the scalar variable x_t) and has exactly one output (represented by the scalar variable y_t). Adding more data channels always increases the set of possible outputs and the row dimension of the output matrix H. In addition, adding data channels may increase the set of states, which increases the dimensions of the transition matrix Φ and the column dimension of H. But increasing the number of data channels does not increase the number of state or output variables in the model.

Nonstationary time series may be modeled by adding a data channel that indicates the time of each observation. For example, the cloud-ceiling data have values in the set {C, N/C}, and the observation times (which occur every 12 hours) have values in the set {AM, PM}. The data on cloud-ceiling may therefore be represented as a two-channel multivariate time series. The four possible outputs of the model can be defined as follows:

$y_t = 1$ means (C, AM)

$y_t = 2$ means (C, PM)

$y_t = 3$ means (N/C, AM)

$y_t = 4$ means (N/C, PM).

Another way of modeling nonstationary time series is to let the output matrix H and the transition matrix Φ be time varying (i.e., using the notation H_t and Φ_t). A disadvantage of using a time-varying Φ is that the extensive theory for Markov chains (Kemeny and Snell, 1960; Meyer, 1975) is no longer applicable for insight into model behavior. This is analogous to the impact of a time-varying transition matrix in a Kalman filter; the extensive spectral theory of time-invariant linear systems is then no longer applicable for insight into model behavior.

3. CATEGORICAL FILTERS

A *categorical filter* for a discrete-state model $(m, n, \mathbf{X}_1, \Phi, H)$ is an algorithm for computing the conditional probability distributions of the current state variable x_t, the next state variable x_{t+1}, and the next output symbol y_{t+1}, given the outputs $\{y\}_t$ and the initial state distribution $\mathbf{X}_1$. In Section 3.1 a recursive categorical filtering algorithm is derived using a matrix notation that parallels the conventional notation for Kalman filtering. The parallel notation is natural, because both the Kalman and the categorical filter are simply recursive implementations of Bayes' rule applied to state models. Since the equations are in matrix form, they are well suited for implementation on vector computers with concurrent processors.

The categorical filtering equations are both formally and numerically much simpler than the Kalman filtering equations, because there are fewer matrices (no means or covariances to update and propagate, but only discrete probability distributions) and no matrix inversions. However, if the categorical filter is used to approximate a nonlinear numerical filter, the dimension of the state vector of the categorical filter is very much larger than the corresponding dimension of the numerical filter. The increase in dimension depends on how finely the numerical state space is quantized (Bucy et al., 1972; Bucy and Mallinckrodt, 1973).

3.1 Derivation of Filtering Equations

Bayesian filtering algorithms for discrete-state models have been previously published (e.g., Bucy et al., 1972; Baum, 1972; Levinson et al., 1983; Castanon and Teneketzis, 1985; White and Janota, 1985; Rabiner and Juang, 1986). The short derivation presented here for a recursive algorithm is intended for readers who are already familiar with the conventions of state-space theory and Kalman filtering.

The inputs to the filtering algorithm are a categorical time series $\{y\}_T$ and its state model $(m, n, \mathbf{X}_1, \Phi, H)$. (The following derivation is valid for time-varying models in which Φ and H are replaced by Φ_t and H_t.) The outputs of the filter are sequences of conditional probability distributions. One of the filter outputs, $\mathbf{Y}_{t+1}[\{y\}_t]$ for $t \in [T]$, is already expressed in terms of the conditional distribution of the state in (2.13). So by substituting the state distribution of (2.14) into (2.13), this output of the filter can be expressed as follows:

$$\mathbf{Y}_{t+1}[\{y\}_t] = H\Phi\mathbf{X}_t[\{y\}_t] \qquad (t \in [T]). \tag{3.1}$$

The conditional probability distribution on the right side of (3.1) is written out as follows:

$$\mathbf{X}_t[\{y\}_t] = \begin{bmatrix} \Pr(x_t = 1|\{y\}_t) \\ \Pr(x_t = 2|\{y\}_t) \\ \vdots \\ \Pr(x_t = n|\{y\}_t) \end{bmatrix}. \tag{3.2}$$

Bayes' rule is now used to compute the conditional probabilities on the right side of (3.2):

$$\begin{aligned} \Pr(x_t|\{y\}_t) &= \Pr(x_t|y_t, \{y\}_{t-1}) \\ &= \frac{\Pr(x_t, y_t|\{y\}_{t-1})}{\Pr(y_t|\{y\}_{t-1})} \\ &= \frac{\Pr(y_t|x_t, \{y\}_{t-1})\Pr(x_t|\{y\}_{t-1})}{\Pr(y_t|\{y\}_{t-1})}. \end{aligned} \tag{3.3}$$

The conditional probability of the output symbol y_t depends only on the value of the current state [as expressed in (2.8)]. That is,

$$\Pr(y_t|x_t, \{y\}_{t-1}) = \Pr(y_t|x_t). \tag{3.4}$$

Therefore, (3.3) can be written more simply as follows:

$$\Pr(x_t|\{y\}_t) = \frac{\Pr(y_t|x_t)\Pr(x_t|\{y\}_{t-1})}{\Pr(y_t|\{y\}_{t-1})}. \tag{3.5}$$

Equation (3.5) expresses Bayes' rule for updating the state distribution, given the observed value of y_t.

The last step in deriving the categorical filter is to express (3.5) using matrix notation. From Section 2.3 the conditional probability distribution of the output y_t, given the state x_t, is represented by the $m \times n$ output matrix H. The element in the ith row and jth column of H is defined as follows [see (2.9) and (2.10)]:

$$H(i, j) = \Pr(y_t = i|x_t = j). \tag{3.6}$$

The equation governing the Bayesian update of the state distribution is conveniently expressed in terms of the n-vectors $\mathbf{h}[k]$ for $k = 1$ to m: By definition, $\mathbf{h}[k] = k$th column of the transposed matrix H^T. These vectors contain the following probabilities:

$$\mathbf{h}[k] = \begin{bmatrix} \Pr(y_t = k|x_t = 1) \\ \Pr(y_t = k|x_t = 2) \\ \vdots \\ \Pr(y_t = k|x_t = n) \end{bmatrix} \qquad (k\text{th column of } H^T). \tag{3.7}$$

The $\mathbf{h}$ vectors enter the filtering equations through a type of vector multiplication called the Hadamard product. This product is also called the Schur product and is defined as follows: Given two n-vectors $\mathbf{a}$ and $\mathbf{b}$, their Hadamard product is the n-vector $\mathbf{c}$:

$$\begin{bmatrix} c(1) \\ c(2) \\ \vdots \\ c(n) \end{bmatrix} = \begin{bmatrix} a(1)b(1) \\ a(2)b(2) \\ \vdots \\ a(n)b(n) \end{bmatrix}. \tag{3.8}$$

The Hadamard product is an element-by-element multiply. Using the symbol "$*$" to denote the Hadamard product, (3.8) may be written as

$$\mathbf{c} = \mathbf{a} * \mathbf{b}. \tag{3.9}$$

By using the $\mathbf{h}$ vectors and the Hadamard product, the Bayes update formula in (3.5) can be rewritten as follows:

$$\mathbf{X}_t[\{y\}_t] = \frac{\mathbf{h}[y_t] * \mathbf{X}_t[\{y\}_{t-1}]}{D_t}, \tag{3.10}$$

where the denominator D_t is the normalizing factor that makes the probabilities sum to unity. This denominator is computed by taking the sum of the elements of the n-vector in the numerator, which is denoted

$\mathbf{v}_t = [v_t(1) \cdots v_t(n)]^T$:

$$\mathbf{v}_t = \mathbf{h}[y_t] * \mathbf{X}_t[\{y\}_{t-1}] \tag{3.11}$$

$$D_t = \sum_{k=1}^{n} v_t(k). \tag{3.12}$$

For simplicity, and to clarify the similarities and differences of categorical and Kalman filtering, the following notations are defined for conditional distributions, where $t = 1, 2, \ldots$:

$$\mathbf{X}_t(-) = \mathbf{X}_t[\{y\}_{t-1}] \tag{3.13}$$

$$\mathbf{X}_t(+) = \mathbf{X}_t[\{y\}_t] \tag{3.14}$$

$$\mathbf{Y}_t = \mathbf{Y}_t[\{y\}_{t-1}]. \tag{3.15}$$

The state distribution $\mathbf{X}_t(-)$ in (3.13) is conditioned on the *past* observations, while (3.14) defines the *updated* distribution $\mathbf{X}_t(+)$ conditioned on the past and *present* observations. The distribution for the observed outputs is represented as simply as possible in (3.15) by dropping the conditioning argument.

3.2 Categorical Filtering Algorithm

The matrix equations for the recursive categorical filtering algorithm can now be written as follows:

Initialize:

$$\mathbf{X}_1(-) = \mathbf{X}_1 \tag{3.16}$$

$$\mathbf{Y}_1 = H\mathbf{X}_1. \tag{3.17}$$

For $t = 1$ to T:

Update:

$$\mathbf{v}_t = \mathbf{h}[y_t] * \mathbf{X}_t(-) \qquad \text{(Hadamard product)} \tag{3.18}$$

$$D_t = \sum_{k=1}^{n} v_t(k) \tag{3.19}$$

$$\mathbf{X}_t(+) = D_t^{-1}\mathbf{v}_t. \tag{3.20}$$

Propagate:

$$\mathbf{X}_{t+1}(-) = \Phi\mathbf{X}_t(+). \tag{3.21}$$

Output:

$$\mathbf{Y}_{t+1} = H\mathbf{X}_{t+1}(-). \tag{3.22}$$

The updating equations (3.18) to (3.20) are a matrix version of (3.5), the

2.1 State Model

The basic state variable is labeled θ_t, where

$$\theta_t^T = (\theta_{tx}, \theta_{ty})$$

with

$$\theta_{tx}^T = (x_t, x_t', x_t'')$$
$$\theta_{ty}^T = (y_t, y_t', y_t''),$$

where the prime sign (′) denotes differentiation with respect to time and the superscript T denotes transpose. For example, θ_{tx} is a 3-vector of position, velocity, and acceleration in the x direction at time t. The basic accelerometer measurements are made every Δ seconds, and we will use a discrete-time analysis with Δ as the sampling interval. In this section a subscript k will denote a multiple of the sampling interval. For example,

$$\theta_k = \theta_{t|t=k\Delta}.$$

If the accelerations in the two dimensions were known at all times, the true position could be obtained by integration. A measurement of the accelerations can be obtained from an accelerometer, which measures forces. If the mass is known, this can be converted to an acceleration. In Section 3 we determine a control strategy that dictates the controlled accelerations. Other forces (outside our control) may act on the system to produce uncontrolled acceleration. The spacecraft's acceleration is assumed to be the sum of these two components. It is convenient to introduce

$$\phi_k^T = (x_k, y_k)$$

and to represent the first and second derivatives (i.e., velocity and acceleration) by ϕ by ϕ' and ϕ'', respectively. By assumption,

$$\phi_k'' = U_k + \eta_k, \tag{2.1}$$

where the controlled acceleration vector is

$$U_k^T = (u_{kx}, u_{ky})$$

and the uncontrolled acceleration vector is

$$\eta_k^T = (n_{kx}, n_{ky}).$$

We will assume that η_k is a white, Gaussian sequence with mean zero and covariance matrix $\sigma_n^2 I$ [denoted by $\eta_k \sim \mathbb{N}(0, \sigma_n^2 I)$], where I is the identity matrix. The controlled portion of (2.1) can be chosen to be constant on each interval of length Δ. A further discussion of the control sequence is given in Section 3. For simplicity, we approximate the uncontrolled portion as a

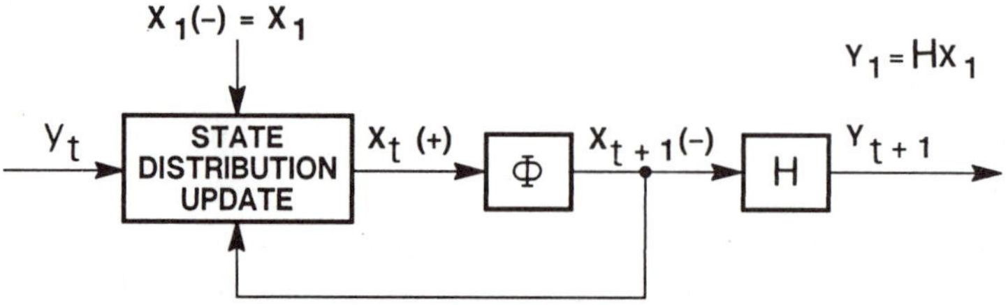

Figure 6. Block diagram of the Bayesian categorical filtering algorithm for state models.

propagation equation (3.21) is equivalent to (2.14), and the output equation (3.22) is equivalent to (2.13).

The categorical filter is depicted in Figure 6, where the Φ and H boxes denote matrix multiplies by the indicated matrix. In contrast, the state distribution update is a nonlinear transformation that maps the observed symbol y_t into the updated conditional probability distribution of the state. The updating operation is defined by (3.18) to (3.20), which correspond to the system shown in Figure 7.

For an example of categorical filtering, consider the model defined in Figure 4. The initial probability distribution of the state variable, $\mathbf{X}_1$, is selected to be the fixed (steady-state) distribution of the Markov chain (i.e., $\mathbf{X}_1 = \Phi\mathbf{X}_1$). In this example, $\mathbf{X}_1 = [0.26 \quad 0.74]^T$. Corresponding to this prior state distribution is the prior distribution of the observable y_1, which is $\mathbf{Y}_1 = H\mathbf{X}_1 = [0.42 \quad 0.58]^T$. According to this prior, there is a 58% probability of observing the second category (cloud ceiling).

Suppose that the first observation turns out to be $y_1 = 1$ (no cloud ceiling). Then $\mathbf{h}[y_1] = \mathbf{h}[1] =$ (first column of the transposed matrix

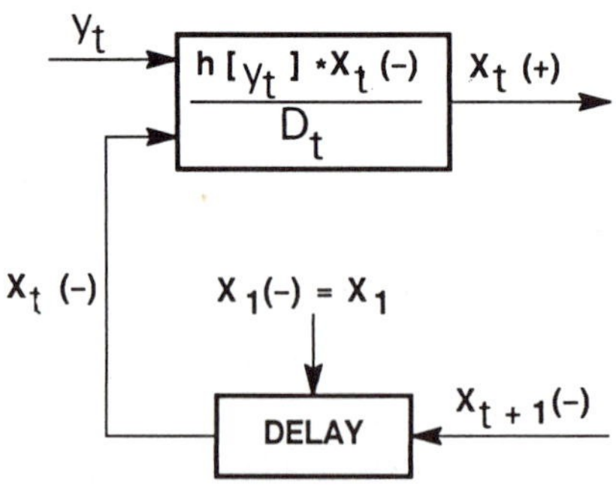

Figure 7. Block diagram of the state distribution update.

$H^T) = [1 \quad 0.21]^T$. The Hadamard product in (3.18) yields $\mathbf{v}_t = [0.26 \quad 0.16]^T$, and (3.19) and (3.20) yield the updated state distribution $\mathbf{X}_1(+) = [0.62 \quad 0.38]^T$. The propagation step, (3.21), yields $\mathbf{X}_2(-) = [0.49 \quad 0.51]^T$, and the output of the filter is the conditional distribution for the next observation, y_2, given by (3.22) as $\mathbf{Y}_2 = H\mathbf{X}_2(-) = [0.60 \quad 0.40]^T$. This distribution states that the probability of a cloud ceiling on the second observation is 40%, given $\mathbf{X}_1$ and the current observation, $y_1 = 1$ (no cloud ceiling).

Continuing this filtering example one more time step, suppose that the second observation is $y_2 = 2$ (cloud ceiling). Then $\mathbf{h}[y_2] = \mathbf{h}[2] =$ (second column of H^T) $= [0 \quad 0.79]^T$. The intermediate vector $\mathbf{v}_2 = [0 \quad 0.40]^T$, and the updated state distribution $\mathbf{X}_2(+) = [0 \quad 1]^T$, which states that the chain is in state 2 with probability 1. [Examination of Figure 4 confirms that "cloud ceiling" ($y_t = 2$) is observed only when the chain is in state 2.] The propagated state distribution is $\mathbf{X}_3(-) = [0.11 \quad 0.89]^T$, and the conditional probability of the next observation is $\mathbf{Y}_3 = [0.30 \quad 0.70]^T$, which indicates a 70% probability of cloud ceiling for the next observation y_3.

3.3 Comparison with Kalman Filtering

In this discussion the reader is assumed to be familiar with Kalman filtering as discussed [e.g., in Gelb (1974) or Harrison and Stevens (1976)]. In Kalman filtering theory, the observed numerical data consist of the m-vectors $\mathbf{y}_t$ for $t = 1, 2, \ldots$, which are related to the n-dimensional state vectors $\mathbf{x}_t$ by the following state-space model:

$$\mathbf{x}_{t+1} = \Phi \mathbf{x}_t + \mathbf{w}_t \tag{3.23}$$

$$\mathbf{y}_t = H\mathbf{x}_t + \mathbf{v}_t. \tag{3.24}$$

In (3.23) $\mathbf{w}_t$ is Gaussian white noise with covariance matrix Q, the initial state $\mathbf{x}_1$ has zero mean and covariance matrix $P_1(-)$, and Φ is the transition matrix. In (3.24) $\mathbf{v}_t$ is Gaussian white noise with covariance matrix R, and H is the output matrix. Comparing (3.23) and (3.24) with (2.15) and (2.16) discloses a *formal* similarity between linear Gaussian state-space models and nonlinear discrete-state models.

The notation $\mathbf{x} \sim N(\mathbf{m}, P)$ is used in the following discussion to indicate that the vector $\mathbf{x}$ is normally distributed with mean vector $\mathbf{m}$ and covariance matrix P. Equations (3.23) and (3.24), plus the initial condition, can now be reexpressed as

$$\mathbf{x}_1 \sim N(\mathbf{0}, P_1(-)) \tag{3.25}$$

$$\mathbf{x}_{t+1} | \mathbf{x}_t \sim N(\Phi \mathbf{x}_t, Q) \tag{3.26}$$

$$\mathbf{y}_t | \mathbf{x}_t \sim N(H\mathbf{x}_t, R). \tag{3.27}$$

For comparison, the distributional equations (2.17) to (2.19) for categorical state models are reproduced with the conditioning explicitly indicated:

$$x_1 \sim \mathbf{X}_1 \tag{3.28}$$

$$x_{t+1} | x_t \sim \mathbf{\Phi}[x_t] \tag{3.29}$$

$$y_t | x_t \sim \mathbf{H}[x_t]. \tag{3.30}$$

Although a formal similarity between (3.25) to (3.27) and (3.28) to (3.30) is apparent, there are also essential differences: (1) in the Gaussian state-space model, the state and output variables are vectors over the real numbers, whereas in the categorical model these variables are scalar quantities taking values in finite sets having no algebraic structure; and (2) the Φ and H matrices in the Gaussian model are linear transformations of the state vector, whereas in the categorical model $\mathbf{\Phi}$ and H are arrays of conditional probability distributions, each column being conditioned on a different value of the state variable.

The Kalman filter computes two conditional means and covariances of the state vector, given an initial state covariance matrix and observations of $\mathbf{y}$ up to and including time t:

1. $\mathbf{x}_t(+)$ and $P_t(+)$, which are the updated (posterior) mean and covariance of $\mathbf{x}_t$.
2. $\mathbf{x}_{t+1}(-)$ and $P_{t+1}(-)$, which are the extrapolated (prior) conditional mean and covariance of $\mathbf{x}_{t+1}$ before $\mathbf{y}_{t+1}$ is observed.

In contrast, the categorical filter computes two discrete probability distributions for its state variable conditioned on an initial state distribution and observations of y up to and including time t:

1. $\mathbf{X}_t(+)$, which is the updated (posterior) distribution of x_t.
2. $\mathbf{X}_{t+1}(-)$, which is the extrapolated (prior) distribution of x_{t+1} before y_{t+1} is observed.

This comparison of Kalman and categorical filtering is summarized in Table 1. An important conclusion from this comparison is that the categorical filter does not converge to the same algorithm as the Kalman filter when the observed symbols are progressively more finely quantized elements of a vector space generated by a linear state-space model. The reason is that these two filters are operating in different coordinate systems: The categorical filter computes discrete probability distributions directly; in contrast, the Kalman filter computes two sufficient statistics (means and covariances) for normal densities.

Table 1. Comparison of Bayesian Filters

Topic	Kalman filter	Categorical filter
Observables	Real vectors	Categories (arbitrary symbols)
	$\mathbf{y}_t \in R^m$	$y_t \in$ finite set
Stochastic model	Linear finite-dimensional system driven by Gaussian white noise	Random function of Markov chain
	$\mathbf{x}_{t+1} \mid \mathbf{x}_t \sim N(\Phi \mathbf{x}_t, Q)$ $\mathbf{y}_t \mid \mathbf{x}_t \sim N(H\mathbf{x}_t, R)$	$x_{t+1} \mid x_t \sim \mathbf{\Phi}[x_t]$ $y_t \mid x_t \sim \mathbf{H}[x_t]$
Filter variables	Conditional means $\mathbf{x}_t(\pm)$ and covariances $P_t(\pm)$	Conditional probability distributions $\mathbf{X}_t(\pm)$
Filter outputs	Conditional means and covariances of $\mathbf{x}_t$ and $\mathbf{y}_{t+1}$	Conditional distributions of x_t and y_{t+1}
Structure of filter algorithm	$(\mathbf{x}_t(-), \mathbf{y}_t) \xrightarrow{\text{update}} \mathbf{x}_t(+)$	$(\mathbf{X}_t(-), y_t) \xrightarrow{\text{update}} \mathbf{X}_t(+)$
	$\mathbf{x}_t(+) \xrightarrow{\text{extrapolate}} \mathbf{x}_{t+1}(-)$	$\mathbf{X}_t(+) \xrightarrow{\text{extrapolate}} \mathbf{X}_{t+1}(-)$
	$\mathbf{x}_{t+1}(-) \xrightarrow{\text{output}} H\mathbf{x}_{t+1}(-)$	$\mathbf{X}_{t+1}(-) \xrightarrow{\text{output}} H\mathbf{X}_{t+1}(-)$
	$P_t(-) \xrightarrow{\text{update}} P_t(+)$	Not applicable
	$P_t(+) \xrightarrow{\text{extrapolate}} P_{t+1}(-)$	Not applicable

4. MODEL INACCURACY

In most applications, a fundamental measure of inaccuracy for stochastic discrete-state models is essential. For example, an inaccuracy measure is needed for fitting model parameters to empirical time series. Even after an accurate reference model has been established, an inaccuracy measure is needed for reduced-order modeling (i.e., for selecting a simpler model that is an accurate approximation to the more complicated reference model). Conventional measures of model inaccuracy for numerical times series, such as the rms error of one-step-ahead predictions, are usually not even defined for categorical models, because there is no metric or arithmetic defined on the set of observable categories. Therefore, Section 4.1 defines a fundamental measure of model inaccuracy called discrimination information (DI), which is consistent with Shannon information theory and which reduces to familiar error measures for numerical time series.

Within the DI framework, log-likelihoods are key statistics for quantifying model inaccuracies. They are also central to using categorical models for Bayesian and maximum-likelihood classification and detection algorithms (Juang and Rabiner, 1985). To support these diverse applications, a recursive algorithm is derived in Section 4.2 for computing estimates of DI from empirical time series. The computationally intensive part of this DI algorithm is the computation of model likelihood statistics. It is shown in Section 4.2 how the Bayesian filtering algorithm in Section 3.2 can be used to compute these likelihoods exactly for finite time series, which is analogous to the use of Kalman filters for computing the likelihoods of Gaussian state-space models for numerical time series. Section 4.3 discusses the practical application of DI to a Bayesian comparison of model accuracies and describes how the Akaike information criterion (AIC) can be applied using the categorical filtering algorithm in Section 3.2.

4.1 Discrimination Information

Consider two state models, A and B, for a categorical time series $\{y\}_T$. If the probability of the observed sequence $\{y\}_T$ is $\Pr_A(\{y\}_T)$ according to model A and $\Pr_B(\{y\}_T)$ according to model B, the log likelihoods of these models are defined as follows:

$$L_A(T) = \log_2 \Pr_A(\{y\}_T) \tag{4.1}$$

$$L_B(T) = \log_2 \Pr_B(\{y\}_T). \tag{4.2}$$

Suppose that both models are ergodic. [A state model is said to be ergodic if the matrices Φ and H are time invariant, and it is possible to go from every state to every other state in a finite number of steps (Kemeny and Snell, 1960).] Suppose, also, that the following limits exist:

$$R_A = \frac{\lim_{T \to \infty} L_A(T)}{T} \tag{4.3}$$

$$R_B = \frac{\lim_{T \to \infty} L_B(T)}{T} \tag{4.4}$$

$$D = R_A - R_B. \tag{4.5}$$

These limits have the units of information rates (expressed in the units of bit per unit time). The difference, D, determines which model is more accurate: If $D > 0$, model A is more accurate; if $D < 0$, model B is more accurate; and if $D = 0$, the models cannot be distinguished. The magnitude of D measures the average rate at which information is available in observations of y_t for discriminating between A and B.

Suppose that model A is the true model, while model B is inaccurate. Then D equals the discrimination information rate, DI, of model B. This is the average rate at which information is observed in y_t for discriminating against model B and recognizing the correctness of model A, and as such, is the fundamental measure of model inaccuracy.

If models A and B are not ergodic, or if time series of a finite length T are being considered, the DI cannot be expressed using limits of time averages. Instead, the fundamental definition of discrimination information (Kullback, 1959) in terms of expected values is invoked. In the general case, with A denoting the true model, the discrimination information rate $\mathrm{DI}(A{:}B;\, T)$ is defined to be proportional to the expected log likelihood ratio of model A to model B, where the expectation is taken with respect to model A:

$$\mathrm{DI}(A{:}B;\, T) = \frac{1}{T} \sum_{s \in S(T)} \mathrm{Pr}_A(s) \log_2 \frac{\mathrm{Pr}_A(s)}{\mathrm{Pr}_B(s)} \tag{4.6}$$

$$S(T) = \text{set of all admissible } \{y\}_T. \tag{4.7}$$

4.2 Likelihood Computations

To compare the inaccuracies of two-state models, given an observed time series $\{y'\}_T$, their log likelihoods are computed and compared. (The symbol y'_t denotes a realization of the random variable y_t.) In this section an efficient recursive algorithm for computing likelihoods is described for state models, which is based on the categorical filtering algorithm in Section 3.2.

Recall that the categorical filter computes the conditional probability distribution of the next symbol, y_{t+1}, given the past and present observations

$\{y'\}_t$. This distribution, $\Pr(y_{t+1}|\{y'\}_t)$, can be used as a building block to compute the log likelihood, $\log_2(\Pr(\{y'\}_T))$, from the entire sequence $\{y'\}_T$. To understand this, note that the probability of observing any admissible sequence $\{y\}_T$ can be computed from the conditional probabilities of observing the individual y_t, as the following calculations indicate:

$$\begin{aligned}\Pr(\{y\}_T) &= \Pr(y_T, y_{T-1}, \ldots, y_2, y_1)\\ &= \Pr(y_T|\{y\}_{T-1})\Pr(\{y\}_{T-1})\\ &= \Pr(y_T|\{y\}_{T-1})\Pr(y_{T-1}|\{y\}_{T-2})\Pr(\{y\}_{T-2})\\ &\ \vdots\\ &= \Pr(y_T|\{y\}_{T-1})\Pr(y_{T-1}|\{y\}_{T-2})\cdots\Pr(y_2|\{y\}_1)\Pr(y_1). \qquad (4.8)\end{aligned}$$

Equation (4.8) for the probability of a sequence $\{y\}_T$ has T factors. The factor $\Pr(y_1)$ on the right end is the prior probability of the first observation. The other factors are conditional probabilities. All of the factors in (4.8) are evaluated on the sample path $\{y'\}_T$ by the categorical filter, as explained in the following.

An output of the filter at time t, given that the partial sequence $\{y'\}_t$ has been processed, is the m-vector $\mathbf{Y}_{t+1}$ [equation (3.22)]. The kth element of this vector is

$$Y_{t+1}(k) = \Pr(y_{t+1} = k|\{y'\}_t). \qquad (4.9)$$

Using the notation of (4.9) allows (4.8) to be reexpressed in terms of the filter output:

$$\Pr(\{y'\}_T) = Y_T(y'_T)Y_{T-1}(y'_{T-1})\cdots Y_2(y'_2)Y_1(y'_1). \qquad (4.10)$$

Therefore, the log-likelihood of the model, $L(T) = \log_2 \Pr\{y'\}_T$, can be computed recursively from the outputs of the categorical filter, $Y_t(y'_t)$:

$$L(0) = 0 \qquad (4.11)$$

$$L(t) = L(t-1) + \log_2 Y_t(y'_t) \qquad (t \in [T]). \qquad (4.12)$$

4.3 Comparing Models

Suppose that an observed categorical time series $\{y'\}_T$ is given and that we wish to determine which of two state models for $\{y'\}_T$ is more accurate, model 1 or model 2. Furthermore, we wish to quantify the difference between the accuracies of these models. Both models may be inaccurate, and the actual model for $\{y'\}_T$ is not given.

Before addressing this comparison problem, consider the idealized case where the actual model for $\{y'\}_T$ is known. Let it be model A, which is different from either model 1 or 2. In principle, the information rates, $\mathrm{DI}(A, 1)$

for model 1 and DI(A, 2) for model 2, could be computed from the stochastic models and compared without even using the series $\{y'\}_T$. That model, 1 or 2, which has the smaller DI is the more accurate, and the difference between the two DIs quantifies the difference between their accuracies. The value of each DI indicates the expected rate of information provided by $\{y'\}_T$ for discriminating against each candidate model, in favor of model A.

The comparison problem cannot be solved using actual DI values, because model A is not known and only a time series $\{y'\}_T$ is given. Instead, the comparison problem is solved by using $\{y'\}_T$ to estimate the difference of the DI values for the two candidate models, as explained in the following discussion.

The definition of DI in (4.6) can be reexpressed in terms of expected log likelihoods (E_A denotes the expectation operator with respect to the correct model):

$$\mathrm{DI}(A\!:\!B;\, T) = \frac{1}{T}\,[E_A L_A(T) - E_A L_B(T)]. \tag{4.13}$$

Therefore, by successively substituting models 1 and 2 for model B in (4.13), the difference between the DI values for the two candidate models can be expressed as follows:

$$\mathrm{DI}(A\!:\!2;\, T) - \mathrm{DI}(A\!:\!1;\, T) = \frac{1}{T}\,[E_A L_1(T) - E_A L_2(T)]. \tag{4.14}$$

The proposed solution to the comparison problem is to estimate the DI difference in (4.14) by replacing the expected log likelihoods on the right side with the sample log likelihoods computed from the series $\{y'\}_T$:

$$\mathrm{DI}(A\!:\!2;\, T) - \mathrm{DI}(A\!:\!1;\, T) = \frac{1}{T}\,[L_1(T) - L_2(T)]. \tag{4.15}$$

If the models A, 1, and 2 are ergodic and the expected log likelihoods in (4.14) exist, the approximation in (4.15) converges with probability 1 as T goes to infinity. The precision of the estimate in (4.15) may be monitored in practice by computing the log likelihoods using the following algorithm based on (4.11) and (4.12). [In the following algorithm, the outputs of the two filters for models 1 and 2 are denoted $Y_t(y'_t, 1)$ and $Y_t(y'_t, 2)$, respectively.]

Initialize:

$$L_1(0) = 0 \tag{4.16}$$

$$L_2(0) = 0 \tag{4.17}$$

$$S(0) = 0. \tag{4.18}$$

For $t = 1$ to T:

$$L_1(t) = L_1(t-1) + \log_2 Y_t(y'_t, 1) \tag{4.19}$$

$$L_2(t) = L_2(t-1) + \log_2 Y_t(y'_t, 2) \tag{4.20}$$

$$S(t) = S(t-1) + [\log_2 Y_t(y'_t, 1) - \log_2 Y_t(y'_t, 2)]^2. \tag{4.21}$$

In (4.21) the sum of squares $S(t)$ of the differences between the incremental log likelihood contributions is computed. After processing the series $\{y'\}_T$ using (4.16) to (4.21), the following summary statistics are computed:

$$d = \frac{1}{T}[L_2(T) - L_1(T)] \tag{4.22}$$

$$v = \frac{S(T)}{T} - d^2 \tag{4.23}$$

$$\sigma_d = \left(\frac{v}{T-1}\right)^{1/2}. \tag{4.24}$$

The number d in (4.22) is the estimate of the DI difference indicated in (4.15), v in (4.23) is the sample variance of the differences between the incremental log likelihoods, and σ_d in (4.24) is the sample standard error of d with $T - 1$ degrees of freedom. The standard error σ_d is an estimate of the standard deviation of the error in using d to estimate the DI difference.

In the following discussion, m denotes the unknown difference in the information rates of the two models [i.e., $m = \mathrm{DI}(A{:}2; T) - \mathrm{DI}(A{:}1; T)$]. By modeling the estimate d as a noisy measurement of m, a Bayesian analysis can be performed to determine the posterior density $p(m|d, \sigma_d^2)$. Specifically, if the prior density of m is normal [$m \sim N(0, \sigma_m^2)$] and the conditional density of d given m and σ_d is normal ($d|m, \sigma_d \sim N(m, \sigma_d^2)$, which is reasonable when the number of observations $T \geqslant 20$), then as the prior becomes noninformative (σ_m goes to infinity), the posterior density $p(m|d, \sigma_d)$ converges to $N(d, \sigma_d^2)$.

The calculations above should be modified when the models being compared are dependent on the time series used for the comparison (Akaike, 1973, 1974, 1976; Larimore and Mehra, 1985). For example, suppose that both models are derived from a single time series, and suppose that this time series is also used to compare their accuracies. Then the model having the larger number of states will usually have the larger likelihood, because the additional parameters can be fit to the unique characteristics of the single time series. In other words, the log likelihood $L_1(T)$ is a biased estimate of the expected value $E_A L_1(T)$ evaluated on independent time series. Akaike developed an asymptotic estimate of this bias for maximum likelihood models developed using long time series. Let $L_1'(T)$ denote the log likelihood of model 1 evaluated on the time series $\{y'\}_T$, and let $L_1(T)$ denote the log likelihood of this model evaluated on an independent realization $\{y\}_T$. Assume that model 1 was derived using maximum likelihood parameter estimation on the time series $\{y'\}_T$. Then according to Akaike's estimate of the bias (modified for base 2 logarithms),

$$E_A L_1(T) = E_A L_1'(T) - \frac{2N}{\log_e 2}, \tag{4.25}$$

where N is the number of independently adjustable parameters in the model. [More precisely, for categorical models N is the least upper bound on the number of real parameters needed to distinguish between all ergodic processes $\{y\}$ having models of the form $(m, n, \mathbf{X}_1, \Phi, H)$ for specified values of m and n as defined in Section 2.4.] For an ergodic model, with $\mathbf{X}_1$ being the fixed state distribution (i.e., $\mathbf{X}_1 = \Phi\mathbf{X}_1$), the number of independent parameters is shown in the Appendix to be

$$N = (3n - 2)(m - 1). \tag{4.26}$$

The total Akaike bias correction for estimating log likelihoods on a sample of length T is $-2N/\log_e 2$, which is incorporated into (4.19) to (4.21) by subtracting the appropriate per sample (incremental) bias from each logarithmic term as shown in the following:

$$\log_2 Y_t(y_t', 1) \rightarrow \log_2 Y_t(y_t', 1) - \frac{2N_1}{T\log_e 2} \tag{4.27}$$

$$\log_2 Y_t(y_t', 2) \rightarrow \log_2 Y_t(y_t', 2) - \frac{2N_2}{T\log_e 2}. \tag{4.28}$$

The T factor appears in the denominators of the bias terms in (4.27) and (4.28), because these are incremental log likelihoods. When T of these incremental log likelihoods are added together to form the total log likelihood, the total bias correction agrees with Akaike's value in (4.25). Parameters N_1 and N_2 are the number of independent parameters in the two models. Basing model comparisons on log likelihoods computed using (4.19) and (4.20), together with the substitutions in (4.27) and (4.28), is equivalent to using the Akaike information criterion (AIC). Moreover, the statistic σ_d, defined by (4.24), provides a measure of the precision of the comparison that is missing when only the AIC is used. More precisely, the previous analysis indicates that the posterior density of $m|d, \sigma_d$, with a noninformative prior and $T \geqslant 20$, is $N(d, \sigma_d^2)$.

5. SUMMARY

The principal contributions of this chapter are summarized in the following:

1. A family of stochastic models is defined for discretely valued (i.e., categorical) time series. According to these models, an observed time series is a random function of an underlying finite Markov chain. The family of models is indexed by the number of states in the chain. This family is analogous to the family of linear state-space models for numerical Gauss-Markov time series.
2. A recursive Bayesian filtering algorithm is derived for categorical time series, which is analogous to the Kalman filter for Gauss-Markov time series. To facilitate comparisons with Kalman filtering, a matrix notation is established for categorical filtering that parallels a widely used notation for Kalman filtering.
3. Discrimination information (DI) is introduced as the fundamental measure of inaccuracy for stochastic models of categorical time series. For estimating DI and comparing the inaccuracies of candidate models, the recursive filtering algorithm is used to compute the exact log likelihoods of state models on finite data sets; a Bayesian analysis then yields the posterior density of DI differences, which measure relative model inaccuracies. When the parameters of two or more candidate models are fitted to a single data set, and the inaccuracies of these models are being compared using this single data set, the likelihood computations are modified to yield the Akaike information criterion (AIC) for categorical models.

ACKNOWLEDGMENTS

The author gratefully acknowledges technical contributions to this study by Steve Baumgartner and Jake Goldstein and to The Analytic Sciences Corporation for supporting this research.

APPENDIX

In this appendix we outline a derivation of N, which is the least upper bound on the number of real parameters needed to distinguish between all ergodic processes $\{y\}$ having state models of the form $(m, n, \mathbf{X}_1, \Phi, H)$, where $\mathbf{X}_1$ is the fixed (steady-state) state distribution (i.e., $\mathbf{X}_1 = \Phi\mathbf{X}_1$). The parameters of interest are the $n \times n$ matrix Φ and the $m \times n$ matrix H. The columns of these matrices are probability distributions. Therefore, these columns sum to unity, and the total number of free parameters in Φ and H is $(n^2 - n) + (mn - n)$. For many models this number is larger than N, because different choices of Φ and H can lead to the same output process $\{y\}$.

An expression for N is derived in the following by assuming that it can be expressed as a multinomial in m and n:

$$N = A + Bm + Cn + Dmn + En^2. \tag{A.1}$$

Higher powers of m and n are not needed, because the number of free parameters in Φ and H is an upper bound on N, and the number of these parameters cannot grow faster than the multinomial in (A.1). The coefficients A, B, C, D, and E are determined by imposing the following boundary conditions on (A.1):

1. If $m = 1$, then $N = 0$. (If there is only one possible value for the output, no parameters are needed to distinguish between different random processes.)
2. If $n = 1$, then $N = m - 1$. (If the chain has exactly one state, the output probabilities are specified by the free parameters in the $m \times 1$ matrix H.)
3. If $m = 2$ and $n = 2$, then $N = 4$. (If the model is as simple as possible without being trivial, N equals the number of free parameters in Φ and H.)

Applying condition 1 to (A.1) yields the equation

$$0 = A + B + Cn + Dn + En^2. \tag{A.2}$$

Because the number of states in the chain, n, can be any positive integer, it follows from (A.2) that $B = -A$, $D = -C$, and $E = 0$. Substituting these results in (A.1) and applying condition 2 yields the following equation for $m \geqslant 1$:

$$m - 1 = A - Am + C - Cm, \tag{A.3}$$

from which it follows that $C = -1 - A$. By substituting this expression for C and the previous expressions for B, D, and E in (A.1) and applying condition 3, it can be shown that $A = 2$. Substituting this value in the previously derived expressions then yields $B = -2$, $C = -3$, and $D = 3$. Finally, these values, when substituted in (A.1), yield the equation

$$N = 2 - 2m - 3n + 3mn, \tag{A.4}$$

which can be factored to yield the final result:

$$N = (3n - 2)(m - 1). \tag{A.5}$$

REFERENCES

Aczel, J., and Z. Daroczy (1975). *On Measures of Information and Their Characterizations*, Academic Press, New York.

Akaike, H. (1973). Information Theory and an Extension of the Maximum Likelihood Principle, in *Second International Symposium on Information Theory* (B. N. Petrov and F. Csaki, eds.), Akadémiai Kiadó, Budapest, pp. 267–281.

Akaike, H. (1974). A new look at the statistical model identification, *IEEE Trans. Autom. Control* AC-19, 716–723.

Akaike, H. (1976). *System Identification: Advances and Case Studies* (R. K. Mehra and D. G. Lainiotis, eds.), Academic Press, New York, pp. 27–96.

Bucy, R. S. (1969). Bayes theorem and digital realizations for nonlinear filters, *J. Astronaut. Sci.* 17, 80–94.

Bucy, R. S., and A. J. Mallinckrodt (1973). An optimal phase demodulator, *Stochastics* 1, 3–23.

Bucy, R. S., and K. D. Senne (1971). Digital synthesis of nonlinear filters, *Automatica* 7, 287–298.

Bucy, R. S., and K. D. Senne (1980). Nonlinear filtering algorithms for vector processing machines, *Comput. Math. Appl.* 6, 317–338.

Bucy, R. S., C. Hecht, and K. D. Senne (1972). *An Engineer's Guide to Building Nonlinear Filters*, Report AD-746921, Vols. I and II, F. J. Seiler Research Laboratory, United States Air Force Academy, Colorado Springs, Colo.

Bucy, R. S., R. Ghovanlou, J. M. F. Moura, and K. D. Senne (1983). Nonlinear filtering and array computation, *IEEE Comput.* 51–61, June.

Castanon, D. A., and D. Teneketzis (1985). Distributed estimation algorithms for nonlinear systems, *IEEE Trans. Autom. Control* AC-30, 418–425.

Gelb, A., ed. (1974). *Applied Optimal Estimation*, MIT Press, Cambridge, Mass.

Harrison, P. J., and C. F. Stevens (1976). Bayesian forecasting, *J. R. Stat. Soc. Ser. B* 38, 205–247.

Juang, B. H., and L. R. Rabiner (1985). A probabilistic distance measure for hidden markov models, *AT&T Tech. J.* 64, 391–408.

Kannappan, P. (1972). On Shannon's entropy, directed divergence and inaccuracy, *Z. Wahrsch.* 22, 95–100.

Kemeny, J. G., and J. L. Snell (1960). *Finite Markov Chains*, D. Van Nostrand, Princeton, N.J.

Kullback, S. (1959). *Information Theory and Statistics*, Dover, New York.

Larimore, W. E., and R. K. Mehra (1985). The problem of overfitting data, *Byte* 10, 167–180.

Levinson, S. E., L. R. Rabiner, and M. M. Sondhi (1983). An introduction to the application of the theory of probabilistic functions of a markov process to automatic speech recognition, *Bell Syst. Tech. J.* 62, 1035–1074.

Matsuoka, T., and T. J. Ulrych (1986). Information theory measures with application to model identification, *IEEE Trans. Acoust. Speech Signal Process. ASSP*-34, 511–517.

Meyer, C. D., Jr. (1975). The role of the group generalized inverse in the theory of finite Markov chains, *SIAM Rev.* 17, 443–464.

Perrenod, S. C., and R. S. Bucy (1985). Supercomputer Performance for Convolution, in *Proceedings of the First International Conference on Supercomputing Systems*, St. Petersburg, Fl., pp. 301–309.

Rabiner, L. R., and B. H. Juang (1986). An introduction to hidden Markov models, *IEEE Acoust. Speech Signal Process. Mag.* 3, 4–16.

Shore, J. E., and R. W. Johnson (1980). Axiomatic derivation of the principle of maximum entropy and the principle of minimum cross-entropy, *IEEE Trans. Inf. Theory* IT-26, 26–36.

U.S. Dept. of Commerce (1984). *Surface Synoptic Codes*, Federal Meteorological Handbook 2, National Oceanic and Atmospheric Administration, Washington, D.C.

White, J. V., and P. Janota (1985). Stochastic Modeling and Forecasting for Categorical Time Series, in *Proceedings of the 9th Conference on Probability and Statistics in Atmospheric Sciences*, Virginia Beach, Va., pp. 163–170.

11

Use of Image Data in a Control Problem

WILLIAM W. DAVIS* Lockheed Missiles & Space Company, Inc., Austin, Texas

1. INTRODUCTION

In this chapter we consider the motion of a spacecraft, which is modeled as a point and is assumed to be moving at a constant altitude. The vehicle possesses two measurement devices (an accelerometer and a radiometer) and a control system that can impart accelerations to the vehicle.

A discrete-time model is given for the dynamical system of the spacecraft. The model utilizes six states consisting of the position, velocity, and acceleration in two dimensions (labeled x and y). A Kalman filter model (e.g., Kalman, 1960) is used for the dynamical system. This model is commonly used in aerospace applications, because it allows a good approximation to the actual system while allowing parsimony in computation and storage.

The Kalman filter model is specified by state equations that describe the change over time of state variables, and measurement equations that relate the measurement sources to the system states. Updating equations for the distributions of the states are derived from the state and the measurement equations. From the Bayesian viewpoint the state updates provide a prior distribution of the state vector. This prior distribution is revised using the measurement equation to obtain a posterior distribution.

*Present address: Statistical Research Division, Bureau of the Census, Washington, D.C.

The Kalman algorithm provides optimal state estimates under different conditions. If the loss function is quadratic, the algorithm is optimal among linear estimators for all wide-sense error distributions (i.e., those distributions having specified means and covariances). If the errors distributions are Gaussian, the Kalman algorithm generates the posterior mean so that the estimator is optimal for a variety of loss functions. Duncan and Horn (1972) described the wide-sense optimality, while Meinhold and Singpurwalla (1983) gave a Bayesian derivation.

In various control problems the optimal control is a function of the current state, which is estimated optimally by the Kalman filter recurrence relations. Probably the most important example is a linear dynamical system with a quadratic cost criterion and Gaussian noise. This is sometimes referred to as the LQG problem (e.g., Whittle, 1983, Chap. 11). For an example with an exponential cost criterion see Chen (1985, Chap. 5).

In Section 2 we review the general Kalman methodology and apply the method to the control of the spacecraft. The system state model and the measurement equation for the accelerometer are given. It is shown that the accuracy of the position estimate degrades over time if only acceleration measurements are utilized. It is shown that this undesirable feature can be eliminated through the use of radiometer measurements.

In Section 3 we give a brief discussion of mission planning for the spacecraft. The premission analysis selects location, called waypoints, where radiometer measurements will be obtained. At each waypoint the position is revised and a course change may be made.

A constant-acceleration turn is made between each pair of waypoints. The control depends only on the current state estimate and the position of the next waypoint. A multiperiod control solution would consider the position of all future waypoints. If the LQG conditions were satisfied, the optimal multiperiod control could be calculated using backward induction (see, e.g., Bertsekas, 1976, Sec. 4.3). Unfortunately, it is difficult to determine a priori the covariance matrix of the radiometer measurement equation, which is a requirement of the LQG problem.

The unique single-period constant-acceleration control which provides mean-zero (Bayesian) errors is given in Section 3. The Kalman recurrence relations are used to update the position estimate at each waypoint.

To update the position at each waypoint, the measurement equation of the radiometer is required. In Section 4 a digital model is developed for the radiometer measurements, and an iterative technique is proposed to calculate the maximum likelihood estimate (MLE) of the parameters. The Fisher information matrix is derived and is used to derive the precision matrix of the measurement equation. We give an example which shows that the MLE is not necessarily asymptotically Gaussian. In such a case one would not expect the covariance estimate to perform well.

For one material map and filter the true covariance matrix is calculated. It is compared by simulation with the approximate covariance matrix and the agreement is shown to be good.

The results of Section 4 on the radiometer measurement equation are related to the image registration literature. In this problem the measurements are assumed to be generated by a portion of a scene, and the goal is to determine the location of the image generator. Good summaries of previously published algorithms for the image registration problem are given by Pratt (1978, Chap. 19) and by Hall (1979, Chap. 7). Techniques such as maximum correlation (Pratt, 1974), minimum absolute deviation (Barnea and Silverman, 1972), and minimum variance (Kuglin, 1981) have been proposed to register the image, but these papers do not contain theoretical results on the performance of the techniques.

In this chapter a model-based approach is used, and the optimal estimator of the parameters is derived. The optimal algorithm reduces to Kuglin's minimum variance algorithm in a special case (no filtering). The model-based approach allows theoretical properties (such as the precision of the estimator) to be derived.

Baird (1984) gave a model-based approach for the image registration problem. His approach determines the optimal registration using linear programming. Noise is modeled using constraints, so it is difficult to quantify the estimate's precision.

The model given in this chapter is not meant to be completely accurate for any spacecraft. A real-world model would include a third dimension as well as other state variables, and would explicitly incorporate gravitational forces. The general conclusions of this paper, however, can be demonstrated adequately using such an example.

2. SYSTEM EQUATIONS

In this section we give the model and analysis for the spacecraft. The description of the state equations is given in Section 2.1. In Section 2.2 the accelerometer measuremeter equation is given. The discrete-time Kalman updating equations are given in Section 2.3 and specialized to the accelerometer measurement equation in Section 2.4. In Section 2.5 the radiometer measurement equation is given; a complete derivation of this equation is deferred to Section 4.

constant on each interval also. With this assumption

$$\phi_t'' = \phi_k'' \qquad (k-1)\Delta \leqslant t \leqslant k\Delta. \tag{2.2}$$

We define ϕ_k to be the position at the start of the $(k+1)$st epoch (i.e., $t = k\Delta$) and similarly, ϕ_k' to be the velocity at that time. Using (2.2) gives

$$\begin{aligned}\phi_{k\Delta}' &= \phi_{(k-1)\Delta}' + \int_{(k-1)\Delta}^{k\Delta} \phi_t''\,dt \\ &= \phi_{(k-1)\Delta}' + \phi_{k\Delta}''\Delta.\end{aligned} \tag{2.3}$$

It follows from (2.3) and the definition of ϕ' that

$$\phi_{k+1}' = \phi_k' + \phi_k''\Delta. \tag{2.4}$$

Similarly, one can show

$$\phi_{k+1} = \phi_k + \Delta\phi_k' + \frac{\Delta^2\phi_k''}{2}. \tag{2.5}$$

Equations (2.1), (2.4), and (2.5) are the basis for the Kalman state equations. In fact, we can write

$$\theta_k = A\theta_{k-1} + u_{k-1} + n_{k-1}, \tag{2.6}$$

where

$$A = \begin{bmatrix} B & 0 \\ 0 & B \end{bmatrix}$$

with

$$B = \begin{bmatrix} 1 & \Delta & \dfrac{\Delta^2}{2} \\ 0 & 1 & \Delta \\ 0 & 0 & 0 \end{bmatrix}$$

$$u_k^T = (0, 0, u_{kx}, 0, 0, u_{ky})$$

$$n_k^T = (0, 0, n_{kx}, 0, 0, n_{ky}),$$

where the sequence $n_k \sim \mathbb{N}(0, M)$ with $M = \operatorname{diag}(0, 0, \sigma_n^2, 0, 0, \sigma_n^2)$.

2.2 Accelerometer Measurement Equation

The accelerometer is assumed to give unbiased estimates of the true accelerations. For the kth time interval we assume that

$$z_k = \phi_k'' + e_k, \tag{2.7}$$

where $z_k^T = (z_{kx}, z_{ky})$ are the measured accelerations and the measurement errors

$$e_k^T = (e_{kx}, e_{ky}) \sim \mathbb{N}(0, N) \tag{2.8}$$

with $N = \sigma_e^2 I$. Equation (2.7) can be written in the form of the measurement equation

$$z_k = C\theta_k + e_k, \tag{2.9}$$

where

$$C = \begin{bmatrix} 0 & 0 & 1 & 0 & 0 & 0 \\ 0 & 0 & 0 & 0 & 0 & 1 \end{bmatrix}.$$

2.3 Kalman Updating Methodology

In this section we give the Kalman recurrence relations for the state [(2.6)] and measurement [(2.9)] equations. We will use p as a generic symbol for a probability density (or distribution) with the argument of p defining the random variable (e.g., Lindley, 1968). For example, $p(\theta_1|z_1)$ denotes the density of θ_1 given z_1.

We label $z^{(k)} = (z_1, z_2, \ldots, z_k)$, where the kth value is obtained at time $t = k\Delta$. Then if $\theta_0 \sim \mathbb{N}(\theta_0^*, \Sigma_0)$, it follows that

$$\{\theta_k|z^{(k)}\} \sim \mathbb{N}(\theta_k^*, \Sigma_k) \tag{2.10}$$

and the Kalman recurrence relations can be used to calculate the mean θ_k^* and the covariance matrix Σ_k. The distribution (2.10) can be viewed as the posterior distribution of θ_k. The Kalman equations can be partitioned into state and measurement updates. The state updates give the parameters of

$$p(\theta_{k+1}|z^{(k)}) = \mathbb{N}(\theta_{k+1|k}^*, \Sigma_{k+1|k})$$

as

$$\begin{aligned} \theta_{k+1|k}^* &= A\theta_k^* + u_k \\ \Sigma_{k+1|k} &= A\,\Sigma_k\,A^T + M. \end{aligned} \tag{2.11}$$

Using the standard Bayesian, Gaussian algebra (e.g., Raiffa and Schlaifer, 1961, Sec. 13.2) applied to

$$p(\theta_{k+1}|z^{(k+1)}) \propto p(z_{k+1}|\theta_{k+1})p(\theta_{k+1}|z^{(k)}), \tag{2.12}$$

the update of (2.10) is completed with

$$\begin{aligned} \theta_{k+1}^* &= \theta_{k+1|k}^* + K_k(z_{k+1} - C\theta_{k+1|k}^*) \\ \Sigma_{k+1} &= \Sigma_{k+1|k} - K_k C\,\Sigma_{k+1|k}, \end{aligned} \tag{2.13}$$

where the Kalman gain matrix is

$$K_k = \Sigma_{k+1|k} C^T (C \Sigma_{k+1|k} C^T + N)^{-1}. \tag{2.14}$$

2.4 Accelerometer Measurement Updating

In this section we discuss the performance of the Kalman filter estimate of the states when only accelerometer measurements are available. We introduce the notation

$$\Sigma_k = \begin{bmatrix} \Sigma_{k,11} & \Sigma_{k,12} \\ \Sigma_{k,21} & \Sigma_{k,22} \end{bmatrix},$$

where each of the submatrices are 3×3. That is, if $\Sigma_k = (\sigma_{ijk})$ for $1 \leqslant i, j \leqslant 6$, then $\Sigma_{k,11} = (\sigma_{ijk})$ for $1 \leqslant i, j \leqslant 3$. A similar representation is assumed for $\Sigma_{k+1|k}$.

We assume that the positions, velocities, and accelerations are independent a priori. That is, Σ_0 is a diagonal matrix. With this assumption it follows that $\sigma_{13k} = \sigma_{23k} = 0$ and $\sigma_{33k} = \sigma_e^2 \sigma_n^2 / (\sigma_e^2 + \sigma_n^2) = \lambda$ (say) for $k > 0$.

A sketch of the recursive calculation of the covariance sequence is given now. From (2.11) we can show that for $k \geqslant 0$,

$$\Sigma_{k+1|k,11} = \Sigma_{k,11} + \begin{bmatrix} H_k & 0 \\ 0^T & \lambda - \sigma_{33k} \end{bmatrix}, \tag{2.15}$$

where

$$H_k = \begin{bmatrix} 2\Delta\sigma_{12k} + \Delta^2\sigma_{22k} + 0.25\Delta^4\sigma_{33k} & \Delta\sigma_{22k} + 0.5\Delta^3\sigma_{33k} \\ \Delta\sigma_{22k} + 0.5\Delta^3\sigma_{33k} & \Delta^2\sigma_{33k} \end{bmatrix}.$$

It follows from (2.13)–(2.15) that the Kalman gain matrix is given by

$$K_k^T = \frac{\sigma_e^2}{\sigma_e^2 + \sigma_n^2} C$$

$$\Sigma_{k+1,11} = \Sigma_{k,11} + \begin{bmatrix} H_k & 0 \\ 0^T & 0 \end{bmatrix}. \tag{2.16}$$

Summation of (2.16) can be used to obtain σ_{ijk} for $1 \leqslant i, j \leqslant 2$. In fact, we have that the 2×2 submatrix of $\Sigma_{k+1,11}$ differs from $\Sigma_{0,11}$ by

$$\sum_{j=0}^{k} H_j = \begin{bmatrix} 2k\Delta\sigma_{220} + k^2\Delta^2\sigma_{220} + k(k+1)(2k+1)\Delta^4 \dfrac{\lambda}{6} & k\Delta\sigma_{220} + \Delta^3 k^2 \dfrac{\lambda}{2} \\ k\Delta\sigma_{220} + \Delta^3 k^2 \dfrac{\lambda}{2} & k\Delta^2\lambda \end{bmatrix}. \tag{2.17}$$

For a fixed sampling interval Δ with $t = k\Delta$, equation (2.17) shows that the standard deviation of the position grows at rate $t^{1.5}$ while the velocity standard deviation grows at rate $t^{0.5}$. In the following section we show that the position standard deviation can be reduced through the use of radiometer measurements taken at selected times.

An alternative look at location prediction can be made using the methodology of Box and Jenkins (1970). Using equations (2.1), (2.3), and (2.4), we have

$$\phi_k = c_k + a_k, \tag{2.18}$$

where $(1 - B)^2 c_k = 0.5\Delta^2 B(1 + B)u_k$ and $(1 - B)^2 a_k = 0.5\Delta^2 B(1 + B)v_k$, where v_k is white noise and B is the backshift operator. Equation (2.18) expresses the (unobservable) position as the sum of controllable and uncontrollable portions. This equation shows that the uncontrollable portion follows an integrated moving average model [IMA (2, 2); see, e.g., Box and Jenkins (1970)], where the moving average parameters are known. If ϕ_0 and $u_1, u_2, \ldots, u_k$ are known, it is easy to show that the prediction variance of each component of ϕ_k has standard deviation of order $k^{1.5}$. (This is the same as the standard deviation of the state estimate of position after k accelerometer measurements.)

2.5 Radiometer Measurement Updating

A small number of radiometer measurements are made throughout the flight to maintain an accurate estimate of position. A complete explanation of the radiometer measurements and the estimation procedure is given in Section 3. In this section we describe the measurement update for this data source.

We assume the (dwell) time Δ_1 necessary to obtain the radiometer data is much smaller than the sampling interval Δ. The mission is designed so that the radiometer measurements are obtained at the end of a sampling interval. Thus they are obtained at $t = k\Delta$ for some integer k.

We show in Section 3 that the radiometer measurements can be summarized in terms of an estimate ϕ_{*k} of the position ϕ_k with

$$\phi_{*k} \sim \mathbb{N}(\phi_k, \Gamma_k). \tag{2.19}$$

There are two equivalent ways of including this information. One is to use the methodology of Sections 2.3 and 2.4 to carry out a state update and a measurement update for the accelerometer measurement. Then the distribution is summarized as

$$\{\theta_k | z^{(k)}\} \sim \mathbb{N}(\theta_k^*, \Sigma_k^*). \tag{2.20}$$

We combine (2.19) and (2.20) using the standard Bayesian methodology.

Since (2.20) depends on θ_k only through ϕ_k, only the distribution of ϕ_k is modified. If the distribution of ϕ_k is

$$\{\phi_k | z^{(k)}\} \sim \mathbb{N}(\phi_k^*, \mathbb{R}_k^*),$$

then after the radiometer data are included,

$$\{\phi_k | z^{(k)}, \phi_{*k}\} \sim \mathbb{N}(\bar{\phi}_k, \bar{\mathbb{R}}_k),$$

where

$$\bar{\mathbb{R}}_k = ((\mathbb{R}_k^*)^{-1} + (\Gamma_k)^{-1})^{-1}$$
$$\bar{\phi}_k = \bar{\mathbb{R}}_k((\Gamma_k)^{-1}\phi_{*k} + (\mathbb{R}_k^*)^{-1}\phi_k^*).$$

The alternative method is to use a measurement update that includes both accelerometer and measurements simultaneously. Incorporating (2.19) in the measurement equation (2.9), we obtain

$$\begin{bmatrix} z_k \\ \phi_{*k} \end{bmatrix} = \begin{bmatrix} 0 & 0 & 1 & 0 & 0 & 0 \\ 0 & 0 & 0 & 0 & 0 & 1 \\ 1 & 0 & 0 & 0 & 0 & 0 \\ 0 & 0 & 0 & 1 & 0 & 0 \end{bmatrix} \theta_k + e_{*k},$$

where the 4×4 dimensional covariance matrix of e_{*k} is $\operatorname{diag}(\sigma_n^2 I, \Gamma_k)$.

The method proposed here involves estimation of the covariance matrix for the radiometer measurement equation. Since the estimated matrix Γ_k is not necessarily diagonal, the use of radiometer measurements introduces correlation between x_k and y_k.

If the precision of the radiometer measurements is high compared to the current estimate of position, the revised precision matrix of the position will be essentially equal to the precision matrix of the radiometer measurement. Stated in terms of covariances: If $\Gamma_k = (\gamma_{ijk})$ for $1 \leqslant i, j \leqslant 2$ and $\max\{\gamma_{11k}, \gamma_{22k}\} \ll \Delta^4 k^3 \lambda/3$, then $\bar{\mathbb{R}}_k \doteq \Gamma_k$.

3. MISSION PLANNING

In this section we give a brief discussion of the premission planning and the in-flight modifications to the plan. The mission plan develops an initial control structure $\{u_k\}$ in equation (2.6). In this section we give a continuous time control sequence $\{u_t\}$, which is approximated by using a constant control sequence on the kth time interval.

The control plan is developed as a function of waypoints w_i, which are selected in advance. The constant-acceleration turn from w_{i-1} to w_i is derived in Section 3.1. Prior to the flight the expected path can be calculated. Minor modifications to the premission plan are made in flight by using the procedure of Section 3.1 with w_{i-1} replaced by $\bar{\phi}_{i-1}$. This control provides the only constant-acceleration turn with no mean error at the next waypoint, w_i.

The waypoints are selected to satisfy the sometimes conflicting requirements of minimization of flight time and maximization of the radiometer measurement's precision. A separate paper could be written on waypoint selection; here we assume that they are given. The next section gives a solution to the guidance problem between waypoints.

3.1 Guidance Between Waypoints

In this section we give the equations for a constant-acceleration turn from initial position w_0 with velocity vector ϕ' to first waypoint w_1. For a constant-acceleration turn the magnitude of the velocity remains constant and

$$a = \frac{v^2}{r}, \tag{3.1}$$

where $a = \|\phi_0''\|$, $v = \|\phi_0'\|$, r is the turning radius, and $\|\cdot\|$ denotes the norm of a vector. For circular motion the acceleration vector is normal to the velocity vector and points toward the center of the circle. Thus the center of the circle ϕ_c is of the form

$$\phi_c = \phi_0 + r\alpha_0,$$

where α_0 is a unit vector in the direction of ϕ_0'',

$$\alpha_0 = \operatorname{sign}(\psi)\left(-\frac{y_0'}{v}, \frac{x_0'}{v}\right),$$

where ψ is the angle between the line-of-sight vector $w_1 - w_0$ and the velocity vector ϕ_0'. Since w_1 is on the circle, one can show that

$$r = \frac{d}{2|\sin\psi|}, \tag{3.2}$$

where

$$d = \|w_1 - w_0\|.$$

The acceleration follows from (3.1) and (3.2):

$$a = \frac{2v^2|\sin\psi|}{d}.$$

The geometry of the path is shown in Figure 1. Since the length of the path is

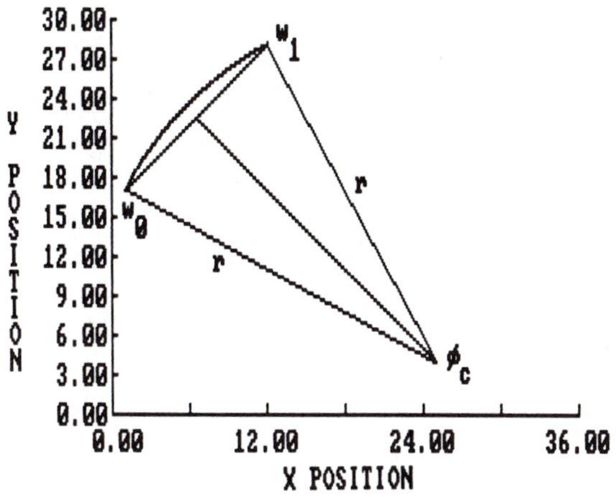

Figure 1. Geometry of one mission segment.

$2r\psi$, the time spent during this segment is

$$T = \frac{d\psi}{v \sin\psi} \doteq \frac{d(1 + \psi^2/6)}{v} \qquad \text{for small } \psi.$$

The time of this segment is d/v if there is no initial heading error, and the percentage difference is approximately $\psi^2/6$ for small ψ.

The position, velocity, and acceleration vector can be expressed in terms of the orthogonal matrix P and the vector u_t as in Table 1, where

$$P = \frac{1}{v}\begin{bmatrix} x'_0 & y'_0 \operatorname{sign}(\psi) \\ y'_0 & x'_0 \operatorname{sign}(\psi) \end{bmatrix}$$

$$u_t = \begin{bmatrix} \cos\left(\dfrac{vt}{r} - \dfrac{\pi}{2}\right) \\ \sin\left(\dfrac{vt}{r} - \dfrac{\pi}{2}\right) \end{bmatrix}.$$

3.2 Example

In this section we discuss a mission with three segments, which are defined in Table 2. We compare the three-segment mission to a mission without any waypoints. The initial velocity is 500 mi/h and the initial velocity vector is (500, 0). In this example the initial line of sight to the target coincides with the initial velocity vector.

Table 1. States During Circular Turn

Quantity	Expression
Position	$\phi_t = \phi_c + rPu_t$
Velocity	$\phi_t' = rPu_t'$
Acceleration	$\phi_t'' = -aPu_t$

Table 2. Premission Planning

Segment	1	2	3
Initial x position (mi)	0	250	500
Initial y position (mi)	0	20	50
Final x position (mi)	250	500	820
Final y position (mi)	20	50	0
Radius of curvature (mi)	1572.5	2908.3	722.8
Initial x velocity (mi/h)	500	493.6	498.7
Initial y velocity (mi/h)	0	80	35.5
Angle subtended (rad)	0.16	0.086	0.44
Time of flight (h)	0.503	0.503	0.653
Final x velocity (mi/h)	493.6	498.7	463
Final y velocity (mi/h)	80	35.5	−187.5

The equations of Section 3.1 can be used to give the expected flight path for the three-segment mission. A typical flight path is shown in Figure 2 and Table 2 gives characteristics of the three segments. The expected mission time is 1.76 h or 7% longer than the constant-heading mission, which can be completed in 1.64 h.

If the waypoints are selected so that an accurate estimate of position is obtained at each one, the only source of error at the final waypoint will be the error incurred in the third segment. We compare the standard deviations at the destination for the two missions. From (2.17) the ratio of the standard deviations is equal to the ratio of the mission times raised to the 3/2 power.

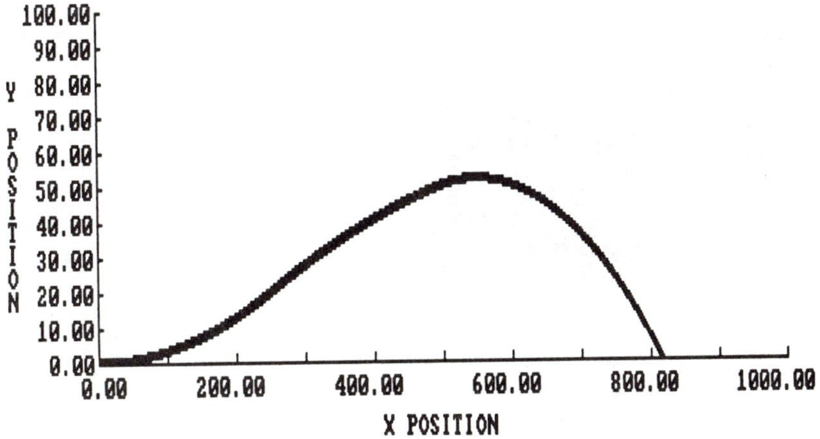

Figure 2. Typical path for the mission of Table 2.

Thus we have

$$\frac{\text{SD(2-waypoint mission)}}{\text{SD(0-waypoint mission)}} = \left[\frac{0.653}{1.64}\right]^{1.5} = 0.25.$$

The introduction of two waypoints has decreased the standard error in each dimension by a factor of 4.

Targeting analyses are frequently summarized in terms of the circular error probable (CEP), which is the median of the radial error. For the circular Gaussian distribution CEP is proportional to the standard deviation, so the CEP is reduced by a factor of 4.

A typical flight path for this mission is shown in Figure 2. The heading deviations from the premission plan are too small to be detected on this scale.

4. IMAGE REGISTRATION

In this section we give a methodology that allows the radiometer measurements to be included in the dynamic model. That is, we derive equation (2.19). The analysis of this section is based on a fixed time, so the time subscript (t or k) is not used here. If the time subscript is ignored, equation (2.19) can be

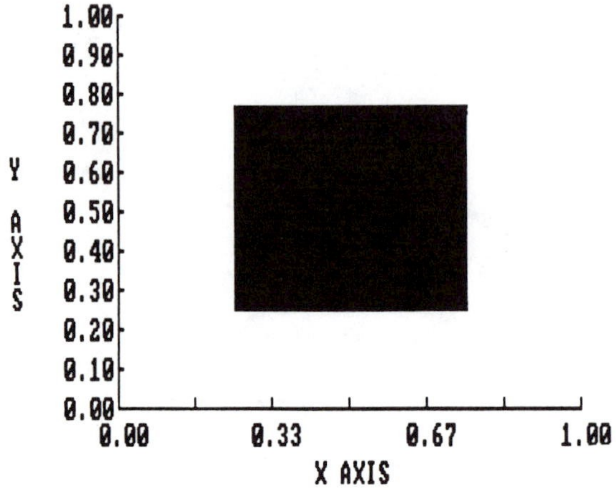

Figure 3. Material map.

written

$$\phi_* \sim \mathbb{N}(\phi, \Gamma). \tag{4.1}$$

The joint asymptotic distribution of an ancillary statistic and ϕ_* is derived in Section 4.3, and the marginal distribution of ϕ_* is used as (4.1). For an alternative approach of utilizing ancillary statistics in dynamical systems, see Spall (1985).

As stated previously, the waypoints are selected so that an accurate estimate of position can be made there. We assume that each scene is a square with sides of length $L(s)$. Thus the scene is composed of $\phi \in [0, L(s)] \times [0, L(s)]$. We assume that each point in the scene can be classified with respect to its material (e.g., land or water). We define a material function $m(\phi)$, where $m(\phi) = k$ if ϕ is material k. We assume only two materials here, but the extension to a finite number is apparent. A material map shown in Figure 3 would imply that

$$m(\phi) = \begin{cases} 1, & |x - 0.5| < 0.25, \ |y - 0.5| < 0.25 \\ 2, & \text{otherwise.} \end{cases}$$

4.1 Continuous Model for Intensities

We assume that each point on the earth radiates energy and that a particular time the amount of energy radiated is a function only of the material type. The amount of energy radiated is called the radiometric temperature and is labeled $f(\phi)$. The radiometric temperatures of a material vary with time and with environmental conditions, so we will not assume that they are known. The radiometric temperature of material k is labeled μ_k for $k = 1, 2$. Under these assumptions the radiometric temperature can be expressed in terms of the material function as

$$f(\phi) = \sum_{k=1}^{2} \mu_k I(m(\phi) = k),$$

where I denotes the indicator function.

The radiometer points at a plot of ground on the earth for a short period of time Δ_1 and collects the energy that is generated. We will assume that radiometer collects energy from a square with sides of length L, where $L < L(s)$. This square is called the image generator, and the image registration problem is to determine its location. We assume in this work that a portion of the image generator is within the scene.

It is convenient in the analysis to have the sides of the image generator parallel to the sides of the scene. This can be accomplished by rotation of the radiometer (with respect to the spacecraft) in two dimensions. The correct rotation can be estimated using the current heading (velocity vectors in x and y). With this assumption the image generator is a square centered at $\phi + 0.5L\underline{1}$, where $\underline{1}^T = (1, 1)$, and the image registration problem is to determine ϕ. The geometry is shown in Figure 4, where the scene of Figure 3 is used.

Much of the image-processing literature utilizes the equation

$$O(\phi_*) = \iint W(\phi_0) f(\phi_* - \phi_0)\, d\phi_0, \tag{4.2}$$

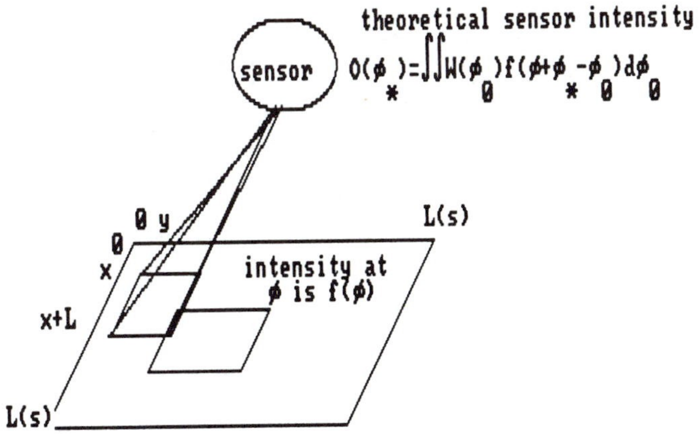

Figure 4. Geometry of the scene and the sensor.

where $W(\cdot)$ is the two-dimensional filter function of the radiometer and $O(\cdot)$ is the measured intensity. For example, one important image-processing problem is to use the measured output O to restore the input f (see, e.g., Hunt, 1983). Typically, the filter function is treated as known since it can be measured in controlled experiments. Equation (4.2) is based on the assumption that the filter is linear and shift invariant. That is, the output from multiple materials is the sum of the individual outputs, and a translation in the materials causes a corresponding translation in the output.

Another important image-processing problem is registration. We assume the model

$$O(\phi_*) = \iint W(\phi_0) f(\phi + \phi_* - \phi_0)\, d\phi_0 \tag{4.3}$$

and the registration problem is to estimate the translation ϕ. The effect of noise and digitization are ignored in (4.3), but they are incorporated in the next section.

4.2 Digital Model for Intensities

Now we carry out a digital analysis of the radiometric temperatures. For a digital analysis the scene is divided into an $N \times N$ matrix of pixels (picture elements), which are each $\ell \times \ell$ squares with $N\ell = L(s)$. The ijth pixel is $P_{i,j} = \{(x, y): (i-1)\ell \leqslant x \leqslant i\ell,\ (j-1)\ell \leqslant y \leqslant j\ell\}$, and the center of $P_{i,j}$ is $\phi_{i,j} = ((i-0.5)\ell,\ (j-0.5)\ell)$. We define an $(N \times N)$-dimensional material matrix (m_{ij}) where $m_{ij} = k$ if material k is dominant in pixel (i, j). For example, $m_{ij} = 1$ if the Lebesgue measure of $\{(u, v) \in P_{ij}: m(u, v) = 1\}$ is greater than $\ell^2/2$. For the scene of Figure 3 with $N = 32$ the material map is given in Figure 5.

The digital analysis is based on an $(M \times M)$-dimensional sensed matrix $S = (s_{ij})$, where $M\ell = L$. The value s_{ij} is the measured radiometric temperature of a pixel centered at $\phi_{ij} + \phi$. The assumed digital model is

$$s_{i,j} = O(\phi_{i,j}) + \varepsilon_{i,j}, \tag{4.4}$$

where $O(\cdot)$ is given by (4.3) and $\varepsilon_{i,j}$ are independent and identically distributed (i.i.d.) measurement errors with the $\mathbb{N}(0, \sigma^2)$ distribution (precision $\tau = \sigma^{-2}$). The parameters of the model are $\beta^T = (\phi^T, \gamma^T)$, where the nuisance parameters are $\gamma^T = (\mu_1, \mu_2, \sigma^2) = (\mu^T, \sigma^2)$. The likelihood function $\ell(\beta|s)$ is

$$\ell(\beta|s) = c\sigma^{-n} \exp\left[-\frac{F(\beta)}{2\sigma^2}\right] \tag{4.5}$$

where $n = M^2$ and

$$F(\beta) = \sum_i \sum_j (s_{i,j} - O(\phi_{i,j}))^2. \tag{4.6}$$

2	2	2	2	2	2	2	2	2	2	2	2	2	2	2	2	2	2	2	2	2	2	2	2	2	2	2	2	2	2	2
2	2	2	2	2	2	2	2	2	2	2	2	2	2	2	2	2	2	2	2	2	2	2	2	2	2	2	2	2	2	2
2	2	2	2	2	2	2	2	2	2	2	2	2	2	2	2	2	2	2	2	2	2	2	2	2	2	2	2	2	2	2
2	2	2	2	2	2	2	2	2	2	2	2	2	2	2	2	2	2	2	2	2	2	2	2	2	2	2	2	2	2	2
2	2	2	2	2	2	2	2	2	2	2	2	2	2	2	2	2	2	2	2	2	2	2	2	2	2	2	2	2	2	2
2	2	2	2	2	2	2	2	2	2	2	2	2	2	2	2	2	2	2	2	2	2	2	2	2	2	2	2	2	2	2
2	2	2	2	2	2	2	2	2	2	2	2	2	2	2	2	2	2	2	2	2	2	2	2	2	2	2	2	2	2	2
2	2	2	2	2	2	2	2	2	2	2	2	2	2	2	2	2	2	2	2	2	2	2	2	2	2	2	2	2	2	2
2	2	2	2	2	2	2	1	1	1	1	1	1	1	1	1	1	1	1	1	1	1	1	2	2	2	2	2	2	2	2
2	2	2	2	2	2	2	1	1	1	1	1	1	1	1	1	1	1	1	1	1	1	1	2	2	2	2	2	2	2	2
2	2	2	2	2	2	2	1	1	1	1	1	1	1	1	1	1	1	1	1	1	1	1	2	2	2	2	2	2	2	2
2	2	2	2	2	2	2	1	1	1	1	1	1	1	1	1	1	1	1	1	1	1	1	2	2	2	2	2	2	2	2
2	2	2	2	2	2	2	1	1	1	1	1	1	1	1	1	1	1	1	1	1	1	1	2	2	2	2	2	2	2	2
2	2	2	2	2	2	2	1	1	1	1	1	1	1	1	1	1	1	1	1	1	1	1	2	2	2	2	2	2	2	2
2	2	2	2	2	2	2	1	1	1	1	1	1	1	1	1	1	1	1	1	1	1	1	2	2	2	2	2	2	2	2
2	2	2	2	2	2	2	1	1	1	1	1	1	1	1	1	1	1	1	1	1	1	1	2	2	2	2	2	2	2	2
2	2	2	2	2	2	2	1	1	1	1	1	1	1	1	1	1	1	1	1	1	1	1	2	2	2	2	2	2	2	2
2	2	2	2	2	2	2	1	1	1	1	1	1	1	1	1	1	1	1	1	1	1	1	2	2	2	2	2	2	2	2
2	2	2	2	2	2	2	1	1	1	1	1	1	1	1	1	1	1	1	1	1	1	1	2	2	2	2	2	2	2	2
2	2	2	2	2	2	2	1	1	1	1	1	1	1	1	1	1	1	1	1	1	1	1	2	2	2	2	2	2	2	2
2	2	2	2	2	2	2	1	1	1	1	1	1	1	1	1	1	1	1	1	1	1	1	2	2	2	2	2	2	2	2
2	2	2	2	2	2	2	1	1	1	1	1	1	1	1	1	1	1	1	1	1	1	1	2	2	2	2	2	2	2	2
2	2	2	2	2	2	2	1	1	1	1	1	1	1	1	1	1	1	1	1	1	1	1	2	2	2	2	2	2	2	2
2	2	2	2	2	2	2	1	1	1	1	1	1	1	1	1	1	1	1	1	1	1	1	2	2	2	2	2	2	2	2
2	2	2	2	2	2	2	2	2	2	2	2	2	2	2	2	2	2	2	2	2	2	2	2	2	2	2	2	2	2	2
2	2	2	2	2	2	2	2	2	2	2	2	2	2	2	2	2	2	2	2	2	2	2	2	2	2	2	2	2	2	2
2	2	2	2	2	2	2	2	2	2	2	2	2	2	2	2	2	2	2	2	2	2	2	2	2	2	2	2	2	2	2
2	2	2	2	2	2	2	2	2	2	2	2	2	2	2	2	2	2	2	2	2	2	2	2	2	2	2	2	2	2	2
2	2	2	2	2	2	2	2	2	2	2	2	2	2	2	2	2	2	2	2	2	2	2	2	2	2	2	2	2	2	2
2	2	2	2	2	2	2	2	2	2	2	2	2	2	2	2	2	2	2	2	2	2	2	2	2	2	2	2	2	2	2
2	2	2	2	2	2	2	2	2	2	2	2	2	2	2	2	2	2	2	2	2	2	2	2	2	2	2	2	2	2	2
2	2	2	2	2	2	2	2	2	2	2	2	2	2	2	2	2	2	2	2	2	2	2	2	2	2	2	2	2	2	2

Figure 5. Digitized material map of Figure 3.

If each pixel is composed of only one material, we show in the Appendix that

$$O(\phi_{i,j}) = \sum_k W_k(i, j)\mu_k, \tag{4.7}$$

where

$$W_k(i, j) = \sum_{(u,v)\in T_k}\sum \iint_{P_{u,v}} W(\phi + \phi_{i,j} - \phi_0)\, d\phi_0 \tag{4.8}$$

$$T_k = \{(u, v) : m_{uv} = k\}.$$

Combining (4.6)–(4.8) yields

$$F(\beta) = \|s - W\mu\|^2,$$

where $s = \text{vec}(S)$ is an n-dimensional vector and $W = W(\phi)$ is an $(n \times 2)$-dimensional matrix whose elements are given by (4.8). With this formulation the registration problem is the linear model

$$s = W\mu + e$$

with the added complication that the design matrix depends on unknown parameters.

4.3 Maximum Likelihood Estimation

Maximization of the likelihood function is similar to the case of the Box and Cox transformation (1964). The conditional MLE of the nuisance parameters (given the registration) can be obtained in closed form, but an iterative procedure is necessary to estimate the parameter of interest. In fact, the conditional MLE given ϕ of the nuisance parameters is

$$\mu_* = (W^T W)^{-1} W^T s$$

$$\sigma_*^2 = \frac{\|s - W\mu_*\|^2}{n} \equiv \frac{\text{SS}(\phi)}{n},$$

where $\text{SS}(\phi) = F(\phi, \gamma_*)$. Since

$$-2 \ln \ell(\phi, \gamma_* | s) = n(\ln \sigma_*^2 + 1), \tag{4.9}$$

the MLE of ϕ_* is any value that minimizes σ_*^2 or $\text{SS}(\phi)$. We will denote this value by ϕ_* even though it may not be unique.

The uniqueness of the MLE depends on a variety of factors, such as the structure of the reference map, the position and the size of the image, the filter characteristics, and the measurement noise. An example, where the MLE is not unique is sketched now. Suppose that the scene is composed of blocks placed in a checkerboard design (e.g., Figure 6), the filter is a Dirac delta function, and there is no measurement noise. For various true locations and sizes of the image, one would be able to identify the position only up to an equivalence class.

A variety of recursive methods could be used to estimate ϕ. The procedure used here was an initial search over a coarse grid and then a fine tuning starting from the best point. For the initial search $\sigma_*^2(\phi)$ was calculated for $\phi = (i\ell k_1, j\ell k_1)$ for $1 \leqslant i, j \leqslant [N/k_1]$, where $[\cdot]$ denotes the greatest integer function. The minimum of these values was used as a starting point for the recursion.

In the recursion a quadratic approximation to the log likelihood

$$-2 \ln \ell(\phi, \gamma_* | s) = c + \alpha^T(\phi - \phi_*) + (\phi - \phi_*)^T Q(\phi - \phi_*) \tag{4.10}$$

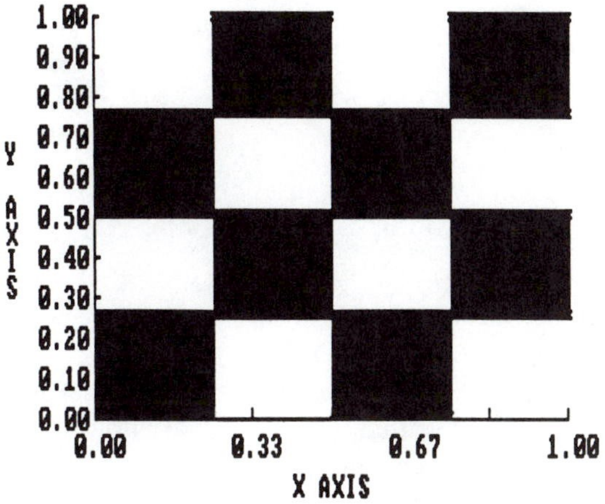

Figure 6. Checkerboard material map.

was assumed, where $\alpha^T = (\alpha_1, \alpha_2)$ and Q is a 2×2 symmetric, positive definite matrix. Completing the square and exponentiating (4.10) we have the familiar Gaussian approximation to the likelihood function (e.g., Box and Jenkins, 1970, Sec. 7.1.7)

$$\ell(\phi, \gamma_* | s) = c_0 \exp\left[-\frac{(\phi - \phi_{**})^T Q(\phi - \phi_{**})}{2} \right]$$

where $\phi_{**} = \phi_* - 0.5Q^{-1}\alpha$.

If ϕ_{i-1} denotes the optimal value after the $(i - 1)$st stage, then the function $-2 \ln \ell(\phi, \gamma_* | s)$ was evaluated from (4.9) at a grid of nine points centered at ϕ_{i-1}. Then (4.10) was used with $\phi_* = \phi_{i-1}$ to obtain least squares parameter estimate (α_i, Q_i), and the new optimum

$$\phi_i = \phi_{i-1} - 0.5Q_i^{-1}\alpha_i \tag{4.11}$$

was calculated. The algorithm terminated when $\|\alpha_i\|^2$ was sufficiently small. The validity of this approach for parameter estimation is demonstrated by simulation in Section 4.5.

Under standard regularity conditions the large-sample distribution of the MLE β_* is Gaussian with mean β and precision matrix I given by the Fisher information matrix

$$I = I(\beta) = E\left(-\frac{\partial^2}{\partial\beta\, \partial\beta^T} \ln \ell(\beta | s) \right) \tag{4.12}$$

(e.g., Rao, 1965, Sec. 6e). Some routine calculations show that in this case

$$I = \begin{bmatrix} \tau K & 0 \\ 0^T & \dfrac{n}{2\tau^2} \end{bmatrix} \tag{4.13}$$

where

$$K = (K_{i,j}) \tag{4.14}$$

and the (2 × 2)-dimensional matrices $K_{i,j} = V_i^T V_j$ for $1 \leqslant i,\ j \leqslant 2$ with $V_1 = \partial(W\mu)/\partial\phi$ and $V_2 = W$. In (4.12), τ rather than σ^2 is used as the fifth component of β. It follows from (4.13) that (ϕ_*, μ_*) is asymptotically independent of τ_*, but ϕ_* and μ_* are dependent, in general.

A formal proof of the asymptotic normality would require additional assumptions about the filter, the number and the size of the pixels, and the composition of the material map. No proof is given, but the simulation results of Section 4.5 indicate that the approximation is good in one special case.

From (4.13) and (4.14) we have the approximation

$$\phi_* \sim \mathbb{N}(\phi, (\tau K_{11.2})^{-1}), \tag{4.15}$$

where

$$K_{11.2} = V_1^T(I - V_2(V_2^T V_2)^{-1} V_2^T)V_1. \tag{4.16}$$

Equation (4.15) is used as (4.1) with

$$\Gamma = (\tau K_{11.2})^{-1}|_{\beta = \beta_*}.$$

4.4 Comparison to Previous Work

Now we assume that the translation is an integer multiple of the pixel length. That is,

$$\phi = (r\ell, s\ell) \tag{4.17}$$

where r and s are the integers such that $0 \leqslant r, s \leqslant N - M$. Thus there are $(N - M + 1)^2$ possible values for ϕ. If the true values of ϕ are uniformly distributed over the square, the (quantization) root mean square error caused by assumption (4.17) is $6^{-0.5}\ell$. The iterative approach of Section 4.3 eliminates this term. In the example of Section 4.5, assumption (4.17) was satisfied, but the performance of the algorithm does not rest on this assumption.

The expressions of Section 4.3 can be simplified considerably when the filter is the Dirac delta. This could be obtained by prefiltering the image, but this induces correlation in the noise (e.g., Hunt, 1983). The previously published algorithms are based on this assumption and (4.17). By (4.3)

$$O(\phi_{ij}) = f(\phi_{ij} + \phi) \tag{4.18}$$
$$W_k(i, j) = I(m_{i+r,j+s} = k).$$

It follows that $SS(\phi)$ can be expressed

$$SS(\phi) = \sum_{k=1}^{2} \sum_{p=1}^{n_k} (s_{kp} - \bar{s}_k)^2, \tag{4.19}$$

where

$$\{s_{kp}\} = \{s_{ij} : m_{i+r,j+s} = k\} \tag{4.20}$$

and $\bar{s}_k$ is the average of $\{s_{kp}\}$.

The translation (4.18) and the material matrix M partition the sensed array indices into two groups [defined by (4.20)], which correspond to the two materials. The sum of squares can be interpreted as the within-groups sum of squares for a one-way analysis of variance (ANOVA). The "minimum variance" algorithm (e.g., Kuglin, 1981) calculates $SS(\phi)$ over the grid of $(N - M + 1)^2$, and the minimum value determines the estimate of the position. Thus this technique finds the translation with the smallest within-materials variance. Equation (4.9) shows that the "minimum variance" algorithm is the MLE for the model (4.4).

The same technique using the L_1 rather than the L_2 norm in (28) was proposed by Barnea and Silverman (1972). This technique should be nearly optimal if the measurement noise in (4.4) has the double exponential rather than the Gaussian distribution.

Pratt (1974) proposed a similar algorithm using maximum correlation rather than minimum variance. In this technique the maximum value of the Pearson correlation coefficient based on $\{s_{kp}\}$ and $\{k\}$ is used to determine the optimal registration.

Now we illustrate these algorithms in a simple case with $N = 3$ and $M = 2$. The material matrix is given in Diagram 1 while the sensed matrix is given in Diagram 2. Diagram 3 gives the values of $SS(\phi)$ while, Diagram 4 gives the values of the Pearson correlation coefficient for the four possible

Diagram 1: Material Matrix M

$$M = \begin{bmatrix} 1 & 2 & 2 \\ 2 & 1 & 1 \\ 1 & 2 & 1 \end{bmatrix}$$

Diagram 2: Sensed Matrix S

$$S = \begin{bmatrix} 4.0 & 4.1 \\ 6.2 & 6.0 \end{bmatrix}$$

values of ϕ of (4.17). Both of these techniques produce the same estimate (0, 1) in this case. The algorithm defined by (4.11) could be used to obtain a more precise estimate, but the large-sample approximation would not be valid in this case.

Diagram 3: Minimum Variance Matrix

$$\begin{array}{cc} & \quad s \\ & \begin{array}{cc} 0 & 1 \end{array} \\ r: \begin{array}{c} 0 \\ 1 \end{array} & \begin{bmatrix} 4.0 & 0.04 \\ 4.0 & 2.55 \end{bmatrix} \end{array}$$

Diagram 4: Correlation Matrix

$$\begin{array}{cc} & \quad s \\ & \begin{array}{cc} 0 & 1 \end{array} \\ r: \begin{array}{c} 0 \\ 1 \end{array} & \begin{bmatrix} 0.10 & 0.995 \\ 0.0 & 0.55 \end{bmatrix} \end{array}$$

4.5 Simulation Results

In this section we study the validity of the approximation for the scene of Figure 3. Several alternatives are available to the variance estimate obtained in Section 4.3. One alternative is simulation, but this approach is not well suited to real-time control problem applications. Another method, which is feasible for simple maps, at least, is to calculate the theoretical asymptotic variance. The asymptotic variance depends on β, but in some problems the variation may be negligible over the range of likely values so that an approximate value can be given in advance. In this section a theoretical asymptotic variance matrix is calculated using the true value of β as known. To make this technique functional, the robustness of this value with respect to parameter variation would have to be demonstrated.

In the simulation the filter function was assumed to be

$$W(u, v) = \frac{I(|u| \leqslant \ell_f) I(|v| \leqslant \ell_f)}{(4\ell_f^2)}.$$

In this case the filter averages over a square with sides of length $2\ell_f$ and (4.8) is given by

$$W_k(i, j) = \sum_{(u,v)\in T_k}\sum g(x, i-u)g(y, j-v), \tag{4.21}$$

where

$$g(x, i) = \frac{0.5\max(0, [\min\{x+(i+0.5)\ell, \ell_f\} - \max\{x+(i-0.5)\ell, -\ell_f\}])}{\ell_f}. \tag{4.22}$$

We assume that $0 \leqslant 0.5\ell \leqslant \ell_f \leqslant \ell$, which implies that the filter averages over an area larger than one pixel but smaller than four. Under these assumptions (4.22) simplifies to $g(i_1\ell, i) = g_0(|i_1 + i|)$, where $g_0(0) = \rho$, $g_0(1) = (1-\rho)/2$, $g_0(\text{i}) = 0$ for $i \geqslant 2$, and $\rho = 1/2\ell_f$. In the special case $\ell = 2\ell_f$ the filter averages over pixels, $\rho = 1$, and the only contribution to the sum in (4.21) is for $(u, v) = (i_1 + 1, j_1 + j)$, and

$$W_k(i, j) = I(m(i + i_1, j + j_1) = k). \tag{4.23}$$

Equation (4.23) gives $W_1(i, j) = I(i \geqslant 5, j \geqslant 5)$ for the map of Figure 3 under the assumption that $M = 16$ and $(i_1, j_1) = (4, 4)$. It is clear from (4.23) that $W_2(i, j) = 1 - W_1(i, j)$.

The derivatives of $W\mu$ are easily obtained from its 16×16 matrix form. The derivatives were approximated using divided differences; for example,

$$\frac{\partial}{\partial x} m(i, j) = \frac{m(i, j) - m(i-1, j)}{\ell}. \tag{4.24}$$

The matrix V_1 is derived as follows. The elements of the matrix of the partials with respect to x of $W\mu$ are

$$\left[\frac{\partial}{\partial x}(W\mu)_{i,j}\right] = qI(i \geqslant 5, j = 5) \tag{4.25}$$

with $q = (\mu_1 - \mu_2)\ell^{-1}$, and the matrix of partials of $W\mu$ with respect to y is the transpose of (4.25). From (4.14) and (4.24) we obtain

$$K_{1,1} = q^2\begin{bmatrix} 12 & 1 \\ 1 & 12 \end{bmatrix}$$

$$K_{2,1} = 12q\begin{bmatrix} 1 & 1 \\ 0 & 0 \end{bmatrix}$$

$$K_{2,2} = \operatorname{diag}(144, 112).$$

The adequacy of the approximation was tested with a simulation with

$n_r = 100$ replications. For each replication the parameters were $N = 32$, $\ell = \frac{1}{32}$, $\ell_f = \frac{1}{16}$, $i_1 = 4$, $j_1 = 4$, $\mu_1 = 6$, $\mu_2 = 3$, and $\tau = 100$ and "grid search" parameter $k_1 = 3$. For each replication random numbers were generated for the errors of (4.4) using the IMSL subroutine GGNPM.

With the parameter values given above,

$$\sigma^2 K_{11.2}^{-1} = (9.86 \times 10^{-8})I. \tag{4.26}$$

For each replication (4.11) was used to estimate the parameters; the estimate of position of the ith replication is labeled ϕ_i and the mean is labeled $\bar{\phi}$. The two components of the bias vector $\bar{b} = \bar{\phi} - \phi$ are given in Table 3. From (4.26) the theoretical standard error of each component of the bias is $(9.86 \times 10^{-8})^{1/2}$. We normalize the biases of Table 3 by dividing by the standard error. Theoretically, these normalized deviates should have an approximate $\mathbb{N}(0, 1)$ distribution. The values given in Table 3 provide no cause to doubt the adequacy of the simulation or the recursive algorithm.

Two estimates of the covariance matrix of the position error are

$$S^* = \frac{\Sigma(\phi_i - \bar{\phi})(\phi_i - \bar{\phi})^T}{n_r} \tag{4.27}$$

$$\bar{S} = \frac{\Sigma Q_i^{-1}}{n_r}, \tag{4.28}$$

where Q_i is the final value in equation (4.11). Since $\bar{S}$ could be used with $n_r = 1$, it does not require a simulation. The values of (4.26)–(4.28) are given in Table 4. Since $\bar{S}$ is the average of independent estimates, the estimated standard deviation of each component can be calculated and are also given in Table 4.

A further test of the simulation can be made by assuming that $\{\phi_i\}$ is a random sample from a $\mathbb{N}(\phi, \Sigma)$ distribution and testing the null hypothesis $\Sigma = \sigma^2 K_{11.2}^{-1}$ [see equation (4.15)]. The likelihood ratio test (e.g., Anderson, 1957, Sec. 10.8) is based on

$$-2 \ln \lambda^* = n_r(\operatorname{tr}(\psi^*) - 2 - \ln |\psi^*|), \tag{4.29}$$

Table 3 Estimation Bias in the Simulation

Parameter	Bias ($\bar{b}$)	Normalized bias [$\bar{b}/\sigma(\bar{b})$]
x	6.8×10^{-6}	0.002
y	-4.4×10^{-5}	-0.15

Table 4. Estimation of the Covariance Matrix ($\times 10^{-8}$)

Parameter	Theoretical value from (4.26)	Estimate from (4.27)	Estimate from (4.28)
$\mathrm{var}(x)$	9.86	9.14	10.70 ± 0.05
$\mathrm{var}(y)$	9.86	10.57	10.11 ± 0.05
$\mathrm{cov}(x, y)$	0.00	-0.15	0.032 ± 0.003

where

$$\psi^* = \tau K_{11.2}(S^*)^{-1}. \tag{4.30}$$

The observed significance level is obtained by comparison of (4.30) with the chi-square distribution with 3 degrees of freedom. Using the values in Table 4, we find $-2 \ln \lambda^* = 0.38$, which is an extremely good fit. Although the validity of the test is less clear, the same analysis can be repeated with S^* replaced by $\bar{S}$; the result is nearly identical in this case.

In summary, for this simple case the asymptotic approximations seem quite adequate. Also, the study indicates that a reasonable variance estimate can be obtained from the recursion based on (4.11). This is useful for real-time control applications since it eliminates the need for a simulation.

APPENDIX

Since each pixel is composed of a single material, we have

$$f(\phi) = \sum_u \sum_v \mu_{m_{uv}} I(\phi \in P_{uv}) \tag{A.1}$$

$$\mu_{m_{uv}} = \sum_k \mu_k I\{(u, v) \in T_k\}. \tag{A.2}$$

To derive (4.7) we use (4.3) with $\phi_* = \phi_{ij}$ and change variables to obtain

$$O(\phi_{ij}) = \iint W(\phi + \phi_{ij} - \phi_0) f(\phi_0)\, d\phi_0. \tag{A.3}$$

Applying (A.1) to (A.3) and then interchanging integration and summation, we obtain

$$O(\phi_{ij}) = \iint W(\phi + \phi_{ij} - \phi_0) \sum_u \sum_v \mu_{m_{uv}} I(\phi_0 \in P_{uv})\, d\phi_0$$

$$= \sum_{u}\sum_{v} \mu_{m_{uv}} \iint_{P_{uv}} W(\phi + \phi_{ij} - \phi_0)\, d\phi_0. \tag{A.4}$$

Applying (A.2) to (A.4), we obtain

$$O(\phi_{ij}) = \sum_{u}\sum_{v}\sum_{k} \mu_k I\{(u, v) \in T_k\} \iint_{P_{uv}} W(\phi + \phi_{ij} - \phi_0)\, d\phi_0$$

$$= \sum_{k} \mu_k \sum\sum_{(u,v)\in T_k} \iint_{P_{uv}} W(\phi + \phi_{ij} - \phi_0)\, d\phi_0,$$

which yields (4.7).

REFERENCES

Anderson, T. W. (1958). *An Introduction to Multivariate Statistical Analysis*, Wiley, New York.

Baird, H. S. (1984). *Model-Based Image Matching Using Location*, MIT Press, Cambridge, Mass.

Barnea, D. T., and H. F. Silverman (1972). A class of algorithms for fast image registration, *IEEE Trans. Comput.* C-21(2), 179–186.

Bertsekas, D. P. (1976). *Dynamic Programming and Stochastic Control*, Academic Press, New York.

Box, G. E. P., and D. R. Cox (1964). An analysis of transformations, *J. R. Stat. Soc. Ser. B* 26, 211–243.

Box, G. E. P., and G. M. Jenkins (1970). *Time Series Analysis: Forecasting and Control*, Holden-Day, San Francisco.

Chen, H. F. (1985). *Recursive Estimation and Control for Stochastic Systems*, Wiley, New York.

Duncan, D. B., and S. D. Horn (1972). Linear dynamic recursive estimation from the viewpoint of regression analysis, *J. Am. Stat. Assoc.* 67, 815–821.

Hall, E. L. (1979). *Computer Image Processing and Recognition*, Academic Press, New York.

Hunt, B. R. (1983). Image Restoration, in *Digital Image Processing Techniques* (M. P. Ekstrom, ed.), Academic Press, Orlando, Fla., pp. 52–76.

Kalman, R. E. (1960). A new approach to linear filtering and prediction problems, *Trans. ASME J. Basic Eng.* 82D, 34–45.

Kuglin, C. D. (1981). Histogram-Based Algorithms for Scene Matching, in *Proceedings of the Photo Optical Institute of Engineers Symposium on Infrared Technology for Target Detection and Classification* (P. Narendra, ed.), pp. 99–107.

Lindley, D. V. (1968). The choice of variables in multiple regression, *J. R. Stat. Soc. Ser. B* 30, 31–66.

Meinhold, R. J., and Singpurwalla, N. D. (1983). Understanding the Kalman filter, *Am. Stat.* 37, 123–127.

Pratt, W. K. (1974). Correlation techniques for image registration, *IEEE Trans. Aerosp. Electron. Syst.* AES-10(3), 353–358.

Pratt, W. K. (1978). *Digital Image Processing*, Wiley, New York.

Raiffa, H., and R. Schlaifer (1961). *Applied Statistical Decision Theory*, Harvard Business School, Boston.

Rao, C. R. (1965). *Linear Statistical Inference and Its Application*, Wiley, New York.

Spall, J. C. (1985). Effects of Uncertain Ancillary Parameters on Maximum likelihood Estimates in Dynamic Models, *Proceedings of the 24th IEEE Conference on Decision and Control*, Ft. Lauderdale, Fla., pp. 1920–1925.

Whittle, P. (1983). *Prediction and Regulation*, 2nd ed., University of Minnesota Press, Minneapolis, Minn.

12

Structure Determination of Regression-Type Models for Adaptive Prediction and Control

MIROSLAV KÁRNÝ and RUDOLF KULHAVÝ Institute of Information Theory and Automation, Czechoslovak Academy of Sciences, Prague, Czechoslovakia

1. INTRODUCTION

The chapter deals with the problem of choosing the best structure of the linear (in parameters) regression-type model of a predicted and/or controlled process from a set of competitive structures. The model parameters may be time varying, but no model of parameter variations is assumed available. The optimal structure is assumed time invariant.

The problem formulation above naturally arises, for instance, when the behavior of a technological process is to be predicted and/or controlled. Here a fixed and relatively simple structure of the model is commonly required in order that the resulting prediction and/or control may be as easy as possible.

The structure needed can be simplified substantially if we admit time variations of model parameters. Recursive estimation of time-varying parameters enables us to describe just the current process behavior. However, if the parameters vary relatively slowly, the quality of prediction and/or control is comparable with the quality achieved in the "full-structure" case. This is the reason why adaptive prediction and/or adaptive control have become so popular.

Most adaptive controllers are based on a version of the multivariate ARX (autoregressive with exogenous variables) model

$$y(t) = \sum_{i=1}^{\ell_A} A_i y(t-i) + \sum_{i=0}^{\ell_B} B_i u(t-i-d_u) + \sum_{i=1}^{\ell_D} D_i v(t-i-d_v) + K + e(t) \qquad (1.1)$$

relating the system output y at time t to its weighted (by A's) lagged values, m_u-dimensional weighted (by B's) lagged system inputs, m_v-dimensional weighted (by D's) lagged exogenous variables v, and white-noise term e (for details, see Kárný et al., 1985). Parameters of such a model are estimated in real time, and the parameter estimates are used for the choice of controller actions.

Complete structure determination in the example above represents the choice of the "orders" ℓ_A, ℓ_B, ℓ_C, common transport delays d_u, d_v, the absolute term K and the selection of proper inputs and exogenous variables. Hence, even for medium-sized control problems, the number of competitive models is excessive.

Obviously, an appropriate choice of the structure of the regression-type model with time-varying, recursively updated parameters substantially influences the resulting control efficiency. For this reason we have attempted to develop a feasible (from an algorithmic viewpoint) procedure to solve it.

A number of results have been published which deal with different aspects of the problem. Kashyap and Rao (1976) review basic approaches to structure determination (for more recent results see, e.g., Akaike, 1979; Neftci, 1982; Neill and Johnson, 1984; Ronchetti, 1985; Poskitt, 1986). The problem of estimating time-varying quantities in the absence of a model of their variations is dealt with, for example, by Jazwinski (1970), Peterka (1981), and Ljung and Söderström (1983).

In comparison with the references cited, the procedure presented below tries to solve the problem in a more general framework. We have attempted to be maximally consistent with the Bayesian solution offered by Peterka (1981). Nevertheless, we have not avoided several concessions in order to handle excessive complexity of the task.

Thus we have been forced to:

1. Assume normality of the noise term e (consequently, robustness with respect to outliers is not attacked).
2. Split the whole task into two related subtasks:
   a. The choice of the best structure under the assumption of time-invariant parameters
   b. The choice of the optimal (regarding parameter variations) structure nested in the structure selected in the first step

   (We make use of the version of the parsimony principle which states that the "best" model structure obtained assuming time-invariant parameters should include the "best" structure of time-varying model.)
3. Use in the step 2a only a suboptimal way of searching for the structure maximizing the posterior probability on the set of all competitive structures (the procedure has been proposed by Kárný, 1983)

4. Cope with the absence of a model of parameter variations by means of a heuristic but well-founded method of forgetting obsolete information (designed by Kulhavý, 1985).

The resulting algorithm exhibits the following distinguished features.

1. It enables effective solution of the cases with about 100 possible entries of regressor.
2. It takes into account possible, incompletely described time variations of model parameters.
3. As a result of rather weak assumptions concerning the regressor, it covers a variety of subcases, such as
   a. Determination of the AR model order
   b. Determination of the number of terms needed in the ARX model
   c. Proper choice of exogenous variables
   d. Proper choice of manipulated variables (inputs)

In spite of the above-mentioned control engineering motivation, we believe that the designed procedure can serve in a diversity of other areas, including economics, medicine, biology, and experimental physics.

The chapter is organized as follows. After presenting the necessary theoretical background in Section 2, the theory is specialized to linear normal regression-type models in Section 3. The main tools for effective numerical implementation of results are treated in Section 4. Section 5 contains an illustrative example. Section 6 completes the chapter with the discussion of open problems and of hitherto obtained practical experience with real-life data analysis (e.g., when modeling a cold-rolling mill or a cement kiln).

Note that readers interested mainly in the final algorithm can look immediately at the results of Section 3 and the example in Section 5 (the notation used is introduced in Section 2.1).

2. UNDERLYING THEORY

The theoretical background of structure determination and of related tasks are sketched in this section. The explanation starts with output prediction (needed for adaptive control, which is of our primary interest). A necessary prerequisite—estimation of time-varying parameters—is discussed in detail. A special formulation of structure determination (relevant to our final goal) and its conceptual solution concludes the section.

2.1 Basic Notions and Notations

Let us assume that we are able to measure a finite collection of data $d(t)$ at discrete-time instants labeled $t = 1, 2, 3, \ldots$ on a system (a part of the real world). The subset of $d(t)$ which can be directly manipulated (possibly empty) will be called the input $u(t)$ of the system. The rest of $d(t)$ which can be manipulated at most indirectly will be called the (extended) output $y(t)$ of the system. Formally, we will arrange data items and their parts into column vectors with entries $d_i(t)$, $i = 1, 2, \ldots, m_d$, $y_i(t)$, $i = 1, 2, \ldots, m_y$, and so on.

The task of control arises when we want to influence, in some definite way, future outputs by choosing an input. Clearly, control (like any other task of rational decision making) requires good output prediction, that is, the ability to forecast future values of the output

$$y(t\,..\,N) = (y(t),\ y(t+1), \ldots, y(N)), \qquad 1 \leqslant t \leqslant N,$$

up to some horizon N. Prediction can be based on the current input $u(t)$, the past data $d(1\,..\,t-1) = (d(1), d(2), \ldots, d(t-1))$ (no data available for $t = 1$), and any other information at our disposal.

Two facts complicate output prediction. First, the system usually exhibits random behavior. Thus the observed past does not determine the observed future in a unique way, because not all influencing "forces" are measured at least. Second, we are, as a rule, uncertain as to which sequence of models of the set of candidates describes best the relations among data items.

The subjective Bayesian approach we apply here deals with uncertainty and randomness in the same way. Both these aspects are described by probabilistic notions in a unified manner, namely by the joint probability of entities under consideration. We restrict ourselves to cases when the Radon-Nikodym derivatives of a probability distribution with respect to the Lebesgue measure (i.e., the probability density function, p.d.f.) or with respect to the counting measure (i.e., the probability function, p.f.) exist. The conditional version is also assumed to exist and will be denoted in a usual, but rather simplified way: $p(A|B)$ stands for the probability (density) function of a random variable A (at the point A) conditioned on a random variable B (having the value B). The above-used abbreviations p.d.f. (p.f.) will be extended to c.p.d.f. (c.p.f.), respectively, for their conditional versions. Moreover, we abbreviate

$$p(A|d(1\,..\,t-1),\ B) = p(A|t-1;\ B) \qquad \text{for } t = 2, 3, \ldots. \tag{2.1}$$

The abbreviation (2.1) will be extended to $t = 1$ with the meaning

$$p(A\,|\,d(1\,..\,0),\ B) = p(A\,|\,B) = p(A\,|\,0;\ B). \tag{2.2}$$

A similar notation will be used for the related expectation

$$\begin{aligned} \mathscr{E}(A\,|\,d(1\,..\,t-1), B) &= \int Ap(A\,|\,d(1\,..\,t-1), B)\, dA \\ &= \mathscr{E}(A\,|\,t-1;\ B) \quad \text{for } t = 1, 2, 3, \ldots, \end{aligned} \tag{2.3}$$

where integration performed over the entire range of A reduces to summation when dA is not the Lebesgue, but the counting measure.

For our purposes it is sufficient to assume that there is a one-to-one correspondence between the set of the considered models and some time-invariant finite-dimensional parametric space $\mathscr{X}$ in the sense that any model at time t is uniquely determined by $X(t) \in \mathscr{X}$.

With the notations above, the complete description of uncertain (unknown parameter) and random (data) entities within the assumed horizon is given by the joint p.d.f.

$$p(d(1\,..\,N), X(1\,..\,N)). \tag{2.4}$$

2.2 Adaptive Prediction and Identification

Adaptive prediction builds up, or at least approximates, the data-related predictive c.p.d.f. $p(d(t\,..\,N)|t-1)$, which can be expressed as

$$p(d(t\,..\,N)|t-1) = \prod_{\tau=t}^{N} p(d(\tau)|\tau-1). \tag{2.5}$$

The evaluation of $p(d(t\,..\,N)|t-1)$ is straightforward after resolving one-step-ahead prediction, that is, after determining the so-called predictive c.p.d.f.'s

$$p(d(t)|t-1) \qquad \text{for } t = 1, 2, \ldots, N. \tag{2.6}$$

Why the adjective "adaptive" has been used will be apparent later.

Elementary rules for c.p.d.f.'s and the definition of $d(t)$ as a pair $(y(t), u(t))$ imply that

$$p(d(t)|t-1) = \int p(d(t)|t-1; X(t))p(X(t)|t-1)\,dX(t) \tag{2.7}$$

$$p(d(t)|t-1; X(t)) = p(y(t)|t-1; u(t), X(t))p(u(t)|t-1; X(t)). \tag{2.8}$$

The first factor in (2.8) describes how the output $y(t)$ responds to the past history $d(1\,..\,t-1)$ and the current input $u(t)$ for given parameters $X(t)$. It represents the (input-output probabilistic) system model. Clearly, in order to solve the prediction, and consequently, a control problem, the parameters $X(t)$ have to determine the system model in the above sense completely.

Similarly, the second factor in (2.8) describes the strategy of the input generator for known parameters $X(t)$. In the rest of the chapter we restrict ourselves to a class of input generators fulfilling natural conditions of control according to Peterka (1981),

$$p(u(t)|t-1; X(t)) = p(u(t)|t-1). \tag{2.9}$$

Input generators in this class can use no additional knowledge about the system model other than that contained in the past data. Validity of (2.9) for adaptive controllers is clear; standard feedback, open-loop input generators as well as person-machine interactions usually meet (2.9).

The left-hand side of (2.7) can be factorized to the system and input generator describing parts as in (2.8). The latter is canceled under (2.9), so that the output prediction reads

$$p(y(t)|t-1; u(t)) = \int p(y(t)|t-1; u(t), X(t))p(X(t)|t-1))\, dX(t), \tag{2.10}$$

where the integral sign denotes integration over the total parameter space $\mathscr{X}$ (the same simplification is used throughout the chapter).

The second factor in (2.7) and (2.10) describes uncertainty about the parameters $X(t)$ that remains after observing data $d(1\,..\,t-1)$. The evaluation of this c.p.d.f. and its characteristics is called either filtering or (in a different context) identification. We will use the latter term, which interprets the unknown entity $X(t)$ as the parameters of the system model.

Data updating

The c.p.d.f. $p(X(t)|t-1)$ can be updated by measured data $d(t)$ according to the Bayes rule

$$p(X(t)|t) = \frac{p(d(t)|t-1;\ X(t))p(X(t)|t-1)}{\int p(d(t)|t-1;\ X(t))p(X(t)|t-1)\, dX(t)} \tag{2.11a}$$

$$= \frac{p(y(t)|t-1;\ u(t),\ X(t))p(X(t)|t-1)}{\int p(y(t)|t-1;\ u(t),\ X(t))p(X(t)|t-1)\, dX(t)} \tag{2.11b}$$

$$\propto p(y(t)|t-1;\ u(t),\ X(t))p(X(t)|t-1). \tag{2.11c}$$

The expression in (2.11b), which needs the value of the input $u(t)$ but no knowledge of the input generator strategy $p(u(t)|t-1)$, is the direct consequence of the natural conditions of control: $p(u(t)|t-1)$ cancels after applying the chain rule (2.8). The expression (2.11c) with the sign "$\propto$" instead of "$=$" stresses the proportionality up to the $X(t)$-independent factor. This notation will be used repeatedly.

Time updating

To complete the recursion

$$p(X(t)|t-1) \xrightarrow[\text{updating}]{\text{data}} p(X(t)|t) \xrightarrow[\text{updating}]{\text{time}} p(X(t+1)|t), \tag{2.12}$$

time updating $p(X(t)|t) \to p(X(t+1)|t)$, which respects time variations of the parameters X, is needed.

To simplify the exposition, we will assume that the conditional distribution of $X(t+1)$ for the given past is fully specified by the preceding value $X(t)$, in other words, $X(1..t-1)$ brings no additional information. Under this condition, we have, when applying elementary calculus,

$$p(X(t+1)|t) = \int p(X(t+1)|t; X(t))p(X(t)|t)\,dX(t). \tag{2.13}$$

The relation (2.13) explicates the intuitively clear need for a model $p(X(t+1)|t; X(t))$ of the evolution of time-varying parameters $X(\cdot)$.

The computational complexity of the full recursion involving the time-update part is usually prohibitive and/or the construction of a proper model $p(X(t+1)|t; X(t))$ difficult. In such cases the time updating is often performed by some heuristic technique. The common root of a great deal of such solutions consists in suppressing information from old data items, information which is believed to be obsolete. Exponential forgetting (Peterka, 1981) serves as a typical example.

Kulhavý (1985) proposed an advanced technique that forms the cornerstone of our treatment of parameter variations in the structure determination problem.

2.3 Estimation of Time-Varying Parameters Using Restricted Exponential Forgetting

Standard techniques of estimating (slowly) time-varying parameters are based on weighting observed data by "windows" of a suitable shape (e.g., exponential or rectangular). However, this simple way of treating parameter changes has its drawbacks. First, old information does not automatically mean obsolete (no longer valid) information. Hence it can happen that a piece of useful information is forgotten, although no new information is gained. This fact is dangerous when data do not carry sufficient information for a long time as the system can become nonidentifiable (Zellner, 1971). Second, a mere data weighting does not fully utilize available prior information about parameter variations, which can (sometimes considerably) improve the quality of estimation.

Requirements on time updating

It is common experience that gathering information is an expensive task. Time updating of the posterior c.p.d.f. $p(X(t)|t)$ which respects this experience should fulfill two (at this point vaguely formulated) requirements:

1. Any available information about parameter changes should be utilized.
2. Only obsolete information should be modified.

To avoid constructing an explicit model of the parameter evolution $X(t) \to X(t+1)$, we will try to build up directly a sufficiently general model of the time updating $p(X(t)|t) \to p(X(t+1)|t)$.

Consistent with the Bayesian standpoint, alternatives for deciding about the optimal $p(X(t+1)|t)$ will be given by the pair of alternative c.p.d.f.'s

$$\{p_i(X(t+1)|t),\ i = 0, 1\}, \tag{2.14}$$

where the "zero" *basic alternative* will correspond to the case of no parameter changes [$p(X(t+1)|t; X(t))$ is the Dirac function $\delta(\cdot)$ model implied in (2.13)]

$$p_0(X(t+1)|t) = \int \delta(X(t+1) - X(t))p(X(t)|t)\, dX(t), \tag{2.15}$$

while the "unit" *reference alternative* should cover possible parameter changes.

The construction of time updating rests heavily on the operational definition of obsolete information. Keeping in mind the price of information, we use a rather conservative policy; we take as obsolete only the piece of information updated by the latest data. The idea is elaborated in detail below.

To fulfill requirements 1 and 2, we perform time updating conceptually in three related steps.

1. *Restriction.* The obsolete piece of basic information and the corresponding piece of reference information are extracted [initial information is described by the basic and reference c.p.d.f.'s (2.14), respectively].
2. *Generalized exponential forgetting.* The obsolete piece of basic information is modified taking into account the extracted piece of reference information.
3. *Extension.* The updated piece of information is complemented by the remaining basic information.

REMARK 1 The reference alternative $p_1(X(t+1)|t)$ is a flexible tool for expressing incomplete knowledge about time variations of parameters. It can be chosen, for example,

a. A fixed p.d.f. assigning to different possible values of $X(t+1)$ the expected degree of occurrence
b. A recursively updated c.p.d.f. describing a "pessimistic" evolution of $X(\cdot)$ (e.g., random walk changes) as a counterpart of the "optimistic" hypothesis $p(X(t+1)|t) = p_0(X(t+1)|t)$ specified by (2.15)
c. A posterior-fitted c.p.d.f. coinciding with the basic c.p.d.f. (2.15) in the conditional expectation

A thorough discussion and examples of all the cases above can be found in Kulhavý (1986).

REMARK 2 The whole explanation can be generalized in a straightforward way to include the case of an arbitrary, but finite number of reference alternatives.

Particular steps of the construction of the time-update model sketched above will now be presented in detail.

Restriction

In the first step of our procedure we are to specify and extract the obsolete piece of basic and the corresponding piece of reference information. We take such a measurable mapping $X^*(t) = T(X(t)|t)$ of unknown parameters which enables us to factorize the prior and posterior c.p.d.f.'s in an analogous way:

$$p(X(t)|t-1) = p(X^*(t)|t-1)q(X(t))$$
$$p(X(t)|t) = p(X^*(t)|t)q(X(t)) \tag{2.16}$$

using a nonnegative measurable function $q(\cdot)$. It is obvious that the mapping $T(\cdot|t)$ with this property distinguishes the prior and posterior c.p.d.f.'s as exactly as the original parameters do it. It is the reason why every mapping fulfilling the equations (2.16) is called sufficient with respect to the pair of the prior and posterior c.p.d.f.'s (Kullback and Leibler, 1951).

Note that to remove all details that do not help distinguishing the prior and posterior c.p.d.f.'s, we should choose the minimal sufficient mapping, that is, such a sufficient mapping T_0 which can be produced by a composition $T_0 = F(T)$ with any other sufficient mapping T. Taking into account the form of Bayes' rule (2.11), it is possible to prove (Csiszár, 1967) that the minimal sufficient mapping can be specified by means of the c.p.d.f. $p(y(t)|t-1; u(t), X(t))$ taken as a function of the unknown parameters $X(t)$:

$$T_0(X(t)|t) = \begin{cases} p(y(t)|t-1; u(t), X(t)) & \text{if } p(X(t)|t-1) > 0 \\ -1 & \text{if } p(X(t)|t-1) = 0. \end{cases} \tag{2.17}$$

After choosing a suitable sufficient mapping $T(\cdot|t)$, we are able to extract the appropriate pieces of basic and reference information simply by evaluating the marginal c.p.d.f.'s of the parameters $X^*(t+1) = T(X(t+1)|t)$:

$$p_0(X(t+1)|t) \to p_0(X^*(t+1)|t)$$
$$p_1(X(t+1)|t) \to p_1(X^*(t+1)|t).$$

Generalized exponential forgetting

In the second step of our procedure we are to choose a c.p.d.f. $p(X^*(t+1)|t)$ using the alternative pieces of knowledge described by the c.p.d.f.'s $p_i(X^*(t+1)|t)$, $i = 0, 1$. This choice under uncertainty will be formalized as a statistical decision problem.

For this reason, let us assume that the probability assigned to the basic alternative (the hypothesis of constant parameters) is $\phi(t+1|t)$. Clearly, the probability of the reference alternative is then $1-\phi(t+1|t)$.

The c.p.d.f. $p(X^*(t+1)|t)$ is searched to minimize the expected loss

$$\phi(t+1|t)L(p(X^*(t+1)|t),\ p_0(X^*(t+1)|t)) \\ +(1-\phi(t+1|t))L(p(X^*(t+1)|t),\ p_1(X^*(t+1)|t)), \qquad (2.18)$$

where the Kullback-Leibler distance (sometimes called the Shannon relative entropy) is selected as a loss functional on the space of all possible c.p.d.f.'s:

$$L(p(X^*(t+1)|t), p_i(X^*(t+1)|t)) = \int p(X^*(t+1)|t) \ln\left(\frac{p(X^*(t+1)|t)}{p_i(X^*(t+1)|t)}\right) dX^*(t+1). \qquad (2.19)$$

It is a relatively easy exercise of variational calculus to find the optimal [unique under rather weak conditions (Kulhavý, 1985)] form of $p(X^*(t+1)|t)$:

$$p(X^*(t+1)|t) \propto [p_0(X^*(t+1)|t)]^{\phi(t+1|t)}[p_1(X^*(t+1)|t)]^{1-\phi(t+1|t)}. \qquad (2.20)$$

The composition rule (2.20), replacing the time-update step (2.13), has been called a generalized exponential forgetting because of its close connection to the standard exponential forgetting (see Remark 3 below).

REMARK 1 The probability $\phi(t+1|t)$ can be chosen

a. A priori (fixed) using its subjective probability interpretation. Note that the selected probability value must necessarily be a compromise suitable for all possible restrictions; however, the algorithm can be further tailored by a careful specification of the reference alternative (see Kulhavý, 1986).
b. By solving a properly stated max-min problem (Kulhavý, 1985); the full Bayesian solution is not tractable.

To simplify the presentation, we will restrict ourselves to the simpler case (a).

REMARK 2 The Shannon relative entropy as the loss functional has been selected because of its excellent properties as a dissimilarity measure and because of the feasibility of the solution gained (see Perez, 1984).

REMARK 3 Taking the special case with a fixed $\phi(t+1|t)=\phi$ and $p_1(X^*(t+1)|t) \propto 1$, the formula (2.20) is reduced to

$$p(X^*(t+1)|t) \propto [p_0(X^*(t+1)|t)]^{\phi}, \qquad (2.21)$$

which is just a Bayesian interpretation of the standard exponential forgetting as given in Peterka (1981).

REMARK 4 It should be noted that the resulting solution is invariant with respect to one-to-one transformations of the parametric space.

REMARK 5 The simplicity of the exponential forgetting is the main reason for its widespread use. This simplicity is, however, paid for by a poor performance under lack of information about estimated parameters in observed data (Åström, 1980; Kárný et al., 1985). This drawback is of critical importance when determining the model structure because overparametrized models have to be handled. Similar problems, however, arise as a result of rate changes of exogenous variables, or of linear feedback in the closed control loop. Note that these problems are effectively overcome by restricting the (generalized) exponential forgetting in the above-mentioned way (see analyse in Kulhavý and Kárný, 1984; Kulhavý, 1985; Kárný et al., 1985).

Extension

It remains to extend the solution $p(X^*(t+1)|t)$ to the final c.p.d.f. $p(X(t+1)|t)$. In order that the piece of information which has been removed by the restriction through the parameter mapping $T(\cdot|t)$ may be restored without change, we also require the mapping $T(\cdot|t)$ to be sufficient with respect to the pair of the posterior $p_0(X(t+1)|t)$ and time-updated $p(X(t+1)|t)$ c.p.d.f.'s. It can be proved from the definitions of the sufficient mapping and conditional probability that the sufficiency requirement implies that

$$p(X(t+1)|t) = \frac{p(T(X(t+1)|t)|t)}{p_0(T(X(t+1)|t)|t)}\, p_0(X(t+1)|t) \tag{2.22}$$

at each point $X(t+1)$ where $p_0(T(X(t+1)|t)|t) > 0$. Under weak conditions (Kulhavý, 1985), the extension (2.22) is unique. Now we are ready to summarize:

Conceptual algorithm for estimation of time-varying parameters

1. For $t = 1$ specify the prior p.d.f. $p(X(1)|0)$.
2. Perform the data updating $p(X(t)|t-1) \rightarrow p(X(t)|t)$ for measured data $u(t)$, $y(t)$ according to (2.11) gaining at the same time the basic alternative $p_0(X(t+1)|t)$ (2.15).
3. Specify (e.g., update by new data) the reference alternative $p_1(X(t+1)|t)$.
4. Select a parameter mapping $T(\cdot|t)$ that is sufficient (preferably minimal sufficient) with respect to the pair of the prior and posterior c.p.d.f.'s $\{p(X(t)|t-1), p(X(t)|t)\}$.
5. Specify the probability of the posterior alternative ϕ.

6. Evaluate the time-updated c.p.d.f. according to the complete rule

$$p(X(t+1)|t) \propto \left[\frac{p_1(T(X(t+1)|t)|t)}{p_0(T(X(t+1)|t)|t)}\right]^{1-\phi} p_0(X(t+1)|t). \tag{2.23}$$

7. Return to point 2 with $t = t + 1$ while data are available.

2.4 Structure Determination

The central problem of the chapter arises whenever there are alternative descriptions of the same process. A choice of the "best" one is often required. The general theory of such a choice can be found in Peterka (1981).

We will restrict ourselves to a special version. A large set of variables that can be used for output prediction is assumed to be at our disposal and the question of which of them can be omitted is asked.

Available information about unknown parameters of the "maximal" model can be gathered in the way sketched in the preceding section. However, for the economy of the application (to save computation time, needed memory, and often, expensive measurements), a preliminary choice is usually inevitable.

Formally, the model is said to have the structure $k(t)$ (at time t) if

$$X(t) \in A_{k(t)} \subset \mathcal{X}, \qquad k(t) \in \{1, 2, \ldots, n\}. \tag{2.24}$$

The sets $A_{k(\cdot)}$ and the range of possible structures describe competitive models. An omission of some available variables corresponds to the special class of sets in (2.24)—some prescribed entries of $X(t)$ are equal to zero in particular A_k. Selecting proper variables used in the model, we are faced with a more-or-less classical decision problem known as testing of hypotheses.

For computational reasons, we will restrict even more. Long-term experience and theoretical results of particular fields have demonstrated that a proper structure of the system model can be and in some cases even must be (to avoid the need for controllers with a variable structure, for instance) chosen beforehand even if parameters vary. Hence it is sensible to assume the proper model structure time invariant, that is,

$$k(t+1) = k(t) = k. \tag{2.25}$$

Bayesian inference

It is a characteristic feature of the Bayesian approach that uncertainty of the model structure k can be taken into account in the same way as uncertainty of the model parameters $X(t)$.

Let prior uncertainty of k at the time instant t be described by the c.p.f.

$p(k|t-1)$. After measuring new data $u(t)$, $y(t)$, the posterior c.p.f. $p(k|t)$ can be evaluated according to the Bayes rule. It is reduced under the natural conditions of control, which in this particular case read

$$p(u(t)|t-1;\ k) = p(u(t)|t-1), \tag{2.26}$$

to a simple formula

$$p(k|t) \propto p(y(t)|t-1;\ u(t),\ k)p(k|t-1). \tag{2.27}$$

To compute the predictive c.p.d.f.

$$p(y(t)|t-1;\ u(t),\ k) = \int p(y(t)|t-1;\ u(t),\ X(t),\ k)p(X(t)|t-1;\ k)\, dX(t), \tag{2.28}$$

we must determine the c.p.d.f.'s of unknown parameters $X(t)$ conditioned on particular hypotheses $k = 1, 2, \ldots, n$:

$$p(X(t)|t-1;k) \propto \begin{cases} p(X(t)|t-1) & \text{for } X(t) \in A_k \\ 0 & \text{for } X(t) \notin A_k. \end{cases} \tag{2.29}$$

Choice of optimal structure

The optimal structure of the system model should be determined by solving a suitably stated decision problem. It can be shown that in the case of models with a time-invariant structure, to which we have restricted ourselves, the posterior c.p.f. $p(k|t)$ concentrates under rather general conditions (Schwartz, 1965; Yashin, 1985) on a single point. Then the maximum a posteriori probability estimate of k will be close to the optimal decisions for a broad class of loss functions. Hence choosing maximum a posteriori likelihood structure, we will not lose too much generality, but achieve a substantial decrease of computational complexity.

In this way the conceptual algorithm for determining the time-invariant structure (within classes embedded into the common parametric space) is ready to be summarized.

Conceptual algorithm for structure determination

1. Set $t = 1$.
2. Specify the prior p.f. $p(k) = p(k|0)$ on the space of possible structures.
3. Condition $p(X(t)|t-1) \rightarrow p(X(t)|t-1;k)$ for $k = 1, 2, \ldots, n$ according to (2.29).
4. Evaluate $p(y(t)|t-1;\ u(t),\ k)$ for $k = 1, 2, \ldots, n$ and measured data $u(t)$, $y(t)$ according to (2.28).
5. Update $p(k|t-1) \rightarrow p(k|t)$ according to (2.27).

6. Update $p(X(t)|t-1) \rightarrow p(X(t+1)|t)$ using the conceptual algorithm for estimation of time-varying parameters.
7. Return to point 3 with $t = t + 1$ while data are available.

2.5 Decomposition of Large-Scale Structure Determination Task

Most of the practical difficulties in the conceptual algorithm of structure determination are related to the "simple" step 5 because the cardinality of the space of possible structure hypotheses is too high to admit storing and evaluating the complete c.p.f. $p(k|t)$. Special measures have to be applied.

Our treatment is based on the following (supported by experience) conjectures:

1. The "best" model structure found under the assumption of time-invariant parameters contains the optimal structure corresponding to the model that "tracks" parameter changes [the A_k chosen under the complementary conditions $X(t) =$ const. contains the best A_{k^*}]
2. The proper structures are simple

We have coped with the curse of dimensionality by the following two-stage procedure which is substantiated when our conjectures are valid.

1. A set of competitive structures is determined under the assumption of time-invariant parameters. This special case can be and will be treated by a sufficiently effective technique (Kárný, 1983) described in detail in Section 3.3.
2. The complete application of the conceptual algorithm for structure determination is performed solely on the set found in the stage 1.

3. APPLICATION TO REGRESSION-TYPE MODELS

The above-outlined theory will now be elaborated on for a special, yet important class of linear (in parameters) normal regression-type models.

3.1 Linear Normal Regression-Type Model

The input-output model of the controlled system, called the linear normal regression-type model, takes the form

$$p(y(t)|t-1;\, u(t),\, P(t),\, W(t)) \\ = (2\pi)^{-m_y/2}|W(t)|^{1/2} \exp\{-\tfrac{1}{2}(y(t)-P^T(t)z(t))^T W(t)(y(t)-P^T(t)z(t))\}. \tag{3.1}$$

The conditional expectation of $y(t)$ is thus assumed to be a linear combination of entries of a m_z-dimensional vector $z(t)$. The regressor $z(t)$ can be any known, possibly nonlinear function of past data $d(1..t-1)$ and of the current input $u(t)$. The (m_z, m_y)-matrix of the regression coefficients $P(t)$ and the positive definite (m_y, m_y)-matrix of precision $W(t)$ (the inverse of the covariance matrix; see DeGroot, 1970, Sec. 5.5) are the unknown parameters $X(t)$ of the model.

The wide-ranging applicability of the model (3.1) is given by its numerical feasibility and by the freedom to choose the regressor $z(t)$. The standard static (e.g., polynomial) regression as well as the description of a dynamic system (with lagged data items in z) are covered by it.

For later use the system model will be expressed in terms of the m_h-dimensional data vector related to the current time instant

$$h^T(t) = [z^T(t), y^T(t)]. \tag{3.2}$$

With the definition (3.2) the model (3.1) can be rewritten as

$$p(y(t)|t-1; u(t), P(t), W(t))$$
$$= (2\pi)^{-m_y/2}|W(t)|^{1/2} \exp\left\{-\frac{1}{2}\operatorname{tr}\left(W(t)\begin{bmatrix}P(t)\\-I\end{bmatrix}^T h(t)h^T(t)\begin{bmatrix}P(t)\\-I\end{bmatrix}\right)\right\} \tag{3.3}$$

where I denotes the unit matrix of the (m_y, m_y)-type.

3.2 Algorithm for Estimation of Time-Varying Parameters

The estimation algorithm described below serves as an intermediate step, but the results presented are of interest in their own right.

Gauss-Wishart prior distribution

The choice of the prior distribution of unknown parameters $(P(t), W(t))$ has been influenced by the following observations. If the prior is chosen in the Gauss-Wishart (GW) form, the posterior c.p.d.f.—after the data updating with the system model (3.1)—will be again in the GW form (Peterka, 1981). Even the time-updated c.p.d.f.—after the application of the restricted forgetting—remains in the GW form if we take a suitable (nonminimal) sufficient mapping. Thus, using the GW form of c.p.d.f.'s, we can reduce the general functional recursions (2.11), (2.23) for parameter estimation to algebraic ones.

At the same time the family of GW priors is rich enough to express specific prior knowledge of unknown parameters. Several cases of specifying the GW priors are discussed by Kárný (1984).

The Gauss-Wishart p.d.f. can be written in the following equivalent forms for $t = 1, 2, 3, \ldots$ [recall (2.2) for $t = 1$].

FORM 1

$$p(P(t), W(t)|t-1) = p_{\mathrm{GW1}}(P(t), W(t)|V, v:(t|t-1))$$

$$p_{\mathrm{GW1}}(P, W|V, v) = J_1^{-1}(V, v)|W|^{(v+m_z-m_y-1)/2} \times \exp\left\{-\frac{1}{2}\mathrm{tr}\left(W\begin{bmatrix} P \\ -I \end{bmatrix}^T V \begin{bmatrix} P \\ -I \end{bmatrix}\right)\right\}. \tag{3.4}$$

This form simplifies the derivation of the fundamental data-update formulas. It also facilitates our explanation of relationships among different structures treated later. Note that we have used the shorthand notation $:(t|t-1)$ for extracting the common index from the statistics $V(t|t-1)$, $v(t|t-1)$.

The symmetric (m_h, m_h)-matrix V has to be positive definite

$$V > 0 \tag{3.5}$$

and the scalar v (the number of degrees of freedom) has to fulfill the inequality

$$v > m_y - 1 \tag{3.6}$$

to guarantee the properness of the corresponding p.d.f. (DeGroot, 1970, Secs. 5.4 and 5.5).

Splitting V into submatrices in correspondence with the definition of the data vector (3.2),

$$V = \begin{bmatrix} V_z & V_{zy} \\ V_{zy}^T & V_y \end{bmatrix} \begin{matrix} \}m_z \\ \}m_y, \end{matrix} \tag{3.7}$$

the normalizing factor reads (Anderson, 1958, Secs. 3.2 and 7.2)

$$J_1(V, v) = f(v, m_z, m_y) \prod_{j=1}^{m_y} \Gamma\left(\frac{v+1-j}{2}\right) |V_z|^{(v-m_y)/2} |V|^{-v/2} \tag{3.8}$$

with the gamma function $\Gamma(\cdot)$ and the scalar $f(\cdot)$ defined by

$$f(v, m_z, m_y) = (2^{v+m_z}\pi^{m_z+m_y-1})^{m_y/2}. \tag{3.9}$$

We can also write (3.4) in an equivalent form which factorizes the GW p.d.f. into the Wishart and Gauss (normal) parts, respectively, and relates the GW statistic to the well-known least squares (LS) quantities. In fact, we have

FORM 2

$$p(P(t), W(t)|t-1) = p_{\mathrm{GW2}}(P(t), W(t)|\hat{P}, C, S, v:(t|t-1))$$

$$p_{\mathrm{GW2}}(P, W|\hat{P}, C, S, v) = J_2^{-1}(\hat{P}, C, S, v)|W|^{(v-m_y-1)/2} \exp\{-\tfrac{1}{2}\mathrm{tr}(WS)\} \times |W|^{m_z/2} \exp\{-\tfrac{1}{2}\mathrm{tr}(W(P-\hat{P})^T C^{-1}(P-\hat{P}))\} \tag{3.10}$$

A direct comparison of (3.4) and (3.10) with the help of (3.7) implies that

$$\hat{P} = V_z^{-1} V_{zy} \qquad \text{(LS parameter estimates)} \tag{3.11}$$

$$C = V_z^{-1} \qquad \text{(imprecision of LS estimates)} \tag{3.12}$$

$$S = V_y - V_{zy}^T V_z^{-1} V_{zy} \qquad \text{(rest of LS).} \tag{3.13}$$

Combining (3.8) and (3.12) and noticing that

$$|S| = \frac{|V|}{|V_z|}, \tag{3.14}$$

we find

$$J_2(\hat{P}, C, S, v) = f(v, m_z, m_y) \prod_{j=1}^{m_y} \Gamma\left(\frac{v+1-j}{2}\right) |C|^{m_y/2} |S|^{-v/2}. \tag{3.15}$$

Data updating

Using the product form of Bayes' rule (2.11), and the system model (3.3) in conjunction with the GW1 prior (3.4), we find immediately that

$$p(P(t), W(t)|t) = p_{\mathrm{GW1}}(P(t), W(t)|V, v:(t|t)), \tag{3.16}$$

where

$$V(t|t) = V(t|t-1) + h(t)h^T(t), \qquad v(t|t) = v(t|t-1) + 1. \tag{3.17}$$

In the LS terms (3.11)–(3.13) the recursion (3.17) changes with the help of the matrix inversion lemma (Rao, 1973, Chap. 1b) into the recursive least squares

$$\hat{P}(t|t) = \hat{P}(t|t-1) + \frac{g(t|t-1)\hat{e}^T(t|t-1)}{1 + \zeta(t|t-1)} \tag{3.18}$$

$$C(t|t) = C(t|t-1) - \frac{g(t|t-1)g^T(t|t-1)}{1 + \zeta(t|t-1)} \tag{3.19}$$

$$S(t|t) = S(t|t-1) + \frac{\hat{e}(t|t-1)\hat{e}^T(t|t-1)}{1 + \zeta(t|t-1)}, \tag{3.20}$$

where the auxiliary quantities (useful later, too) are

$$g(t|t-1) = C(t|t-1)z(t) \tag{3.21}$$

$$\zeta(t|t-1) = g^T(t|t-1)z(t) = z^T(t)C(t|t-1)z(t) \tag{3.22}$$

$$\hat{e}(t|t-1) = y(t) - \hat{P}^T(t|t-1)z(t). \tag{3.23}$$

Specification of time updating

The restricted exponential forgetting will be used for updating the posterior c.p.d.f. $p(P(t), W(t)|t)$. Let us specify the version used.

1. The reference alternative is chosen in the GW form to keep computations at the "algebraic" level: the manipulations with functions (c.p.d.f.'s) are reduced to those with finite-dimensional statistics.

   The structure determination task arises when little is a priori known about the model parameters; therefore, the "noninformative" fixed alternative described by

$$V = \varepsilon I, \qquad v = m_y - 1 + \varepsilon \qquad \text{for } \varepsilon \to 0^+ \tag{3.24}$$

   is of direct use, keeping the presentation simple enough.
2. The parameter mapping sufficient with respect to the pair of the c.p.d.f.'s $\{p(P(t), W(t)|t-1), p(P(t), W(t)|t)\}$ is selected as follows:

$$T(P, W|t) = (y(t) - P^T z(t), W). \tag{3.25}$$

   The mapping is clearly nonminimal [see (2.17)]. However, owing to its linearity [the Gaussian part is mapped by a linear projector into the direction of $z(t)$], it preserves the GW form throughout the time updating.
3. The time-invariant probability ϕ of the basic alternative (of the parameter time-invariance hypothesis) is assumed.

Evaluating the marginal alternative c.p.d.f.'s of the parameter mapping (3.25) and substituting them into the formula (2.23), the time-updated density can be determined in a straightforward way. Joining the results of the data and time updating, we get the following algorithm.

Complete updating

The functional form of the c.p.d.f. of unknown parameters is

$$p(P(t), W(t)|t-1) = p_{\mathrm{GW}2}(P(t), W(t)|\hat{P}, C, S, v:(t|t-1)) \qquad \text{for } t = 1, 2, \ldots. \tag{3.26}$$

The statistics of the prior p.d.f. (initial conditions) are

$$\hat{P}(1|0) \text{ arbitrary}, \quad C(1|0) > 0, \quad S(1|0) > 0, \quad v(1|0) > m_y - 1 \tag{3.27}$$

The full-update recursions read

For $z^T(t)z(t) > 0$ (the regular branch):

$$\hat{P}(t+1|t) = \hat{P}(t|t-1) + \frac{g(t|t-1)\hat{e}^T(t|t-1)}{1 + \zeta(t|t-1)} \tag{3.28a}$$

$$C(t+1|t) = C(t|t-1) - \frac{g(t|t-1)g^T(t|t-1)}{\gamma^{-1}(t) + \zeta(t|t-1)} \tag{3.29a}$$

where quantities (3.21)–(3.23) are used and

$$\gamma(t) = \phi - \frac{1-\phi}{\zeta(t|t-1)}. \tag{3.30}$$

For $z^T(t)z(t) = 0$ (the singular branch):

$$\hat{P}(t+1|t) = \hat{P}(t|t-1) \tag{3.28b}$$

$$C(t+1|t) = C(t|t-1). \tag{3.29b}$$

Independently of $z^T(t)z(t)$ (the common branch):

$$S(t+1|t) = \phi\left(S(t|t-1) + \frac{\hat{e}(t|t-1)\hat{e}^T(t|t-1)}{1+\zeta(t|t-1)}\right) \tag{3.31}$$

$$v(t+1|t) = \phi(v(t|t-1)+1). \tag{3.32}$$

REMARK. The substantial difference of the foregoing algorithm in comparison with the exponentially forgotten least squares lies in the relation (3.29). Note that instead of the whole matrix $C(t|t)$ only the last data dyad $z(t)z^T(t)$ is weighted as a result of forgetting (see a similar result by Hägglund, 1985). Thus the matrix $C^{-1}(t+1|t)$ cannot become singular even for linearly dependent regressors (at least prior information is saved). This fact considerably improves the reliability of parameter estimation in case the data do not contain sufficient information.

3.3 Algorithm for Structure Determination

In the sequel we restrict ourselves to a special class of hypotheses about the model structure:

$$P_{ij}(t) = 0 \qquad \text{for } i \in \mathcal{I}_k \subset \{1, 2, \ldots, m_z\}, k = 1, 2, \ldots, n$$
$$\text{and } j = 1, 2, \ldots, m_{\mathcal{I}}, t = 1, 2, \ldots. \tag{3.33}$$

Thus each hypothesis k is represented by the index set y_k of zero rows of the matrix $P(t)$. This corresponds with our goal (see Section 2.4): the possibility to omit some entries of a "maximal" $z(t)$ is tested.

Conditioning

The evaluation of the output predictions (2.28) requires the c.p.d.f.'s $p(P(t), W(t)|t-1; k)$ to be known for all the hypotheses in (3.33) with $k = 1, 2, \ldots, n$. To make the situation more transparent, let us arrange the relevant and spare entries of P for each hypothesis k into the blocks Q_r^k and Q_s^k, respectively, using a suitable permutation matrix E^k:

$$E^k P = \begin{bmatrix} E_r^k \\ E_s^k \end{bmatrix} P = Q^k = \begin{bmatrix} Q_r^k \\ Q_s^k \end{bmatrix} \begin{matrix} \}\, m_k \\ \}\, m_z - m_k. \end{matrix} \tag{3.34}$$

Then the evaluation of the predictive c.p.d.f. (2.28) can be specialized as follows:

$$p(y(t)|t-1; u(t), k) = \int p(y(t)|t-1; u(t), Q_r^k(t), Q_s^k(t)=0, W(t)) \cdot p(Q_r^k(t), W(t)|t-1; Q_s^k(t)=0)\, dQ_r^k(t)\, dW(t). \tag{3.35}$$

It is easy to show that if the original c.p.d.f. has the GW1 form

$$p(P(t), W(t)|t-1) = p_{\mathrm{GW1}}(P(t), W(t)|V, v:(t|t-1)),$$

then the c.p.d.f.'s corresponding to particular structure hypotheses $k = 1, 2, \ldots, n$ are of the same type

$$p(Q_r^k(t), W(t)|t-1; Q_s^k(t)=0) = p_{\mathrm{GW1}}(Q_r^k(t), W(t)|V, v:(t|t-1; k)) \tag{3.36}$$

with the statistics

$$V(t|t-1; k) = \begin{bmatrix} E_r^k & 0 \\ 0 & I \end{bmatrix} V(t|t-1) \begin{bmatrix} E_r^k & 0 \\ 0 & I \end{bmatrix}^T \tag{3.37}$$

$$\underbrace{}_{m_k} \quad \underbrace{}_{m_y}$$

$$v(t|t-1; k) = v(t|t-1) + (m_z - m_k). \tag{3.38}$$

Output predictions

Evaluating the predictive p.d.f. (3.35) with the c.p.d.f. of unknown parameters in the GW1 form (3.36), we find

$$\begin{aligned} p(y(t)|t-1; u(t), k) &= \frac{(2\pi)^{-m_y/2} J_1(V, v:(t|t; k))}{J_1(V, v:(t|t-1; k))} \\ &= \pi^{-m_y/2} \frac{\Gamma(v(t|t; k)/2)}{\Gamma((v(t|t-1; k) + 1 - m_y)/2)} \\ &\quad \times \left(\frac{|V_z(t|t; k)|}{|V_z(t|t-1; k)|} \right)^{-m_y/2} \left(\frac{|V(t|t; k)|}{|V_z(t|t; k)|} \right)^{-v(t|t; k)/2} \\ &\quad \times \left(\frac{|V(t|t-1; k)|}{|V_z(t|t-1; k)|} \right)^{v(t|t-1; k)/2}. \end{aligned} \tag{3.39}$$

Utilizing the one-to-one correspondence of the joint statistic V to the LS statistics $\hat{P}$, C, S given by the relations (3.11)–(3.13) and taking into account the recursions (3.18)–(3.20) together with the definition of the related quantities (3.21)–(3.23), we get, after necessary algebraic manipulations (Rao,

1973, Chap. 1b), that

$$p(y(t)|t-1; u(t), k) = \frac{(2\pi)^{-m_y/2} J_2(\hat{P}, C, S, v:(t|t; k))}{J_2(\hat{P}, C, S, v:(t|t-1; k))}$$

$$= \pi^{-m_y/2} \frac{\Gamma((v(t|t-1; k)+1)/2)}{\Gamma((v(t|t-1; k)+1-m_y)/2)}$$

$$\times |S(t|t-1; k)|^{-1/2} (1+\zeta(t|t-1; k))^{-m_y/2}$$

$$\times \left(1 + \frac{\eta(t|t-1; k)}{1+\zeta(t|t-1; k)}\right)^{-(v(t|t-1; k)+1)/2}, \tag{3.40}$$

where we have denoted

$$\eta(t|t-1; k) = \hat{e}^T(t|t-1; k) S^{-1}(t|t-1; k) \hat{e}(t|t-1; k). \tag{3.41}$$

REMARK Notice that the LS statistics evaluated for each competitive model are sufficient for the Bayesian model comparison, specifically, for evaluating the c.p.f. $p(k|t)$ according to the Bayes rule (2.27).

Organization of computations

A plain parallel computation of the output predictions in (3.40) is possible, but then the computational burden increases linearly with the number of hypotheses involved. This observation excludes the possibility to check fully the significance of particular entries and their combinations even for medium-sized regressors as the number of possible combinations grows exponentially with the dimension of the maximal admissible regressor.

The practical problems met have forced us to give up an exact solution because of the curse of dimensionality (strengthened by the adaptivity needed) when choosing among tens (even hundreds) of potential regressor entries. Instead, relying on conjectures formulated in Section 2.5, we have proposed the following two-stage algorithm.

Stage 1: preliminary structure determination with constant parameter supposed

At this preparatory stage the time variations of model parameters are neglected; hence the subjective probability of the basic alternative (the forgetting factor) is set equal to the unity. In this case the time updating is trivial

$$V(t+1|t) = V(t|t), \qquad v(t+1|t) = v(t|t) \tag{3.42}$$

and the predictive c.p.d.f. (3.39) takes the form

$$p(y(t)|t-1; u(t), k) = (2\pi)^{-m_y/2} J_1(V, v:(t+1|t; k)) / J_1(V, v:(t|t-1; k)). \tag{3.43}$$

The repetitive use of the Bayes rule (2.27) in the time range $1, 2, \ldots, t$ results, because of canceling intermediate factors, in

$$p(k|t) \propto l(k|t)p(k|0) \tag{3.44}$$

with the "likelihood" function

$$l(k|t) \propto \frac{J_1(V, v:(t+1|t;\, k))}{J_1(V, v:(1|0;\, k))} \times \prod_{j=1}^{m_y} \frac{\Gamma((v(t+1|t;\, k)+1-j)/2)}{\Gamma((v(1|0;\, k)+1-j)/2)} \left(\frac{|V_z(t+1|t;\, k)|}{|V_z(1|0;\, k)|}\right)^{-m_y/2} \times \left(\frac{|V(t+1|t;\, k)|}{|V_z(t+1|t;\, k)|}\right)^{-v(t+1|t;\, k)/2} \left(\frac{|V(1|0;\, k)|}{|V_z(1|0;\, k)|}\right)^{v(1|0;\, k)/2} \tag{3.45}$$

The first stage is used when little is a priori known, so that the selection of the uniform prior p.f. $p(k|0)$ is fully justified. As we search for the maximum a posteriori probability structure (recall the discussion of Section 2.4), the preliminary structure determination is then reduced to finding the maximum of the function (3.45). The problem is by no means simple because the function (3.45) is defined on a high-dimensional discrete grid and has been found multimodal.

We have succeeded in developing a special algorithm for this task. Insufficient space prevents us from describing it in detail. Instead, a commented "flowchart" is presented. A full description with a nontrivial example is given in Kárný (1983).

Stage 1 flow chart

1. A flat GW1 prior p.d.f. with statistics of the type (3.24) is chosen with a small, but positive ε.

COMMENT Notice that owing to this choice, the values of determinants $|V(1|0)|$ and $|V_z(1|0)|$ present no computational burden.

2. The statistics for the most complex structure

$$V(t+1|t) = \sum_{\tau=1}^{t} h(\tau)h^T(\tau) + \varepsilon I, \qquad v(t+1|t) = t + v(1|0) \tag{3.46}$$

are collected. The form (3.46) follows from (3.17) and (3.42).

COMMENT The evaluation of (3.46) can be a time-consuming task for long data sets and large m_z, but it is performed only once. A special shifted structure of the data vector $h(t)$ (typical for dynamic models) can be used to decrease the computational burden.

3. The user specifies which entries must be, which must not be, which are expected to be, and which are not expected to be in the final model. The initial arrangement (3.34) (the initial structure k_0) is defined in this way. The counter, say j, is set to zero at the same time.
4. The statistics $V(t + 1|t; k_0)$, $v(t + 1|t; k_0)$ corresponding to the initial structure chosen in the step 3 are generated according to the relations (3.37), (3.38).

COMMENT The step is again relatively time consuming, but it is performed once for each initial guess made in the step 3. Only several guesses suffice as a rule (typically, 3).

5. An initial value of the function (3.45) is computed. It is advantageous to use a logarithmic version of (3.45) throughout to keep a proper numerical level.
6. The full trial search is performed in a suitable neighborhood of the previous best guess of the optimal structure. The neighborhood of the structure z_j, given by k_j, contains those regressors arising from z_j either by omitting or adding just one admissible entry of z_j. The local maximizer is taken as a new "center" for the next step. The counter j is increased.

COMMENT Computational requirements of this local search essentially determine the upper bound on dimensionality of the problems which can be handled at reasonable computational demands. Effective tools proposed in Kárný (1983) are discussed in Section 4.

7. The search is stopped if either too many steps have been performed (rarely met), or if no increase of value of the function (3.45) in a new local neighborhood is detected. If the search is not stopped, step 6 is repeated. If the search is stopped, the procedure can be repeated starting from point 3 to assure that the global maximum has been reached; in the opposite case, step 8 follows.

COMMENT Usually, it suffices to compare runs starting with the empty and full regressor.

8. A final arrangement of the best regressor z_j is performed so that its entries are placed closer to the tail end, so they influence less the maximal achieved value of the function (3.45).

COMMENT This step extends the algorithm in Kárný (1983). The chosen regressor is arranged in such a way that a "natural" nesting of order determination problems is imitated. Thus the initial most complex structure for the second stage of the entire algorithm is prepared.

The following discussion brings additional insight into the first stage and its connection with the second stage.

Expressing the function (3.45) in the LS statistics

$$l(k|t) \propto \frac{J_2(\hat{P}, C, S, v:(t+1|t;\, k))}{J_2(\hat{P}, C, S, v:(1|0;\, k))} \times \prod_{j=1}^{m_y} \frac{\Gamma((v(t+1|t;\, k)+1-j)/2)}{\Gamma((v(1|0;\, k)+1-j)/2} \times \frac{|C(t+1|t;\, k)|^{m_y/2}}{|C(1|0;\, k)|^{m_y/2}} \frac{|S(t+1|t;\, k)|^{-v(t+1|t;\, k)/2}}{|S(1|0;\, k)|^{-v(1|0;\, k)/2}}, \tag{3.47}$$

we can find that the choice of an optimum structure is influenced by the following items:

1. The quality of prediction measured by the determinant values of the rest-of-LS matrices $S(t+1|t;k)$; this is the first-rate item due to high powers of determinants $v(t+1|t;k)$ equal approximately to the number of data t.
2. Parameter uncertainty measured by the determinant values of the corresponding matrices $C(t+1|t;k)$ and by the gamma functions of the corresponding numbers of freedom $v(t+1|t;k)$; this item starts being significant if the determinants of some rest-of-LS matrices are essentially the same.
3. Prior information supplied through $S(1|0;k)$, $C(1|0;k)$, and $v(1|0;k)$.

Forgetting changes the relations mentioned. It essentially limits the amount of stored information. Consequently, instead of describing the whole observed process by a unique complex model, we are building and adapting a simple model to involve only a limited memory range. This feature is broadly used for adaptive prediction and control and the main purpose of the second stage is to "tailor" the chosen structure to this fact.

Stage 2: nested structure determination with time-varying parameters supposed

This stage essentially belongs to the class of algorithms that compare a sufficiently narrow set, say K, of competitive models, on the basis of the c.p.f. $p(k|t)$ recursively evaluated according to the Bayes formula (2.27). In contrast to the first stage, the time variations of parameters are taken into account.

The set K used here is formed (relying on the conjectures of Section 2.5) by nested submodels found in the first stage. It should be noticed, however, that the algorithm can be used directly in real time for any, not too large, set of competitors with nested structures (e.g., for model order determination).

The description of the algorithm is given again in terms of a commented flowchart.

Stage 2 flowchart

1. The set K of competitors with a nested structure is defined. It is taken either from the first stage, or specified by the user. Operationally, it means choosing the maximal regressor that has to contain the regressors of competitors in the leading entries.
2. The prior probabilities $p(k|0)$ of particular structures are chosen (no restriction is imposed on them).

COMMENT The full power of the restricted forgetting can be used, including the possibility of data-dependent forgetting and a broader set of alternatives. This makes sense mainly when this stage is directly used in real time (not as a part of our overall algorithm). For the necessary background, see Kulhavý (1986).

4. The prior statistics (3.27) of the GW2 p.d.f. of unknown parameters for the most complex model are specified.

COMMENT Close attention has to be paid to the choice in order to be sure that the prior p.d.f.'s of parameters induced by nesting really express available knowledge.

5. After measuring data the statistics of the most complex model are updated using the algorithm (3.28)–(3.32).
6. The statistics of all the nested models are computed using the relations (3.37)–(3.38) and (3.11)–(3.13) together with the quantities $\zeta(t|t-1;k)$ (3.22) and $\eta(t|t-1;k)$ (3.41) needed for evaluating the output predictions (3.40).

COMMENT This step represents, as in the first stage, the crucial point of computations. See Section 4 for an efficient solution.

7. The predictive c.p.d.f.'s of the form (3.40) are evaluated with quantities precomputed in the preceding step and the c.p.f. of nested structures $p(k|t-1)$ is updated according to the Bayes rule (2.27).
8. Step 5 is repeated while data are available.

4. NUMERICAL IMPLEMENTATION

Extensive computational demands of the structure determination task have forced us to pay great attention to efficient numerical implementation of all the above presented algorithms. Due to the limited space, only essential points are discussed below. For details, see Kárný (1983) and Peterka (1986); for basic ideas of the L-D factorization, see Bierman (1977).

4.1 Matrix Factorization

The core of implementation lies in the L-D factorization of the matrix statistic $V(t|t-1) = V$. Any positive definite matrix V can be uniquely factorized into the product

$$V = LDL^T, \tag{4.1}$$

where L denotes a lower triangular matrix with the unit diagonal and D a diagonal matrix with positive entries. Splitting the matrices L, D in accordance with the structure of the data vector (3.2),

$$L = \begin{bmatrix} L_z & 0 \\ L_{yz} & L_y \end{bmatrix}, \qquad D = \begin{bmatrix} D_z & 0 \\ 0 & D_y \end{bmatrix}, \tag{4.2}$$

we easily derive the corresponding expressions of the LS statistics from the relations (3.11)–(3.13):

$$\hat{P} = L_z^{-T} L_{yz}^T \tag{4.3}$$

$$C = L_z^{-T} D_z^{-1} L_z^{-1} \tag{4.4}$$

$$S = L_y D_y L_y^T. \tag{4.5}$$

In the first stage, only the terminal statistic V is factorized into the form (4.1). The evaluations are substantially reduced in this way whenever the regressor has the shifted structure (Kárný, 1983).

In the second stage, the L, D factors of V are updated recursively [instead of the statistics $\hat{P}$, C, S, (3.28)–(3.31), using so-called dyadic reductions (Peterka, 1986)].

The L-D factorization guarantees excellent numerical properties necessary for poorly conditioned structure determination tasks with overparametrized models. Even more important, however, is the fact that the factorization enables us to simplify considerably the evaluation of output predictions for the compared structures. This aspect is analyzed in the next section.

4.2 Factorization-Based Conditioning

To find the conditional statistic $V(t|t-1; k) = V^k$ (3.37), we first apply the permutation (3.34) to appropriate rows and columns of the L, D factors of the statistic V:

$$H(E^k) = \begin{bmatrix} E_r^k & 0 \\ 0 & I \\ E_s^k & 0 \end{bmatrix} L \begin{bmatrix} E_r^k & 0 \\ 0 & I \\ E_s^k & 0 \end{bmatrix}^T, \qquad G(E^k) = \begin{bmatrix} E_r^k & 0 \\ 0 & I \\ E_s^k & 0 \end{bmatrix} D \begin{bmatrix} E_r^k & 0 \\ 0 & I \\ E_s^k & 0 \end{bmatrix}^T. \tag{4.6}$$

Clearly, the product $H(E^k)G(E^k)H^T(E^k)$ does not represent the L-D factorization of the permuted statistic $V(E^k)$ as the matrix $H(E^k)$ is generically nontriangular. The L-D form $V(E^k) = L(E^k)D(E^k)L^T(E^k)$ can be restored in a computationally nondemanding way (e.g., using dyadic reductions mentioned above). Splitting the factors $L(E^k)$, $D(E^k)$ in accordance with (4.6),

$$L(E^k) = \begin{bmatrix} L_r^k & 0 & 0 \\ L_{yr}^k & L_y^k & 0 \\ L_{sr}^k & L_{sy}^k & L_s^k \end{bmatrix}, \qquad D(E^k) = \begin{bmatrix} D_r^k & 0 & 0 \\ 0 & D_y^k & 0 \\ 0 & 0 & D_s^k \end{bmatrix}, \tag{4.7}$$

we easily find out that the conditional statistic V^k is determined by the left upper (nested) submatrices of the factors $L(E^k)$, $D(E^k)$ as follows:

$$V^k = \begin{bmatrix} L_r^k & 0 \\ L_{yr}^k & L_y^k \end{bmatrix} \begin{bmatrix} D_r^k & 0 \\ 0 & D_y^k \end{bmatrix} \begin{bmatrix} L_r^k & 0 \\ L_{yr}^k & L_y^k \end{bmatrix}^T. \tag{4.8}$$

The statistics $\hat{P}^k$, C^k, S^k are related to appropriate submatrices of V^k by the relations (4.2)–(4.5).

The nesting property just illustrated can be exploited for an effective solution of structure determination problems which have "naturally" nested regressors. The problem of order determination (Kárný, 1980) serves as a typical case of this type. The second stage of the structure determination algorithm described in Section 3.3 also belongs to this class.

As is obvious from (4.8), the evaluation of the L^k, D^k factors for nested structures with a shorter regressor requires only the recomputation of the L_y^k, D_y^k factors. Usually, the output dimension m_y is much smaller than the maximal regressor dimension m_z, thus the needed triangulization is computationally unburdensome. Moreover, it is always possible to solve m_y one-output problems independently, still gaining generality since each output can have its own regressor.

The systematic use of the L-D factorization makes high-dimensional problems feasible by considerably simplifying necessary repetitive evaluations of the normalizing factors. The complexity of J_1 and J_2 [equations (3.8) and (3.15)] lies in the need for the determinants of $V_z(t|t-1;k) = V_r^k$ and $V(t|t-1;k) = V^k$, or $S(t|t-1;k) = S^k$. It follows immediately from (4.8) and (4.4)–(4.5) that

$$|V_r^k| = |D_r^k| \tag{4.9}$$

$$|S^k| = \frac{|V^k|}{|V_r^k|} = |D_y^k|. \tag{4.10}$$

The overall computational burden is strongly influenced by a suitable organization of computations. For instance, much can be saved if adjoint rows of P are permuted and the L-D factorization is immediately restored. The required factors can be found by a sequence of such elementary steps. Throughout, the quantities $\zeta(t|t-1;k)$ and $\eta(t|t-1;k)$ [equations (3.22)–(3.41)] for nested submodels can be gained as by-products.

5. ILLUSTRATIVE EXAMPLE

The behavior of the designed algorithm is illustrated by a simulated example. We demonstrate that

1. Even simple cases lead to regressors with tens of entries.
2. The achieved speed of evaluation is high enough.
3. The second stage properly simplifies the chosen structure.

The algorithm is a member of the package for simulation, identification, and adaptive control (SIC) operating under the interactive monitor KOS (see Kárný et al., 1985, for details). This portable package is running on the IBM 370/135 computer under the OS/VS1 operating system in a partition with 256 kB. Information about the CPU time given below includes simulations, person-machine interactions, and so on.

The simulated system had the form of a multivariate ARMAX model excited by the normally distributed white noise input $u(t) \sim N(0, I)$ in an open loop

$$y(t) = \begin{bmatrix} 0.004 & 0.8 & 0.1 \\ 0. & 1.5 & 0.1 \end{bmatrix} u(t-1) + \begin{bmatrix} 0.04 & -0.82 & 0 \\ 0.20 & -1.10 & 0 \end{bmatrix} u(t-2)$$
$$+ \begin{bmatrix} 0.98 & -0.1 \\ 0 & 0.6 \end{bmatrix} y(t-1) + \begin{bmatrix} 0.18 \\ -0.02 \end{bmatrix} + \begin{bmatrix} 1 & 0.02 \\ 0 & 1.00 \end{bmatrix} e(t)$$
$$+ \begin{bmatrix} 0.995 & 0.8 \\ 0.300 & -0.5 \end{bmatrix} e(t-1).$$

The white noise $e(t) \sim N(0, I)$ was independent of $u(t)$.

The ARMAX system model having zeros of the stochastic part close to the stability boundary was chosen to imitate an infinite-order ARX model. This could be described successfully only by time-varying finite-order ARX models in order that mismodeling errors might be handled (the typical reason for adaptivity in practice!). A stable autoregression with some interactions between the outputs was selected with one step of the common transport delay, some other zero entries at the inputs, and roughly the same static gains of input-output and noise-output transfer functions.

A collection of 1600 data items were processed and analyzed in the preliminary stage with $\varepsilon = 0.01$ in (3.24). The empty regressor and the full regressor cases were used as the initial guesses of the optimal structure. Models up to the tenth order were compared; thus the maximal admissible regressor contained 54 variables. Both the initial guesses gave the same maximizing solution, including only 22 variables. The preliminary choice of the optimal structure (for each guess) took less than 12 minutes of CPU time. The chosen structure had the following properties:

1. The transport delay and the absolute term were correctly recognized.
2. All data items having nonzero coefficients in the autoregression and regression parts of the simulated model were found significant with the exception of $u_1(t-1)$, which influenced both outputs weakly.
3. The order of the autoregression part was increased to 5 (but not more!) due to the high noise correlation.
4. The orders of the regression part were also increased because of the noise correlation, but again did not go beyond 5 [(3, 5, 1) for (u_1, u_2, u_3)].
5. The entries compensating for the noise correlation were recognized as less important (according to their contribution to the likelihood function); thus they were placed among the tail entries of the resulting regressor.

The second stage was realized under the same conditions starting from the best solution of the first stage. The forgetting factors 0.9, 0.8, 0.7, and 0.6 were used to demonstrate the influence of the chosen value. Every complete run required about 18 minutes of CPU time. It was found that

1. The final structure contained 21, 17, 11, 11 entries for the decreasing values of the forgetting factor; from this viewpoint, the expected monotonic dependence of the required model complexity on the forgetting factor value was demonstrated.
2. A substantial reduction of the initial structure was achieved only for relatively low values of the forgetting factor [this fact is not surprising if the probabilistic interpretation of forgetting (see Section 2.3) is taken into account].
3. The most lagged input-output data items in regressor, compensating for the noise correlation, were more or less regularly discarded when the simpler model was chosen; the original variables in the regression and autoregression parts remained in the final regressor.

6. CONCLUDING REMARKS

The proposed formulation of the structure determination task takes account of:

Stochastic nature of the system
Available, but incomplete knowledge of the system
Excessive dimensionality of real-life problems caused by interdependence of influencing factors
Variations of estimated parameters

The coupling of Bayesian analysis and decision theory has resulted in a feasible algorithm that in a unified manner takes into account:

Ability of the model to fit data
Precision of parameter estimates
Complexity of the model
Number of available data items
Prior information about the identified system

Naturally, much should be done to broaden applicability of the algorithm.

1. Like other existing techniques, the algorithm is unable to recognize when a particular data item does not influence the predicted output. The most extreme example we have encountered was a temperature value in a cement kiln recognized as independent of a constant (during the accumulation of data) fuel flow. Clearly, the necessary prerequisite to structure determination is a suitable, if not optimal, design of the experiment.
2. Robustness of the algorithm should be treated. Many particular results are available since least squares forms the core of evaluations. The analysis is, however, incomplete. We plan to address the problem by realistic modeling of the analyzed situations. This, of course, leads beyond the boundaries of linear normal models, to the problems for which a feasible approximation of Bayesian solution is inevitable. A promising step in this direction has been taken by Kárný and Hangos (1987).
3. An asymptotic analysis is another unfinished task. With no forgetting, general favorable properties of Bayesian identification are applicable (Schwartz, 1965; Yashin, 1985). Almost nothing is available in the adaptive case with forgetting; this is similar to most of the truly adaptive algorithms.

The above-mentioned open questions, however, do not represent a serious restriction to immediate applicability. Extensive simulations as well as several practical tests have confirmed the usefulness of the procedure. Poorly conditioned cases have been safely managed irrespectively of the source of data "noninformativity" (overparametrization, linear feedback, rare changes of some entries of the regressor, etc.).

In simulated cases, the correct structure has been found almost without exception when multimodality of the likelihood function has been accounted for.

For real-life data, the chosen structures have usually been accepted by experts as reasonable. In several cases the results were surprising at first sight, but an explanation was always found afterward. One example was the cement kiln case mentioned above. Another typical example was a production quality index detected as independent of all 60 tested variables in spite of evident relevance of many of them. An unrealistic definition of the index, verified independently later, was indicated in this way. It is not surprising that such a successful analysis, showing mistakes in the solution of the particular technical problem, has not been published.

In a few previously implemented applications the solution of preliminary structure determination has resulted in good controllers. Adaptive control of a cold-rolling mill (Ettler, 1986) serves as an example where preliminary structure determination contributed to overall success: Unnecessary measurements of some variables were avoided and the computational time, important for the high sampling rates needed, was substantially saved.

At present we are preparing an analysis of nontechnical data related to Hodgkin's disease. We hope to answer in the Bayesian way the questions usually solved by the analysis of variance.

REFERENCES

Akaike, H. (1979). A Bayesian extension of the minimum AIC procedure of autoregressive model fitting, *Biometrika* 66, 237–242.

Anderson, T. W. (1958). *An Introduction to Multivariate Statistical Analysis*, Wiley, New York.

Åström, K. J. (1980). Design Principles for Self-Tuning Regulators, in *Methods and Applications in Adaptive Control* (H. Unbehauen, ed.), Springer-Verlag, Berlin, pp. 1–20.

Bierman, G. J. (1977). *Factorization Methods for Discrete Sequential Estimation*, Academic Press, New York.

Csiszár, I. (1967). Information-type measures of difference of probability distributions and indirect observations, *Stud. Sci. Math. Hung.* 2, 299–318.

DeGroot, M. H. (1970). *Optimal Statistical Decisions*, McGraw-Hill, New York.

Ettler, P. (1986). An adaptive controller for Škoda twenty-rolls cold rolling mills, *Preprints of the 2nd IFAC Workshop on Adaptive Systems in Control and Signal Processing*, Lund, Sweden, pp. 277–280.

Hägglund, T. (1985). Recursive Estimation of Slowly Time-Varying Parameters, in *Preprints of the 7th IFAC Symposium on Identification and System Parameter Estimation*, University of York, England, Vol. 2, pp. 1137–1142.

Jazwinski, A. H. (1970). *Stochastic Processes and Filtering Theory*, Academic Press, New York.

Kárný, M. (1980). Bayesian estimation of model order, *Probl. Control Inf. Theory* 9, 33–46.

Kárný, M. (1983). Algorithms for determining the model structure of a controlled system, *Kybernetika* 19, 164–178.

Kárný, M. (1984). Quantification of prior knowledge about global characteristics of linear model, *Kybernetika* 20, 376–385.

Kárný, M., and K. Hangos (1987). One-sided Approximation of Bayes Rule: Theoretical Background, in *Preprints of the 10th IFAC World Congress*, Munich, FRG, Vol. 10, pp. 312–317.

Kárný, M., A. Halousková, J. Böhm, R. Kulhavý, and P. Nedoma (1985). Design of linear quadratic adaptive control: theory and algorithms for practice, *Kybernetika* 21, supplement to numbers 3, 4, 5, 6.

Kashyap, R. L., and A. R. Rao (1976). *Dynamic Stochastic Models from Empirical Data*, Academic Press, New York.

Kulhavý, R., and M. Kárný (1984). Tracking of Slowly Varying Parameters by Directional Forgetting, in *Preprints of the 9th IFAC World Congress*, Budapest, Hungary, Vol. X, pp. 178–183.

Kulhavý, R. (1985). Restricted Exponential Forgetting in Real-Time Identification, in *Preprints of the 7th IFAC Symposium on Identification and System Parameter Estimation*, University of York, England, Vol. 2, pp. 1143–1148. A revised version published in *Automatica* 23 (1987), 589–600.

Kulhavý, R. (1986). Directional Tracking of Regression-Type Model Parameters, in *Preprints of the 2nd IFAC Workshop on Adaptive Systems in Control and Signal Processing*, Lund, Sweden, pp. 97–102.

Kullback, S., and R. Leibler (1951). On information and sufficiency, *Ann. Math. Stat.* 22, 79–87.

Ljung, L., and T. Söderström (1983). *Theory and Practice of Recursive Identification*, MIT Press, Cambridge, Mass.

Neill, J. W., and D. E. Johnson (1984). Testing for the lack of fit in regression—a review, *Commun. Stat. Theory Methods* 13, 485–511.

Neftci, S. N. (1982). Specification of economic time series models using Akaike's criterion, *J. Am. Stat. Assoc.* 77, 531–549.

Perez, A. (1984). "Barycenter" of a Set of Probability Measures and Its Application in Statistical Decision, in *Proceedings of the 6th Symposium on Computational Statistics*, Prague, Czechoslovakia, pp. 154–159.

Peterka, V. (1981). Bayesian Approach to System Identification, in *Trends and Progress in System Identification* (P. Eykhoff, ed.), Pergamon Press, Oxford.

Peterka, V. (1986). Control of uncertain processes: applied theory and algorithms, *Kybernetika* 22, supplement to numbers 3, 4, 5, 6.

Poskit, D. S. (1986). A Bayes procedure for the identification of univariate time series models, *Ann. Stat.* 14, 502–506.

Rao, C. R. (1973). *Linear Statistical Inference and Its Applications*, Wiley, New York.

Ronchetti, E. (1985). Robust model selection in regression, *Stat. Probab. Lett.* 3, 21–23.

Schwartz, L. (1965). On Bayes procedure, *Z. Wahrsch.* 4, 10–26.

Yashin, Y. (1985). Martingale Approach to Identification of Stochastic Systems, in *Preprints of the 7th IFAC Symposium on Identification and System Parameter Estimation*, University of York, England, Vol. 2, pp. 1755–1761.

Zellner, A. (1971). *An Introduction to Bayesian Inference in Econometrics*, Wiley, New York.

13

Bayesian Analysis of Regression Models with Time Series Errors

JOAQUIN DIAZ Department of Decision and Information Sciences, University of Houston, Houston, Texas

1. INTRODUCTION

A problem that has interested several authors recently is the analysis of time series when the model that generated the series could have changed during the recording of the data. In this regard, the work of Salazar (1982) and Cook (1983) study the case in which the level of the series remains constant, but the parameters of the ARMA (autoregressive moving average) model change at an unknown point in time.

In the area of intervention analysis, the point in time at which the change could have taken place is known, and if in fact there was a change, it affected the level of the series but not the noise structure. Probably the first paper related to the topic is the one by Box and Tiao (1965), in which they study the problem of analyzing the change in level of an integrated moving average process. A Bayesian analysis of the problem was included there. Box and Tiao (1975) proposed the use of difference equation models to represent the dynamic intervention effect and noise. Their analysis uses maximum likelihood procedures, and although the authors mention that the Bayesian analysis can be done after approximating the posterior density, they do not report any detailed results.

The intervention analysis approach has been used widely in different fields. As examples we can mention the papers by Stoline and Huitema (1978) in the social sciences, Shahabudin (1980) in the area of finance, and Lasarre and Tau

(1982) in the area of public safety, among others. It is important to mention here that the time series models, through the so-called state-space representation, correspond to special cases of linear dynamic system models. From this point of view, the problems described above are connected to those in the areas of control and fault detection in linear dynamic systems. These are very active research areas, and in particular the fault detection methods have been widely applied in monitoring the performance of aircraft and power plants. The papers by Willsky (1976), Willsky and Jones (1976), Isermann (1984), and Spall (1988) contain very valuable information for those interested in this subject.

In Diaz-Saiz (1985), a Bayesian procedure is proposed to analyze time series with interventions, in which a large-sample approximation to the likelihood is used. It is shown there that the procedure eliminates some of the nonlinearities and produces posterior distributions which, at least conditionally, correspond to multivariate t distributions, for suitable prior information.

In this chapter we propose the use of these procedures to analyze a linear regression model with ARMA errors described by

$$y_t = \mathbf{X}_t^T \boldsymbol{\beta} + N_t \qquad t = 1, 2, \ldots, n, \tag{1.1}$$

where

$$N_t = \frac{\theta(B)}{\phi(B)} a_t \qquad t = 1, 2, \ldots, n. \tag{1.2}$$

$\theta(B)$ and $\phi(B)$ are polynomials in the backshift operator B, and a_t, $t = 1, 2, \ldots, n$, are assumed to be independent normal random variables with mean zero and unknown variance σ^2.

Notice that the model described by (1.1) and (1.2) includes as particular cases some of the intervention models described in Box and Tiao (1975). It also includes the models used in Glass et al. (1975) in the area of analysis of time series experiments. Another important case of interest is that of regression models with autocorrelated errors (see Zellner, 1971).

In the next section we include a description of the procedure to approximate the likelihood function together with the derivation of the conditional posterior distribution of the regression parameters given the noise parameters, and the marginal posterior distributions of the noise parameters.

Although it is not difficult to derive the conditional posterior distribution of the noise parameters, we will not include it because the main purpose of this chapter is to show that at least through conditional procedures, the estimation and hypothesis testing steps usually followed in standard regression modeling can be implemented.

In Section 3 this procedure is applied to two numerical examples. The first corresponds to a series of monthly averages of the level of O_3 in the atmosphere of downtown Los Angeles that was analyzed by Box and Tiao (1975). The second is an artificially generated data set.

2. BAYESIAN ANALYSIS

In this section we describe the procedure to approximate the likelihood function, and since one is most likely interested in determining the regression variables in the model, we derive the expression for the conditional posterior distribution of the regression parameters, given the noise parameters. The use of the conditional distribution is proposed due to the existence of nonlinearities that cannot be eliminated except for some particular cases, as when the order of the autoregressive part in the noise model is equal to zero.

The model described by (1.1) and (1.2) can be written as

$$\phi(B)y_t = \phi(B)\mathbf{X}_t^T\boldsymbol{\beta} + \theta(B)a_t \tag{2.1}$$

and the residuals at time t can be given in terms of the observations $y_1, y_2, \ldots, y_t$ and the residuals $a_1, a_2, \ldots, a_{t-1}$ by the equation

$$a_t = \phi(B)y_t - \phi(B)\mathbf{X}_t^T\boldsymbol{\beta} - \sum_{i=1}^{q} \theta_i B^i a_t. \tag{2.2}$$

Now, since the Jacobian of the transformation, assuming the first p observations as fixed and the first p residuals with the value of zero, is equal to 1, the likelihood can be written as

$$\ell(\boldsymbol{\beta}, \boldsymbol{\eta}, \sigma^2|\mathbf{y}) \propto (\sigma^2)^{-(n-p)/2} \exp\left(\frac{-\sum_{t=p+1}^{n} a_t^2}{2\sigma^2}\right), \tag{2.3}$$

where $\boldsymbol{\beta}^T = (\beta_1, \beta_2, \ldots, \beta_m)$, $\boldsymbol{\eta}^T = (\boldsymbol{\theta}^T, \boldsymbol{\phi}^T)$, and a_t, $t = p+1, \ldots, n$, are given by (2.2).

To eliminate some of the nonlinearities, we propose to substitute for the unobservable variables a_t in the right-hand side of (2.2) the observable variables $\hat{a}_t$, $t = p+1, \ldots, n$, calculated recursively from (2.2) using the maximum likelihood estimators as the parameter values. This approach was used successfully by Broemeling and Shaarawy (1984, 1986), Shaarawy (1984), and Shaarawy and Broemeling (1984, 1985) to develop a complete Bayesian time series analysis procedure. See Chapter 1 of this volume for a detailed presentation of the subject.

The approximate likelihood function corresponds then to

$$\ell^*(\boldsymbol{\beta}, \boldsymbol{\eta}, \sigma^2|\mathbf{y}) \propto (\sigma^2)^{-(n-p)/2} \exp\left[-\frac{(\mathbf{Z} - U\boldsymbol{\beta})^T(\mathbf{Z} - U\boldsymbol{\beta})}{2\sigma^2}\right], \tag{2.4}$$

where the components of the vector $\mathbf{Z}$ and the rows of the matrix U are given by

$$Z_t = \phi(B)y_t - \sum_{i=1}^{q} \theta_i B^i \hat{a}_t \tag{2.5}$$

$$\mathbf{U}_t^T = \phi(B)\mathbf{X}_t^T, \qquad t = p+1, \ldots, n. \tag{2.6}$$

Notice that Z_t and $\mathbf{U}_t$, $t = p + 1, \ldots, n$. depend on the noise parameters.

For the prior information we propose the use of a multivariate normal

$$\pi(\boldsymbol{\beta}, \boldsymbol{\eta}|\sigma^2) \propto (\sigma^2)^{-(m+p+q)/2} \times \exp\left[-\frac{(\boldsymbol{\beta} - \boldsymbol{\beta}_0(\boldsymbol{\eta}))^T \Sigma_{11.2}^{-1} (\boldsymbol{\beta} - \boldsymbol{\beta}_0(\boldsymbol{\eta})) + (\boldsymbol{\eta} - \boldsymbol{\eta}_0)^T \Sigma_{22}^{-1} (\boldsymbol{\eta} - \boldsymbol{\eta}_0)}{2\sigma^2} \right]. \tag{2.7}$$

for the conditional distribution of the model parameters given σ^2, and an inverted gamma

$$\pi(\sigma^2) \propto (\sigma^2)^{-(k+1)} \exp\left(-\frac{\alpha}{2\sigma^2} \right) \tag{2.8}$$

for the marginal of σ^2. This family of priors is flexible enough to be used under a wide variety of conditions and facilitates the mathematical calculations. From (2.7) and (2.8), the joint prior distribution can be written as

$$\pi(\boldsymbol{\beta}, \boldsymbol{\eta}, \sigma^2) \propto (\sigma^2)^{-((m+p+q+2k)/2+1)} \times \exp\left[-\frac{(\boldsymbol{\beta} - \boldsymbol{\beta}_0(\boldsymbol{\eta}))^T \Sigma_{11.2}^{-1} (\boldsymbol{\beta} - \boldsymbol{\beta}_0(\boldsymbol{\eta})) + (\boldsymbol{\eta} - \boldsymbol{\eta}_0)^T \Sigma_{22}^{-1} (\boldsymbol{\eta} - \boldsymbol{\eta}_0) + \alpha}{2\sigma^2} \right]. \tag{2.9}$$

where $\sigma^2\Sigma_{11.2}$ and $\boldsymbol{\beta}_0(\boldsymbol{\eta})$ are the parameters of the conditional distribution of $\boldsymbol{\beta}$ given σ^2 and $\boldsymbol{\eta}$, and $\sigma^2\Sigma_{22}$ is the variance covariance matrix of the marginal distribution of $\boldsymbol{\eta}$ given σ^2. Also, k and α are the parameters of the marginal distribution of σ^2. Notice that the approximate likelihood, as a function of σ^2, has a form which coincides with that of (2.8). Also, (2.4) as a function of $\boldsymbol{\beta}$ can be transformed to have the form of a multivariate normal as in (2.7). Therefore, at least conditionally, (2.9) is a conjugate prior for (2.4). As for the marginal prior distribution of $\boldsymbol{\eta}$, any distribution can be used, except that if we want to work with its conditional posterior distribution, given $\boldsymbol{\beta}$, the use of a normal prior facilitates the calculations.

From (2.4) and (2.9), the joint posterior distribution of $\boldsymbol{\beta}$, $\boldsymbol{\eta}$, and σ^2 is

$$\pi(\boldsymbol{\beta}, \boldsymbol{\eta}, \sigma^2|\mathbf{y}) \propto (\sigma^2)^{-((n+m+q+2k)/2+1)} \times \exp\left[-\frac{(\mathbf{Z} - U\boldsymbol{\beta})^T(\mathbf{Z} - U\boldsymbol{\beta}) + (\boldsymbol{\beta} - \boldsymbol{\beta}_0(\boldsymbol{\eta}))^T \Sigma_{11.2}^{-1} (\boldsymbol{\beta} - \boldsymbol{\beta}_0(\boldsymbol{\eta})) + (\boldsymbol{\eta} - \boldsymbol{\eta}_0)^T \Sigma_{22}^{-1} (\boldsymbol{\eta} - \boldsymbol{\eta}_0) + \alpha}{2\sigma^2} \right]. \tag{2.10}$$

Now, after integrating out σ^2, the conditional posterior distribution of $\boldsymbol{\beta}$ given $\boldsymbol{\eta}$, and of course the prior parameters, is given by

$$\pi(\boldsymbol{\beta}|\boldsymbol{\eta}, \mathbf{y}) \propto \left[1 + \frac{(\boldsymbol{\beta} - \boldsymbol{\beta}_*)^T \Pi (\boldsymbol{\beta} - \boldsymbol{\beta}_*)}{R} \right]^{-(n+m+q+2k)/2}, \tag{2.11}$$

while the marginal posterior distribution of $\boldsymbol{\eta}$ is

$$\pi(\boldsymbol{\eta}|\mathbf{y}) \propto |\Pi|^{-1/2} R^{-(n+q+2k)/2}, \tag{2.12}$$

where

$$\Pi = U^T U + \Sigma_{11.2}^{-1}$$

$$\boldsymbol{\beta}_* = \Pi^{-1}(U^T U \boldsymbol{\beta}_e + \Sigma_{11.2}^{-1} \boldsymbol{\beta}_0(\boldsymbol{\eta})) \tag{2.13}$$

$$\boldsymbol{\beta}_e = (U^T U)^{-1} U^T \mathbf{Z} \tag{2.14}$$

$$\begin{aligned} R = {} & \boldsymbol{\beta}_e^T (U^T U) \boldsymbol{\beta}_e + \boldsymbol{\beta}_0(\boldsymbol{\eta})^T \Sigma_{11.2}^{-1} \boldsymbol{\beta}_0(\boldsymbol{\eta}) - \boldsymbol{\beta}_*^T \Pi \boldsymbol{\beta}_* \\ & + (\mathbf{Z} - U\boldsymbol{\beta}_e)^T (\mathbf{Z} - U\boldsymbol{\beta}_e) + (\boldsymbol{\eta} - \boldsymbol{\eta}_0)^T \Sigma_{22}^{-1} (\boldsymbol{\eta} - \boldsymbol{\eta}_0) + \alpha. \end{aligned} \tag{2.15}$$

Notice that while (2.11) corresponds to a multivariate t-distribution, (2.12) in general does not correspond to any standard distribution and its analysis has to be done numerically. As mentioned before, if $p = 0$, the conditional analysis is not required. A more detailed derivation of formulas (2.11) and (2.12) appears in Appendix A.

Now, if the vague prior $\pi(\boldsymbol{\beta}, \boldsymbol{\eta}, \sigma^2) \propto 1/\sigma^2$ is used, equations (2.11) and (2.12) become

$$\pi(\boldsymbol{\beta}|\boldsymbol{\eta}, \mathbf{y}) \propto \left[1 + \frac{(\boldsymbol{\beta} - \boldsymbol{\beta}_e)^T U^T U (\boldsymbol{\beta} - \boldsymbol{\beta}_e)}{R_0} \right]^{-(n-p)/2} \tag{2.16}$$

$$\pi(\boldsymbol{\eta}|\mathbf{y}) \propto |U^T U|^{-1/2} R_0^{-(n-m-p)/2} \tag{2.17}$$

with $R_0 = (\mathbf{Z} - U\boldsymbol{\beta}_e)^T (\mathbf{Z} - U\boldsymbol{\beta}_e)$ and $\boldsymbol{\beta}_e$ defined as in (2.14).

Formulas (2.11) and (2.16) can be used to obtain conditional point estimators, as well as highest posterior density (HPD) regions which can also be used for hypothesis-testing procedures on the regression parameters. The conditioning values to be used can be the maximum likelihood estimators. If there is a large set of observations for which the values of the regression variables are all equal to zero, the point estimators obtained through the time series analysis procedures given in the series of papers by Broemeling and Shaarawy, mentioned in the preceding section, can be used. Another alternative would be to calculate the mode or the expected value of the noise parameters ($\boldsymbol{\eta}$) using (2.12) or (2.17).

It is worth mentioning here that if the noise model does not include seasonal factors, the conditional posterior distribution of the noise parameters given the regression parameters can easily be obtained and corresponds to a multivariate t. An iterative procedure, using the two conditional distributions, can be implemented to obtain the joint mode for the vector of all the parameters in the model. The values for the noise parameters obtained through this procedure can be used as the conditioning values mentioned above. For a detailed derivation of the conditional distribution of the noise parameters in the case of an intervention model, see Diaz-Saiz (1985).

In the next section we apply the results obtained here to two numerical examples.

3. NUMERICAL EXAMPLES

We apply here the results in Section 2 to analyze two data sets. The first corresponds to one of the examples in Box and Tiao (1975) and the second is an artificially generated data set.

EXAMPLE 1 Box and Tiao (1975) analyzed a series of monthly averages of oxidant (O_3) level in downtown Los Angeles from January 1955 to December 1972. The reader should refer to that paper for a more detailed justification of the model used and a description of the nature of the interventions involved. We assume here that no prior information is available and will use the vague prior $\pi(\boldsymbol{\beta}, \boldsymbol{\eta}, \sigma^2) \propto 1/\sigma^2$.

The model proposed corresponds to

$$y_t = \omega_{01}\xi_{t,1} + \frac{\omega_{02}\xi_{t,2}}{1 - B^{12}} + \frac{\omega_{03}\xi_{t,3}}{1 - B^{12}} + \frac{(1 - \theta_1 B)(1 - \theta_2 B^{12})}{1 - B^{12}} a_t, \tag{3.1}$$

where

$$\xi_{t,1} = \begin{cases} 0, & t < \text{January, 1960} \\ 1, & t \geqslant \text{January, 1960} \end{cases}$$

$$\xi_{t,2} = \begin{cases} 1, & \text{“summer” months June–October beginning 1966} \\ 0, & \text{otherwise} \end{cases}$$

$$\xi_{t,3} = \begin{cases} 1, & \text{“winter” months November–May beginning 1966} \\ 0, & \text{otherwise} \end{cases}$$

and a_t, $t = 1, 2, \ldots, 216$, are assumed to be independent identically distributed normal random variables with mean zero and variance σ^2.

After multiplying both sides of (3.1) by $(1 - B^{12})$, the model corresponds to a particular case of (1.1). The conditional posterior distribution of the intervention parameters given θ_1 and θ_2 corresponds to

$$\pi(\omega_{01}, \omega_{02}, \omega_{03} | \theta_1, \theta_2, \mathbf{y}) \propto \left[1 + \frac{(\boldsymbol{\beta} - \boldsymbol{\beta}_e)^T U^T U (\boldsymbol{\beta} - \boldsymbol{\beta}_e)}{rs^2}\right]^{-(r+3)/2}, \tag{3.2}$$

where

$$\boldsymbol{\beta}^T = (\omega_{01}, \omega_{02}, \omega_{03})$$

$$U = \begin{bmatrix} \xi_{13,1} - \xi_{1,1} & \xi_{13,2} & \xi_{13,3} \\ \xi_{14,1} - \xi_{2,1} & \xi_{14,2} & \xi_{14,3} \\ \cdots & \cdots & \cdots \\ \cdots & \cdots & \cdots \\ \xi'_{n,1} - \xi_{n-12,1} & \xi_{n,2} & \xi_{n,3} \end{bmatrix}$$

$$U^T U = \begin{bmatrix} 12 & 0 & 0 \\ 0 & 35 & 0 \\ 0 & 0 & 45 \end{bmatrix}$$

$$Z_t = y_t - y_{t-12} + \theta_1 \hat{a}_{t-1} + \theta_2 \hat{a}_{t-12} - \theta_1 \theta_2 \hat{a}_{t-13}, \qquad t = 13, \ldots, n$$

$$\boldsymbol{\beta}_e = (U^T U)^{-1} U^T \mathbf{Z}$$

$$s^2 = \frac{(\mathbf{Z} - U\boldsymbol{\beta}_e)^T(\mathbf{Z} - U\boldsymbol{\beta}_e)}{r} \tag{3.3}$$

and $r = n - 15$.

For the data taken from the SAS/ETS User's Guide (page 174 of Version 5 Edition), $\boldsymbol{\beta}_e^T = (-1.1297, \ -0.2320, \ -0.06456)$, $r = 201$, and $s^2 = 0.69015$, if the conditional maximum likelihood estimators for θ_1 and θ_2, -0.2998 and 0.5923, respectively, are used as conditioning values.

The conditional point estimator for $\boldsymbol{\beta}$ is given by $\boldsymbol{\beta}_e$ if a quadratic loss function is considered. Also, from (3.2) HPD regions can be obtained and used to test some hypotheses of interest. For example, to test the null hypothesis that $\omega_{03} = 0$ (i.e., to test whether there is any effect of the winter intervention) we can use the fact that conditionally, $(\omega_{03} + 0.06456)/0.11867$ has a t distribution with 211 degrees of freedom, and that the 95% HPD region corresponds to $-0.2972 < \omega_{03} < 0.16801$.

Since the value of zero is included in the region, we can conclude that there was no effect of the winter intervention. We checked the sensitivity of the test for changes in the conditioning values and, at least for the values in a region of probability close to 1, calculated from the marginal posterior distribution

$$\pi(\theta_1, \theta_2) \propto [rs^2]^{-r/2}$$

with r and s^2 as in (3.3), the results of the test did not change.

Marginal expectations were obtained, using numerical integration procedures, and produced the following results:

$$E[\omega_{03}] = -1.1343$$
$$E[\omega_{03}] = -0.2309$$
$$E[\omega_{03}] = -0.0667$$
$$E[\theta_1] = -0.2725$$
$$E[\theta_1] = 0.5824,$$

which do not differ much from the conditional maximum likelihood estimators. Other parameters of interest were calculated and reported by Diaz-Saiz (1985).

EXAMPLE 2 In this case we generated $n = 200$ observations with the model

$$y_t = 3X_{1,t} - X_{2,t} + \frac{1 - 0.25B}{1 + 0.9B} a_t,$$

where

$$X_{1,t} = \begin{cases} 0, & t \leqslant 100 \\ 1, & \text{otherwise} \end{cases}$$
$$X_{2,t} = \begin{cases} 0, & t \leqslant 150 \\ 1, & \text{otherwise} \end{cases}$$

and a_t, $t = 1, 2, \ldots, 200$, distributed normally with mean zero and variance $\sigma^2 = \frac{1}{4}$.

Naturally, the model used in the analysis was

$$y_t = \beta_1 X_{1,t} + \beta_2 X_{2,t} + \frac{1 + \theta B}{1 + \phi B} a_t$$

and the parameters of the conditional posterior distribution (2.14), using the maximum likelihood estimators, based on the first 100 observations, of θ and ϕ ($\theta = -0.27$ and $\phi = 0.88$) as the conditioning values, correspond to

$$U^T U = \begin{bmatrix} 352.57 & 175.89 \\ 175.89 & 175.01 \end{bmatrix},$$

$\boldsymbol{\beta}_e^T = (3.04, -1.02)$, and $R_0 = 50.56$. The number of degrees of freedom for the multivariate t distribution is equal to $n - 3 = 197$.

Notice that the quadratic form defined by

$$Q(\boldsymbol{\beta}) = \frac{(\boldsymbol{\beta} - \boldsymbol{\beta}_e)^T U^T U (\boldsymbol{\beta} - \boldsymbol{\beta}_e)}{2R_0/197}$$

has an F distribution with 2 and 197 degrees of freedom (see Box and Tiao, 1973, p. 117), and the 95% HPD region for $\boldsymbol{\beta}$ given by $Q(\boldsymbol{\beta}) \leqslant 1.39$ does not contain the vector $\boldsymbol{\beta} = \mathbf{0}$. Therefore, the hypothesis that the two regression parameters are equal to zero can be rejected.

The marginal HPD regions for β_1 and β_2 correspond to the intervals [2.99; 3.12] and [−1.13; −0.91], respectively, and the value of zero is not included in either one. We can conclude in this case that the two parameters are different from zero.

4. COMMENTS AND CONCLUSIONS

We have proposed an approximate conditional Bayesian procedure to analyze a class of models that includes very useful cases in terms of applications. The implementation of the procedure is not difficult and can be done using the SAS statistical packages. A FORTRAN program to produce automatically results like the ones described in the examples is being developed.

It is important to stress the fact that since conditional distributions are used, a sensitivity analysis should be performed before jumping to conclusions about the behavior of the regression parameters.

An important problem that can be solved using a procedure similar to the one presented here is the prediction of a value of y. The predictive distribution is not difficult to derive, but it will be reported elsewhere.

ACKNOWLEDGMENT

The author would like to thank a referee for valuable comments and suggestions leading to the improved presentation and correctness of this chapter.

APPENDIX: MATHEMATICAL DERIVATIONS

The purpose of this appendix is to give the definition of the distribution used for the prior (2.8) of σ^2, and to show the calculations that led to equations (2.11) and (2.12). Derivations like these are common in the area of Bayesian analysis of linear models, and the reader should refer to Broemeling (1985) for a more detailed coverage of this and other related topics.

First we define what we mean by an inverse gamma distribution (see Berger, 1985, p. 561) as follows:

DEFINITION A random variable X has an inverse gamma distribution if and only if its probability density function is given by

$$g(x) = \lambda^r(\Gamma(r))^{-1}x^{-(r+1)}\exp\left(-\frac{\lambda}{x}\right) \qquad (x > 0) \tag{A.1}$$

with $\lambda > 0$ and $r > 0$.

Notice that (2.8) corresponds to (A.1) with $r = k$, $\lambda = \alpha/2$, and $x = \sigma^2$.

LEMMA A.1

$$\int_0^\infty x^{-(r+1)}\exp\left(-\frac{\lambda}{x}\right)dx = \lambda^{-r}\Gamma(r) \tag{A.2}$$

Proof It is trivial due to the fact that (A.1) is a density function and its integral from zero to infinity is equal to 1.

LEMMA A.2 Equation (2.10) can be written as

$$\pi(\boldsymbol{\beta}, \boldsymbol{\eta}, \sigma^2|\mathbf{y}) \propto (\sigma^2)^{-((n+m+q+2k)/2+1)} \times \exp\left[-\frac{R + (\boldsymbol{\beta} - \boldsymbol{\beta}_*)^T\Pi(\boldsymbol{\beta} - \boldsymbol{\beta}_*)}{2\sigma^2}\right] \tag{A.3}$$

with R, Π, and $\boldsymbol{\beta}_*$ as in (2.13) and (2.15).

Proof: The proof is not difficult but requires a series of manipulations which we prefer not to include here.

Now we are ready to derive equations (2.11) and (2.12) from (2.10).

THEOREM A.1 The conditional posterior distribution of $\boldsymbol{\beta}$ given $\boldsymbol{\eta}$ and $\mathbf{y}$ is given by (2.11).

Proof: From (A.3) the conditional posterior distribution of $\boldsymbol{\beta}$ and σ^2, given $\boldsymbol{\eta}$ and $\mathbf{y}$, corresponds to

$$\pi(\boldsymbol{\beta}, \sigma^2|\boldsymbol{\eta}, \mathbf{y}) \propto (\sigma^2)^{-((n+m+q+2k)/2+1)} \times \exp\left[-\frac{R + (\boldsymbol{\beta} - \boldsymbol{\beta}_*)^T\Pi(\boldsymbol{\beta} - \boldsymbol{\beta}_*)}{2\sigma^2}\right]. \tag{A.4}$$

Now since (A.4) as a function of σ^2 corresponds to the integrand in (A.2) with $r = (n + m + q + 2k)/2$ and $\lambda = [R + (\boldsymbol{\beta} - \boldsymbol{\beta}_*)^T\Pi(\boldsymbol{\beta} - \boldsymbol{\beta}_*)]/2$, it is easy to see that integrating out σ^2, the conditional distribution of $\boldsymbol{\beta}$ given $\boldsymbol{\eta}$ and $\mathbf{y}$ is given by

$$\pi(\boldsymbol{\beta}|\boldsymbol{\eta}, \mathbf{y}) \propto [R + (\boldsymbol{\beta} - \boldsymbol{\beta}_*)^T \Pi (\boldsymbol{\beta} - \boldsymbol{\beta}_*)]^{-(n+m+q+2k)/2}, \tag{A.5}$$

which can be written as

$$\pi(\boldsymbol{\beta}|\boldsymbol{\eta}, \mathbf{y}) \propto \left[1 + \frac{(\boldsymbol{\beta} - \boldsymbol{\beta}_*)^T \Pi (\boldsymbol{\beta} - \boldsymbol{\beta}_*)}{R}\right]^{-(n+m+q+2k)/2}. \tag{A.6}$$

Now, (A.6) coincides with (2.11).

THEOREM A.2 The posterior distribution of $\boldsymbol{\eta}$ given $\mathbf{y}$ corresponds to (2.12).

Proof: As function of $\boldsymbol{\beta}$, (A.3) is proportional to a normal density function. We can easily integrate out $\boldsymbol{\beta}$ and obtain

$$\pi(\boldsymbol{\eta}, \sigma^2|\mathbf{y}) \propto |\Pi|^{-1/2} (\sigma^2)^{-((n+q+2k)/2+1)} \exp\left(-\frac{R}{2\sigma^2}\right). \tag{A.7}$$

Now, using (A.2) to integrate out σ^2, we have that

$$\pi(\boldsymbol{\eta}|\mathbf{y}) \propto |\Pi|^{-1/2} R^{-(n+q+2k)/2}, \tag{A.8}$$

which corresponds to (2.12).

Equations (2.16) and (2.17) are obtained using basically the same procedures.

REFERENCES

Berger, J. O. (1985). *Statistical Decision Theory and Bayesian Analysis, 2nd ed.*, Springer-Verlag, New York.

Box, G. E. P., and G. C. Tiao (1965). A change in level of a non-stationary time series, *Biometrika* 52, 181–192.

Box, G. E. P., and G. C. Tiao (1973). *Bayesian Inference in Statistical Analysis*, Addison-Wesley, Reading, Mass.

Box, G. E. P., and G. C. Tiao (1975). Intervention analysis with applications to economic and environmental problems, *J. Am. Stat. Assoc.* 70, 70–79.

Broemeling, L. D. (1985). *Bayesian Analysis of Linear Models*, Marcel Dekker, New York.

Broemeling, L. D., and S. M. Shaarawy (1984). Bayesian identification of time series, unpublished manuscript.

Broemeling, L. D., and S. M. Shaarawy (1986). A Bayesian Analysis of Time Series, in *Bayesian Inference and Decision Techniques with Applications:*

Essays in Honor of Bruno de Finetti (P. Goel and A. Zellner, eds.), Elsevier, New York, pp.337–354.

Cook, P. J. (1983). A Bayesian analysis of autoregressive processes: time and frequency domain, Ph.D. dissertation, Oklahoma State University.

Diaz-Saiz, J. (1985). Time series analysis with interventions: a Bayesian approach, Ph.D. dissertation, Oklahoma State University.

Glass, G. V., V. L. Willson, and J. M. Gottman (1975). *Design and Analysis of Time Series Experiments*, Colorado Associated University Press, Boulder, Colo.

Isermann, R. (1984). Process fault detection based on modeling and estimation—a survey, *Automatica* 20, 387–404.

Lasarre, S., and S. H. Tau (1982). Evaluation of Safety Measures on the Frequency and Gravity of Traffic Accidents in France by Means of Intervention Analysis, in *Time Series Analysis: Theory and Practice 1* (O. D. Anderson, ed.), North-Holland, Amsterdam.

Salazar, D. (1982). Structural changes in time series models, *J. Econometrics* 19, 147–163.

Shaarawy, S. M. (1984). Bayesian analysis of moving average processes, Ph.D. dissertation, Oklahoma State University.

Shaarawy, S. M., and L. D. Broemeling (1984). Bayesian inferences and forecasts with moving average processes, *Commun. Stat. Theory Methods* 13, 1871–1888.

Shaarawy, S. M., and L. D. Broemeling (1985). Bayesian identification of multiple ARMA models, unpublished manuscript.

Shahabudin, S. (1980). Analysis of Stock Prices Using Time Series with Intervention Analysis, in *Proceedings of the American Statistical Association, Business and Economic Statistics Section*, pp. 311–314.

Spall, J. C. (1988). Bayesian Error Isolation for Models of Large-Scale Systems, *IEEE Trans. Autom. Control* AC-33, 341–347.

Stoline, M. R., and B. E. Huitema (1978). Aspects of the Intervention Time Series Analysis of the Pilot Programs of Michigan Public Act 339: An Act Decriminalizing Drunkenness, in *Proceedings of the American Statistical Association, Social Statistics Section*, pp. 397–402.

Willsky, A. S. (1976). A survey of design methods for failure detection in dynamic systems, *Automatica* 12, 601–611.

Willsky, A. S., and H. L. Jones (1976). A generalized likelihood ratio approach to the detection and estimation of jumps in linear systems, *IEEE Trans. Autom. Control* AC-21, 108–112.

Zellner, A. (1971). *An Introduction to Bayesian Inference in Econometrics*, Wiley, New York.

14

Modeling and Monitoring Discontinuous Changes in Time Series

K. GORDON* and ADRIAN F. M. SMITH Department of Mathematics, University of Nottingham, Nottingham, England

1. INTRODUCTION

In this chapter we examine the use of the linear dynamic model (LDM) for modeling time series which are subject to various kinds of abrupt change in pattern. Throughout, we emphasize the problem of on-line monitoring of such time series, both to detect changes and to identify the particular kind of change that has occurred. Recursive Bayesian updating procedures are developed, which extend the standard Kalman filter techniques to deal with the multiprocess models used to represent the various kinds of potential change in the series.

The following three time series typify the kinds of problems we shall consider. Figure 1 shows a series of blood creatinine measurements from a kidney transplant recipient (Smith and West, 1983). There are physiological reasons for believing that the series behaves rather like a sequence of reversing linear trends, but with additional perturbations in the form of outliers or changes in level of the series. Figure 2 consists of 4-hourly measurements of peak expiratory flow rate (a measure of airway capacity) from an asthmatic patient (Gordon, 1986). In this case the underlying behavior seems to be rhythmic in nature, but with perturbations to these

*Present address: Mathematical Applications, CIBA-GEIGY AG, Basel, Switzerland.

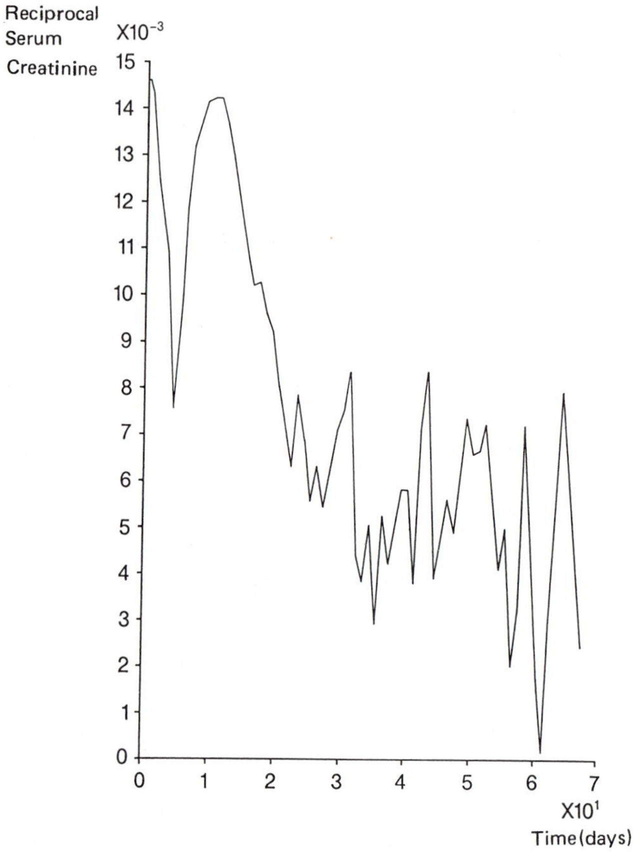

Figure 1. Daily blood creatinine measurements from a kidney transplant recipient.

stable characteristics taking the form of changes of level and amplitude of the cyclical variation. Figure 3 shows a simulated time series of 100 observations, Gordon (1986), for which the typical behavior is autoregressive but with several change-points induced in the series.

Clearly, in practice there are many possible forms of unperturbed underlying behavior for time series, together with many different forms of potentially interesting changes in pattern. We require, therefore, a flexible modeling framework that can incorporate a wide variety of such patterns and also allow for the possibility of various discontinuities such as those presented in Figures 1 to 3.

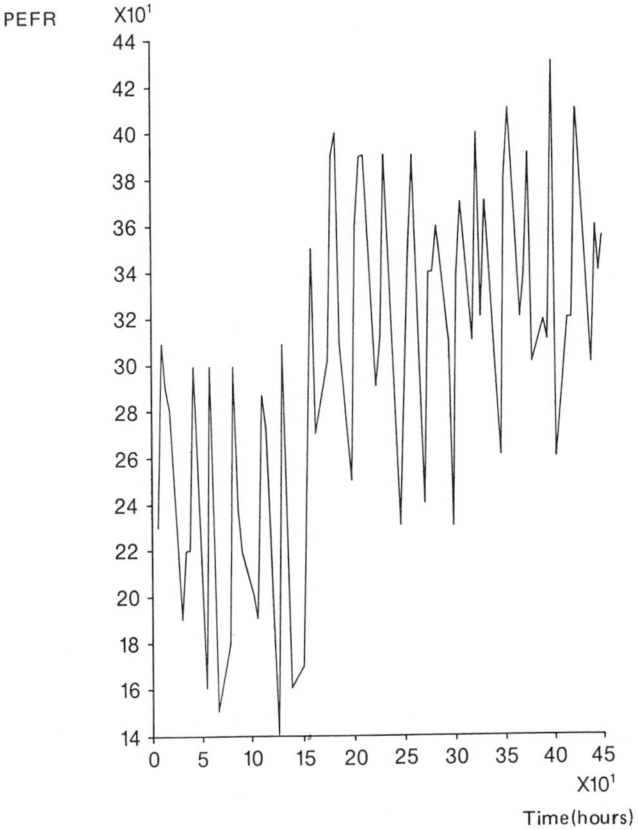

Figure 2. Four-hourly measurements of peak expiratory flow rate from an asthmatic patient.

In Section 2 we examine a modeling and recursive learning framework based around linear dynamic models and present a number of special cases. In Section 3 we describe the extension of this framework to a multistate structure well suited to change-point detection. In Section 4 we provide an illustration of the monitoring procedure.

The methods we present in this chapter are developed within the unified framework of the linear dynamic model with multiprocess Kalman filter updating motivated and utilized from a Bayesian perspective. However, the problem of detecting abrupt changes in system behavior has been studied extensively in the control theory literature and a number of alternative approaches and "fault detection" algorithms have been developed. The

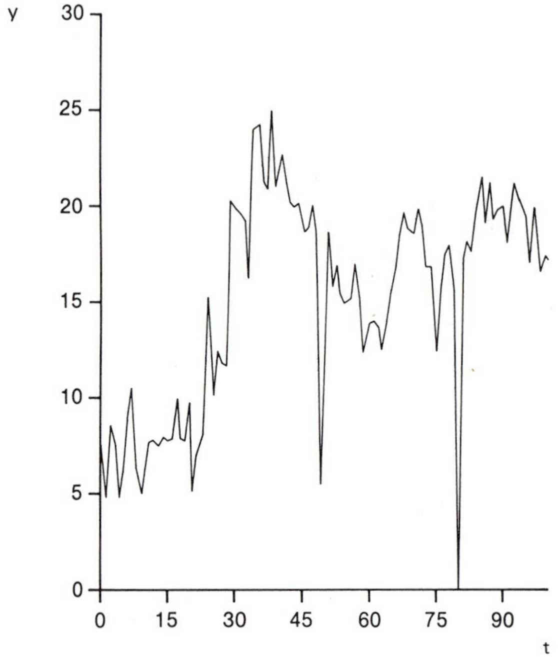

Figure 3. Simulated autoregressive series with superimposed discontinuities.

survey articles by Willsky (1976), Isermann (1984), and Kerr (1983) contain substantial bibliographies relating to a wide range of procedures and further recent results are reported in Basseville and Benveniste (1986). Most of this literature is non-Bayesian, but a specifically Bayesian approach is adopted in Chow and Willsky (1984). A systematic comparison of alternative methods is beyond the scope of this chapter, however, and we shall confine ourselves here to presenting our version of the Bayesian methodology.

2. LINEAR DYNAMIC MODEL

2.1 Basic Model

All our developments are based on the (state space) general linear dynamic model

$$\mathbf{y}_t = \mathbf{H}_t\boldsymbol{\theta}_t + \boldsymbol{\varepsilon}_t \tag{2.1}$$

$$\boldsymbol{\theta}_t = \mathbf{G}_t\boldsymbol{\theta}_{t-1} + \boldsymbol{\omega}_t, \tag{2.2}$$

where

$$\begin{aligned} \mathbf{y}_t &= \text{vector of observations made at time } t \\ \boldsymbol{\theta}_t &= \text{vector of system parameters at time } t \\ \mathbf{H}_t &= \text{known regression matrix at time } t \\ \mathbf{G}_t &= \text{known transition matrix at time } t \\ \boldsymbol{\varepsilon}_t, \boldsymbol{\omega}_t &= \text{zero-mean, random vectors at time } t. \end{aligned} \tag{2.3}$$

For the basic LDM we make the following assumptions about the quantities outlined in (2.3):

(i) $\boldsymbol{\varepsilon}_t$ is independent of $\boldsymbol{\varepsilon}_s$, $\forall s \neq t$

(ii) $\boldsymbol{\varepsilon}_t$, $\boldsymbol{\omega}_t$ are independent of $\boldsymbol{\theta}_{t-1}$, $\forall t$, given the past observations $y_1, \ldots, y_{t-1}$ (denoted by $\mathbf{D}_{t-1}$) (2.4)

(iii) $\boldsymbol{\varepsilon}_t$ is independent of $\boldsymbol{\omega}_t$, $\forall t$

(iv) $\boldsymbol{\varepsilon}_t \sim N(\mathbf{O}, \mathbf{E}_t)$; $\boldsymbol{\omega}_t \sim N(\mathbf{O}, \mathbf{W}_t)$.

The LDM described above has two components: (2.1) is the *observation* equation, describing the measuring process which relates the current system parameters to the resulting observations; (2.2) is the *system* (or *state*) equation, describing the Markovian process by which the system parameters evolve in time.

There have been a number of papers describing various extensions of this standard LDM framework. For example, Sorenson and Alspach (1971), Masreliez (1975), and West (1981, 1982) discuss robustification of the algorithm by dropping assumption (iv) above and West et al. (1985) go on to describe nonlinear versions of such models, thus formulating a framework for *general* dynamic models. We work with the standard normality assumptions in this chapter, but the methods we describe could certainly be generalized using the kinds of techniques discussed by Masreliez and West.

2.2 Recursive Estimation

If the assumptions of (2.4) are maintained and if it is further assumed that, initially,

$$\boldsymbol{\theta}_0 \sim N(\mathbf{m}_0, \mathbf{C}_0), \tag{2.5}$$

then at time $t - 1$ the distribution of $\boldsymbol{\theta}_{t-1}$, given $\mathbf{D}_{t-1}$, the data up to that point, is described by

$$(\boldsymbol{\theta}_{t-1}|\mathbf{D}_{t-1}) \sim N(\mathbf{m}_{t-1}, \mathbf{C}_{t-1}), \tag{2.6}$$

where the values of $\mathbf{m}$ and $\mathbf{C}$ are obtained recursively from the Kalman filter algorithms (Kalman and Bucy, 1961) as described below.

Following the notation of (2.3), let

$$\begin{aligned} \mathbf{f}_t &= \mathbf{H}_t\mathbf{G}_t\mathbf{m}_{t-1} \\ \mathbf{e}_t &= \mathbf{y}_t - \mathbf{f}_t \\ \mathbf{P}_t &= \mathbf{G}_t\mathbf{C}_{t-1}\mathbf{G}_t^T + \mathbf{W}_t \\ \mathbf{F}_t &= \mathbf{H}_t\mathbf{P}_t\mathbf{H}_t^T + \mathbf{E}_t \\ \mathbf{S}_t &= \mathbf{P}_t\mathbf{H}_t^T(\mathbf{F}_t)^{-1}; \end{aligned} \tag{2.7}$$

then

$$\mathbf{m}_t = \mathbf{G}_t\mathbf{m}_{t-1} + \mathbf{S}_t\mathbf{e}_t \tag{2.8}$$

$$\mathbf{C}_t = \mathbf{P}_t - \mathbf{S}_t\mathbf{F}_t\mathbf{S}_t^T. \tag{2.9}$$

To facilitate the interpretation of these quantities, note, for example, that for time t, $\mathbf{f}_t$ is the one-step-ahead predictor, $\mathbf{e}_t$ the corresponding prediction error, $\mathbf{P}_t$ the predictive variance of the system parameter $\boldsymbol{\theta}_t$, and $\mathbf{F}_t$ the predictive variance of the observation $\mathbf{y}_t$.

To calculate the quantities involved, one more assumption has been made: namely, that both the observation and system variances are known for each time t. A number of approaches to the problem of unknown variances in dynamic linear models have been suggested. Harrison and Stevens (1976) suggested the use of discrete-valued grids, covering a range likely to include plausible values for the variances (an approach related to the ideas of Sorenson and Alspach, 1971), but the "inefficiency" of this approach has been criticized by Stoodley and Mirnia (1979). Goodrich and Caines (1979) and Tsang et al. (1981) suggest more efficient procedures for on-line variance updating, as does West (1982), whose procedure, based on the idea of a joint conjugate prior distribution for the normal distribution with unknown mean and variance (see, e.g., DeGroot, 1970), is the one we adopt in this chapter.

We assume that $\operatorname{var}(\boldsymbol{\varepsilon}_t) = \mathbf{E}_t = \sigma^2\mathbf{R}_\varepsilon$, $\operatorname{var}(\boldsymbol{\omega}_t) = \mathbf{W}_t = \sigma^2\mathbf{R}_\omega$, and that

$$\boldsymbol{\theta}_0 \sim N(\mathbf{m}_0, \sigma^2\mathbf{C}_0) \tag{2.10}$$

$$(\boldsymbol{\theta}_{t-1}|\mathbf{D}_{t-1}) \sim N(\mathbf{m}_{t-1}, \sigma^2\mathbf{C}_{t-1}) \tag{2.11}$$

[replacing (2.5) and (2.6), respectively].

If σ^2 is known, an obvious modification of the recursion described by (2.7)–(2.9) above applies. However, when σ^2 is unknown, we can use the following procedure, due to West (1982). Let $\lambda^{-1} = \sigma^2$. We replace (2.11) by

$$(\boldsymbol{\theta}_{t-1}|\mathbf{D}_{t-1}, \lambda) \sim N(\mathbf{m}_{t-1}, \lambda^{-1}\mathbf{C}_{t-1}) \tag{2.12}$$

and

$$(\lambda|\mathbf{D}_{t-1}) \sim G(\tfrac{1}{2}n_{t-1}, \tfrac{1}{2}r_{t-1}) \tag{2.13}$$

where $G(\alpha, \beta)$ denotes the gamma distribution. We also assume that, initially,

$$\lambda \sim G(\tfrac{1}{2}n_0, \tfrac{1}{2}r_0). \tag{2.14}$$

Standard Bayesian analysis (see, e.g., DeGroot, 1970) shows that

$$(\boldsymbol{\theta}_t|\mathbf{D}_t, \lambda) \sim N(\mathbf{m}_t, \lambda^{-1}\mathbf{C}_t) \tag{2.15}$$

and

$$(\lambda|\mathbf{D}_t) \sim G(\tfrac{1}{2}n_t, \tfrac{1}{2}r_t) \tag{2.16}$$

where $\mathbf{m}_t$ and $\mathbf{C}_t$ are defined by (2.8) and (2.9), and where

$$n_t = n_{t-1} + 1 = n_0 + t \tag{2.17}$$

$$r_t = r_{t-1} + \mathbf{e}_t^T(\mathbf{F}_t)^{-1}\mathbf{e}_t \tag{2.18}$$

with $\mathbf{e}_t$ and $\mathbf{F}_t$ as defined by (2.7). This procedure has been demonstrated (Smith and West, 1983; Trimble et al., 1983) to be an improvement on the method of Harrison and Stevens (1976).

Although the general framework of a linear dynamic model incorporates the possibility of multiple observations, we shall restrict ourselves for the present to the case of a univariate observation, y_t. In this case it can easily be seen that

$$(y_t|\mathbf{D}_{t-1}, \lambda) \sim N(f_t, \lambda^{-1}F_t), \tag{2.19}$$

where f_t and F_t are defined in (2.7). It follows that

$$p(y_t|\mathbf{D}_{t-1}) = \int_0^\infty p(y_t|\mathbf{D}_{t-1}, \lambda)p(\lambda|\mathbf{D}_{t-1})\,\mathrm{d}\lambda,$$

and after some routine manipulation, this yields

$$p(y_t|\mathbf{D}_{t-1}) \propto F_t^{-1/2} r_{t-1}^{(1/2)n_t - 1} r_t^{-(1/2)n_t}, \tag{2.20}$$

which has the form of a t-density (see, e.g., Aitchison and Dunsmore, 1975). This predictive density is useful not only for providing predictions but also in the derivation of multiprocess probabilities (see Section 3.3).

2.3 Nuisance Parameters

When the model depends upon either one or more nuisance parameters, $\boldsymbol{\phi}$, we adopt the following procedures. First, we assign a probability distribution to $\boldsymbol{\phi}$ over a sensibly defined range using a discrete-valued grid. For instance, if the system model is a first-order autoregressive process and the autoregressive parameter is considered as a nuisance parameter, a suitable range would be $(-1, 1)$ and some kind of distribution would be specified over this range. In this case we have

$$(\boldsymbol{\theta}_{t-1}|\mathbf{D}_{t-1}, \lambda, \boldsymbol{\phi}) \sim N(\mathbf{m}_{t-1}, \lambda^{-1}\mathbf{C}_{t-1}) \tag{2.21}$$

$$(\lambda|\mathbf{D}_{t-1}, \boldsymbol{\phi}) \sim G(\tfrac{1}{2}n_{t-1}, \tfrac{1}{2}r_{t-1}), \tag{2.22}$$

where, now, $\mathbf{m}_{t-1}$, $\mathbf{C}_{t-1}$ and r_{t-1} are dependent on $\boldsymbol{\phi}$ and will typically have different values at each point on the $\boldsymbol{\phi}$-grid. Therefore, replacing (2.15) and (2.16), we have

$$(\boldsymbol{\theta}_t|\mathbf{D}_t, \lambda, \boldsymbol{\phi}) \sim N(\mathbf{m}_t, \lambda^{-1}\mathbf{C}_t) \tag{2.23}$$

$$(\lambda|\mathbf{D}_t, \boldsymbol{\phi}) \sim G(\tfrac{1}{2}n_t, \tfrac{1}{2}r_t) \tag{2.24}$$

where $\mathbf{m}_t$ and $\mathbf{C}_t$ are defined by (2.8) and (2.9); and n_t and r_t are defined by (2.17) and (2.18), except that each of the quantities in (2.7) now depends on $\boldsymbol{\phi}$.

Let the probability distribution of $\boldsymbol{\phi}$ at time $t-1$, over a suitable grid $\boldsymbol{\Phi}$, be represented by

$$p(\boldsymbol{\phi}|\mathbf{D}_{t-1}) \approx K_{t-1}(\boldsymbol{\phi}). \tag{2.25}$$

Then

$$p(\boldsymbol{\theta}_t|\mathbf{D}_t, \lambda) \approx \sum_{\boldsymbol{\Phi}} p(\boldsymbol{\theta}_t|\mathbf{D}_t, \lambda, \boldsymbol{\phi})p(\boldsymbol{\phi}|\mathbf{D}_t, \lambda)$$

$$= \sum_{\boldsymbol{\Phi}} p(\boldsymbol{\theta}_t|\mathbf{D}_t, \lambda, \boldsymbol{\phi})K_t(\boldsymbol{\phi}), \tag{2.26}$$

where the first term in the summation is defined by (2.23); this replaces (2.15) when $\boldsymbol{\phi}$ is present. Also,

$$p(\lambda|\mathbf{D}_t) = \sum_{\boldsymbol{\Phi}} p(\lambda|\mathbf{D}_t, \boldsymbol{\phi})p(\boldsymbol{\phi}|\mathbf{D}_t)$$

$$= \sum_{\boldsymbol{\Phi}} p(\lambda|\mathbf{D}_t, \boldsymbol{\phi})K_t(\boldsymbol{\phi}), \tag{2.27}$$

where the first term in the summation is defined by (2.24); this replaces (2.16) when $\boldsymbol{\phi}$ is present. Moreover,

$$p(y_t|\mathbf{D}_{t-1}) = \sum_{\boldsymbol{\Phi}} p(y_t|\mathbf{D}_{t-1}, \boldsymbol{\phi})p(\boldsymbol{\phi}|\mathbf{D}_{t-1})$$

$$= \sum_{\boldsymbol{\Phi}} p(y_t|\mathbf{D}_{t-1}, \boldsymbol{\phi})\mathbf{K}_{t-1}(\boldsymbol{\phi}), \tag{2.28}$$

where the first term in the summation is defined by (2.20) for each grid point; this replaces (2.20) when $\boldsymbol{\phi}$ is present.

Note that (2.26) represents a mixture of normal distributions. We use the minimum Kullback-Leibler divergence criterion (Kullback and Leibler, 1951) to approximate (2.26) with a single normal distribution having mean vector and covariance matrix

$$\mathbf{m}_t = \sum_{\boldsymbol{\Phi}} \mathbf{m}_t(\boldsymbol{\phi}) K_t(\boldsymbol{\phi}) \tag{2.29}$$

$$\mathbf{C}_t = \sum_{\boldsymbol{\Phi}} \{\mathbf{C}_t(\boldsymbol{\phi}) + (\mathbf{m}_t(\boldsymbol{\phi}) - \mathbf{m}_t)(\mathbf{m}_t(\boldsymbol{\phi}) - \mathbf{m}_t)^T\} K_t(\boldsymbol{\phi}), \tag{2.30}$$

respectively, where $\mathbf{m}_t(\boldsymbol{\phi})$, $\mathbf{C}_t(\boldsymbol{\phi})$ denote the means and covariances conditional on the $\boldsymbol{\phi}$-grid points. Similarly, (2.27) is a mixture of gamma distributions which we replace by a single gamma density with parameters n_t and r_t, where

$$(r_t)^{-1} = \sum_{\boldsymbol{\Phi}} (r_t(\boldsymbol{\phi}))^{-1} K_t(\boldsymbol{\phi}), \tag{2.31}$$

again derived by minimizing the Kullback-Leibler divergence (see West, 1982, for details).

To calculate the quantities in (2.29)–(2.31), we need to update the grid weights iteratively, so that

$$K_t(\boldsymbol{\phi}) = p(\boldsymbol{\phi}|\mathbf{D}_t) = \frac{p(y_t|\mathbf{D}_{t-1}, \boldsymbol{\phi})p(\boldsymbol{\phi}|\mathbf{D}_{t-1})}{\sum_{\boldsymbol{\Phi}} p(y_t|\mathbf{D}_{t-1}, \boldsymbol{\phi})p(\boldsymbol{\phi}|\mathbf{D}_{t-1})}$$

(using Bayes' theorem), that is,

$$K_t(\boldsymbol{\phi}) = \frac{p(y_t|\mathbf{D}_{t-1}, \boldsymbol{\phi})K_{t-1}(\boldsymbol{\phi})}{\sum_{\boldsymbol{\Phi}} p(y_t|\mathbf{D}_{t-1}, \boldsymbol{\phi})K_{t-1}(\boldsymbol{\phi})}, \tag{2.32}$$

where $p(y_t|\mathbf{D}_{t-1}, \boldsymbol{\phi})$ is calculated from (2.20) for each grid point in $\boldsymbol{\Phi}$.

2.4 Examples of LDMs

Linear growth

For series whose underlying behavior consists of straight-line trends, such as that shown in Figure 1, the linear growth model may be appropriate. This model can be written as

$$\begin{aligned} y_t &= \mu_t + \varepsilon_t \\ \mu_t &= \mu_{t-1} + \beta_t + \delta\mu_t \\ \beta_t &= \beta_{t-1} + \delta\beta_t, \end{aligned} \tag{2.33}$$

where μ_t is usually interpreted as the system level at time t and β_t is the incremental growth (i.e., slope) at time t. Furthermore, it is assumed that

$$\varepsilon_t \sim N(0, \lambda^{-1}R_\varepsilon), \qquad \delta\mu_t \sim N(0, \lambda^{-1}R_\mu), \qquad \delta\beta_t \sim N(0, \lambda^{-1}R_\beta),$$

and that these perturbations are independent of one another. It is important to note that although no suffix t has been attached to the variances, they *are* assumed to be time dependent.

Using the LDM representation, we may write

$$y_t = [1 \quad 0]\begin{pmatrix}\mu_t\\ \beta_t\end{pmatrix} + \varepsilon_t \tag{2.34}$$

$$\begin{pmatrix}\mu_t\\ \beta_t\end{pmatrix} = \begin{pmatrix}1 & 1\\ 0 & 1\end{pmatrix}\begin{pmatrix}\mu_{t-1}\\ \beta_{t-1}\end{pmatrix} + \begin{pmatrix}\delta\mu_t + \delta\beta_t\\ \delta\beta_t\end{pmatrix}, \tag{2.35}$$

that is,

$$\mathbf{H}_t = \mathbf{H} = [1 \quad 0], \qquad \boldsymbol{\theta}_t = \begin{pmatrix}\mu_t\\ \beta_t\end{pmatrix}, \qquad \mathbf{G}_t = \mathbf{G} = \begin{pmatrix}1 & 1\\ 0 & 1\end{pmatrix},$$

$$\boldsymbol{\omega}_t = \begin{pmatrix}\delta\mu_t + \delta\beta_t\\ \delta\beta_t\end{pmatrix} \tag{2.36}$$

and

$$\mathbf{R}_\varepsilon = R_\varepsilon, \qquad \mathbf{R}_\omega = \begin{pmatrix}R_\mu + R_\beta & R_\beta\\ R_\beta & R_\beta\end{pmatrix}. \tag{2.37}$$

The procedure for updating λ is defined by (2.14) and (2.16) to (2.18).

A Sinusoidal model

In order to capture rhythmic characteristics, such as those exhibited by the series in Figure 2, we consider a simple cyclical model, the sinusoidal waveform:

$$\begin{aligned} y_t &= \mu_t + c_t\alpha_t + \varepsilon_t\\ \mu_t &= \mu_{t-1} + \delta\mu_t\\ \alpha_t &= \alpha_{t-1} + \delta\alpha_t, \end{aligned} \tag{2.38}$$

where $c_t = \cos(2\pi\omega t + \phi)$, $\omega =$ the rhythm frequency, $\phi =$ the phase, $\varepsilon_t \sim N(0, \lambda^{-1}R_\varepsilon)$, $\delta\mu_t \sim N(0, \lambda^{-1}R_\mu)$, $\delta\alpha_t \sim N(0, \lambda^{-1}R_\alpha)$, and where μ_t can be interpreted as the *level* of the series at time t, with α_t representing the rhythm *amplitude* at time t.

We will often be able to assume that ω is fixed and known (e.g., when a periodicity of 24 hours is specified), whereas it is unlikely that we will be able to stipulate an accurate value for ϕ. We therefore adopt the procedure described in Section 2.3, and use a discrete-valued grid to update recursively our beliefs about ϕ. In the absence of further information, the range adopted for this purpose is $[0, 2\pi)$, representing all possible values for ϕ. Writing this model in LDM form, we have

$$y_t = [1 \quad c_t]\begin{pmatrix} \mu_t \\ \alpha_t \end{pmatrix} + \varepsilon_t \tag{2.39}$$

$$\begin{pmatrix} \mu_t \\ \alpha_t \end{pmatrix} = \begin{pmatrix} 1 & 0 \\ 0 & 1 \end{pmatrix}\begin{pmatrix} \mu_{t-1} \\ \alpha_{t-1} \end{pmatrix} + \begin{pmatrix} \delta\mu_t \\ \delta\alpha_t \end{pmatrix}, \tag{2.40}$$

so that

$$\mathbf{H}_t = [1 \quad c_t] = [1 \quad \cos(2\pi\omega t + \phi)] \tag{2.41}$$

$$\boldsymbol{\theta}_t = \begin{pmatrix} \mu_t \\ \alpha_t \end{pmatrix}, \qquad \mathbf{G}_t = \mathbf{G} = \begin{pmatrix} 1 & 0 \\ 0 & 1 \end{pmatrix}, \qquad \boldsymbol{\omega}_t = \begin{pmatrix} \delta\mu_t \\ \delta\alpha_t \end{pmatrix} \tag{2.42}$$

and

$$\mathbf{R}_\varepsilon = R_\varepsilon, \qquad \mathbf{R}_\omega = \begin{pmatrix} R_\mu & 0 \\ 0 & R_\alpha \end{pmatrix}. \tag{2.43}$$

AR(1)

Time series can often be adequately modeled by assuming a low-order ARMA model for the system evolution, together with a simple additive error structure for the observation equation. For instance, a suitable description of the series given in Figure 3 may be the following first-order autoregressive model:

$$\begin{aligned} y_t &= \mu_t + \varepsilon_t \\ \mu_t - v_t &= \phi(\mu_{t-1} - v_{t-1}) + \delta\mu_t \\ v_t &= v_{t-1} + \delta v_t, \end{aligned} \tag{2.44}$$

where v_t represents the base level of the system at time t; $\mu_t - v_t$ represents the deviation of the actual system level, μ_t, from the base level and ϕ is the autoregressive parameter, which, for this model, is considered as a nuisance parameter. Also, we assume that $\varepsilon_t \sim N(0, \lambda^{-1}R_\varepsilon)$, $\delta\mu_t \sim N(0, \lambda^{-1}R_\mu)$, and $\delta v_t \sim N(0, \lambda^{-1}R_v)$.

The grid method is used to update ϕ, using the "natural" range $(-1, 1)$ corresponding to the stationarity region.

In LDM form we have

$$y_t = [1 \quad 0]\begin{pmatrix}\mu_t \\ \nu_t\end{pmatrix} + \varepsilon_t \tag{2.45}$$

$$\begin{pmatrix}\mu_t \\ \nu_t\end{pmatrix} = \begin{pmatrix}\phi & 1-\phi \\ 0 & 1\end{pmatrix}\begin{pmatrix}\mu_{t-1} \\ \nu_{t-1}\end{pmatrix} + \begin{pmatrix}\delta\mu_t + \delta\nu_t \\ \delta\nu_t\end{pmatrix}, \tag{2.46}$$

so that

$$\mathbf{H}_t = \mathbf{H} = [1 \quad 0], \qquad \boldsymbol{\theta}_t = \begin{pmatrix}\mu_t \\ \nu_t\end{pmatrix}, \qquad \mathbf{G}_t = \mathbf{G} = \begin{pmatrix}\phi & 1-\phi \\ 0 & 1\end{pmatrix}, \tag{2.47}$$

$$\boldsymbol{\omega}_t = \begin{pmatrix}\delta\mu_t + \delta\nu_t \\ \delta\nu_t\end{pmatrix} \tag{2.48}$$

and

$$\mathbf{R}_\varepsilon = R_\varepsilon, \qquad \mathbf{R}_\omega = \begin{pmatrix}R_\mu + R_\nu & R_\nu \\ R_\nu & R_\nu\end{pmatrix}. \tag{2.49}$$

Should we wish to allow for nonstationarities, we might easily add, say, a trend component to the system parameters, thus restricting the range for ϕ even for nonstationary time series.

3. MODELING DISCONTINUITIES

3.1 Change-Points and the Multistate Structure

Consider the linear growth model of Section 2.4 and the series shown in Figure 1. What kind of sudden departures from stability can we expect to see? From the point of view of time series monitoring one must recall that observations are processed on a sequential basis. In this case if the "current" observation is not part of an existing trend, it may have any one of a number of possible interpretations. It may, for instance, be a single anomolous observation after which the previous trend continues exactly as it had before this outlying value. On the other hand, it may be that the current observation defines the beginning of a new process level, in which case the subsequent observations will display the same slope as beforehand, but with this trend displaced to the new level. The current observation could also be the beginning of a new slope, in which case the following observations will indicate a total change in direction of the overall trend.

These simple change-point types (states), along with the possibility that the current observation is, indeed, in line with the previous trend, are shown in stylized form in Figure 4 and are referred to as the multistate structure for the

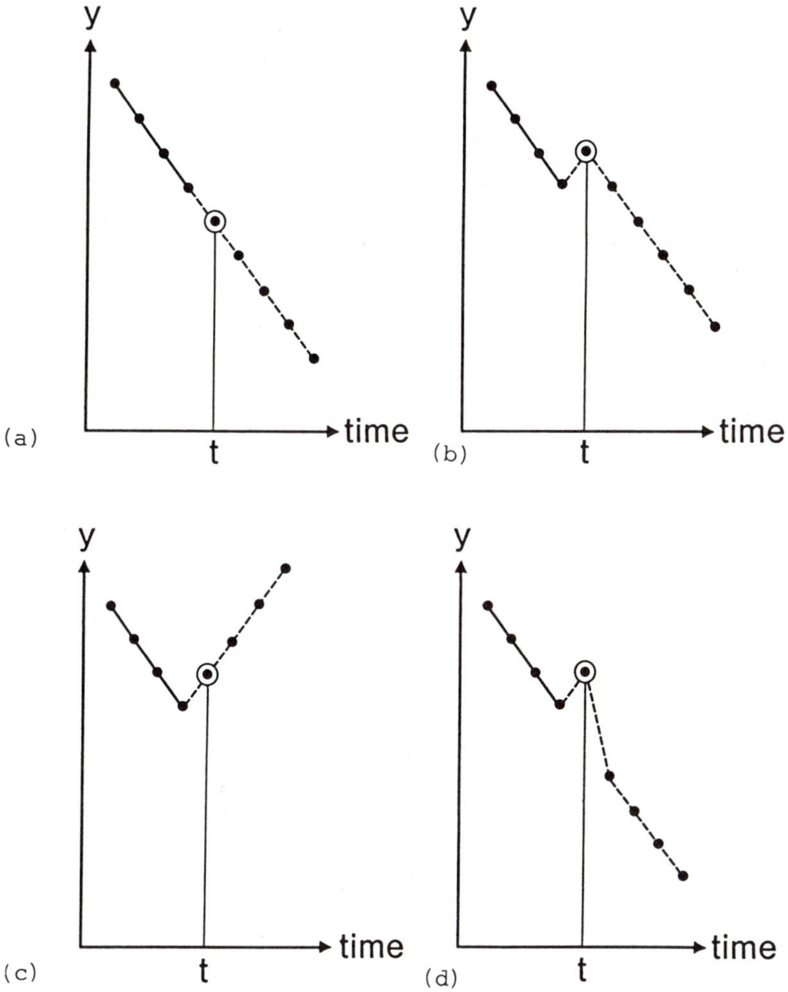

Figure 4. Stylized multistate structure for a linear growth model: (a) $j = 1$: Steady state: (b) $j = 2$: level change: (c) $j = 3$: slope change: (d) $j = 4$: transient.

linear growth model. Figure 5 shows the stylized multistate structure for the sinusoidal model of Section 2.4. In this case, departures from stability include a change in the rhythm amplitude. The stylized multistate structure for the AR(1) model of Section 2.4 is shown in Figure 6. The inclusion, into the multistate structure, of an impulse, which takes on a different form depending on whether the autoregressive parameter is positive or negative, will be more

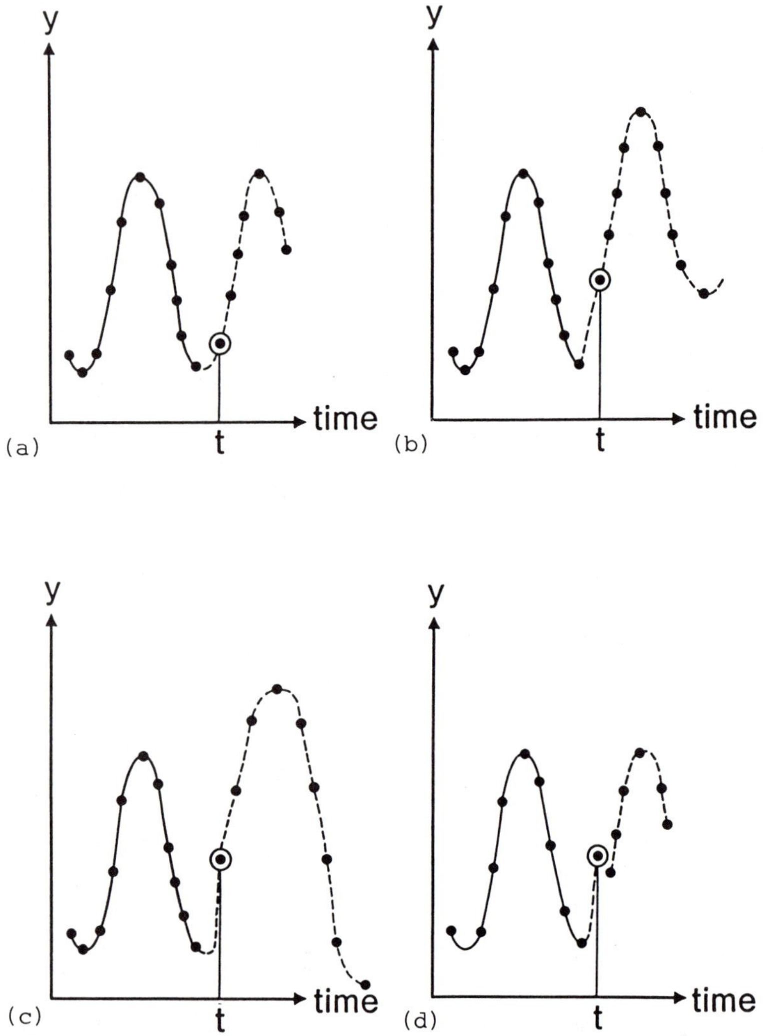

Figure 5. Stylized multistate structure for a sinusoidal model: (a) $j = 1$: steady state: (b) $j = 2$: level change: (c) $j = 3$: amplitude change: (d) $j = 4$: transient.

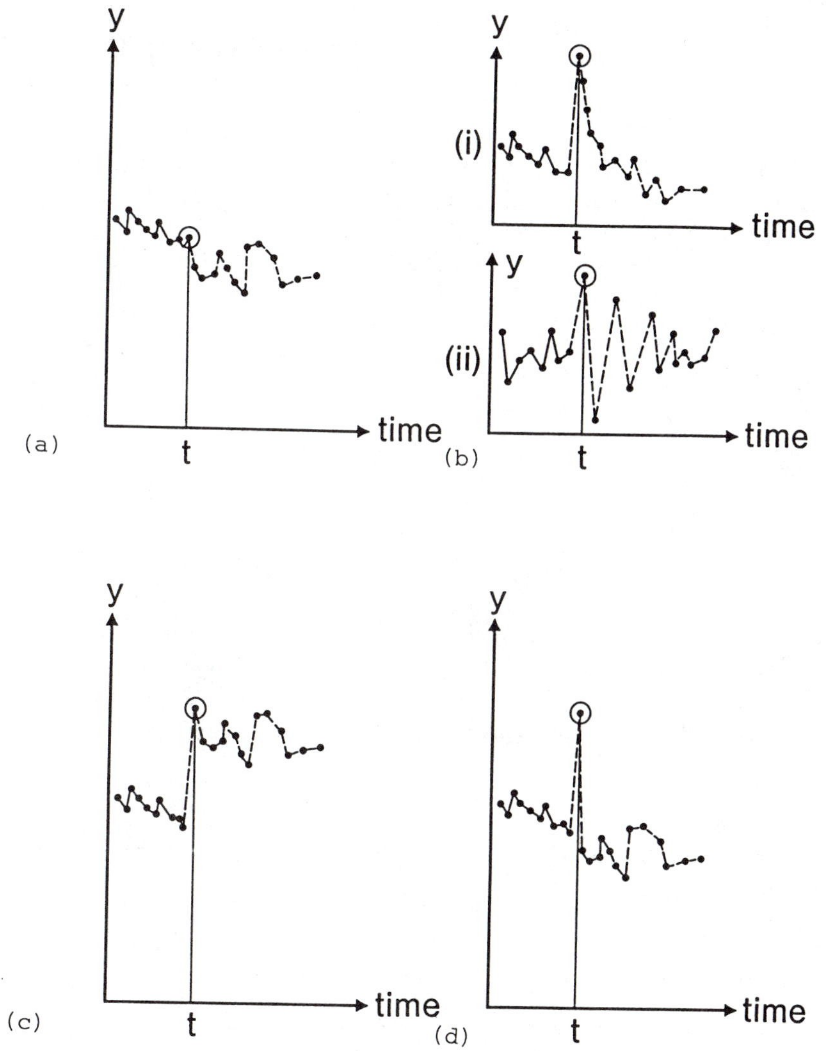

Figure 6. Stylized multistate structure for an autoregressive model: (a) $j = 1$: steady state: (b) $j = 2$: impulse: (i) $\phi > 0$, (ii) $\phi < 0$; (c) $j = 3$: level change: (d) $j = 4$: transient.

fully understood when, in the next section, we examine how we might combine this change-point structure with the linear dynamic model.

3.2 Multistate Modeling

Let us now turn our attention to the addition of a multistate structure to the normal linear dynamic model framework, as described by Harrison and Stevens (1976). Define the multistate structure by $\{M_t^{(j)}, p_0^{(j)}; j = 1, \ldots, J\}$, where $M_t^{(j)}$ indicates that model j obtains at time t, and $p_0^{(j)}$ is the a priori probability that model j obtains. Then the linear dynamic model given by (2.1) and (2.2) can be extended to

$$\mathbf{y}_t = \mathbf{H}_t\boldsymbol{\theta}_t + \boldsymbol{\varepsilon}_t^{(j)} \tag{3.1}$$

$$\boldsymbol{\theta}_t = \mathbf{G}_t\boldsymbol{\theta}_{t-1} + \boldsymbol{\omega}_t^{(j)} \tag{3.2}$$

for $j = 1, \ldots, J$, where

$$\mathbf{E}_t^{(j)} = \text{var}\,(\boldsymbol{\varepsilon}_t^{(j)}) = \lambda^{-1}\mathbf{R}_\varepsilon^{(j)} \tag{3.3}$$

$$\mathbf{W}_t^{(j)} = \text{var}\,(\boldsymbol{\omega}_t^{(j)}) = \lambda^{-1}\mathbf{R}_\omega^{(j)} \tag{3.4}$$

so that (3.1) and (3.2) represent J possible models, differing only through the elements of $\mathbf{R}_\varepsilon$ and $\mathbf{R}_\omega$. It will be shown in Section 3.4 how particular choices of $\mathbf{R}_\varepsilon$ and $\mathbf{R}_\omega$ can result in a variety of change-point models, such as those presented in Figures 4 to 6.

As well as the assumptions given by (2.4), it will also be assumed here that

$$p(M_t^{(j)}|H) = p_0^{(j)}, \qquad j = 1, \ldots, J, \forall t \tag{3.5}$$

where H is the process history *prior* to time t. We may now rewrite the Kalman filter recursion, given in (2.10) to (2.18), as follows (it being understood throughout that i, j run through the range $1, \ldots, J$):

$$\boldsymbol{\theta}_0 \sim N(\mathbf{m}_0, \lambda^{-1}\mathbf{C}_0). \tag{3.6}$$

Equation (2.12) is replaced by

$$p(\boldsymbol{\theta}_{t-1}|\mathbf{D}_{t-1}, \lambda, M_{t-1}^{(i)}) \sim N(\mathbf{m}_{t-1}^{(i)}, \lambda^{-1}\mathbf{C}_{t-1}^{(i)}), \tag{3.7}$$

and equation (2.13) is replaced by

$$p(\lambda|\mathbf{D}_{t-1}, M_{t-1}^{(i)}) \sim G(\tfrac{1}{2}n_{t-1}, \tfrac{1}{2}r_{t-1}^{(i)}), \tag{3.8}$$

where, again, $U \sim G(\alpha, \beta)$ means that U has a gamma distribution. We assume that the initialization of (3.8) is given by (2.14).

Upon receipt of $\mathbf{y}_t$, we can update (3.7) and (3.8) to give

$$p(\boldsymbol{\theta}_t|\mathbf{D}_t, \lambda, M_{t-1}^{(i)}, M_t^{(j)}) \sim N(\mathbf{m}_t^{(ij)}, \lambda^{-1}\mathbf{C}_t^{(ij)}) \tag{3.9}$$

$$p(\lambda|\mathbf{D}_t, M_{t-1}^{(i)}, M_t^{(j)}) \sim G(\tfrac{1}{2}n_t, \tfrac{1}{2}r_t^{(ij)}), \tag{3.10}$$

where $\mathbf{m}_t^{(ij)}$ and $\mathbf{C}_t^{(ij)}$ are given by the Kalman filter recursions

$$\mathbf{m}_t^{(ij)} = \mathbf{G}_t\mathbf{m}_{t-1}^{(i)} + \mathbf{S}_t^{(ij)}\mathbf{e}_t^{(i)} \tag{3.11}$$

$$\mathbf{C}_t^{(ij)} = \mathbf{P}_t^{(ij)} - \mathbf{S}_t^{(ij)}\mathbf{F}_t^{(ij)}(\mathbf{S}_t^{(ij)})^T, \tag{3.12}$$

and where, by standard Bayesian conjugate analysis, n_t and $r_t^{(ij)}$ are given by

$$n_t = n_{t-1} + 1 \tag{3.13}$$

$$r_t^{(ij)} = r_{t-1}^{(i)} + (\mathbf{e}_t^{(i)})^T(\mathbf{F}_t^{(ij)})^{-1}\mathbf{e}_t^{(i)} \tag{3.14}$$

with

$$\begin{aligned}
\mathbf{f}_t^{(i)} &= \mathbf{H}_t\mathbf{G}_t\mathbf{m}_{t-1}^{(i)} \\
\mathbf{e}_t^{(i)} &= \mathbf{y}_t - \mathbf{f}_t^{(i)} \\
\mathbf{P}_t^{(ij)} &= \mathbf{G}_t\mathbf{C}_{t-1}^{(i)}\mathbf{G}_t^T + \mathbf{R}_\omega^{(j)} \\
\mathbf{F}_t^{(ij)} &= \mathbf{H}_t\mathbf{P}_t^{(ij)}\mathbf{H}_t^T + \mathbf{R}_\varepsilon^{(j)} \\
\mathbf{S}_t^{(ij)} &= \mathbf{P}_t^{(ij)}\mathbf{H}_t^T(\mathbf{F}_t^{(ij)})^{-1}.
\end{aligned} \tag{3.15}$$

It is important to realize that whereas (3.7) and (3.8) describe J models, (3.9) and (3.10) describe J^2 models. In other words, each iteration produces a J-fold increase in the number of models under consideration. Clearly, this will soon lead to a combinatorial explosion even when the number of states, J, is small (e.g., when $J = 2$ there would be over 1000 models by the time $t = 10$). To avoid this problem we will therefore have to approximate the forms of (3.9) and (3.10), so that they resemble (3.7) and (3.8). In the next section we describe a pragmatic algorithm for the general class of multistate linear dynamic models.

3.3 Recursive Estimation

As in Section 2, we restrict ourselves, for ease of exposition, to a univariate observation y_t. From Equations (3.1), (3.2), and (3.15), we have

$$p(y_t|\mathbf{D}_{t-1}, \lambda, M_{t-1}^{(i)}, M_t^{(j)}) \sim N(f_t^{(i)}, \lambda^{-1}F_t^{(ij)}), \tag{3.16}$$

so that

$$p(y_t|\mathbf{D}_{t-1}, M_{t-1}^{(i)}, M_t^{(j)}) = \int_0^\infty p(y_t|\mathbf{D}_{t-1}, M_{t-1}^{(i)}, M_t^{(j)}, \lambda) \times p(\lambda|\mathbf{D}_{t-1}, M_{t-1}^{(i)}, M_t^{(j)})\,d\lambda,$$

and following the arguments of Section 2.2, we can use (3.16) and (3.8) to show that

$$Z_t^{(ij)} = p(y_t|\mathbf{D}_{t-1}, M_{t-1}^{(i)}, M_t^{(j)}) \propto (F_t^{(ij)})^{-1/2}(r_{t-1}^{(i)})^{(1/2)n_{t-1}}(r_t^{(ij)})^{-(1/2)n_t}. \tag{3.17}$$

If we let

$$p_t^{(j)} = p(M_t^{(j)}|\mathbf{D}_t) \tag{3.18}$$

$$p_t^{(ij)} = p(M_{t-1}^{(i)}, M_t^{(j)}|\mathbf{D}_t), \tag{3.19}$$

then, using (3.18),

$$\begin{aligned} p(y_t|\mathbf{D}_{t-1}) &= \sum_{i=1}^{J} p(y_t|\mathbf{D}_{t-1}, M_{t-1}^{(i)})p(M_{t-1}^{(i)}|\mathbf{D}_{t-1}) \\ &= \sum_{i=1}^{J}\left[\sum_{j=1}^{J} p(y_t|\mathbf{D}_{t-1}, M_{t-1}^{(i)}, M_t^{(j)})p(M_t^{(j)}|\mathbf{D}_{t-1}, M_{t-1}^{(i)})\right]p_{t-1}^{(i)}. \end{aligned}$$

Using (3.5), we see that

$$p(y_t|\mathbf{D}_{t-1}) = \sum_{i=1}^{J}\sum_{j=1}^{J} Z_t^{(ij)}p_0^{(j)}p_{t-1}^{(i)}. \tag{3.20}$$

From Equation (3.19), using Bayes' theorem it follows that

$$\begin{aligned} p_t^{(ij)} &= p(M_{t-1}^{(i)}, M_t^{(j)}|\mathbf{D}_t) \\ &= \frac{p(y_t|\mathbf{D}_{t-1}, M_{t-1}^{(i)}, M_t^{(j)})p(M_{t-1}^{(i)}, M_t^{(j)}|\mathbf{D}_{t-1})}{p(y_t|\mathbf{D}_{t-1})} \\ &= \frac{Z_t^{(ij)}p(M_t^{(j)}|\mathbf{D}_{t-1}, M_{t-1}^{(i)})p(M_{t-1}^{(i)}|\mathbf{D}_{t-1})}{p(y_t|\mathbf{D}_{t-1})}, \end{aligned}$$

so that

$$p_t^{(ij)} = \frac{Z_t z_t^{(ij)}p_0^{(j)}p_{t-1}^{(i)}}{p(y_t|\mathbf{D}_{t-1})}. \tag{3.21}$$

Completing the recursion

1. *Collapsing procedures.* To close the recursion, we need to approximate (3.9) and (3.10) so that they take the forms of (3.7) and (3.8), respectively. The normal approximation used is that proposed by Harrison and Stevens (1976), which fits a mixture of Gaussians by a single Gaussian distribution having the same mean vector and covariance matrix. West (1982) pointed out that this form of "collapsing" the mixture to a single distribution corresponds to minimizing the well-known Kullback-Leibler divergence. This suggests using a gamma approximation which also minimizes the Kullback-Leibler divergence (see West, 1982, for details). We therefore make the following assumptions.

We approximate the J^2 normal distributions $N(\mathbf{m}_t^{(ij)}, \lambda^{-1}\mathbf{C}_t^{(ij)})$, $i = 1, \ldots, J$,

$j = 1, \ldots, J$, by the J distributions: $N(\mathbf{m}_t^{(j)}, \lambda^{-1}\mathbf{C}_t^{(j)})$, $j = 1, \ldots, J$, where

$$\mathbf{m}_t^{(j)} = \sum_{i=1}^{J} \frac{p_t^{(ij)}}{p_t^{(j)}} \mathbf{m}_t^{(ij)} \tag{3.22}$$

$$\mathbf{C}_t^{(j)} = \sum_{i=1}^{J} \frac{p_t^{(ij)}}{p_t^{(j)}} \{\mathbf{C}_t^{(ij)} + (\mathbf{m}_t^{(ij)} - \mathbf{m}_t^{(j)})(\mathbf{m}_t^{(ij)} - \mathbf{m}_t^{(j)})^T\}. \tag{3.23}$$

We approximate the J^2 gamma distributions: $G(\frac{1}{2}n_t, \frac{1}{2}r_t^{(ij)})$, $i = 1, \ldots, J$, $j = 1, \ldots, J$, by the J distributions $G(\frac{1}{2}n_t, \frac{1}{2}r_t^{(j)})$, $j = 1, \ldots, J$, where

$$(r_t^{(j)})^{-1} = \sum_{i=1}^{J} \frac{p_t^{(ij)}}{p_t^{(j)}} (r_t^{(ij)})^{-1} \tag{3.24}$$

For some reported studies of the effectiveness of this approximation, see Smith and Makov (1980).

2. *Updating procedures.* The foregoing procedure for updating the parameter estimates and covariances from (3.7) and (3.8) given the latest observation y_t is completed by specifying the form of $p_t^{(j)}$. From Bayes' theorem and equation (3.5),

$$\begin{aligned} p_t^{(j)} &= p(M_t^{(j)}|\mathbf{D}_t) \\ &= p(y_t|\mathbf{D}_{t-1}, M_t^{(j)})p(M_t^{(j)}|\mathbf{D}_{t-1})/p(y_t|\mathbf{D}_{t-1}) \\ &= \frac{[\sum_{i=1}^{J} p(y_t|\mathbf{D}_{t-1}, M_{t-1}^{(i)}, M_t^{(j)})p(M_{t-1}^{(i)}|\mathbf{D}_{t-1}, M_t^{(j)})]p_0^{(j)}}{p(y_t|\mathbf{D}_{t-1})}, \end{aligned}$$

so that

$$p_t^{(j)} = \frac{\sum_{i=1}^{J} Z_t^{(ij)} p_{t-1}^{(i)} p_0^{(j)}}{p(y_t|\mathbf{D}_{t-1})}, \tag{3.25}$$

where $Z_t^{(ij)}$ is given by (3.17) and where the denominator is given by (3.20).

State probabilities

As a by-product of the parameter-updating process, we have calculated the quantities $p_t^{(j)}$, $j = 1, \ldots, J$. It can easily be seen that $p_t^{(j)} = p(M_t^{(j)}|\mathbf{D}_t)$ denotes the probability that process (or state) j obtains at time t, given all the data up to and including time t. So that, for instance, if state j represents the change in level model (see Section 3.4), $p_t^{(j)}$ is the probability of a change in level at time t, and therefore can be used to indicate the timing of the change-point. However, when some change in pattern occurs at time t, it may not be readily apparent which particular one of several alternative types of change of pattern has obtained until further information is available. It may be essential, therefore, to be able to update our beliefs about the state obtaining at time t having received observations $y_{t+1}, y_{t+2}, \ldots$, in addition to those up to time t.

We therefore consider $p(M_{t-1}^{(i)}|\mathbf{D}_t)$, which denotes the *one-step-back* probability of state i obtaining at time $t-1$, and $p(M_{t-2}^{(h)}|\mathbf{D}_t)$, which denotes the *two-steps-back* probability of state h obtaining at time $t-2$. For example, if h is the level change model, $p(M_{t-2}^{(h)}|\mathbf{D}_t)$ denotes the probability of a level change at time $t-2$, given all the data up to and including time $t-2$ *and* the additional observations y_{t-1} and y_t.

We use Bayes' theorem in order to calculate these quantities:

$$p(M_{t-1}^{(i)}|\mathbf{D}_t) = \frac{p(y_t|\mathbf{D}_{t-1}, M_{t-1}^{(i)})p(M_{t-1}^{(i)}|\mathbf{D}_{t-1})}{p(y_t|\mathbf{D}_{t-1})}$$
$$= \frac{\sum_{j=1}^{J} Z_t^{(ij)} p_0^{(j)} p_{t-1}^{(i)}}{p(y_t|\mathbf{D}_{t-1})}, \tag{3.26}$$

with $Z_t^{(ij)}$ specified by (3.17) and the denominator by (3.20), and similarly

$$p(M_{t-2}^{(h)}|\mathbf{D}_t) = \frac{p(y_t|\mathbf{D}_{t-1}, M_{t-2}^{(h)})p(M_{t-2}^{(h)}|\mathbf{D}_{t-1})}{p(y_t|\mathbf{D}_{t-1})}$$
$$= \frac{\sum_{j=1}^{J} p(y_t|\mathbf{D}_{t-1}, M_{t-2}^{(h)}, M_t^{(j)})p_0^{(j)}p(M_{t-2}^{(h)}|\mathbf{D}_{t-1})}{p(y_t|\mathbf{D}_{t-1})}$$
$$= \frac{\sum_{j=1}^{J}\sum_{i=1}^{J} Z_t^{(ij)}p(M_{t-1}^{(i)}|\mathbf{D}_{t-1}, M_{t-2}^{(h)}, M_t^{(j)})p_0^{(j)}p(M_{t-2}^{(h)}|\mathbf{D}_{t-1})}{p(y_t|\mathbf{D}_{t-1})}$$
$$= \frac{\sum_{j=1}^{J}\sum_{i=1}^{J} Z_t^{(ij)}[Z_{t-1}^{(hi)}p_0^{(i)}/\sum_{i=1}^{J} Z_{t-1}^{(hi)}p_0^{(i)}]p_0^{(j)}p(M_{t-2}^{(h)}|\mathbf{D}_{t-1})}{p(y_t|\mathbf{D}_{t-1})} \tag{3.27}$$

where $Z_t^{(ij)}$ is specified by (3.17), $Z_{t-1}^{(hi)}$ has been calculated at time $t-1$ from (3.17), $p(M_{t-2}^{(h)}|\mathbf{D}_{t-1})$ is given by (3.26), and the denominator is given by (3.20). Similar results can, of course, be obtained for $P(M_{t-3}^{(i)}|\mathbf{D}_t)$, and so on, if required.

The use of posterior probabilities on alternative models as the basis for model choice or comparison was first popularized by Jeffreys (1931) as a systematic Bayesian alternative to significance tests. For more recent references, see, for example, Prabhu (1970), Harrison and Stevens (1976), Kashyap (1977), Smith and Spiegelhalter (1980), Poskitt and Tremayne (1983), Chow and Willsky (1984), and Spall (1988).

Nuisance parameters

If an LDM and its corresponding "change-of-state" processes depend on one or more nuisance parameters, $\boldsymbol{\phi}$, we adopt the procedure outlined in Section 2 and specify a probability distribution for $\boldsymbol{\phi}$ over a suitably chosen discrete grid. We replace (3.7) and (3.8) by

$$p(\boldsymbol{\theta}_{t-1}|\mathbf{D}_{t-1}, \lambda, M_{t-1}^{(i)}, \boldsymbol{\phi}) \sim N(\mathbf{m}_{t-1}^{(i)}, \lambda^{-1}\mathbf{C}_{t-1}^{(i)}) \tag{3.28}$$

$$p(\lambda|\mathbf{D}_{t-1}, M_{t-1}^{(i)}, \boldsymbol{\phi}) \sim G(\tfrac{1}{2}n_{t-1}, \tfrac{1}{2}r_{t-1}^{(i)}) \tag{3.29}$$

where $\mathbf{m}_{t-1}^{(i)}$, $\mathbf{C}_{t-1}^{(i)}$, and $r_{t-1}^{(i)}$ now all depend on $\boldsymbol{\phi}$, so that if there are N nodes in the $\boldsymbol{\phi}$ grid, $\boldsymbol{\Phi}$, (3.28) and (3.29) represent N possible normal-gamma distributions.

It can readily be seen that

$$p(\boldsymbol{\theta}_t|\mathbf{D}_t, \lambda, M_{t-1}^{(i)}, M_t^{(j)}, \boldsymbol{\phi}) \sim N(\mathbf{m}_t^{(ij)}, \lambda^{-1}\mathbf{C}_t^{(ij)}) \tag{3.30}$$

$$p(\lambda|\mathbf{D}_t, M_{t-1}^{(i)}, M_t^{(j)}, \boldsymbol{\phi}) \sim G(\tfrac{1}{2}n_t, \tfrac{1}{2}r_t^{(ij)}), \tag{3.31}$$

where $\mathbf{m}_t^{(ij)}$, $\mathbf{C}_t^{(ij)}$, n_t, and $r_t^{(ij)}$ are defined by (3.11) to (3.15) for *each* node in $\boldsymbol{\Phi}$.

Let

$$p(\boldsymbol{\phi}|\mathbf{D}_{t-1}, M_{t-1}^{(i)}) = K_{t-1}^{(i)}(\boldsymbol{\phi}) \tag{3.32}$$

$$p(\boldsymbol{\phi}|\mathbf{D}_t, M_{t-1}^{(i)}, M_t^{(j)}) = K_t^{(ij)}(\boldsymbol{\phi}). \tag{3.33}$$

Then we replace (3.9) by

$$\begin{aligned} p(\boldsymbol{\theta}_t|\mathbf{D}_t, \lambda, M_{t-1}^{(i)}, M_t^{(j)}) &= \sum_{\boldsymbol{\Phi}} p(\boldsymbol{\theta}_t|\mathbf{D}_t, \lambda, M_{t-1}^{(i)}, M_t^{(j)}, \boldsymbol{\phi})p(\boldsymbol{\phi}|\mathbf{D}_t, \lambda, M_{t-1}^{(i)}, M_t^{(j)}) \\ &= \sum_{\boldsymbol{\Phi}} p(\boldsymbol{\theta}_t|\mathbf{D}_t, \lambda, M_{t-1}^{(i)}, M_t^{(j)}, \boldsymbol{\phi})K_t^{(ij)}(\boldsymbol{\phi}), \end{aligned} \tag{3.34}$$

where the first of the two terms in each summand is given by (3.30). In practice, (3.34) is a mixture of normals and we approximate by an $N(\mathbf{m}_t^{(ij)}, \lambda^{-1}\mathbf{C}_t^{(ij)})$ distribution, where

$$\mathbf{m}_t^{(ij)} = \sum_{\boldsymbol{\Phi}} \mathbf{m}_t^{(ij)}(\boldsymbol{\phi})K_t^{(ij)}(\boldsymbol{\phi}) \tag{3.35}$$

$$\mathbf{C}_t^{(ij)} = \sum_{\boldsymbol{\Phi}} [\mathbf{C}_t^{(ij)}(\boldsymbol{\phi}) + (\mathbf{m}_t^{(ij)}(\boldsymbol{\phi}) - \mathbf{m}_t^{(ij)})(\mathbf{m}_t^{(ij)}(\boldsymbol{\phi}) - \mathbf{m}_t^{(ij)})^T]K_t^{(ij)}(\boldsymbol{\phi}). \tag{3.36}$$

Similarly, we replace (3.10) by

$$p(\lambda|\mathbf{D}_t, M_{t-1}^{(i)}, M_t^{(j)}) = \sum_{\boldsymbol{\Phi}} p(\lambda|\mathbf{D}_t, M_{t-1}^{(i)}, M_t^{(j)}, \boldsymbol{\phi})K_t^{(ij)}(\boldsymbol{\phi}), \tag{3.37}$$

where the first of the two terms in each summand is given by (3.31). Since (3.37) is a mixture of gamma distributions, we approximate it by

$$(r_t^{(ij)})^{-1} = \sum_{\boldsymbol{\Phi}} (r_t^{(ij)}(\boldsymbol{\phi}))^{-1}K_t^{(ij)}(\boldsymbol{\phi}). \tag{3.38}$$

We now replace (3.17) by

$$Z_t^{(ij)}(\phi) = p(y_t|\mathbf{D}_{t-1}, M_{t-1}^{(i)}, M_t^{(j)}, \phi)$$
$$\propto (F_t^{(ij)})^{-1/2}(r_{t-1}^{(i)})^{(1/2)n_{t-1}}(r_t^{(ij)})^{-(1/2)n_t} \tag{3.39}$$

where $F_t^{(ij)}$, $r_{t-1}^{(i)}$, and $r_t^{(ij)}$ now depend on ϕ.

To calculate (3.35), (3.36), and (3.38) we need to derive $K_t^{(ij)}(\phi)$, where, using Bayes' theorem,

$$\begin{aligned} K_t^{(ij)}(\phi) &= p(\phi|\mathbf{D}_t, M_{t-1}^{(i)}, M_t^{(j)}) \\ &= \frac{p(y_t|\mathbf{D}_{t-1}, M_{t-1}^{(i)}, M_t^{(j)}, \phi)p(\phi|\mathbf{D}_{t-1}, M_{t-1}^{(i)}, M_t^{(j)})}{\sum_{\Phi} p(y_t|\mathbf{D}_{t-1}, M_{t-1}^{(i)}, M_t^{(j)}, \phi)p(\phi|\mathbf{D}_{t-1}, M_{t-1}^{(i)}, M_t^{(j)})} \\ &= \frac{Z_t^{(ij)}(\phi)K_{t-1}^{(i)}(\phi)}{\sum_{\Phi} Z_t^{(ij)}(\phi)K_{t-1}^{(i)}(\phi)}, \end{aligned} \tag{3.40}$$

where $Z_t^{(ij)}(\phi)$ is given by (3.39).

Updating the probabilities

In the presence of nuisance parameters, it can readily be shown, after some manipulation, that

$$p_t^{(j)} = \frac{\sum_{i=1}^{J} \sum_{\Phi} Z_t^{(ij)}(\phi)K_{t-1}^{(i)}(\phi)p_{t-1}^{(i)}p_0^{(j)}}{p(y_t|\mathbf{D}_{t-1})} \tag{3.41}$$

[see equation (3.25)];

$$p(M_{t-1}^{(i)}|\mathbf{D}_t) = \frac{\sum_{j=1}^{J} \sum_{\Phi} Z_t^{(ij)}(\phi)K_{t-1}^{(i)}(\phi)p_{t-1}^{(i)}p_0^{(j)}}{p(y_t|\mathbf{D}_{t-1})} \tag{3.42}$$

[see equation (3.26)];

$$\begin{aligned} p(M_{t-2}^{(h)}|\mathbf{D}_t) = \frac{1}{p(y_t|\mathbf{D}_{t-1})} \Bigg\{ \sum_{j=1}^{J} \sum_{i=1}^{J} \sum_{\Phi} \left[Z_t^{(ij)}(\phi) \frac{Z_{t-1}^{(hi)}(\phi)K_{t-2}^{(h)}(\phi)}{\sum_{\Phi} Z_{t-1}^{(hi)}(\phi)K_{t-2}^{(h)}(\phi)} \right] \\ \times \frac{\sum_{\Phi} Z_{t-1}^{(hi)}(\phi)K_{t-2}^{(h)}(\phi)p_0^{(i)}}{\sum_{i=1}^{J} \sum_{\Phi} Z_{t-1}^{(hi)}(\phi)K_{t-2}^{(h)}(\phi)p_0^{(i)}} \, p_0^{(j)}p(M_{t-2}^{(h)}|\mathbf{D}_{t-1}) \Bigg\} \end{aligned} \tag{3.43}$$

[see equation (3.27)]. To close the recursion, we need to update the ϕ-grid as follows:

$$K_t^{(j)}(\phi) = \frac{\sum_{i=1}^{J} Z_t^{(ij)}(\phi)K_{t-1}^{(i)}(\phi)p_{t-1}^{(i)}}{\sum_{i=1}^{J} \sum_{\Phi} Z_t^{(ij)}(\phi)K_{t-1}^{(i)}(\phi)p_{t-1}^{(i)}}. \tag{3.44}$$

3.4 The Change-Point Models

In this section we examine the implications of attaching a multistate structure to the models described in Section 2.4. As mentioned previously, change point models can be introduced through the adjustment of observation and system variances, without having to change the underlying model structure described by the transition matrix $\mathbf{G}_t$.

Linear growth

Consider the linear growth model of Section 2.4 with a multistate structure defined by

$$y_t = \mu_t + \varepsilon_t^{(j)} \tag{3.45}$$

$$\mu_t = \mu_{t-1} + \beta_t + \delta\mu_t^{(j)} \tag{3.46}$$

$$\beta_t = \beta_{t-1} + \delta\beta_t^{(j)} \tag{3.47}$$

with

$$\begin{aligned} \varepsilon_t^{(j)} &\sim N(0, \lambda^{-1}R_\varepsilon^{(j)}) \\ \delta\mu_t^{(j)} &\sim N(0, \lambda^{-1}R_\mu^{(j)}) \\ \delta\beta_t^{(j)} &\sim N(0, \lambda^{-1}R_\beta^{(j)}). \end{aligned} \tag{3.48}$$

We have four simple states (i.e., $J = 4$):

1. $j = 1$. $R_\varepsilon^{(1)} = 1$, $R_\mu^{(1)} = 0$, $R_\beta^{(1)} = 0$. In this case $\delta\mu_t^{(j)}$ and $\delta\beta_t^{(j)}$ are identically zero, and therefore μ_t and β_t will not be perturbed, which represents the system in the *steady state*.
2. $j = 2$. If $\delta\mu_t^{(j)}$ were nonzero only at time t, this would affect μ_t and, since μ_t is related to μ_{t+1} according to (3.46), μ_{t+1}, μ_{t+2}, and so on, would also be influenced. Therefore, a single nonzero $\delta\mu_t^{(j)}$ results in a *change in the level* of the systems evolution, beginning at time t. This can be achieved by setting: $R_\varepsilon^{(2)} = 1$, $R_\mu^{(2)} =$ positive, $R_\beta^{(2)} = 0$.
3. $j = 3$. If $\delta\beta_t^{(j)}$ were nonzero only at time t, this would affect β_t and, since β_t is related to β_{t+1} according to (3.47), β_{t+1}, β_{t+2}, and so on, would also be influenced. Also, according to (3.46), μ_{t+1} will be influenced by β_{t+1} and μ_t (which is affected by β_t). In other words, a single nonzero $\delta\beta_t^{(j)}$ results in an incremental effect on the level, μ, and therefore produces a *change in the slope* of the systems evolution, beginning at time t. This can be achieved by setting: $R_\varepsilon^{(3)} = 1$, $R_\mu^{(3)} = 0$, $R_\beta^{(3)} =$ positive.
4. $j = 4$. If $\varepsilon_t^{(j)}$ were very large only at time t, this would affect y_t, according to (3.45), but *not* future values of y, since y_{t+1} is not directly related to y_t. Therefore, a single large $\varepsilon_t^{(j)}$ results in a *transient observation* at time t. This can be achieved by setting $R_\varepsilon^{(4)} =$ large positive, $R_\mu^{(4)} = 0$, $R_\beta^{(4)} = 0$.

Clearly, it would be possible to extend the multistate structure by including combinations of these four states in the overall model. For example, with $R_\varepsilon^{(5)} = 1$, $R_\mu^{(5)} =$ positive, $R_\beta^{(5)} =$ positive, we can model the situation where there is a simultaneous level change and slope change.

Figure 4 presents a stylized pictorial display of the linear growth multistate structure where, for clarity of presentation, it is assumed that $\mathbf{C}_t = 0$ for all t, where

$$\lambda^{-1}\mathbf{C}_t = \operatorname{var}\left(\begin{pmatrix}\mu_t\\ \beta_t\end{pmatrix}\middle|\mathbf{D}_t\right).$$

It can be seen from Figure 4 that the current observation, y_t (circled in the diagram), may not be sufficient to discriminate between change-point types, and that it is necessary to receive y_{t+1} in order to be able to attempt to identify a specific change in pattern.

The series shown in Figure 1 exemplifies a situation well modeled by the multiprocess linear growth model and for which a detailed discussion and analysis is provided in Smith and West (1983).

Sinusoidal model

For the case of the sinusoidal model:

$$y_t = \mu_t + c_t\alpha_t + \varepsilon_t^{(j)} \tag{3.49}$$

$$\mu_t = \mu_{t-1} + \delta\mu_t^{(j)} \tag{3.50}$$

$$\alpha_t = \alpha_{t-1} + \delta\alpha_t^{(j)}, \tag{3.51}$$

where

$$c_t = \cos(2\pi\omega t + \phi)$$

and

$$\begin{aligned} \varepsilon_t^{(j)} &\sim N(0, \lambda^{-1}R_\varepsilon^{(j)}) \\ \delta\mu_t^{(j)} &\sim N(0, \lambda^{-1}R_\mu^{(j)}) \\ \delta\alpha_t^{(j)} &\sim N(0, \lambda^{-1}R_\alpha^{(j)}). \end{aligned} \tag{3.52}$$

This multistate structure is demonstrated in stylized form in Figure 5. As in the previous case, $j = 1, 2, 4$ correspond, respectively, to steady state, change in level, and transient, but $j = 3$ now corresponds to a change in amplitude.

Note that there are nuisance parameters in this model since $c_t = \cos(2\pi\omega t + \phi)$ with ω (the frequency) and ϕ (the phase) treated as unknown parameters. The linear dynamic model specified above does not consider sudden changes in ω or ϕ. If this situation were likely to arise, we would need to formulate an alternative linear dynamic model in order to handle it. The series shown in Figure 2 exemplifies a situation well modeled by the multiprocess sinusoidal model and for which a detailed discussion and analysis is provided in Gordon (1986).

AR (1)

For the AR (1) model,

$$y_t = \mu_t + \varepsilon_t^{(j)} \tag{3.53}$$

$$\mu_t - \nu_t = \phi(\mu_{t-1} - \nu_{t-1}) + \delta\mu_t^{(j)} \tag{3.54}$$

$$\nu_t = \nu_{t-1} + \delta\nu_t^{(j)}, \tag{3.55}$$

where

$$\begin{aligned} \varepsilon_t^{(j)} &\sim N(0, \lambda^{-1}R_\varepsilon^{(j)}) \\ \delta\mu_t^{(j)} &\sim N(0, \lambda^{-1}R_\mu^{(j)}) \\ \delta\nu_t^{(j)} &\sim N(0, \lambda^{-1}R_\nu^{(j)}). \end{aligned} \tag{3.56}$$

The multistate structure is shown in Figure 6. Here $j = 1, 3, 4$ correspond, respectively, to steady state, change in level, and transient, and $j = 2$ corresponds to an impulse. Notice that for $j = 2$ the change-point phenomenon arising from a large $\delta\mu_t^{(j)}$ takes on a different appearance depending on whether the autoregressive parameter, ϕ, is positive or negative. In the former case, the effect of the impulse damps out with an exponential decay; in the latter case, there is oscillatory damping. Figure 3 shows a typical AR (1) series with a positive autoregressive parameter, which is analyzed in detail in the next section to illustrate many of the ideas presented in this chapter.

4. ILLUSTRATIVE ANALYSIS

The series shown in Figure 3 was simulated in the following manner from the AR (1) change-point model defined by (3.53)–(3.56). The starting values were taken to be $\nu_0 = \mu_0 = 10$ and the autoregressive parameter was fixed at $\phi = 0.7$. The steady-state measurement error distribution was taken to be $\varepsilon_t^{(1)} \sim N(0, 1.0)$. At $t = 30$ and $t = 35$, *level changes* were simulated by setting $R_\nu = 10.0$, so that $\delta\nu_{30}^{(3)} \sim N(0, 10.0)$ and $\delta\nu_{35}^{(3)} \sim N(0, 10.0)$. At $t = 25$ and $t = 75$, *impulses* were simulated by setting $R_\mu = 20.0$, so that $\delta\mu_{25}^{(2)} \sim N(0, 20.0)$ and $d\mu_{75}^{(2)} \sim N(0, 20.0)$. At $t = 50$ and $t = 80$, *transients* were simulated by setting $R_\varepsilon = 30.0$, so that $\varepsilon_{50}^{(4)} \sim N(0, 30.0)$ and $\varepsilon_{80}^{(4)} \sim N(0, 30.0)$.

One might think (quite reasonably) that some of the change points contained in this series are fairly obvious, even by eye, and that this level of statistical processing is, perhaps, not essential to detect these changes. Certainly, this example has been included more for the sake of a clear demonstration of the ability of these techniques not only to detect but also distinguish between a number of change-point types than for its difficulty. Recall, though, that the observations are processed sequentially; it is always much easier to identify change-points retrospectively when one has the whole data series to examine. If one moves slowly across Figure 3 from left to right, uncovering new observations one at a time, one may get a feel for the difficulty involved in the on-line identification of a specific type of change.

The series shown in Figure 3 was then analyzed on-line using the monitoring procedures described above. The prior values assumed were as follows:

$$\mathbf{m}_0 = \begin{pmatrix} \mu_0 \\ \nu_0 \end{pmatrix} = \begin{pmatrix} 10.0 \\ 10.0 \end{pmatrix}, \qquad \mathbf{C}_0 = \begin{pmatrix} 15.0 & 0 \\ 0 & 15.0 \end{pmatrix}, \qquad \text{[equation (2.10)]}$$

$$n_0 = 5, \qquad r_0 = 3 \qquad \text{[equation (2.14)]}$$

with ϕ [the autoregressive parameter in (3.54)] discretized over a 21-point equally spaced grid on the interval $(-1, 1)$ and $K_0(\phi) = \frac{1}{21}$ for all grid points [equation (2.25)]; the values of $p_0^{(j)}$ [equation (3.5)] and $R_\varepsilon^{(j)}$, $\mathbf{R}_\omega^{(j)}$ [equations (3.3), (3.4)], $j = 1, \ldots, 4$, were taken to be

	$j=1$	$j=2$	$j=3$	$j=4$
$p_0^{(j)}$	0.85	0.06	0.07	0.02
$R_\varepsilon^{(j)}$	1	1	1	30
$R_\mu^{(j)}$	0	20	0	0
$R_\nu^{(j)}$	0	0	10	0

Figure 7 shows the output from the multistate monitoring analysis described in Section 3.3. The upper part of the figure shows the original series, together with the one-step-ahead forecasts, $E(y_t|\mathbf{D}_{t-1})$, based on equation (3.20). The three bar diagrams show the one-step-back probabilities of the various possible types of change-point, calculated from equation (3.42).

It can be seen (see also the first row of the table below) that the level changes simulated at $t = 30$ and $t = 35$ are clearly picked up by $p(M_{30}^{(3)}|\mathbf{D}_{31}) = 0.91$ and $p(M_{35}^{(3)}|\mathbf{D}_{36}) = 0.83$, and that the transient simulated at $t = 50$ and $t = 80$ are clearly picked up by $p(M_{50}^{(4)}|\mathbf{D}_{51}) = 0.55$ and $p(M_{80}^{(4)}|\mathbf{D}_{81}) = 0.97$, although at $t = 50$ the monitoring procedure attached some probability to the possibility of an impulse, $p(M_{50}^{(2)}\mathbf{D}_{51}) \approx 0.42$. The impulse simulated at $t = 75$ is detected, $p(M_{75}^{(2)}|\mathbf{D}_{76}) = 0.27$, but not very confidently.

Clearly, these monitoring procedures are potentially sensitive to the setting

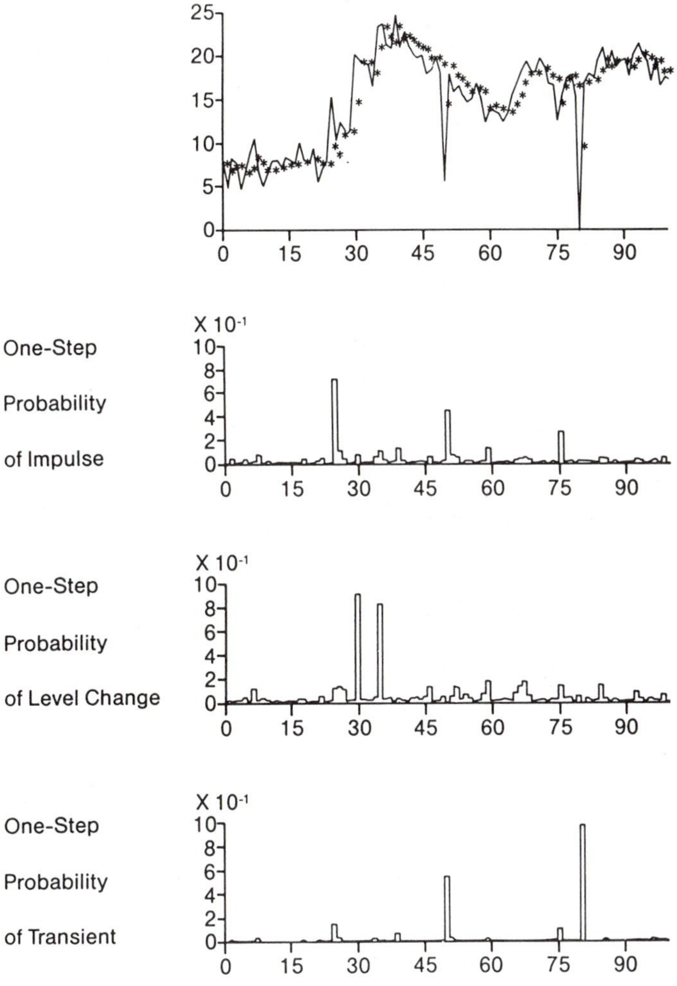

Figure 7. On-line analysis of the simulated series of Figure 3.

of prior parameters, and in implementing an on-line system for routine use in actual applications there is a need for careful initial study of system variation, together with systematic sensitivity studies, in order to tune the system.

As an illustration of the kinds of sensitivity analysis that can be performed, and of the effects of choices of prior parameters on the monitoring process, we present in Table 1 a summary of the consequences of various changes to the initial settings of the prior parameters described above. The changes in the initial setting are indicated in the left-hand column; the headings I_t, L_t, T_t

Table 1 Consequences of Changes[a]

	I_{26}	L_{31}	L_{36}	T_{51}	I_{76}	T_{81}	NFP	$\hat{\phi}$
Initial setting	0.72	0.91	0.83	0.55	0.27	0.97	1	0.65
$n_0 = 25, r_0 = 23$	0.68	0.91	0.75	0.52	0.17	0.97	1	0.68
$n_0 = 5, r_0 = 1$	0.75	0.95	0.93	0.58	0.69	0.96	14	0.37
$n_0 = 5, r_0 = 9$	0.61	0.89	0.66	0.52	0.15	0.96	1	0.68
$\mathbf{R}_\omega^{(j)} = 9 \times$ initial setting of $\mathbf{R}_\omega^{(j)}, j = 1, \ldots, 4$	0.72	0.94	0.82	0.42	0.15	0.93	1	0.69
$p_0^{(1)} = 0.97, p_0^{(2)} = p_0^{(3)} = p_0^{(4)} = 0.1$	0.57	0.90	0.49	0.82	0.03	1.00	1	0.84
$v_0 = \mu_0 = 3.33$	0.72	0.84	0.74	0.76	0.24	0.99	2	0.73
$v_0 = \mu_0 = 30.0$	0.71	0.85	0.74	0.72	0.23	0.99	2	0.75
$K_0(\phi) = \frac{1}{11}$ on 11-point grid	0.72	0.91	0.83	0.55	0.26	0.97	1	0.67

[a] I_t, L_t, T_t denote one-step back probabilities of impulse, level change, and transient, respectively. NFP denotes the number of false positives signaled (i.e., one-step-back probabilities $\geqslant 0.2$). $\hat{\phi}$ is the posterior mean for ϕ derived from (4.1).

denote the one-step-back probabilities of an impulse, level change or transient, respectively; NFP denotes the number of false positives recorded (defined by incorrect one-step-back probability signals greater than 0.2); and $\hat{\phi}$ denotes the mean of the posterior distribution for ϕ, the latter given (with $t = 100$) by

$$\sum_{j=1}^{J} p_t^{(j)} K_t^{(j)}(\phi), \tag{4.1}$$

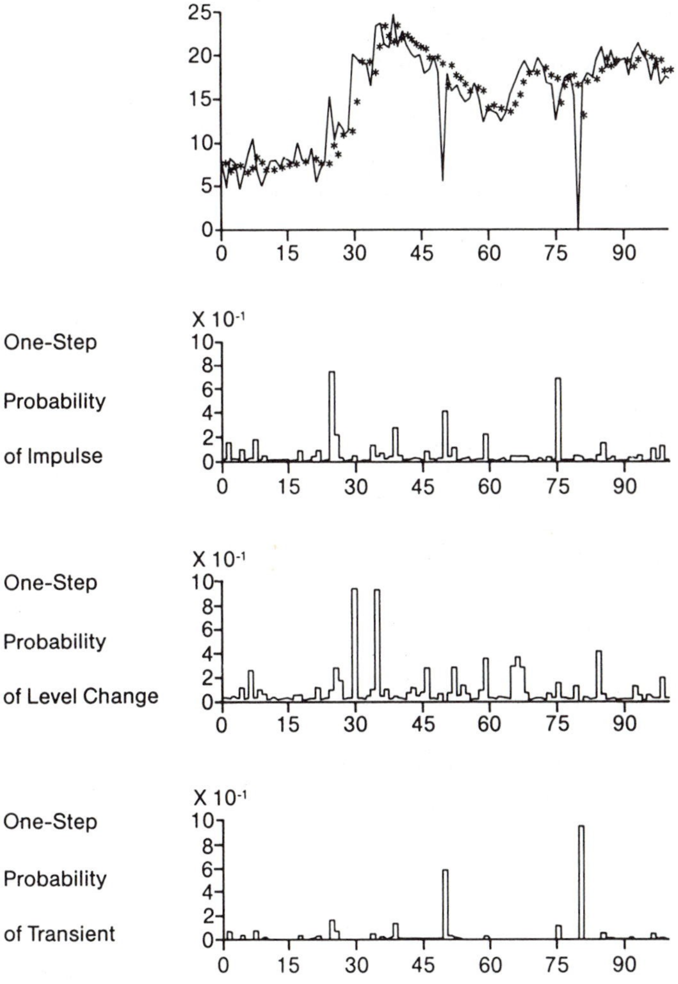

Figure 8. Retuned on-line analysis of the simulated series of Figure 3.

where $p_t^{(j)}$ and $K_t^{(j)}(\phi)$ are defined by equations (3.41) and (3.44), respectively.

It can be seen that neither a sharper (but with the same expected value) prior for the baseline measurement variance σ^2 ($n_0 = 25, r_0 = 23$) nor a prior which specifies a larger expected value of σ^2 ($n_0 = 5, r_0 = 9$) have much effect on the performance of the monitoring system, whereas the specification of a much smaller expected value of σ^2 ($n_0 = 5, r_0 = 1$) produces much higher sensitivity to changes. On the one hand, this manifests itself in sharper values for the one-step-back probabilities correctly identifying changes, but on the other hand produces a large number of false-positive signals. This overestimation of the amount of instability in the system in turn causes a degradation in the learning process for ϕ.

Neither changes in the direction of larger values of $\mathbf{R}_\omega^{(j)}$, nor in the misspecification of prior levels of v and μ, nor in the cruder treatment of the nuisance parameter ϕ (using an 11-point rather than a 21-point grid) causes much deterioration in performance. However, a specification of $p_0^{(j)}$ which asserts that changes are highly unlikely can result in a polarized response to "clear" and "unclear" changes.

Figure 8 shows a rerun of the monitoring process shown in Figure 7, corresponding to the third row of Table 1, where (with $n_0 = 5$, $r_0 = 1$) an inappropriately small value of the baseline measurement variance is specified

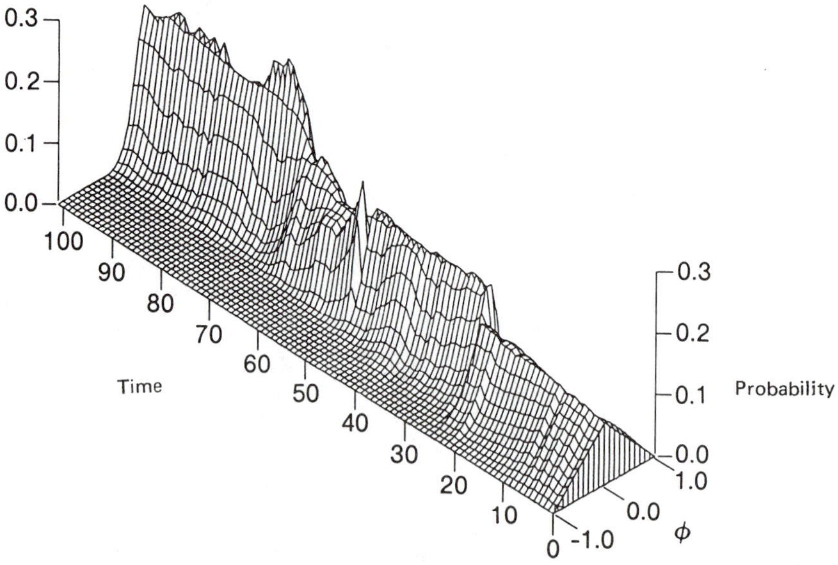

Figure 9. Evolution of the posterior distribution of the autoregressive parameter.

a priori. The bar diagrams clearly show the increase in false-positive responses, and the one-step-ahead forecast pattern shows that in this case there is a tendency to follow the series too closely.

As an illustration of other outputs from the system which may be of independent, or additional, interest, we present in Figure 9 the evolving posterior distribution for ϕ, defined by (4.1) above, as t increases, for the case of the initial setting of prior parameters described above. It can be seen that the distribution has become tightly focused near the correct value ($\phi = 0.7$) by about $t = 30$. In part, this is explained by the clear identification of an impulse around $t = 25$, since the rate of decay of the impulse provides a rather "pure" estimate of ϕ.

We believe that these techniques have tremendous potential. They have been used in a number of areas of application, including kidney transplantation (Smith and West, 1983; Trimble et al., 1983), television advertising (Migon and Harrison, 1983), and fetal heart monitoring (Gordon, 1986). The latter reference also contains far-reaching extensions of the methodology to deal with unequally spaced data and multivariate series.

Editor's Note: For a discussion of this chapter, see J. L. Maryak (1987), *Proc. Am. Statist. Assoc., Business and Econ. Statist. Sect.*, pp. 80–82.

REFERENCES

Aitchison, J., and I. R. Dunsmore (1975). *Statistical Prediction Analysis*, Cambridge University Press, Cambridge.

Basseville, M., and A. Benveniste, eds. (1986). *Detection of Abrupt Changes in Signals and Dynamical Systems*, Vol. 77 of *Lecture Notes in Control and Information Sciences*, Springer-Verlag, New York.

Chow, E. Y., and A. S. Willsky (1984). Bayesian design of decision rules for failure detection, *IEEE Trans. Aerosp. Electron. Syst.* AES-20, 761–773.

DeGroot, M. H. (1970). *Optimal Statistical Decisions*, McGraw-Hill, New York.

Goodrich, R. L., and P. E. Caines (1979). Linear system identification from non-stationary cross-sectional data, *IEEE Trans. Autom. Control* AC-24, 403–411.

Gordon, K. (1986). Modelling and monitoring of medical time series, Ph.D. thesis, University of Nottingham.

Harrison, P. J., and C. F. Stevens (1976). Bayesian forecasting (with discussion), *J. R. Stat. Soc. Ser. B* 38, 205–247.

Isermann, R. (1984). Process fault detection based on modeling and estimation—a survey, *Automatica* 20, 387–404.

Jeffreys, H. (1931). *Theory of Probability*, Cambridge University Press, Cambridge.

Kalman, R. E., and R. S. Bucy (1961). New results in linear filtering and prediction theory, *Trans. ASME J. Basic Eng.* 83D, 95–108.

Kashyap, R. L. (1977). A Bayesian comparison of different classes of dynamic models using empirical data, *IEEE Trans. Autom. Control* AC-22, 715–727.

Kerr, T. H. (1983). The Controversy over Use of SPRT and GLR Techniques and Other Loose Ends in Failure Detection, in *Proceedings of the American Control Conference*, pp. 966–977.

Kullback, S., and R. A. Leibler (1951). On information and sufficiency, *Ann. Math. Stat.* 22, 79–86.

Masreliez, C. J. (1975). Approximate non-Gaussian filtering with linear state and observation relations, *IEEE Trans. Autom. Control* AC-20, 107–110.

Migon, H. S., and P. J. Harrison (1983). An Application of Non-linear Bayesian Forecasting to Television Advertising, in *Bayesian Statistics 2* (J. M. Bernardo et al., eds.), North-Holland, Amsterdam, pp. 681–696.

Poskitt, D. S., and A. R. Tremayne (1983). On the posterior odds of time series models, *Biometrika* 70, 157–162.

Prabhu, K. P. S. (1970). On the detection of a sudden change in system parameters, *IEEE Trans. Inf. Theory* IT-16, 497–500.

Smith, A. F. M., and U. E. Makov (1980). Bayesian Detection and Estimation of Jumps in Linear Systems, in *The Analysis and Optimization of Stochastic Systems* (C. Harris and O. Jacobs, eds.), Academic Press, New York, pp. 333–345.

Smith, A. F. M., and D. J. Spiegelhalter (1980). Bayes factors and choice criteria for linear models, *J. R. Stat. Soc. Ser. B* 42, 213–220.

Smith, A. F. M., and M. West (1983). Monitoring renal transplants: an application of the multiprocess Kalman filter, *Biometrics* 39, 867–878.

Spall, J. C. Bayesian error isolation for models of large-scale systems, *IEEE Trans. Autom. Control* AC-33, 341–347.

Sorenson, H. W., and D. L. Alspach (1971). Recursive Bayesian estimation using Gaussian sums, *Automatica* 7, 465–479.

Stoodley, K. D. C., and M. Mirnia (1979). The automatic detection of transients, step changes and slope changes in the monitoring of medical time series, *Statistician* 28, 163–170.

Trimble, I. M., M. West, M. S. Knapp, R. Pownall, and A. F. M. Smith (1983). Detection of renal allograft rejection by computer, *Br. Med. J.* 286, 1695–1699.

Tsang, W. L., J. D. Glover, and R. E. Bach (1981). Identifiability of unknown noise covariance matrices for some special cases of a linear time-invariant, discrete-time dynamic system, *IEEE Trans. Autom. Control* AC-26, 970–974.

West, M. (1981). Robust sequential approximate Bayesian estimation, *J. R. Stat. Soc. Ser. B* 43, 157–166.

West, M. (1982). Aspects of recursive Bayesian estimation, Ph.D. thesis, University of Nottingham.

West, M., P. J. Harrison, and H. S. Migon (1985). Dynamic generalised linear models and Bayesian forecasting (with discussion), *J. Am. Stat. Assoc.* 80, 73–97.

Willsky, A. S. (1976). A survey of design methods for failure detection in dynamic systems, *Automatica* 12, 601–611.

15

Equivalence Between Bayesian Smoothness Priors and Optimal Smoothing for Function Estimation

ROBERT KOHN and CRAIG F. ANSLEY Australian Graduate School of Management, University of New South Wales, Kensington, New South Wales, Australia

1. INTRODUCTION

The purpose of this chapter is to show that for a general signal plus white noise model, signal (function) estimation by optimal smoothing (also known as penalized least squares smoothing) is equivalent to placing a Gaussian smoothness prior on the function and estimating it by its posterior mean. Our second aim is to present this equivalence and the mathematical theory required to obtain it in an easy and accessible way. Because of the generality of our results, we now motivate them by considering two applications. The first is a structural time series model and the second is function estimation by spline smoothing.

One approach to the problem of modeling time series with trends and seasonal cycles that has received much attention lately is the use of smoothness priors to constrain the form of the components in what is otherwise a tranditional *trend-plus-seasonal-plus-irregular* model. For a sequence of observations $y(t)$, a typical model might be

$$y(t) = T(t) + S(t) + e(t), \tag{1.1}$$

where $T(t)$ represents the trend component, $S(t)$ the seasonal, and $e(t)$ the irregular. The practical problem is how to choose the stochastic models generating the individual components. Usually, $T(t)$ and $S(t)$ are chosen as solutions to nonstationary stochastic difference equations such that the forecast function for $T(t)$ is a polynomial trend, and for $S(t)$ a cycle with period one year. The irregular component $e(t)$ is usually stationary, often white noise, with mean zero. For instance, a very simple model for quarterly data which fits into the class proposed by Kitagawa and Gersch (1984) is

$$T(t) = T(t-1) + w_1(t), \qquad \sum_{j=0}^{3} S(t-j) = w_2(t), \tag{1.2}$$

where $e(t)$, $w_1(t)$, and $w_2(t)$ are independent sequences of independent Gaussian random variables with mean zero and variances σ^2, $\lambda_1\sigma^2$, and $\lambda_2\sigma^2$, respectively. These authors discuss a wide class of models for seasonal processes, and in other papers extend their work to prediction problems (Gersch and Kitagawa, 1983), to autoregressive models with time-varying coefficients (Kitagawa and Gersch, 1985a) and to the estimation of the spectral density of a stationary process (Kitagawa and Gersch, 1985b). The distinguishing characteristic of these models is that the process generating each of the components is assumed to be known a priori; for this reason the authors refer to them as smoothness priors, and the method is essentially Bayesian.

A very similar class of models is discussed by Harvey and Todd (1983) and Harvey (1985), with an interesting application to a multivariate intervention problem by Harvey and Durbin (1986). These authors refer to their models as structural time series models. However, because only the $y(t)$ are observed and not the components $T(t)$ or $S(t)$, the decomposition (1.1) may not be unique. We therefore consider (1.2) a "prior" for the trend and seasonal rather than a "model" for $y(t)$, preferring to reserve the term model for a situation where it can be uniquely determined from the data, at least in large samples.

Given the data and smoothness priors, we can obtain the posterior distributions of the unknown components, and in particular their posterior means and credible regions for the unobserved components. The posterior mean of $T(t) + e(t)$ in (1.1), for example, would be regarded as the seasonally adjusted version of the series $y(t)$. As with any Bayesian approach, the usefulness of the solution depends on the reasonableness of the prior.

An alternative way of estimating the components is to set up a loss function, and then find the sequences $\hat{S}(t)$ and $\hat{T}(t)$ which minimize it (see, e.g., Schlicht, 1981). This is known as optimal smoothing. For example, suppose that for the model (1.1) we want to minimize

$$\sum_{t=1}^{n} \{y(t) - T(t) - S(t)\}^2 + \lambda_1^{-1} \sum_{t=1}^{n} \{\nabla T(t)\}^2 + \lambda_2^{-1} \sum_{t=1}^{n} \left\{ \sum_{j=0}^{3} S(i-j) \right\}^2, \tag{1.3}$$

where $\nabla T(t) = T(t) - T(t-1)$. The first term in (1.3) measures the deviation of $y(t)$ from the signal $T(t) + S(t)$, the second attempts to make the trend as smooth as possible, and the third tries to make the sum $S(t) + S(t-1) + S(t-2) + S(t-3)$ as close to zero as possible, so that the seasonal cycle stays as close as possible to a deterministic cycle. It seems to us that these requirements are quite natural. The relative weights given the three competing requirements are controlled by the parameters λ_1 and λ_2.

What we will show is that if we use the prior (1.2) and estimate $T(t)$ and $S(t)$ by their posterior means and take $T(0)$ and $S(-3), \ldots, S(0)$ as unknown constants or diffuse, then $\hat{S}(t) = E\{S(t)|y\}$ and $\hat{T}(t) = E\{T(t)|y\}$, so that we get the same solution as for the optimal smoothing problem (1.3). In addition to the function estimates $\hat{S}(t)$ and $\hat{T}(t)$, the Bayesian approach also gives us posterior probability regions for $S(t)$ and $T(t)$. By contrast the optimal smoothing approach gives us just the function estimates.

This example illustrates that there are two ways of looking at function estimation and each complements the other. If we take the smoothness priors approach to estimating $T(t)$ and $S(t)$, then, because we only observe $y(t)$ and not the individual components $T(t)$ and $S(t)$, we cannot verify the appropriateness of the prior even in large samples. The fact that we obtain the same answer using a sensible smoothing criterion such as (1.3) adds credibility to our answer. Conversely, estimating $T(t)$ and $S(t)$ by minimizing (1.3) does not give us confidence intervals for the unknown trend and seasonal, but these can be obtained from the smoothness priors approach. In addition, by expressing (1.1) and (1.2) in state-space form, the smoothness priors approach leads to an efficient algorithm for estimating $S(t)$ and $T(t)$. See, for example, Kitagawa and Gersch (1984), who write (1.1) and (1.2) in state-space form and then use the Kalman filter and fixed interval smoothing algorithms to obtain $\hat{T}(t)$ and $\hat{S}(t)$.

Both the smoothness priors and the optimal smoothing approaches require a knowledge of the smoothing parameters λ_1 and λ_2 and the scale parameter σ^2, or estimates of them. By treating the smoothness prior as a stochastic model for the observations, Kitagawa and Gersch (1984) estimate these parameters by maximum likelihood, which are the values at the posterior mode using a uniform prior on λ_1, λ_2, and σ^2. One could also use an informative prior for a full Bayesian analysis. In contrast, the optimal smoothing approach gives no guidance on how to choose the parameters.

The model (1.1) is an example of a signal-plus-noise model with signal $T(t) + S(t)$ and noise $e(t)$. The existence of an equivalent optimal smoothing problem, in this case (1.3), holds for a much wider class of such models, including the more general component models discussed by the authors cited above. In fact, it holds for any signal-plus-noise model where the signal is generated by a stochastic difference or differential equation, typically nonstationary, and the noise is white or stationary, or at least has a known distribution. It even extends to certain nonstationary noise processes, but we will not consider this case here.

Our second example of a signal-plus-noise model arises when we want to estimate an unknown function $g(t)$ $(0 \leqslant t \leqslant 1)$ observed with noise. More specifically, we observe $y = \{y(1), \ldots, y(n)\}^T$ with

$$y(i) = g(t_i) + e(i), \tag{1.4}$$

where $e(i) \sim N(0, \sigma^2)$ is an independent sequence and for convenience we assume that $0 \leqslant t_1 \leqslant \cdots \leqslant t_n \leqslant 1$. One way of estimating $g(t)$ is to place a smooth prior on it and estimate it by the mean of the posterior distribution. One family of smooth priors which is particularly easy to work with is

$$g(t) = \sum_{j=1}^{m} \alpha_j t^{j-1}/(j-1)! + \sigma\sqrt{\lambda} \int_0^t \frac{(t-h)^{m-1}}{(m-1)!} W(dh), \tag{1.5}$$

where $W(h)$ is a zero-mean Wiener process with var$\{W(1)\} = 1$ and $\alpha_1, \ldots, \alpha_m$ are unknown constants. This prior is used by Wecker and Ansley (1983) and is an integrated Wiener process with $m - 1$ continuous derivatives. In Section 7 we give reasons why this is a reasonable prior. The model (1.4) is a signal-plus-noise model with signal $g(t)$ and noise $e(i)$, where the signal is given by (1.5).

An alternative way to estimate $g(t)$ is to minimize

$$\sum_{i=1}^{n} \{y(i) - g(t_i)\}^2 + \lambda^{-1} \int_0^1 \{g^{(m)}(t)\}^2 \, dt \tag{1.6}$$

over all functions that have a square integrable mth derivative $g^{(m)}(t)$ on the interval $[0, 1]$. The loss function (1.6) is a sum of two terms, the first a sum of squared deviations of $y(i)$ from $g(t_i)$, and the second a measure of the smoothness of $g(t)$ over $[0, 1]$. Thus minimizing (1.6) is a trade-off between a sum of squares fit and the smoothness of $g(t)$, with the trade-off being determined by the smoothness parameter λ. If λ is large, the sum of squares part dominates, while if λ is small, the smoothness part dominates. The solution to the minimization problem (1.6) is a polynomial smoothing spline of degree $2m - 1$, that is, a continuous function which is a piecewise polynomial of degree $2m - 1$ between the points $t_1, \ldots, t_n$, and whose first $2m - 2$ derivatives are continuous across these points.

Wahba (1978) showed that the smoothing spline estimated by minimizing (1.6) could be computed equivalently by obtaining the posterior mean of $g(t)$ in (1.4) using the smoothness prior (1.5) with a diffuse prior (in a sense to be defined in Sections 2 and 4) on the α_j's. Thus, as in the structural time series model discussed above, we have a Bayesian approach and an optimal smoothing approach giving essentially the same answer. As before, however, the optimal smoothing approach gives an estimate of the function alone, and to obtain confidence intervals for the unknown function we need to make use of the Bayesian model.

Even if we only want to estimate the function $g(t)$ by minimizing (1.6) and are not interested in obtaining confidence intervals for it, the Bayesian formulation is still valuable in that it leads to efficient computing algorithms for the smoothing spline itself. Wecker and Ansley (1983) and Kohn and Ansley (1987) show how to write the stochastic model (1.4) and (1.5) in state-space form and thus compute the posterior mean and variance for $g(t)$ efficiently with a diffuse prior on the α_j's, and hence obtain confidence intervals also. We note that Weinert et al. (1980) give an alternative state-space algorithm which uses a proper prior on the α_j's but imposes a particular pattern of correlation between the signal $g(t)$ and noise $e(i)$, $i = 1, \ldots, n$. See Ansley and Kohn (1986a) for a discussion of the equivalence between these formulations.

Finally, we usually need to estimate the smoothing parameter λ from the data and here the optimal smoothing approach offers no guidance. The stochastic formulation immediately suggests a number of possible estimation methods. Wecker and Ansley (1983) use maximum likelihood, and discuss a Bayesian estimator in a later paper (Ansley and Wecker, 1984). Craven and Wahba (1979) introduced a method of generalized cross-validation to estimate λ without assuming a stochastic model for $g(t)$, but even here the stochastic model is still useful because it leads to an efficient $O(n)$ algorithm for obtaining the cross-validated estimate of λ using a state-space method (see Ansley and Kohn, 1987).

A particularly simple example that illustrates the usefulness of the Bayesian approach is the function interpolation problem dealt with in Section 7. Here we observe a function at a discrete number of points without noise and want to estimate all its values in an interval. A smoothness prior Bayesian approach will give us an estimate of the unknown function together with a confidence interval, whereas spline interpolation will just give us the same estimate of the function without any measure of uncertainty.

To summarize, the correspondence between Bayesian smoothness priors and optimal smoothing for signal-plus-noise models is important because:

1. It gives us a way of assessing the smoothness priors underlying the Bayesian formulation. Even if the prior models are slightly misspecified, it is likely that we will still get sensible posterior estimates for the components because they correspond to the solution of a sensible smoothing problem.
2. The correspondence gives us two criteria for evaluating a model. For example, a smoothness prior for seasonal adjustment should imply an appropriate smoothing criterion, and vice versa.

3. The optimal smoothing approach does not give any guidance on how to estimate unknown parameters, such as the smoothness parameter in spline smoothing. The stochastic analog, however, immediately suggests Bayesian or maximum likelihood estimation procedures. Even if we have a method of parameter estimation which is not stochastically based, the stochastic analog may suggest an efficient way to implement it, for example, the state-space algorithm for generalized cross-validation of Ansley and Kohn (1987).
4. Given an optimal smoothing problem, its Bayesian equivalent often suggests an efficient algorithm for computing the unknown function.
5. For a given optimal smoothing problem, the equivalent prior distribution allows us to obtain posterior credible regions for the estimated signal.

Our purpose here is to show how the correspondence arises, to link it to previous results in the literature, and to discuss its practical importance for a number of examples. The natural mathematical framework for this discussion is a reproducing kernel Hilbert space. Because these spaces are unfamiliar to many statisticians, we spend some time giving a simple introduction. We really need little more than the definition of such a space, or in other words a notation capturing the basic property, so the investment required to make use of this structure is quite small.

The plan of this chapter is to obtain in Section 2 the basic result for some simple examples without using Hilbert spaces at all. In Section 3 we define a reproducing kernel Hilbert space and discuss some of its properties, in particular its equivalence to the space generated by a stochastic process. In Section 4 we obtain the basic equivalence results between optimal smoothing and function estimation using the mean of the posterior distribution. Section 5 deals with interpolation. Section 6 deals with a Hilbert space of functions where the inner product in the Hilbert space is defined in terms of a differential operator. In Section 7 we consider function interpolation by first imposing a prior on the function and then showing that this is equivalent to an optimal interpolation approach. In Section 8 we consider function estimation in a signal-plus-noise model using both a Bayesian approach and an optimal smoothing approach and show the equivalence between them. In Section 8.3 we relate our work to previous work on spline smoothing, and in Section 8.4 we discuss how to estimate the unknown function efficiently using the modified Kalman filter and fixed interval smoothing algorithms of Kohn and Ansley (1987). Section 9 generalizes the discussion in Section 8 to partial spline models. In Section 10 we consider a Hilbert space for functions defined on the integers, with the inner product defined by a difference operator. Finally, in Section 11 we consider function estimation by both Bayesian and optimal smoothing approaches for discrete time series models.

2. SOME SIMPLE EXAMPLES

In this section we develop the correspondence between optimal interpolation and smoothing and Bayesian smoothness priors models for three simple examples chosen to illustrate the underlying mathematical ideas. The first example deals with function interpolation on a discrete index set, but as we show in Examples 2.2 and 2.3, and more generally in Sections 3 and 4, we can easily rewrite the apparently more difficult smoothing problem as a function interpolation problem and obtain its solution as in Example 2.1. Example 2.2 considers a smoothing problem, and Example 2.3 considers a smoothing problem with independent regressors. We note the Bayesian smoothness prior in Example 2.3 is particularly simple and this simplicity is retained in more general smoothing problems with independent regressors. This provides the key insight into the effectiveness of our theory, as applied for example to the spline smoothing problem. We take a complicated smoothing problem, convert it to a simpler smoothing problem with independent regressors (the underlying Hilbert space is simpler), and then find the Bayesian analog of the latter smoothing problem.

EXAMPLE 2.1 We start with a simple interpolation problem. Suppose that we observe $y(j) = f(i_j), j = 1, \ldots, n$, with $1 \leqslant i_1 < i_2 < \ldots < i_n \leqslant N$ and with $n \leqslant N$, and we wish to estimate the entire vector $f = \{f(1), \ldots, f(N)\}^T$ by minimizing

$$f^T\Omega^{-1}f \quad \text{subject to} \quad y(j) = f(i_j), \quad j = 1, \ldots, n \tag{2.1}$$

over $f \in R^N$, where Ω is an $N \times N$ positive definite matrix and R^N is N-dimensional Euclidean space. We will derive the stochastic analog of this problem, using a method that generalizes immediately to more complicated interpolation problems.

First we define an inner product in R^N as

$$\langle f, g\rangle = f^T\Omega^{-1}g \tag{2.2}$$

with $f, g \in R^N$. This is a proper inner product and the corresponding norm is

$$\|f\| = \{f^T\Omega^{-1}f\}^{1/2}. \tag{2.3}$$

To later generalize the argument it is convenient to write $\mu f = \{f(i_1), \ldots, f(i_n)\}^T$ and $y = \{y(1), \ldots, y(n)\}^T$, and thus we need to minimize $\|f\|$ subject to $\mu f = y$. Now write the ith column of Ω as $\Omega(\cdot, i)$, and let $\rho_j = \Omega(\cdot, i_j)\, j = 1, \ldots,$

n. Note that $\langle \Omega(\cdot, j), \Omega(\cdot, l) \rangle = \Omega(j, l)$, and in particular,

$$\langle \rho_j, \rho_l \rangle = \langle \Omega(\cdot, i_j), \Omega(\cdot, i_l) \rangle = \Omega(i_j, i_l), \qquad j, l = 1, \ldots, n.$$

Define V to be the $n \times n$ matrix whose jlth element is $V(j, l) = \langle \rho_j, \rho_l \rangle = \Omega(i_j, i_l)$, so that V is made up of the rows and columns of Ω corresponding to elements of f which are observed without error. Now any $f \in R^N$ can be written as

$$f = \sum_{j=1}^{n} \delta_j \rho_j + \eta, \tag{2.4}$$

where the first term on the right in (2.4) is the projection of f on $\rho_1, \ldots, \rho_n$ and $\langle \eta, \rho_j \rangle = 0, j = 1, \ldots, n$. Therefore,

$$\|f\|^2 = \|\sum \delta_j \rho_j\|^2 + \|\eta\|^2, \tag{2.5}$$

where all norms are given by (2.3). Because $f(i_j) = \langle f, \rho_j \rangle = y(j)$ and $\langle \rho_l, \rho_j \rangle = \Omega(i_j, i_l)$, by successively taking inner products of both sides of (2.4) with $\rho_1, \ldots, \rho_n$ we obtain $\delta = (\delta_1, \ldots, \delta_n)^T = V^{-1}y$. To minimize $\|f\|$ in (2.5), we choose $\eta = 0$, and the solution to (2.1) or (2.5) is $\hat{f} = (\rho_1, \ldots, \rho_n)\delta = \rho V^{-1}y$, where $\rho = (\rho_1, \ldots, \rho_n)$.

We now set up the equivalent stochastic model. Because Ω is positive definite, we can define an N-dimensional Gaussian random vector $\tilde{f}$ with mean 0 and covariance matrix Ω. Let

$$y(j) = \tilde{f}(i_j) \qquad j = 1, \ldots, n$$

and write $y = \{y(1), \ldots, y(n)\}^T$ as above. Then $\operatorname{cov}\{\tilde{f}, y(j)\} = \Omega(\cdot, i_j) = \rho_j$, so that $\operatorname{cov}(\tilde{f}, y) = \rho$, while $\operatorname{cov}\{y(j), y(l)\} = \Omega(i_j, i_l) = V(j, l)$, so that $\operatorname{var}(y) = V$. Thus

$$E(\tilde{f}|y) = \operatorname{cov}(\tilde{f}, y)\{\operatorname{var}(y)\}^{-1}y = \rho V^{-1}y = \hat{f},$$

establishing the equivalence we require.

We will be interested later in the minimum value of (2.1). In this case, because $\langle \rho_i, \rho_j \rangle = V(i,j)$ it follows from (2.5) with $\eta = 0$ that

$$\|\hat{f}\|^2 = \delta^T V \delta = y^T V^{-1} V V^{-1} y = y^T V^{-1} y. \tag{2.6}$$

To anticipate their generalization in later sections, we now look more closely at the sequence of steps taken to obtain the correspondence above.

1. We rewrite the problem as a minimum norm problem subject to a set of constraints, and minimize over all elements in the vector (Hilbert) space R^N.
2. The constraints on the vector f are $y(j) = \mu_j f$ and the μ_j are bounded linear functionals on R^N because $\mu_j f = \langle f, \rho_j \rangle$ so that $|\mu_j f| \leqslant \|\rho_j\| \, \|f\|$

for all $f \in R^N$. In the general case we will always be dealing with a bounded linear functional and then by the Riesz representation theorem (Simmons, 1963, p. 261) there exists a vector ρ so that $\mu f = \langle f, \rho \rangle$ for all f.

3. We find a Gaussian random vector $\tilde{f}$, that is, a stochastic process with index set $\{1, \ldots, N\}$, such that $\text{cov}\{\tilde{f}(j), \tilde{f}(l)\} = \langle \Omega(\cdot, j), \Omega(\cdot, l) \rangle = \Omega(j, l)$. This is the step that presents problems for a general index set, and is the one that requires the introduction of reproducing kernels. It is discussed in the next section.
4. Next, we define a Gaussian random vector y such that

   $$\text{cov}\{y(j), \tilde{f}(l)\} = \langle \rho_j, \Omega(\cdot, l) \rangle$$

   for each j and l. In Example 2.1 it is easy: $y(j) = \tilde{f}(i_j)$. In general, however, we have to be a little more formal, although in practical examples the model for y is rarely difficult to set up.
5. Finally, we show that the solution to the optimal smoothing problem is just $E(\tilde{f} | y)$.

We go through exactly the same steps in the general case to obtain the solution.

EXAMPLE 2.2 We will now use the same argument to solve a simple smoothing problem and set up the equivalent Bayesian model. Let Ω be an $N \times N$ positive definite matrix, and let V_2 be an $n \times n$ positive definite matrix with $n \leqslant N$. For a vector $f \in R^N$ define $\mu f = \{f(i_1), \ldots, f(i_n)\}^T$ where the i_j are defined as in Example 2.1. Suppose now that we observe $y = \{y(1), \ldots, y(n)\}^T$, and we want to solve the optimal smoothing problem of minimizing

$$f^T \Omega^{-1} f + (y - \mu f)^T V_2^{-1} (y - \mu f). \tag{2.7}$$

Following Weinert et al. (1980) we obtain the solution by rewriting (2.7) as an interpolation problem similarly to (2.4) and so use the same arguments as for Example 2.1. Let $\zeta = \{\zeta(1), \ldots, \zeta(n)\}^T = y - \mu f$ and we will rewrite (2.7) in terms of the ordered pairs $(f^T, \zeta^T)^T$, which can be regarded as vectors in R^{N+n}. For notational convenience we will often write $(f^T, \zeta^T)^T$ as (f, ζ). Note first that for $f, g \in R^N$ and $\zeta, \xi \in R^n$,

$$\langle (f, \zeta), (g, \xi) \rangle = f^T \Omega^{-1} g + \zeta^T V_2^{-1} \xi \tag{2.8}$$

is an inner product on R^{N+n} with corresponding norm $\|(f, \zeta)\| = (f^T \Omega^{-1} f + \zeta^T V_2^{-1} \zeta)^{1/2}$. For $(f, \zeta) \in R^{N+n}$, define the n-dimensional vector functional $\lambda(f, \zeta) = \{f(i_1) + \zeta(1), \ldots, f(i_n) + \zeta(n)\}^T$. Then (2.7) can be rewritten as

$$\|(f, \zeta)\|^2 \tag{2.9}$$

and we minimize (2.9) over $(f, \zeta) \in R^{N+n}$ subject to $\lambda(f, \zeta) = y$. We have now expressed (2.7) in exactly the same form as the minimization problem in Example 2.1 and hence we can solve it in exactly the same way. Let $\Omega(\cdot, j)$ be the jth column of Ω and $V(\cdot, j)$ the jth column of V, and let $\rho_j = \{\Omega(\cdot, i_j), V_2(\cdot, j)\} \in R^{N+n}$. Then for any $(f, \zeta) \in R^{N+n}$,

$$\langle (f, \zeta), \rho_j \rangle = f(i_j) + \zeta(j)$$

and in particular

$$\langle \rho_l, \rho_j \rangle = \Omega(i_l, i_j) + V_2(l, j) = V_1(l, j) + V_2(l, j),$$

where V_1 is the $n \times n$ matrix with ljth element $\Omega(i_l, i_j)$. Now, just as in (2.5), we can write

$$(f, \zeta) = \sum_{j=1}^{n} \delta_j \rho_j + \eta,$$

where $\langle (f, \zeta), \eta \rangle = 0$, so that if $\lambda(f, \zeta) = y$ we must have $\delta = (V_1 + V_2)^{-1} y$, and we take $\eta = 0$ to minimize $\|(f, \zeta)\|$. Thus the optimal solution to (2.7) or (2.9) is

$$(\hat{f}, \hat{\zeta}) = \rho (V_1 + V_2)^{-1} y$$

where ρ is the $(N + n) \times n$ matrix $(\rho_1, \ldots, \rho_n)$. Because $\langle \rho_l, \rho_j \rangle = V_1(l, j) + V_2(l, j)$, it follows from (2.6) that the minimum value of the criterion function is $y^T (V_1 + V_2)^{-1} y$.

It is now very simple to set up the stochastic equivalent to (2.7). Let $\tilde{f}$ be an N-dimensional Gaussian random vector with covariance matrix Ω, and let e be an n-dimensional Gaussian random vector with covariance matrix V_2 and which is independent of $\tilde{f}$. As above we will write the $(N + n)$-dimensional vector $(\tilde{f}^T, e^T)^T$ as $(\tilde{f}, e)$ to simplify the notation. Suppose that the n-dimensional random vector y is generated by the stochastic model

$$y = \mu \tilde{f} + e,$$

where we define $\mu \tilde{f} = \{\tilde{f}(i_1), \ldots, \tilde{f}(i_n)\}^T$. Then it is easy to check that $\operatorname{var}(y) = V_1 + V_2$ and $\operatorname{cov}\{(\tilde{f}, e), y(j)\} = \{\Omega(\cdot, j), V_2(\cdot, j)\}$, so that $\operatorname{cov}\{(\tilde{f}, e), y\} = \rho$. Thus

$$E\{(\tilde{f}, e) | y\} = \operatorname{cov}\{(\tilde{f}, e), y\}\{\operatorname{var}(y)\}^{-1} y = \rho (V_1 + V_2)^{-1} y = (\hat{f}, \hat{\zeta}).$$

which is the equivalence we require.

Example 2.2 is identical to Example 2.1 once the smoothing problem has been rewritten as an interpolation problem. In doing this, note that:

1. We minimize (2.7) over a vector (Hilbert) space R^N with inner product $\langle f, g \rangle_1 = f^T \Omega^{-1} g$. The "residual" vector $\zeta = y - \mu f$ is a vector in the

space R^n with inner product $\langle \zeta, \xi \rangle_2 = \zeta^T V_2^{-1} \xi$. When we recast the problem as an interpolation problem (2.9), we have defined the direct sum $R^N \oplus R^n$ of the two spaces, with inner product

$$\langle (f, \zeta), (g, \xi) \rangle = \langle f, g \rangle_1 + \langle \zeta, \xi \rangle_2.$$

In this case the direct sum is just R^{N+n}.

2. The criterion function in (2.7) involves the coordinates $f(i_j)$ of the vector f, which are in fact linear functions μ_j defined by $\mu_j f = f(i_j)$. The corresponding coordinates $\zeta(j)$ of the residual vector can be regarded as linear functionals v_j on ζ defined by $v_j \zeta = \zeta(j)$. We can then define linear functionals λ_j on the direct sum $R^N \oplus R^n$ by $\lambda_j(f, \zeta) = \mu_j f + v_j \zeta = f(i_j) + \zeta(j)$. The vectors ρ_j are the representers of the λ_j because $\langle (f, \zeta), \rho_j \rangle = \lambda_j(f, \zeta)$. In Example 2.2, it was not necessary to set up all this formal machinery because it was easy to find the vectors ρ_j explicitly, but in general these are the steps that we must go through.

From this point on, Examples 2.1 and 2.2 are formally equivalent. Note, however, that in general we regard the random vector $(\tilde{f}, e)$ as a stochastic process with index set $T_1 \cup T_2$, where T_1 is the index set for $\tilde{f}$, in this case $\{1, \ldots, N\}$, and T_2 is the index set for e, in this case $\{1, \ldots, n\}$.

EXAMPLE 2.3 Finally, we consider a smoothing problem with independent regressors. Using the same notation as in Example 2.2, suppose that instead of minimizing (2.7) we seek to minimize

$$f^T \Omega^{-1} f + (y - \mu f - Z\beta)^T V_2^{-1} (y - \mu f - Z\beta). \tag{2.10}$$

where Z is an $n \times r$ matrix of fixed regressor variables, and β is an $r \times 1$ coefficient vector. We use the solution to Example 2.2 to show that the equivalent Bayesian model is

$$y = u\tilde{f} + Z\tilde{\beta} + e. \tag{2.11}$$

where $\tilde{f} \sim N(0, \Omega)$, $\tilde{\beta} \sim N(0, kI_r)$, and $e \sim N(0, V_2)$, where $\tilde{f}$, $\tilde{\beta}$, and e are mutually independent and k is a positive constant. Then the optimal solution to (2.10) is

$$\hat{f} = \lim_{k \to \infty} E(f \mid y; k), \qquad \hat{\beta} = \lim_{k \to \infty} E(\beta \mid y; k). \tag{2.12}$$

In other words, the solution to (2.11) is given by the posterior means of $\tilde{f}$ and $\tilde{\beta}$ under a diffuse prior on $\tilde{\beta}$.

Suppose that β is known, and we minimize (2.10) over $f \in R^N$. Let $J(f, \beta)$ be the expression in (2.10). From the solution to Example 2.2, the minimum value is

$$J\{\hat{f}(\beta), \beta\} = (y - Z\beta)^T(V_1 + V_2)^{-1}(y - Z\beta), \tag{2.13}$$

where the minimizing value $\hat{f}(\beta)$ is

$$\hat{f}(\beta) = \operatorname{cov}(\tilde{f}, y)(V_1 + V_2)^{-1}(y - Z\beta).$$

Furthermore, $J\{\hat{f}(\beta), \beta\} \leqslant J(f, \beta)$ for all f and β. Let $\hat{\beta}$ minimize (2.13), so that

$$\hat{\beta} = \{Z^T(V_1 + V_2)^{-1}Z\}^{-1}Z^T(V_1 + V_2)^{-1}y. \tag{2.14}$$

Then $J\{\hat{f}(\hat{\beta}), \hat{\beta}\} \leqslant J(f, \beta)$ for all f and β, so that the solution to (2.10) is $\hat{f} = \hat{f}(\hat{\beta})$ with $\hat{\beta}$ given by (2.14). Now it is well known that $\hat{\beta}$ in (2.14) is the limit in (2.12), and thus $\hat{f}$ is also given by (2.12). See also the derivation in Section 4.

3. REPRODUCING KERNEL HILBERT SPACES

The only step in the solutions to the examples in the preceding section that causes any real difficulties is finding a stochastic process with the right covariance properties. A related concern is exactly what these properties should be for the general case. In this section we introduce the notion of a reproducing kernel Hilbert space (RKHS), which is the proper and most general setting for our results, and use it to provide a simple and elegant solution to these problems.

One of the first authors, if not the first, to introduce the ideas of a RKHS into theoretical statistical work was Parzen (1961); see also the collection of papers in Parzen (1967). A fundamental reference to RKHS is Aronszajn (1950). Although RKHS are used extensively by Wahba and her co-workers (see, e.g., Kimeldorf and Wahba, 1970a), and also by Weinert and his co-workers [see, in particular, Weinert et. al. (1980) in connection with spline smoothing], in general the use and understanding of RKHS has been limited and disappointing. Perhaps this is because the Hilbert spaces used by these authors were rather tricky.

Let H be a Hilbert space whose elements are functions from some index set $\mathbf{T}$ to the real numbers. For example, $\mathbf{T}$ may be a subinterval of the real line, such as $[0, 1]$, or a subset of the integers, such as $\{1, \ldots, N\}$. The examples in the preceding section involved an N-dimensional vector f, which we can regard as a function from the index set $\{1, \ldots, N\}$ to the real line. Conversely, a real function defined on the interval $[0, 1]$ could be regarded as a vector of uncountably infinite dimension; this is sometimes a useful way of viewing things when dealing with RKHS. Let $\langle \cdot, \cdot \rangle$ be the inner product on H, and $\|\cdot\|$ the corresponding norm.

Suppose that there exists a function $K(\cdot, \cdot)$ from $\mathbf{T} \times \mathbf{T}$ to the real numbers such that:

1. For each $t \in \mathbf{T}$, $K(\cdot, t)$ belongs to H.
2. For $f \in H$, $\langle f, K(\cdot, t)\rangle = f(t)$ for every $t \in \mathbf{T}$.

Then H is called a RKHS with reproducing kernel K. From (i) and (ii) we have $\langle K(\cdot, s), K(\cdot, t)\rangle = K(t, s)$, and because H is a real Hilbert space, $K(s, t) = K(t, s)$. For $f \in H$, it is useful to regard $f(t)$ as the tth coordinate of f, so that we can view $K(\cdot, t)$ as projecting f onto its tth coordinate. This interpretation of the reproducing kernel will be particularly useful when we discuss the RKHS formed by the direct product of two RKHS.

EXAMPLE 3.1 Consider the Hilbert space R^N under the inner product (2.2), as discussed in Example 2.1. With $\Omega(\cdot, j) \in R^N$ representing the jth column of the positive definite matrix Ω, it is easy to check that for any vector $f \in R^N$,

$$\langle f, \Omega(\cdot, j)\rangle = f^T \Omega^{-1} \Omega(\cdot, j) = f(j),$$

so that R^N is a RKHS with reproducing kernel Ω. Notice that the role of $K(\cdot, j)$ is to project f onto its jth coordinate $f(j)$ as in (2) above.

EXAMPLE 3.2 In Example 2.2 we discussed the direct sum $R^N \oplus R^n$ of R^N and R^n made up of the ordered pairs (f, ζ) with $f \in R^N$ and $\zeta \in R^n$. This space is a Hilbert space under the inner product (2.8). It was convenient in working through Example 2.2 to identify $R^N \oplus R^n$ with R^{N+n} and regard (f, ζ) as a vector in R^{N+n}. However, to place this example in a more general context it is better to regard R^N as a space of functions f on the index set $\mathbf{T}_1 = \{1, \ldots, N\}$ and R^n as a space of functions ζ on the index set $\mathbf{T}_2 = \{1, \ldots, n\}$, and to regard (f, ζ) as a function on the union $\mathbf{T}_1 \cup \mathbf{T}_2$, where we treat the elements of $\mathbf{T}_1$ as being distinct from those of $\mathbf{T}_2$. Now define

$$K(i, j) = \begin{cases} \Omega(i, j), & i, j \in \mathbf{T}_1 \\ V_2(i, j), & i, j \in \mathbf{T}_2 \\ 0, & \text{otherwise,} \end{cases}$$

so that

$$K(\cdot, j) = \begin{cases} (\Omega(\cdot, j), 0), & j \in \mathbf{T}_1 \\ (0, V_2(\cdot, j)), & j \in \mathbf{T}_2 \end{cases} \tag{3.1}$$

and $K(\cdot, j) \in R^N \oplus R^n$ for all $j \in \mathbf{T}_1 \cup \mathbf{T}_2$. Moreover, using (3.1), if $j \in \mathbf{T}_1$,

$$\langle (f, \zeta), K(\cdot, j)\rangle = f^T \Omega^{-1} \Omega(\cdot, j) = f(j),$$

while if $j \in \mathbf{T}_2$,

$$\langle (f, \zeta), K(\cdot, j) \rangle = \zeta^T V_2^{-1} V_2(\cdot, j) = \zeta(j),$$

so that K is a reproducing kernel on $R^N \oplus R^n$.

Returning to the general case we now show that the RKHS H defined above is equivalent to a space generated by a zero-mean Gaussian random function indexed by the set $\mathbf{T}$, with covariance kernel $K(s, t)$. The random function $\tilde{f}(t)$ generating this space forms the prior for the unknown function or sequence.

Note that the kernel $K(s, t)$ is positive semidefinite in the sense that

$$\sum_{i=1}^{n} \sum_{j=1}^{n} \alpha_i \alpha_j K(t_i, t_j) = \left\| \sum_{i=1}^{n} \alpha_i K(\cdot, t_i) \right\|^2 \geqslant 0$$

for arbitrary real numbers $\alpha_1, \ldots, \alpha_n$ and indices $t_1, \ldots, t_n \in \mathbf{T}$. Thus (Doob, 1951, p. 72) there exists a zero-mean Gaussian stochastic process $\{\tilde{f}(t), t \in \mathbf{T}\}$ with covariance function $\operatorname{cov}\{\tilde{f}(t), \tilde{f}(s)\} = K(t, s)$, for $t, s \in \mathbf{T}$. Moreover (Parzen, 1961) the random variables $\{\tilde{f}(t), t \in \mathbf{T}\}$ generate a Hilbert space H_f with the inner product being defined as the covariance, and there is an isometry $\psi: H \to H_f$ such that $\psi K(\cdot, t) = \tilde{f}(t)$ for all $t \in \mathbf{T}$. An isometry is a linear mapping which is one to one, onto, and inner product preserving. Thus each RKHS is equivalent to a space of zero-mean Gaussian random variables. All we need from a reproducing kernel Hilbert space is the property that the reproducing kernel is positive semidefinite, so there is a stochastic process with index set $\mathbf{T}$ and covariance kernel $K(s, t)$.

We note that ψ maps $K(\cdot, t)$, the representer of the tth coordinate functional, to $\tilde{f}(t)$. This suggests the following definition. Suppose that μ is a bounded linear functional on H, with representer ρ (i.e., $\mu f = \langle f, \rho \rangle$ for all $f \in H$). Then for $\tilde{f} \in H_f$, we define the random variable $\mu(\tilde{f})$ by $\mu(\tilde{f}) = \psi(\rho)$. For brevity we often write $\mu(\tilde{f})$ as $\mu\tilde{f}$.

Suppose now that $\mu_i, i = 1, \ldots, n$, are n bounded and linearly independent functionals in H with the corresponding representers $\rho_i, i = 1, \ldots, n$, and let V be the $n \times n$ matrix having ijth element $V_{ij} = \langle \rho_i, \rho_j \rangle$. Let $\tilde{f}(t)$ be the random function defined above and define the $n \times 1$ vector random variable $\mu\tilde{f} = (\mu_1 \tilde{f}, \ldots, \mu_n \tilde{f})^T$. Then $\operatorname{var}(\mu\tilde{f}) = V$, because

$$\operatorname{cov}(\mu_i \tilde{f}, \mu_j \tilde{f}) = \operatorname{cov}(\psi \rho_i, \psi \rho_j) = \langle \rho_i, \rho_j \rangle = V_{ij},$$

recalling that the mapping ψ preserves inner products.

The following result extends the solution given for Example 2.1 to the general interpolation problem.

THEOREM 3.1 Let $y = \{y(1), \ldots, y(n)\}^T$ be a vector of observations and suppose that $\hat{f}$ minimizes

$$\|f\| \quad \text{subject to} \quad \mu f = y.$$

Now suppose that y is generated by the stochastic model $y = \mu\tilde{f}$, where $\tilde{f}(t)$ is the random function and $\mu\tilde{f}$ the random vector defined above. Then

$$\hat{f}(t) = E\{\tilde{f}(t) | y\}. \tag{3.2}$$

Furthermore,

$$\|\hat{f}\|^2 = y^T V^{-1} y. \tag{3.3}$$

Proof: As in Example 2.1, we can show that

$$\hat{f}(t) = \rho(t)^T V^{-1} y, \tag{3.4}$$

where ρ is the vector of functions $\rho = (\rho_1, \ldots, \rho_n)^T$. Now

$$\operatorname{cov}\{\mu_i\tilde{f}, \tilde{f}(t)\} = \operatorname{cov}(\psi\rho_i, \psi K(\cdot, t)) = \rho_i(t),$$

so that

$$E\{\tilde{f}(t) | y\} = \operatorname{cov}\{\tilde{f}(t), \mu\tilde{f}\}\{\operatorname{var}(\mu\tilde{f})\}^{-1} y,$$

as required. Eq. (3.2) follows exactly as in Example 2.1. Similarly, Eq. (3.3) can be obtained as in Example 2.1.

The first part of the proof, leading to (3.4), can be found in Luenberger (1969, p. 65) and showing that (3.2) is equivalent to (3.4) is in Kimeldorf and Wahba (1970a). This theorem is the basis of many results in Kimeldorf and Wahba (1970a, b, 1971) and Weinert et al. (1980).

We now generalize Example 2.2. Suppose that H_1 and H_2 are two RKHSs with index sets $\mathbf{T}_1$ and $\mathbf{T}_2$, reproducing kernels K_1 and K_2, inner products $\langle\cdot,\cdot\rangle_1$ and $\langle\cdot,\cdot\rangle_2$ and the corresponding norms $\|\cdot\|_1$ and $\|\cdot\|_2$, respectively. Now define the index set $\mathbf{T}$ as the index set $\mathbf{T}_2$ appended to the index set $\mathbf{T}_1$, and let H be the direct sum $H_1 \oplus H_2$ of H_1 and H_2 with typical element $h = (f, g)$, where $f \in H_1$ and $g \in H_2$. As in Example 3.2 define the function $h: \mathbf{T} \to R$ as

$$h(t) = \begin{cases} f(t), & t \in \mathbf{T}_1 \\ g(t), & t \in \mathbf{T}_2. \end{cases}$$

To illustrate what we mean by appending $\mathbf{T}_2$ to $\mathbf{T}_1$ suppose that $\mathbf{T}_1 = \{1, 2, 3, 4\}$ and $\mathbf{T}_2 = \{1^*, 2^*, 3^*, 4^*\}$, and we have starred the elements of $\mathbf{T}_2$ to differentiate them from those of $\mathbf{T}_1$. Then $\mathbf{T} = \{1, \ldots, 4, 1^*, \ldots, 4^*\}$. For

$h_1 = (f_1, g_1)$ and $h_2 = (f_2, g_2)$, define the inner product in H as

$$\langle h_1, h_2 \rangle = \langle f_1, f_2 \rangle_1 + \langle g_1, g_2 \rangle_2$$

with corresponding norm

$$\|h\|^2 = \|f\|_1^2 + \|g\|_2^2.$$

Then H is a Hilbert space. We now show that it is also a reproducing kernel Hilbert space. Recalling that a reproducing kernel maps a function into its coordinates, we require that the reproducing kernel K satisfy

$$\langle K(\cdot, t), h \rangle = \begin{cases} f(t), & t \in \mathbf{T}_1 \\ g(t), & t \in \mathbf{T}_2. \end{cases}$$

It is not difficult to check that the required reproducing kernel is

$$K(s, t) = \begin{cases} (K_1(s, t), 0), & t, s \in \mathbf{T}_1 \\ (0, K_2(s, t)), & t, s \in \mathbf{T}_2 \\ 0 & \text{otherwise.} \end{cases}$$

Let $\mu_1, \ldots, \mu_n$ be n bounded linear functionals over H_1, and let $v_1, \ldots, v_n$ be n bounded linear functionals over H_2. Define the n linear functionals $\lambda_1, \ldots, \lambda_n$ over H as

$$\lambda_i(f, g) = \mu_i(f) + v_i(g).$$

Then it is straightforward to show that the λ_i are bounded. As above there exists a random function $\tilde{f}(t)$, $t \in \mathbf{T}_1$, with K_1 as its covariance kernel, and a random function $\tilde{g}(t)$, $t \in \mathbf{T}_2$, with K_2 as its covariance kernel. Furthermore, $\tilde{f}$ and $\tilde{g}$ are defined on the same probability space and can be chosen to be independent. Define the random function $\tilde{h}$ on the composite index set $\mathbf{T}$ as

$$\tilde{h}(t) = \begin{cases} \tilde{f}(t), & t \in \mathbf{T}_1 \\ \tilde{g}(t), & t \in \mathbf{T}_2. \end{cases}$$

Then we can readily check that $\tilde{h}$ has K as its covariance kernel. Let $\mu\tilde{f}$ be the $n \times 1$ random vector with ith element $\mu_i\tilde{f}$ and let $v\tilde{g}$ be the $n \times 1$ random vector with ith element $v_i\tilde{g}$. Let V_1 be the variance covariance matrix of $\mu\tilde{f}$ and let V_2 be the variance covariance matrix of $v\tilde{g}$. Then the argument in Example 2.2 extends immediately to give the following result.

THEOREM 3.2 Suppose that we want to minimize $\|h\|$ over $h \in H$, subject to the n constraints $\lambda_i(h) = y(i)$, $i = 1, \ldots, n$. The solution is given by estimating h from the stochastic model

$$y(i) = \lambda_i(\tilde{h})$$

with $\tilde{h}$ being a random Gaussian function having zero mean and covaraince kernel K. Thus

$$\hat{f}(t) = E\{\tilde{f}(t)|y\}, \quad t \in \mathbf{T}_1, \qquad \hat{g}(t) = E\{\tilde{g}(t)|y\}, \quad t \in \mathbf{T}_2$$

and $\hat{h} = (\hat{f}, \hat{g})$. Furthermore, $\|\hat{h}\|^2 = y^T(V_1 + V_2)^{-1}y$.

Clearly, Theorem 3.2 can be generalized to three or more RKHSs.

4. BASIC EQUIVALENCE RESULT

We now obtain the equivalence between estimating an unknown function by optimal smoothing and the estimation of the function by the posterior mean for a given prior distribution. First we set out the optimal smoothing problem. Let H be a reproducing kernel Hilbert space with index set $\mathbf{T}$ and reproducing kernel $K(s, t)$. We use the same notation for norm and inner product as in Section 3. Suppose that $\mu_1, \ldots, \mu_n$ are n bounded linear functionals on H. We observe $y = \{y(1), \ldots, y(n)\}^T$ and suppose that

$$y(i) = z(i)^T\beta + \mu_i(f) + e(i), \tag{4.1}$$

where $Z = \{z(1), \ldots, z(n)\}^T$ is a sequence of $r \times 1$ independent regressors, β is a $r \times 1$ unknown vector parameter, and $e = \{e(1), \ldots, e(n)\}^T \sim N(0, \sigma^2 I_n)$ is the noise. The optimal smoothing approach estimates $f(t)$ and β by minimizing

$$J(f, \beta) = \sum_{i=1}^{n} \{y(i) - z(i)^T\beta - \mu_i(f)\}^2 + \|f\|^2. \tag{4.2}$$

over $f \in H$ and $\beta \in R^r$.

Turning to the Bayesian approach, we impose a zero-mean Gaussian prior on $\tilde{f}(t)$ with covariance function $\sigma^2 K(s, t)$ and with $\tilde{f}(t)$ independent of e. Let $\mu\tilde{f} = \{\mu_1\tilde{f}, \ldots, \mu_n\tilde{f}\}^T$ and write (4.1) as

$$y = Z\beta + \mu\tilde{f} + e. \tag{4.3}$$

Let $V = \text{var}(\mu\tilde{f})/\sigma^2$ so that V is independent of σ^2. For given β, the posterior mean of $\tilde{f}(t)$ is

$$\hat{f}(t; \beta) = E\{\tilde{f}(t)|y, \beta\} = \text{cov}\{\tilde{f}(t), \mu\tilde{f} + e\}\,\text{var}(y)^{-1}(y - Z\beta), \qquad t \in \mathbf{T}. \tag{4.4}$$

However, β is unknown and so we need to estimate it. Let $\hat{\beta}$ be the generalized least squares estimate of β in (4.3), and suppose that Z is of full column rank. Then

$$\hat{\beta} = \{Z^T(I + V)^{-1}Z\}^{-1}Z^T(I + V)^{-1}y. \tag{4.5}$$

Our estimate of $f(t)$ is $\hat{f}(t;\hat{\beta})$. The basic equivalence result obtained by Kohn and Ansley (1986) is

THEOREM 4.1 The generalized least squares estimate $\hat{\beta}$ of β given in (4.5) and the posterior mean $\hat{f}(t;\hat{\beta})$ of the function $\tilde{f}(t)$ given in (4.4) minimize (4.2) over all β and $f \in H$.

The proof follows from Theorem 3.2 and the argument of Example 2.3. We will use the following extension of Theorem 4.1 in the following sections.

COROLLARY 4.1 Let $\lambda > 0$ be given. The prior corresponding to the smoothing problem of minimizing

$$\sum_{i=1}^{n} \{y(i) - z(i)^T\beta - \mu_i(f)\}^2 + \frac{\|f\|^2}{\lambda}.$$

over $f \in H$ and $\beta \in R^r$ is the same as above except that $\operatorname{cov}\{\tilde{f}(t), \tilde{f}(s)\} = \sigma^2 \lambda K(s, t)$.

Proof: By redefining the inner product as $\langle \cdot, \cdot \rangle / \lambda$ the reproducing kernel becomes $\lambda K(s, t)$ and the result now follows from Theorem 4.1.

We now show that instead of regarding β as an unknown constant vector in (4.3), β can equivalently have a diffuse prior distribution in a sense to be made clear below. This result was first obtained by Wahba (1978) for spline smoothing, and we generalize it to the general smoothing problem (4.2), using a different, simpler proof.

For given k consider the problem of minimizing

$$J_k(f, \beta) = J(f, \beta) + \frac{\|\beta\|^2}{k}, \tag{4.6}$$

where $\|\beta\|^2 = \beta^T\beta$. Using the proof of Corollary 4.1 and a simple extension of Theorem 3.2 this is equivalent to estimating $\tilde{f}$ and $\tilde{\beta}$ from (4.3) with $\tilde{\beta}$ having the prior distribution $N(0, \sigma^2 k I_r)$ independent of e and $\tilde{f}(t)$.

Keeping β fixed and minimizing (4.6) over $f \in H$ gives

$$J_k(\hat{f}(\cdot, \beta, k), \beta) = (y - Z\beta)^T(I + V)^{-1}(y - Z\beta) + \frac{\|\beta\|^2}{k} \tag{4.7}$$

with

$$\hat{f}(t; \beta, k) = E\{\tilde{f}(t)|y, \beta, k) = \sigma^{-2} \operatorname{cov}\{\tilde{f}(t), \mu\tilde{f} + e\}(I + V)^{-1}(y - Z\beta).$$

Let $\hat{\beta}_k$ minimize (4.7). Then $E(\tilde{\beta}|y; k) = \hat{\beta}_k$ and it is not difficult to show that

$$\lim_{k \to \infty} \hat{\beta}_k = \hat{\beta},$$

where $\hat{\beta}$ is the generalized least squares estimate (4.5). Thus

THEOREM 4.2 Consider the model (4.3) with the prior distributions on $\tilde{f}(t)$ and $\tilde{\beta}$ as above. Then

$$\hat{f}(t) = \lim_{k\to\infty} E\{\tilde{f}(t)|y; k\} \quad \text{and} \quad \hat{\beta} = \lim_{k\to\infty} E\{\tilde{\beta}|y; k\}$$

minimize (4.2).

Theorem 4.2 is important because it leads to efficient algorithms for computing $\hat{\beta}$ and $\hat{f}$. In particular, it leads to the efficient spline smoothing algorithm in Kohn and Ansley (1987).

5. EQUIVALENCE FOR AN OPTIMAL INTERPOLATION PROBLEM

Keeping all notation the same as in Section 4 except that we now take $\sigma^2 = 1$, we consider the corresponding interpolation problem, which is to minimize $\|f\|$ subject to

$$y(i) = z(i)^T\beta + \mu_i(f), \qquad i = 1, \ldots, n \tag{5.1}$$

with β an $r \times 1$ vector of unknown parameters. Obtaining the equivalent stochastic model and prior distribution proceeds similarly to Section 4. Thus for fixed β it follows from Section 3 that

$$\hat{f}(t; \beta) = \operatorname{cov}\{\tilde{f}(t), y\}V^{-1}(y - Z\beta)$$

is the function minimizing $\|f\|$ and

$$\|\hat{f}\|^2 = (y - Z\beta)^T V^{-1}(y - Z\beta).$$

Minimizing over β gives

$$\hat{\beta} = (Z^T V^{-1} Z)^{-1} Z^T V^{-1} y \tag{5.2}$$

so that the equivalent stochastic model is (5.1) with the prior for the function $f(t)$ being the same as in Section 4 and $\hat{\beta}$ is the generalized least squares estimate of β in (5.2). The estimate of $f(t)$ is $\hat{f}(t; \hat{\beta})$. Alternatively, we can obtain $\hat{\beta}$ and $\hat{f}$ by taking $\beta \sim N(0, kI_r)$ in the model (5.1), and letting $k \to \infty$.

6. HILBERT SPACE OF FUNCTIONS

We now introduce a space of smooth functions on the interval [0, 1] whose inner product is defined by a differential operator, and obtain the reproducing kernel for it. This space will be central to our discussion of function smoothing and interpolation in the next two sections.

Let the mth-order differential operator L_t be defined by

$$L_t = \frac{d^m}{dt^m} + a_{m-1}(t)\frac{d^{m-1}}{dt^{m-1}} + \cdots + a_0(t), \tag{6.1}$$

where the coefficients $a_i(t)$ are continuous functions of $t \in [0, 1]$.

Let H be the space of functions on $[0, 1]$ having square integrable mth derivative and with initial conditions

$$f^{(j)}(0) = 0, \qquad j = 0, \ldots, m-1. \tag{6.2}$$

Then it is not difficult to check that H is a Hilbert space with inner product

$$\langle f_1, f_2 \rangle = \int_0^1 (L_t f_1)(L_t f_2)\, dt.$$

We now show that H is a RKHS by introducing the Green's function for L_t.

There exists a Green's function $G(t, s)$, $s, t \in [0, 1]$ so that for $f \in H$,

$$f(t) = \int_0^1 G(t, s) L_s f(s)\, ds. \tag{6.3}$$

A nice description of $G(t, s)$ and a proof of its properties is given by Hurewicz (1961, pp. 52–54). Here we just note that for fixed s:

1. $G(t, s) = 0$ for $0 \leqslant t < s$.
2. $L_t G(t, s) = 0$ for $t \geqslant s$.
3. For $j = 0, \ldots, m-2$, $\partial^j G(t, s)/\partial t^j$ is a continuous function of t for all $t \in [0, 1]$, and $\partial^{m-1} G(t, s)/\partial t^{m-1} \to 1$ as $t \to s$ from above.

Thus $G(t, s)$ can be determined uniquely on $[s, 1]$ because of (2) and (3). It is identically zero on $[0, s)$.

In particular, if $L_t = d^m/dt^m$, then

$$G(t, s) = \begin{cases} (t-s)^{m-1}/(m-1)! & t \geqslant s \\ 0 & t < s. \end{cases} \tag{6.4}$$

Now, if a reproducing kernel $K(s, t)$ exists, then

$$f(t) = \langle f(\cdot), K(t, \cdot) \rangle = \int_0^1 L_s f(s) L_s K(t, s)\, ds$$

and (6.3) suggests that $L_s K(t, s) = G(t, s)$. Applying (6.3) again, we obtain

$$K(t, s) = \int_0^1 G(s, v) G(t, v)\, dv. \tag{6.5}$$

We can in fact check that for fixed s, $K(t, s)$, as given by the right side of (6.5), belongs to H and is the reproducing kernel.

Let H^* be the space of functions having square integrable mth derivative on [0, 1], and let $H_0 = \{g \in H^*, L_t g = 0\}$ be the null space of L_t, with $h_1(t), \ldots, h_m(t)$ a basis of H_0. Then any $g \in H^*$ can be written uniquely as

$$g(t) = \sum_{i=1}^{m} \alpha_i h_i(t) + f(t),$$

where $f(t) \in H$ and $L_t g = L_t f$. In particular, if $L_t = d^m/dt^m$, then $1, t, t^2/2!, \ldots, t^{m-1}/(m-1)!$ is a basis for H_0 and any $g(t)$ with square integrable mth derivative can be written as

$$g(t) = \sum_{i=1}^{m} \frac{\alpha_i t^{i-1}}{(i-1)!} + f(t), \tag{6.6}$$

where $\alpha_i = g^{(i)}(0)$ and $f(t)$ satisfies the initial conditions (6.2).

7. FUNCTION INTERPOLATION

Let $g(t)$ be a function on the real line whose values we observe at a finite number of points only, and suppose that we would like to have an estimate of the function at all points in some bounded interval. For ease of exposition, and without loss in generality, suppose that $g(t)$ is a function on [0, 1] and we observe $y(i) = g(t_i)$, $i = 1, \ldots, n$, with $0 \leqslant t_1 < \cdots < t_n \leqslant 1$. We first present a Bayesian approach to computing $g(t)$ and then the equivalent optimal interpolation approach which gives the optimal interpolating spline. For the rest of the chapter we will not make a distinction between a function as a member of a Hilbert space and the corresponding stochastic process. This will simplify the notation and should not cause any confusion. It will also be consistent with the Bayesian view of placing a prior on a function.

7.1 A Bayesian Approach

One way of estimating $g(t)$ is to place a prior on it, and then use the mean of the posterior distribution of $g(t)$ as the estimate. This approach also gives us posterior confidence intervals for $g(t)$. Suppose that we believe that $g(t)$ is smooth. Specifically, suppose that we believe that $g(t)$ is m times differentiable, but that we have little information about the mth derivative, so that it may be reasonable to put a prior with a large variance on $d^m g(t)/dt^m$.

One effective way of doing this is to suppose that the prior for $g(t)$ is generated by the differential equation

$$L_t g(t) = \frac{dW(t)}{dt}, \tag{7.1}$$

where $L_t = d^m/dt^m$, and $W(t)$ is a zero-mean Wiener process with $\text{var}\{W(1)\} = 1$. A discussion of the Wiener process is given in Doob (1953). For our purposes we note that $W(t)$ is a Gaussian process with orthogonal increments, and that for $h > 0$ we have that $\text{var}\{W(t+h) - W(t)\} = h$, so that $\text{var}\{W(t+h) - W(t)\}/h \to \infty$ as $h \to 0$. It is clear from (6.3) and (6.6) that the solution of (7.1) is

$$g(t) = \sum_{j=1}^{m} \frac{\alpha_j}{(j-1)!} t^{j-1} + f(t) \tag{7.2}$$

$$f(t) = \int_0^t G(t, s)W(ds), \tag{7.3}$$

where $G(t, s)$ is the Green's function (6.4) corresponding to the differential operator L_t above, and $\alpha_j = d^j g(0)/dt^j$.

Thus we can consider the prior on $g(t)$ as the sum of a polynomial of degree $m - 1$ with unknown coefficients, together with a function $f(t)$ whose mth derivative has infinite variance, whose lower-order derivatives have well-defined Gaussian distributions, and which satisfies the zero initial conditions (6.2). Because we only observe the function at a discrete number of points, the data do not give us any information about the smoothness of the function, and hence the sample paths generated by the posterior distribution will have the same smoothness properties as the sample paths of the prior. We note, however, that the posterior mean of $g(t)$ has continuous derivatives of order $2m - 2$ and so is smoother than the sample paths generated by the posterior distribution. See Weinert et al. (1979), and for a general treatment, Kohn and Ansley (1983). These results extend to function smoothing, which is dealt with in the next section.

An interesting and appealing feature of the prior above which can be deduced from (8.10) below is that for $h > 0$ is that

$$E\{g(t+h)|g(t), \ldots, g^{(m-1)}(t)\} = g(t) + hg^{(1)}(t) + \cdots + \frac{h^{m-1}}{(m-1)!} g^{(m-1)}(t)$$

and in general for $j = 0, \ldots, m$,

$$E\{g^{(j)}(t+h)|g(t), \ldots, g^{(m-1)}(t)\}$$
$$= g^{(j)}(t) + hg^{(j+1)}(t) + \cdots + \frac{h^{m-j-1}}{(m-j-1)!} g^{(m-1)}(t),$$

so that in particular the prior implies that the best predictor of $g(t+h)$ given $g(t), \ldots, g^{(m-1)}(t)$ is its Taylor series expansion of order $m - 1$, and the best predictor of $g^{(m-1)}(t+h)$ is just $g^{(m-1)}(t)$.

Let $w(t) = \{1, t, \ldots, t^{m-1}/(m-1)!\}$, and $\alpha = (\alpha_1, \ldots, \alpha_m)^T$, so that

$w(t)^T\alpha = \sum \alpha_j t^{j-1}/(j-1)!$. We are given the observations

$$y(i) = w(t_i)^T\alpha + f(t_i), \qquad i = 1, \ldots, n. \tag{7.4}$$

As in Section 4, for given α, let $\hat{f}(t: \alpha)$ be the estimate of $f(t)$ and let $\hat{\alpha}$ be the generalized least squares estimate of α obtained from (7.4). Then the estimate of $g(t)$ is

$$\hat{g}(t) = w(t)^T\hat{\alpha} + \hat{f}(t; \hat{\alpha}).$$

Alternatively, to specify a prior for $g(t)$ fully, we need to impose a prior on α, that is, on the value of g and its first $m - 1$ derivatives at $t = 0$. Often we do not have prior information about α, so that a diffuse prior may be appropriate. We define a diffuse prior on α by a limiting operation as in Sections 4 and 5, and furthermore, we assume that α is independent of f. Thus for given $k > 0$, take $\alpha \sim N(0, kI_m)$ and suppose that α is independent of $f(t)$. Then for fixed k, our estimate of $g(t)$ is the posterior mean $E\{g(t)| y; k\}$ with the posterior variance of our estimate being $\operatorname{var}\{g(t)| y; k\}$. We define the posterior mean and variance of $g(t)$ for a diffuse prior as

$$\hat{g}(t) = \lim_{k\to\infty} E\{g(t)| y; k\}, \qquad \hat{S}_g(t) = \lim_{k\to\infty} \operatorname{var}\{g(t)| y; k\}. \tag{7.5}$$

Thus $\hat{S}_g(t)$ indicates the uncertainty in estimating $g(t)$ given y and the 95% confidence interval for $g(t)$ is $\{\hat{g}(t) - 1.96\sqrt{\hat{S}_g(t)}, \hat{g}(t) + 1.96\sqrt{\hat{S}_g(t)}\}$. At the observed points $\hat{g}(t_i) = y(i)$ and $\hat{S}_g(t_i) = 0$.

The posterior mean $\hat{g}(t)$ and the posterior variance $\hat{S}_g(t)$ can be computed efficiently as in Kohn and Ansley (1987). See Section 8.4 for some remarks.

7.2 Optimal Interpolation Approach

From (7.3), $\operatorname{cov}\{f(t), f(s)\} = K(t, s)$, with $K(t, s)$ given by (7.5). It therefore follows from Section 5 that the smoothing problem corresponding to the prior in (7.1) is to minimize

$$\|f\|^2 = \int_0^1 (L_t f)^2\, dt$$

subject to (7.4). This characterizes an optimal interpolating spline.

Thus our Bayesian approach to interpolation is equivalent to spline interpolation. Note, however, that with the Bayesian approach we also obtain a posterior variance of $g(t)$ to give us some idea of the accuracy of our estimate.

We can interpret the equivalence between the posterior mean of the unknown function and the solution of the optimal smoothing problem as follows. Suppose that we have n observations on the unknown function $g(t)$ and we choose as its estimate the function $\hat{g}(t) \in H^*$ having square integrable mth derivative which minimizes $\int \{\hat{g}^{(m)}(t)\}^2 \, dt$ subject to $y(i) = \hat{g}(t_i)$. What we have shown is that if $g(t)$ has the prior (7.2) and (7.3), then $E\{g(t) | y\}$ belongs to H^* and is the solution to this optimal smoothing problem. There is no implication that the true function $g(t)$ has a square integrable mth derivative. If we know that it does and if this derivative is also continuous, we should incorporate this knowledge in our prior. One way of doing so is to use (7.1) as a prior with $L_t = d^{m+1}/dt^{m+1}$ so that the posterior mean will be an interpolating spline of order $2m + 1$, which will in fact be smoother than the solution to the optimal smoothing problem just discussed. This remark clears up the paradox mentioned in Wahba (1983) and Silverman (1985), who observe that the solution of the optimal smoothing problem is smoother than the sample paths generated by the prior and posterior distributions.

8. ESTIMATING A FUNCTION OBSERVED WITH NOISE

Let $g(t)$ be a function on the real line whose values we observe with noise at a finite number of points only, and suppose that we would like to have an estimate of the function at all points in some bounded interval. As above, suppose that $g(t)$ is a function on $[0, 1]$ and we observe

$$y(i) = g(t_i) + e(i), \qquad i = 1, \ldots, n. \tag{8.1}$$

where the $e(i)$ are $N(0, \sigma^2)$ and independent and $0 \leqslant t_1 \leqslant \cdots \leqslant t_n \leqslant 1$. We note that the t_i need no longer be distinct because it is now possible to have several observations at the one point.

We first present the Bayesian approach to estimating $g(t)$ and then the equivalent optimal smoothing approach. The treatment here is similar to Section 7, because our assumptions on the unknown function will be the same. In Section 8.3 we point out that Sections 4 and 5 allow the generalization of the results in Section 7 and Sections 8.1 and 8.2 above to more general splines called Lg splines. We also relate our results to previous results in the literature. In Section 8.4 we describe briefly the state-space approach of Kohn and Ansley (1987) for computing the smoothing spline efficiently, and in Section 8.5 we discuss how to compute unknown parameters.

8.1 Bayesian Approach

Because our prior knowledge of the unknown function $g(t)$ will be the same whether we observe the function with or without noise, our prior for $g(t)$ will be similar to that of Section 7.1 as described by (7.1)–(7.3). Now, however, we need to differentiate between the variance of the $e(i)$ and the variance of $f(t)$, so that we retain (7.2) while (7.1) and (7.3) become, respectively,

$$L_t g(t) = \sigma\sqrt{\lambda}\,\frac{dW(t)}{dt} \tag{8.2}$$

and

$$f(t) = \sigma\sqrt{\lambda}\int_0^1 G(t, s)W(ds), \tag{8.3}$$

so that σ is the scale of the noise $e(i)$ and $\sigma\sqrt{\lambda}$ is the scale of $f(t)$. The prior on α is $N(0, \sigma^2 k I_m)$ and α is independent of f. We rewrite the observation equation (8.1) as

$$y(i) = w(t_i)^T\alpha + f(t_i) + e(i), \qquad i = 1, \ldots, n,$$

so for given λ and σ^2 the posterior mean and variance of $g(t)$ are given by (7.5), with $\hat{S}_g(t)$ finite if there are at least m distinct points t_i. Equivalently, we can treat α as an unknown parameter.

When λ and σ^2 are given, the conditional distribution of $g(t) = w(t)^T\alpha + f(t)$ given y is Gaussian, and we can obtain its mean and variance, and hence credible regions, as described in Section 8.4.

8.2 Equivalent Optimal Smoothing Problem

From Section 8.1, $\operatorname{cov}\{f(t), f(s)\} = \sigma^2\lambda K(s, t)$ and $\operatorname{cov}\{e(i), e(j)\} = \sigma^2$ if $i = j$ and is zero otherwise. It follows from Corollary 4.1 that the smoothing problem corresponding to (7.2) and (8.3) is to minimize

$$\sum_{i=1}^{n} \{y(i) - w(t_i)^T\alpha - f(t_i)\}^2 + \frac{\|f\|^2}{\lambda} \tag{8.4}$$

over $f \in H$ and α, with H defined as in Section 6. Rewriting (8.4) as

$$\sum_{i=1}^{n} \{y(i) - g(t_i)\}^2 + \frac{1}{\lambda}\int_0^1 (L_t g)^2\, dt.$$

it is readily seen that minimizing over $g \in H^*$ gives the optimal smoothing spline of degree $2m - 1$. The space H^* is described in Section 6.

8.3 Generalization and Relationship to Previous Work

Using the results in Sections 4 and 5, the equivalence between function estimation using a prior for $g(t)$ and function estimation by spline smoothing or interpolation can be generalized to the case where L_t is the mth-order linear differential operator (6.1), with the prior for $g(t)$ given by (8.2) and the linear functionals $g(t_i)$ replaced by general bounded linear functionals. The resulting function estimates $\hat{g}(t)$ are called *Lg* smoothing splines. See Kohn and Ansley (1987) for details.

The equivalence between spline smoothing and the estimation of a stochastic process observed with noise was first obtained for Chebycheffian splines by Kimeldorf and Wahba (1971); see also Wahba (1978). This prior is the same we used in Section 8.1. The equivalence for the most general *Lg* smoothing spline was given by Weinert et al. (1980), who used a somewhat different stochastic process with $g(t)$ correlated with the $e(i)$. For a discussion of the relation between the Wahba and Weinert priors, see Ansley and Kohn (1986a).

Weinert et al. (1980) also presented an efficient $O(n)$ spline smoothing algorithm, given the smoothing parameter λ. They were the first to point out that the stochastic prior could be written in state-space form, and thus the smoothing spline could be obtained efficiently using the Kalman filter and the fixed-interval smoothing algorithm.

It is unfortunate, therefore, that the papers of Weinert and his co-workers, and in particular the important papers by Weinert and Sidhu (1978) and Weinert et al. (1980), have been largely ignored by statisticians.

Our proof of the equivalence between the Bayesian and optimal smoothing approaches for spline smoothing (and the related problem of interpolation) seems significantly simpler than that of either Kimeldorf and Wahba (1971) or Weinert et al. (1980), because the Hilbert space we work with is much simpler. We obtain this simplicity by reducing the mathematical problem to the Hilbert space H defined in Section 6 together with an unknown vector parameter α, or H plus a Hilbert space defined with respect to α. By contrast, both the Weinert and Wahba approaches involve either a separate Hilbert space for H_0 or a Hilbert space on the whole of H^*, where H^* is defined in Section 6.

We want to point out, however, that the results and insights of Kimeldorf and Wahba (1970a, b, 1971), Wahba (1978), and Weinert et al. (1980) were very important in deriving our results.

When λ and σ^2 are known, our method gives $\hat{g}(t) = E\{g(t)|y\}$ and $\hat{S}_g(t) = \text{var}\{g(t)|y\}$ for each t, and we can obtain 95% posterior credible regions. Alternative approaches to computing $\text{var}[g(t)|y]$ are given by Silverman (1985) and Wahba (1983). Silverman (1985) first assumes that $g(t)$ is a polynomial spline of degree $2m - 1$ with knots at the points t_i. Let $g = \{g(t_1), \ldots, g(t_n)\}^T$ and $\hat{g} = \{\hat{g}(t_1), \ldots, \hat{g}(t_n)\}^T$. Silverman first obtains $\text{var}(g|y)$ using the Bayesian model and computes $\text{var}[g(t)|y]$ from $\text{var}(g|y)$ as discussed below. Our conditional variance, which is the same as that in Wecker and Ansley (1983), is at least as big as that obtained by Silverman and is only the same when $g(t)$ is a polynomial spline of degree $2m - 1$. It is instructive to see why this is so.

Suppose first that $\hat{g}(t)$ is the spline interpolating the points $y(i) = g(t_i)$, $i = 1, \ldots, n$. Then from the stochastic model in Section 7.1, we can write

$$\hat{g}(t) = \sum_{i=1}^{n} c_i(t)y(i) = \sum_{i=1}^{n} c_i(t)g(t_i),$$

where the $c_i(t)$ are nonstochastic and depend on t. Put $c(t) = \{c_1(t), \ldots, c_n(t)\}^T$. Then $\hat{g}(t) = c(t)^T g$. If $g(t)$ is a spline, then $g(t) = c(t)^T g$, so that $\hat{g}(t) = g(t)$ for all t and $\operatorname{var}(g(t)|y) = 0$. Hence if $g(t)$ is a spline of a given order, we know it exactly for all t once we observe it at a fixed number of points. By contrast our approach above and that of Wecker and Ansley (1983) does not assume that $g(t)$ is a spline, so we obtain that $\operatorname{var}(g(t)|y) > 0$ for $t \neq t_i$. To see this, note that

$$\operatorname{var}\{g(t)|y\} = E\{\operatorname{var}[g(t)|g]|y\} + \operatorname{var}\{E[g(t)|g]|y\}. \tag{8.5}$$

Now for the interpolation problem $g = y$, so that

$$E[g(t)|g] = c(t)^T g \tag{8.6}$$

and the second term in (8.5) is identically zero. The first term in (8.5) is $E\{[g(t) - c(t)^T g]^2|y\} > 0$ for $t \neq t_i$ $(i = 1, \ldots, n)$ because of the uncertainty about $g(t)$ inherent in the Bayesian model.

For spline smoothing, Silverman (1985) first computes $\hat{g}$ and uses the Bayesian prior to obtain $\operatorname{var}(g|y) = \operatorname{var}(g - \hat{g})$. Because $\hat{g}(t)$ is a smoothing spline, it follows from above that $\hat{g}(t) = c(t)^T\hat{g}$. If, in addition, $g(t)$ is a spline of order $2m - 1$, then $g(t) = c(t)^T g$ and Silverman computes $\operatorname{var}[g(t)|y]$ as

$$\operatorname{var}[g(t) - \hat{g}(t)] = c(t)^T \operatorname{var}(g - \hat{g})c(t). \tag{8.7}$$

Consider now the stochastic model. Equation (8.6) still holds so that the second term in (8.5) is given by (8.7). As in the interpolation case, the first term of (8.5) will be greater than zero except at $t = t_i$ $(i = 1, \ldots, n)$, indicating our inherent uncertainty about $g(t)$.

Wahba (1983, Sect. 2) obtains an expression for $\operatorname{var}[g(t)|y]$ in (8.5) for the case of polynomial splines. However, this expression is rather complicated, and in her examples she restricts herself to periodic cubic splines with observations at $t_i = i/n$ $(i = 1, \ldots, n)$, in which case she is able to obtain a relatively simple expression for $\operatorname{var}(g|y)$. This gives confidence regions for $\hat{g}(t_i)$, $i = 1, \ldots, n$. Confidence bands for intermediate points are plotted in the figures in Section 3 of Wahba (1983), but she notes that strictly speaking these have meaning only at the arguments t_i $(i = 1, \ldots, n)$, and presumably intermediate values are obtained by interpolation.

All these arguments depend on knowing λ and σ^2, and Wahba (1983), Silverman (1985), and Wecker and Ansley (1983) condition their results on the estimated values of these parameters. Ansley and Wecker (1984) give a Bayesian argument that places an improper prior on λ and σ^2 and compute the full posterior distribution of $g(t)$ given y.

8.4 Efficient Computation of Smoothing Splines

It was Weinert et al. (1980) who first pointed out that we can rewrite the observation equation (8.1) and the stochastic prior for $g(t)$ in state-space form, and so obtain an efficient $O(n)$ spline smoothing algorithm for computing the smoothing spline and its posterior variance, given the smoothing parameter λ. This is also the basis of the spline smoothing algorithms of Wecker and Ansley (1983) and Kohn and Ansley (1987). Previously, Reinsch (1967) gave an $O(n)$ cubic spline smoothing algorithm, but unlike the state-space-based methods, Reinsch's algorithm does not give an estimate of the estimation error variance, does not generalize to the more general Lg smoothing spline case discussed in the first paragraph of Section 8.3, and because it requires divided differences of the data, may be numerically more unstable than algorithms based on the state-space approach.

We now follow Kohn and Ansley (1987) and briefly outline the state-space approach to computing the smoothing spline and its conditional mean squared error. Define the $m \times 1$ state vector $x(t) = \{g(t), g^{(1)}(t), \ldots, g^{(m-1)}(t)\}^T$, so that the state vector consists of $g(t)$ and its first $(m-1)$ derivatives. Then the observation equation (8.1) becomes

$$y(i) = x_1(t_i) + e(i), \tag{8.8}$$

where $x_1(t)$ is the first element of $x(t)$, and as in Kohn and Ansley (1987), we have the state transition equation

$$x(t_{i+1}) = F(t_{i+1} - t_i)x(t_i) + u(t_{i+1}, t_i), \tag{8.9}$$

where

$$F(t) = \begin{bmatrix} 1 & t & \cdot & \cdot & \cdot & t^{m-1}/(m-1)! \\ 0 & 1 & t & \cdot & \cdot & t^{m-2}/(m-2)! \\ \cdot & \cdot & \cdot & \cdot & \cdot & \cdot \\ \cdot & \cdot & \cdot & \cdot & \cdot & \cdot \\ \cdot & \cdot & \cdot & \cdot & \cdot & \cdot \\ 0 & 0 & 0 & 0 & 0 & 1 \end{bmatrix}$$

and

$$u(t_{i+1}, t_i) = \sigma\sqrt{\lambda}\int_{t_i}^{t_{i+1}} F_m(t_{i+1} - s)W(ds).$$

The vector $F_m(t)$ is the mth column of $F(t)$. Equations (8.8) and (8.9) are the state-space equations, and together with the initial conditions $x(0) \sim N(0, kI_m)$, make up our state-space model. In the limit as $k \to \infty$ we impose diffuse initial conditions on the state vector.

Let the mean and variance of $x(t)$ conditional on $y(1), y(2), \ldots, y(j)$ and k be, respectively,

$$x(t|j;\, k) = E\{x(t)|y(1), \ldots, y(j)\}, \quad S(t|j;\, k) = \operatorname{var}\{x(t)|\, y(1), \ldots, y(j)\}.$$

Then, by Ansley and Kohn (1985) or Kohn and Ansley (1987),

$$x(t|j;\, k) = x^{(0)}(t|j) + O\left(\frac{1}{k}\right)$$

$$S(t|j;\, k) = kS^{(1)}(t|j) + \sigma^2 S^{(0)}(t|j) + O\left(\frac{1}{k}\right),$$

where $x^{(0)}(t|j)$, $S^{(1)}(t|j)$, and $S^{(0)}(t|j)$ are independent of k and σ^2. Furthermore, $S^{(1)}(t|j) = 0$ for $j \geqslant m$.

For $i = 1, \ldots, n$, $x^{(0)}(t_i|i)$, $S^{(1)}(t_i|i)$, and $S^{(0)}(t_i|i)$ can be computed efficiently using the modified Kalman filter of Ansley and Kohn (1985) and for any $t \in [0, 1]$, $x^{(0)}(t|n)$ and $S^{(0)}(t|n)$ are obtained using the modified fixed-interval smoothing algorithm of Ansley and Kohn (1985). All the details are set out in Kohn and Ansley (1987). We note that

$$\lim_{k\to\infty} x(t|n;\, k) = x^{(0)}(t|n), \qquad \lim_{k\to\infty} S(t|n;\, k) = S^{(0)}(t|n),$$

so that we obtain directly the limiting values of $x(t|n;\, k)$ and $S(t|n;\, k)$ as $k \to \infty$.

The smoothing spline and its conditional mean squared error are given by

$$\hat{g}(t) = x_1^{(0)}(t|j) + O\left(\frac{1}{k}\right)$$

$$\sigma^2 \hat{S}_g(t) = \operatorname{var}\{g(t)|\, y\} = \sigma^2 S_{11}^{(0)}(t|n).$$

When there are repeated observations at the t_i some computational savings may be obtained as outlined by Hutchinson and de Hoog (1985) and Ansley and Kohn (1986b).

8.5 Parameter Estimation

The estimate $\hat{g}(t)$ is a function of the smoothness parameter λ. Usually, λ is unknown and has to be estimated. To obtain the posterior variance $\sigma^2 \hat{S}_g(t)$ of $g(t)$, we also need an estimate of σ^2. Wecker and Ansley (1983) obtained the first $O(n)$ spline smoothing algorithm that estimated both λ and, given λ, the smoothing spline. They did so by treating the prior for $g(t)$ as a stochastic model for the observations y, writing down the likelihood for λ and σ^2, and then estimating these two parameters by maximum likelihood. Neither Reinsch (1967) nor Weinert et al. (1980) provided satisfactory ways to estimate λ. A slightly different likelihood is defined and efficiently computed by Kohn and Ansley (1987) using the modified Kalman filter.

Craven and Wahba (1979) suggested estimating λ by generalized cross-validation (GCV), which is a variant of ordinary cross-validation and does not depend for its justification on the validity of the prior for $g(t)$. Simulation and asymptotic theory is presented by Wahba (1985) suggesting that if the unknown function $g(t)$ is smoother than the prior (8.2) for $g(t)$, then the GCV estimate of λ will be better than the maximum likelihood estimate in the sense of giving a smaller value of

$$E\frac{1}{n}\sum_{i=1}^{n}\{(g(t_i)-\hat{g}(t_i;\lambda)\}^2.$$

For polynomial smoothing splines, Silverman (1984) obtained an approximate $O(n)$ GCV estimate and Hutchinson and de Hoog (1985) obtained the first exact $O(n)$ GCV algorithm. Ansley and Kohn (1987) gave an exact $O(n)$ algorithm for computing GCV in general state-space models, and in particular for general Lg smoothing splines. The Ansley and Kohn GCV algorithm is a direct result of the Bayesian interpretation of smoothing splines.

9. PARTIAL SPLINES

Suppose that we have the following model for the observations:

$$y(i) = v(i)^T\gamma + g(t_i) + e(i), \tag{9.1}$$

where the $e(i)$ are an independent $N(0, \sigma^2)$ sequence, γ is a $p \times 1$ vector of unknown parameters, and $v(i)$ is a $p \times 1$ vector of independent regressors. The function $g(t)$ is an unknown function of the $(p + 1)$st variable t which we observe at the points $t = t_i$. Thus (9.1) represents a regression with $p + 1$ independent variables, p of which enter linearly and the last nonlinearly through $g(t)$.

To estimate both γ and $g(t)$ we can proceed as in Section 8, where without loss of generality we consider t to lie in the interval [0, 1]. Consider first the Bayesian solution. Let $g(t)$ have the same prior as in Section 8.1, so that we can write

$$y(i) = v(i)^T\gamma + w(t_i)^T\alpha + f(t_i) + e(i), \qquad i = 1, \ldots, n.$$

Let $\beta = (\alpha^T, \gamma^T)^T$ be the $r(=m+p)\times 1$ vector of unknown parameters and the prior for $f(t)$ is given by (8.3). Alternatively, either α or γ or both can be taken to have a diffuse prior. The model (9.1) and the stochastic approach to estimating γ and $g(t)$ were first discussed by Ansley and Wecker (1981), who also gave a number of empirical applications of the model (9.1). Ansley and Wecker (1981) suggested estimating the parameters α, γ, λ, σ^2 by maximum likelihood.

Let $z(i) = \{v(i)^T, w(t_i)^T\}^T$, and write

$$y(i) = z(i)^T\beta + f(t_i) + e(i),$$

so that from Section 4 the optimal smoothing problem corresponding to the foregoing prior on $f(t)$ is to minimize

$$\sum_{i=1}^{n} \{y(i) - z(i)^T\beta - f(t_i)\}^2 + \frac{\|f\|^2}{\lambda}$$

over f and β, where $\|f\|$ is defined as in Section 6.

As in Section 8.4, we can use the prior for $g(t)$ to write $f(t_i) + e(i)$ in state-space form. We can now estimate $g(t)$ and γ using the results in Kohn and Ansley (1986).

10. HILBERT SPACE OF FUNCTIONS DEFINED BY A DIFFERENCE OPERATOR

Let $L_t(B)$ be the mth-order difference operator

$$L_t(B) = a_0(t) + a_1(t)B + \cdots + a_m(t)B^m, \tag{10.1}$$

where B is the backshift operator $Bf(t) = f(t-1)$.

Let $N \geqslant m$ be a fixed positive integer, and let H^* be the set of functions defined on the integers $1-m, \ldots, N$. Let H be the subset of H^* consisting of all functions with the zero initial conditions

$$f(1-m) = \cdots = f(0) = 0. \tag{10.2}$$

We show that H is a reproducing kernel Hilbert space with an inner product induced by the difference operator L_t. The Hilbert space H will be used extensively when we discuss structural time series models in the next section. We note that results similar to those below were given in Section 6. Because of the zero initial conditions (10.2), we will often regard a function $f \in H$ as being defined on the integers $1, \ldots, N$ only. This should cause no confusion.

For $f \in H$, let

$$v_f(t) = L_t(B)f(t), \qquad t = 1, \ldots, N. \tag{10.3}$$

Suppose that Ω is a given $N \times N$ positive definite matrix. For $f, g \in H$ define the inner product in H as

$$\langle f, g \rangle = v_f^T \Omega^{-1} v_g, \tag{10.4}$$

where $v_f = \{v_f(1), \ldots, v_f(N)\}^T$ and v_g is defined similarly. Regarding $f \in H$ as an $N \times 1$ vector, there is a unique $N \times N$ lower triangular matrix D with ones on the diagonal so that $f = Dv_f$, so we can write (10.4) as

$$\langle f, g \rangle = f^T (D\Omega D^T)^{-1} g.$$

As in Example 3.1, H is a RKHS with reproducing kernel $D\Omega D^T$.

Now let H_0 be the subspace of H^* consisting of all g such that $L_t(B)g(t) = 0$. Thus H_0 is the null space of L_t. Let $h_1(t), \ldots, h_m(t)$ be a basis of H_0. Then any $g \in H^*$ can be written as

$$g(t) = \sum_{j=1}^{m} \alpha_j h_j(t) + f(t), \tag{10.5}$$

where $f \in H$, and $\alpha_1, \ldots, \alpha_m$ are constants.

11. SIGNAL ESTIMATION IN DISCRETE TIME SERIES MODELS

Consider the signal-plus-noise model

$$y(i) = g(t_i) + e(i) \tag{11.1}$$

with $g(t)$ the unobserved signal which is defined on the integers and $e(i)$ an independent $N(0, \sigma^2)$ sequence. We want to estimate $g(t)$ for integer t. Let $L_t(B)$ be the difference operator (10.1), and suppose that our prior for $g(t)$ can be expressed by the difference equation

$$L_t(B)g(t) = u(t), \qquad t = 1, \ldots, N \tag{11.2}$$

with $u = \{u(1), \ldots, u(N)\}^T \sim N(0, \sigma^2\Omega)$. The initial values $g(1-m), \ldots, g(0)$ are taken as unknown constants or equivalently diffuse. The equivalent optimal smoothing problem is described by the following theorem.

THEOREM 11.1 Suppose that we minimize

$$\sum_{i=1}^{n} \{y(i) - g(t_i)\}^2 + u^T \Omega^{-1} u \tag{11.3}$$

over $g \in H^*$ where the space H^* is defined in Section 10, and $u(t) = L_t(B)g(t)$, $t = 1, \ldots, N$. The solution is the posterior mean of $g(t)$ in the signal-plus-noise model (11.1) with the prior for $g(t)$ given by (11.2) and with the initial values of $g(t)$ unknown constants or diffuse.

If $\Omega = \lambda I_N$, we can write (11.3) as

$$\sum_{i=1}^{n} \{y(i) - g(t_i)\}^2 + \frac{1}{\lambda} \sum_{t=1}^{N} \{L_t g(t)\}^2. \tag{11.4}$$

Proof: Using (10.5), we can write $g(t)$ as $g(t) = w(t)^T \alpha + f(t)$, where $w(t) = \{h_1(t), \ldots, h_m(t)\}^T$ and $\alpha = \{\alpha_1, \ldots, \alpha_m\}^T$. The function $f(t)$ belongs to the Hilbert space H, so that $f(1-m) = \cdots = f(0) = 0$. Because $L_t(B)g(t) = L_t(B)f(t) = u(t)$ and $f(t)$ has zero initial conditions, we can find a lower triangular matrix D so that $f = Du$ where $f = \{f(1), \ldots, f(N)\}^T$. Therefore, we can write (11.3) as

$$\sum_{i=1}^{n} \{y(i) - w(t_i)^T \alpha - f(t_i)\}^2 + f^T (D \Omega D^T)^{-1} f$$

and it follows from Section 10 and Theorem 4.1 that the corresponding signal-plus-noise model is

$$y(t_i) = w(t_i)^T \alpha + f(t_i) + e(i)$$

with the prior for $f(t)$ given by $f = Du \sim N(0, \sigma^2 D \Omega D^T)$. An alternative way of writing this prior is $L_t f(t) = u(t)$ with $u \sim N(0, \sigma^2 \Omega)$. This is equivalent to the prior (11.2) with the initial values of $g(t)$ unknown constants or diffuse.

EXAMPLE 11.1 Suppose that in (11.1) $g(t)$ is the mean of $y(t)$, which we think is changing slowly. A suitable prior for $g(t)$ may be

$$g(t) = g(t-1) + u(t),$$

where $u(t)$ is an independent $N(0, \lambda\sigma^2)$ sequence and $g(0)$ is an unknown constant or diffuse. For small λ this will give a smooth evolution of $g(t)$. It follows from (11.4) that the corresponding optimal smoothing problem estimates $g(t)$ by minimizing

$$\sum_{i=1}^{n} \{y(i) - g(t_i)\}^2 + \frac{\sum_{t=1}^{N} \{g(t) - g(t-1)\}^2}{\lambda} \tag{11.5}$$

over the sequence $\{g(0), \ldots, g(N)\}$. The second term in (11.5) indicates that we do not want the estimate of $g(t)$ to change too quickly, or in other words we want it to be smooth. As in Section 8, λ is called the smoothness parameter because it controls the trade-off between least squares fit as measured by the first term in (11.5), and the smoothness of $g(t)$, which is measured by the second term.

EXAMPLE 11.2 Suppose that we have monthly observations and $g(t)$ is a seasonal signal. A reasonable way to estimate $g(t)$ (see e.g., Schlicht, 1981) is to minimize

$$\sum_{i=1}^{n} \{y(i) - g(t_i)\}^2 + \frac{\sum_{t=1}^{N} \{g(t) + g(t-1) + \cdots + g(t-11)\}^2}{\lambda} \tag{11.6}$$

The second term in (11.6) indicates that we want each sum of 12 successive terms of $g(t)$ to be as close to zero as possible and is a measure of how close we are over all to achieving this.

It follows from (11.4) that the corresponding prior for $g(t)$ is

$$g(t) + g(t-1) + \cdots + g(t-11) = u(t),$$

where $u(t)$ is a $N(0, \lambda\sigma^2)$ independent sequence. Thus the sum of 12 successive lags of $g(t)$ is a zero-mean random variable.

As in Section 8.4, we can write Examples 11.1 and 11.2 in state-space form, take the initial values of $g(t)$ as diffuse and then use the modified Kalman filter and fixed interval smoothing algorithms to obtain the posterior distribution of $g(t)$. See Kitagawa and Gersch (1984) for details.

We now generalize (11.1) to several additive signals. Consider

$$y(t) = g_1(t) + g_2(t) + g_3(t) + v(t)^T\gamma + e(t), \tag{11.7}$$

where $g_1(t)$, $g_2(t)$, $g_3(t)$ are three unknown signals, $v(t)$ is a $p \times 1$ vector of independent regressors, γ is a $p \times 1$ parameter vector, and $e(t)$ is an independent $N(0, \sigma^2)$ sequence as before. Suppose that for $i = 1, 2, 3$ the prior for $g_i(t)$ is described by the difference equation $L_{it}(B)g_i(t) = u_i(t)$. Furthermore, for $i = 1, 2, 3$, $u_i = \{u_i(1), \ldots, u_i(N)\}^T \sim N(0, \sigma^2\Omega_i)$, with $u_1(t)$, $u_2(t)$, $u_3(t)$ independent of each other and of $e(t)$.

By an extension of Theorem 3.2 to three or more Hilbert spaces, and the corresponding extensions of Theorems 4.1 and 4.2, the corresponding smoothing problem is

$$\sum_{i=1}^{n} \{y(t_i) - g_1(t_i) - g_2(t_i) - g_3(t_i) - v(t_i)^T\gamma\}^2$$
$$+ u_1^T\Omega_1^{-1}u_1 + u_2^T\Omega_2^{-1}u_2 + u_3^T\Omega_3^{-1}u_3.$$

An important example of (11.7) is the monthly variance components model of Kitagawa and Gersch (1984):

$$y(t) = v(t)^T\gamma + T(t) + S(t) + \xi(t) + e(t),$$

where $T(t)$ is the trend, $S(t)$ the seasonal, $\xi(t)$ an irregular stationary term, and $v(t)^T$ a trading day adjustment term in which $v(t)$ is a 7×1 vector with $v_1(t)$ the number of Mondays in month t, $v_2(t)$ the number of Tuesdays and so on. The coefficients $\gamma_1, \gamma_2, \ldots, \gamma_7$ represent the relative contributions to $y(t)$ of

Monday to Sunday respectively and Kitagawa and Gersch (1984) impose the normalization $\gamma_1 + \cdots + \gamma_7 = 0$, so that $v(t)^T\gamma = 0$ if the $v_i(t)$ are all equal.

Kitagawa and Gersch (1984) take the priors for $T(t)$ and $S(t)$ to be, respectively,

$$(1 - B)^d T(t) = u_1(t), \qquad (1 + B + \cdots + B^{11})S(t) = u_2(t) \tag{11.8}$$

and $\xi(t)$ is a stationary autoregressive process so that $(1 - \phi_1 B - \cdots - \phi_q B^q)\xi(t) = u_3(t)$. We assume that $u_1(t)$ is an independent $N(0, \lambda_1\sigma^2)$ sequence, $u_2(t)$ is an independent $N(0, \lambda_2\sigma^2)$ sequence, and $u_3(t)$ is an independent $N(0, \lambda_3\sigma^2)$ sequence, and these three sequences are also independent of each other.

The corresponding smoothing problem is to minimize

$$\sum_{i=1}^{n} \{y(i) - T(t_i) - S(t_i) - \xi(t_i) - v(t_i)^T\gamma\}^2 + \frac{\sum_{t=1}^{N} \{(1 - B)^d T(t)\}^2}{\lambda_1}$$
$$+ \frac{\sum_{t=1}^{N} \{S(t) + \cdots + S(t - 11\}^2}{\lambda_2}$$
$$+ \frac{\{\xi(t) - \phi_1\xi(t-1) - \phi_2\xi(t-2) - \cdots - \phi_q\xi(t-q)\}^2}{\lambda_3}.$$

As we have remarked in the introduction, we consider (11.8) as specifying a prior for $T(t)$ and $S(t)$ rather than a model because we only observe $y(t)$ and therefore there will be many possible decompositions of $y(t)$ into components. The prior (11.8) reflects our best knowledge of what these components look like.

12. CONCLUSION

For signal-plus-noise models we showed that signal (function) estimation by a Bayesian smoothness prior is equivalent to function estimation by penalized least squares. This gives us two complementary ways of viewing the function estimates. The first approach puts a prior on the unknown function and estimates it by the mean of the posterior distribution of the function. This approach usually suggests efficient algorithms to obtain the function and statistical methods for estimating any unknown parameters, and furthermore, provides us with posterior confidence intervals for the unknown function. Because we cannot usually check how reasonable the prior on the unknown function is, even in large samples, the optimal smoothing or penalized least squares approach adds credence to the smoothness prior if the smoothness penalty is sensible.

The main mathematical idea in the chapter which makes the derivation of the results straightforward is that of a reproducing kernel Hilbert space, which is just a Hilbert space (inner product space) of functions such that the coordinate (value) functionals are continuous. That is, if two functions are sufficiently close together in the norm of the space, their coordinates (values) will be as well.

ACKNOWLEDGMENT

We would like to thank Jim Spall and two referees for the many suggestions that helped improve the presentation of the chapter.

REFERENCES

Ansley, C. F., and R. Kohn (1985). Estimation, filtering and smoothing in state space models with incompletely specified initial conditions, *Ann. Stat.* 13, 1286–1316.

Ansley, C. F., and R. Kohn (1986a). On the equivalence of two stochastic approaches to spline smoothing, in Essays in Time Series and Allied Processes (J. Gani and M. B. Priestly, eds.), *J. Appl. Probab.* 23A, 391–405.

Ansley, C. F., and R. Kohn (1986b). Spline smoothing with repeated values, *J. Stat. Comput. Simul.* 25, 251–258.

Ansley, C. F., and R. Kohn (1987). Efficient generalized cross-validation for state space models, *Biometrika* 74, 139–148.

Ansley, C. F., and W. E. Wecker (1981). Extensions and Examples of the Signal Extraction Approach to Regression, in *Proceedings of the ASA-CENSUS-NBER Conference on Applied Time Series Analysis of Economic Data*, U.S. Bureau of the Census, Washington, D.C., pp. 181–192.

Ansley, C. F., and W. E. Wecker (1984). A nonparametric Bayesian Approach to the Calibration Problem, in *Proceedings of the American Statistical Association, Business and Economics Statistics Section*, pp. 96–101.

Aronszajn, N. (1950). Theory of reproducing kernels, *Trans. Am. Math. Soc.* 68, 337–404.

Craven, P., and G. Wahba (1979). Smoothing noisy data with spline functions, *Numer. Math.* 31, 377–403.

Doob, J. L. (1953). *Stochastic Processes*, Wiley, New York.

Gersch, W., and G. Kitagawa (1983). The prediction of time series with trends and seasonalities, *J. Bus. Econ. Stat.* 1, 253–264.

Harvey, A. C. (1985). Trends and cycles in macroeconomic time series, *J. Bus. Econ. Stat.* 3, 216–227.

Harvey, A. C., and J. Durbin (1986). The effects of seat belt legislation on

British road casualties: a case study in structural time series modelling, *J. R. Stat. Soc. Ser. A* 149, 187–227.

Harvey, A. C., and P. H. J. Todd (1983). Forecasting economic time series with structural and Box-Jenkins models: a case study (with discussion), *J. Bus. Econ. Stat.* 1, 299–315.

Hurewicz, W. W. (1961). *Lectures on Ordinary differential Equations*, Wiley, New York.

Hutchinson, M., and F. R. de Hoog (1985). Smoothing data with spline functions, *Numer. Math.* 47, 99–106.

Kimeldorf, G. S., and G. Wahba (1970a). A correspondence between Bayesian estimation on stochastic processes and smoothing splines, *Ann. Math. Stat.* 41, 495–502.

Kimeldorf, G. S., and G. Wahba (1970b). Spline functions and stochastic processes, *Sankhyā Ser. A* 32, 173–180.

Kimeldorf, G. S., and G. Wahba (1971). Some results on Tchebycheffian splines, *J. Math. Anal. Appl.* 33, 82–95.

Kitagawa, G., and W. Gersch (1984). A smoothness priors-state space modeling of times series with trend and seasonality, *J. Am. Stat. Soc.* 79, 378–389.

Kitagawa, G., and W. Gersch (1985a). A smoothness priors time-varying AR coefficient modeling of nonstationary covariance time series, *IEEE Trans. Autom. Control* AC-30, 48–56.

Kitagawa, G., and W. Gersch (1985b). A smoothness priors long AR model method for spectral estimation, *IEEE Trans. Autom. Control* AC-30, 57–65.

Kohn, R., and C. F. Ansley (1983). On the smoothness properties of the best linear unbiased estimate of a stochastic process observed with noise, *Ann. Stat.* 11, 1011–1017.

Kohn, R., and C. F. Ansley (1985). Efficient estimation and prediction in time series regression models, *Biometrika* 694–697.

Kohn, R., and C. F. Ansley (1986). *On the Equivalence Between Optimal Smoothing and Signal Extraction*, Technical Report, University of New South Wales, Kensington, Australia.

Kohn, R., and C. F. Ansley (1987). A new algorithm for spline smoothing and interpolation based on smoothing a stochastic process, *SIAM J. Sci. Stat. Comput.* 8, 33–48.

Luenberger, B. G. (1969). *Optimization by Vector Space Methods*, Wiley, New York.

Parzen, E. (1961). An approach to time series analysis, *Ann. Math. Stat.* 32, 951–989.

Parzen, E. (1967). *Time Series Analysis Papers*, Holden-Day, San Franscisco.

Reinsch, C. H. (1967). Smoothing by spline functions, *Numer. Math.* 10, 177–183.

Schlicht, E. (1981). A seasonal adjustment principle and a seasonal adjustment method derived from this principle, *J. Am. Stat. Assoc.* 76, 374–378.

Silverman, B. W. (1984). A fast and efficient cross-validation method for smoothing parameter choice in spline regression, *J. Am. Stat. Assoc.* 79, 584–589.

Silverman, B. W. (1985). Some aspects of the spline smoothing approach to nonparametric regression curve fitting (with discussion), *J. R. Stat. Soc. Ser. B* 47, 1–52.

Simmons, G. F. (1963). *Topology and Modern Analysis*, McGraw-Hill, New York.

Wahba, G. (1978). Improper priors, spline smoothing and the problem of guarding against model errors in regression, *J. R. Stat. Soc. Ser. B* 40, 364–372.

Wahba, G. (1983). Bayesian confidence intervals for the cross-validated smoothing spline, *J. R. Stat. Soc. Ser. B* 45, 133–150.

Wahba, G. (1985). A comparison of GCV and GML for choosing the smoothing parameter in the generalized spline smoothing problem, *Ann. Stat.* 13, 1378–1402.

Wecker, W. E., and C. F. Ansley (1983). The signal extraction approach to nonlinear regression and spline smoothing, *J. Am. Stat. Soc.* 78, 81–89.

Weinert, H. L., and G. S. Sidhu (1978). A stochastic framework for recursive computation of spline functions: Part I. Interpolating splines, *IEEE Trans. Inf. Theory* IT-24, 45–50.

Weinert, H. L., U. B. Desai, and G. S. Sidhu (1979). ARMA splines, system inverses, and least squares estimation, *SIAM J. Control Optimization* 17, 525–536.

Weinert, H. L., R. H. Byrd, and G. S. Sidhu (1980). A stochastic framework for recursive computation of spline functions: Part II. Smoothing splines, *J. Optimization Theory Appl.* 30, 255–268.

16

Smoothness Priors in Time Series

WILL GERSCH Department of Information and Computer Science, University of Hawaii, Honolulu, Hawaii

GENSHIRO KITAGAWA The Institute of Statistical Mathematics, Tokyo, Japan

1. INTRODUCTION

Several different kinds of stationary and nonstationary time series modeling problems are considered here from a Bayesian smoothness priors approach. The smoothness priors specify the prior distribution of the time series model parameters.

The term "smoothness priors" is very likely due to Shiller (1973). Shiller modeled the distributed lag (impulse response) relationship between the input and output of economic time series under difference equation "smoothness" constraints on the distributed lags. A trade-off of the goodness of fit of the solution to the data and the goodness of fit of the solution to a smoothness constraint was determined by a single smoothness trade-off parameter. Shiller did not offer an objective method of choosing the smoothness trade-off parameter. Akaike (1980) completed the analysis initiated by Shiller. Akaike developed and exploited the concept of the likelihood of the Bayesian model and used a maximization of the likelihood procedure for determining the smoothness trade-off parameter. [In Bayesian terminology, the smoothness trade-off parameter is referred to as the "hyperparameter" (Lindley and Smith, 1972).] The origin of Shiller-Akaike smoothness priors can be seen in a smoothing problem posed by Whittaker (1923). The smoothing problem context is now understood to be common to a variety of other statistical data analysis problems, including density estimation and image analysis (Titterington, 1985).

In the problem treated by Whittaker, the observations y_n, $n = 1, \ldots, N$ are given. They are assumed to consist of the sum of a "smooth" function and observation noise,

$$y_n = f_n + \varepsilon_n. \tag{1.1}$$

The problem is to estimate the unknown f_n, $n = 1, \ldots, N$. In a time series interpretation of this problem, f_n, $n = 1, \ldots, N$ is the trend of a nonstationary mean time series. A typical approach to this problem is to use a class of parametric models. The quality of the analysis is completely dependent on the appropriateness of the assumed model class. A flexible model is desirable. In this context, Whittaker suggested that the solution balance a trade-off of goodness of fit to the data and goodness of fit to a smoothness criterion. This idea was realized by minimizing

$$\left[\sum_{n=1}^{N} (y_n - f_n)^2 + \mu^2 \sum_{n=1}^{N} (\nabla^k f_n)^2\right] \tag{1.2}$$

for some appropriately chosen smoothness trade-off parameter μ^2. In (1.2) $\nabla^k f_n$ expresses a kth-order difference constraint on the solution f, with $\nabla f_n = f_n - f_{n-1}$, $\nabla^2 f_n = \nabla(\nabla f_n)$, and so on. [Whittaker's original solution was not expressed in a Bayesian context. Whittaker and Robinson (1924) does invoke a Bayesian interpretation of this problem.]

The properties of the solution to the problem in (1.1)–(1.2) are clear. If $\mu^2 = 0$, $f_n = y_n$ and the solution is a replica of the observations. As μ^2 becomes increasingly large, the smoothness constraint dominates the solution and the solution satisfies a kth-order constraint. For large μ^2 and $k = 1$, the solution is a constant; for $k = 2$, it is a straight line; and so on. Whittaker left the choice of μ^2 to the investigator.

Kohn and Ansley (Chapter 15, this volume) demonstrate that the signal extraction problem of (1.1) and the smoothing problem of (1.2) are equivalent problem statements. The equivalence also holds for broad variations of signal extraction and smoothing problems. All of the time series analysis problems that we treat here are variations of the signal extraction–smoothing problem in (1.1) and (1.2).

An implication of Akaike (1980) is that a Bayesian interpretation of the smoothing problem in (1.2) implies that the difference equation constraint is a stochastically perturbed zero-mean unknown variance difference equation. The stochastically perturbed difference equation constraint in the trend estimation problem is a smoothness priors constraint in the time domain. Akaike (1980) considered other time domain smoothness priors constraint problems, including the Shiller distributed lag problem and the seasonal adjustment of time series. Ishiguro et al. (1981) used time domain smoothness priors constraints and fixed-effects regression in an analysis of tidal effects.

Akaike (1979) employed a frequency-domain smoothness priors constraint on the distributed lag parameters in the Shiller problem. Gersch and Kitagawa (1984) and Kitagawa and Gersch (1985a) are other frequency-domain smoothness priors time series problem analyses.

Shiller (1973), Akaike (1980), and all of the aforementioned smoothness priors analyses are Bayesian analyses of the linear model with Gaussian stochastic constraints and Gaussian disturbances. The critical ideas in smoothness priors are the likelihood of the Bayesian model and the use of likelihood as a measure of the goodness of fit of the model. In our analysis, hyperparameters have interpretations as noise-to-signal ratios and they have a remarkable role in the analysis. The maximization of the likelihood of a small number of hyperparameters permits the robust modeling of a time series with relatively complex structure and a very large number of implicitly inferred parameters. When we consider alternative smoothness priors models, with different distributional assumptions or different numbers of parameters to model the same data, we use the Akaike AIC statistic (Akaike 1973) to choose between candidate models. Kitagawa (1987) is a smoothness priors state-space modeling of nonstationary time series in which neither the system noise nor the observation noise are necessarily Gaussian distributed.

The original Whittaker problem has also given rise to work on splines in numerical analysis and to related smoothing problem analysis, particularly by Wahba (1977, 1982). Smoothness priors relate to the ill-posed problems and problems of statistical regularization that have been considered extensively in the Soviety Union by Tikhonov (1965) and his associates. Also related are the "bump hunting"-penalized likelihood methods (Good and Gaskins, 1980; Wecker and Ansley, 1983; O'Sullivan et al., 1986). Vigorous work, primarily at the Institute of Statistical Mathematics, Tokyo, has resulted in the application of smoothness priors methods to a variety of applications other than the ones we discuss here. These applications include the seasonal adjustment of time series (Akaike, 1980b), tidal analysis (Ishiguro et al., 1981), binary regression (Ishiguro and Sakamoto, 1983), cohort analysis (Nakamura, 1986), and density estimation (Tanabe and Sagae, 1987).

Smoothness priors problems that are amenable to analysis by least squares algorithms are treated in Section 2. The likelihood of the Bayesian model, as done by Akaike, is presented in Section 2.1. The Bayesian solution to the smoothing problem originally posed by Whittaker is also shown there. In that problem, the smoothness priors constraints are time-domain constraints. The priors are expressed as zero-mean unknown variance stochastically perturbed kth-order random walk difference equations. In Section 2.2, the estimation of the power spectral density from short-duration stationary time series illustrates the use of frequency-domain smoothness priors constraints. In Section 3 several smoothness priors nonstationary time series problems, which are amenable to a Kalman filter state-space method of analysis, including examples of the modeling of nonstationary mean and nonstationary covariance time series, are shown. All of the aforementioned treat a linear model, Gaussian distributions situation. That method is generalized in Section 4. There we show a smoothness priors state space not necessarily Gaussian nonstationary time series analysis method. Finally, Section 5 is a summary.

2. SMOOTHNESS PRIORS MODELING: LEAST SQUARES ANALYSIS

In Section 2.1 we review the concept of smoothness priors Bayesian modeling as introduced in Akaike (1980). That method is applied to the problem addressed by Whittaker (1923), the estimation of a trend in white noise. The smoothness priors constraint is expressed as a kth-order random walk with a zero-mean, unknown variance perturbation. The variance is a hyperparameter of the prior distribution. The constraint is a time-domain constraint on the priors. In Section 2.2 we introduce the notion of frequency-domain constraint on the priors. That method is used in the estimation of the power spectral density of a stationary time series. Section 2.3 is a discussion.

The frequency-domain smoothness priors method is particularly suited for the situation in which only a short span of data is available for analysis. In that case, the results of conventional parametric model analysis methods are particularly sensitive to the choice of model order. We circumvent that problem, using the frequency-domain smoothness priors, by tending to fit models that are "too long." Those priors reflect the integrated squared kth derivative with respect to frequency of the departure from model smoothness. The estimation of the model parameters and an additional small number of hyperparameters is required. The maximization of the likelihood of the hyperparameters is the critical computation.

2.1 Smoothness Priors Bayesian Modeling

Consider the linear regression model subject to Bayesian-stochastic constraints

$$\begin{pmatrix} y \\ \theta \end{pmatrix} \sim N\left(\begin{bmatrix} X\theta \\ 0 \end{bmatrix}, \begin{bmatrix} \sigma^2 I & 0 \\ 0 & \sigma^2\lambda^{-2}D^{-1}D^{-T} \end{bmatrix}\right). \tag{2.1}$$

The dimensions of the matrices in (2.1) are $y: n \times 1$; $X: n \times p$; $\theta: p \times 1$. σ^2 and λ^2 are unknown, y is the vector of observed data, and X and D are assumed known. θ is the normally distributed prior parameter vector. The observation noise variance is σ^2. In this conjugate family Bayesian situation (Berger, 1985), the mean of the posterior normal distribution of the parameter vector θ minimizes

$$\|y - X\theta\|^2 + \lambda^2\|D\theta\|^2. \tag{2.2}$$

If λ^2 were known, the computational problem in (2.2) could be solved by an ordinary least squares computation. The solution for $\hat{\theta}$, the posterior mean, is the minimizer of

$$\left\|\begin{pmatrix} y \\ 0 \end{pmatrix} - \begin{pmatrix} X \\ \lambda D \end{pmatrix}\theta\right\|^2. \tag{2.3}$$

That solution is

$$\hat{\theta} = [X^T X + \lambda^2 D^T D]^{-1} X^T y \tag{2.4}$$

with the residual sum of squares,

$$\text{SSE}(\hat{\theta}, \lambda^2) = y^T y - \hat{\theta}^T [X^T X + \lambda^2 D^T D]\hat{\theta}. \tag{2.5}$$

For a smoothness priors interpretation of the problem in (2.1) and (2.2), multiply (2.2) by $-1/2\sigma^2$ and exponentiate. Then the θ that minimizes (2.2) also maximizes

$$\exp\left\{-\frac{1}{2\sigma^2}\|y - X\theta\|^2\right\}\exp\left\{-\frac{\lambda^2}{2\sigma^2}\|D\theta\|^2\right\}. \tag{2.6}$$

In (2.6), the posterior distribution interpretation of the parameter vector θ is that it is proportional to the product of the conditional data distribution (likelihood), $p(y \mid X, \theta, \sigma^2)$, and a prior distribution, $\pi(\theta \mid \lambda^2, \sigma^2)$ on θ,

$$\pi(\theta \mid y, \lambda^2, \sigma^2) \propto p(y \mid X, \theta, \sigma^2)\pi(\theta \mid \lambda^2, \sigma^2). \tag{2.7}$$

The integration of (2.7) yields $L(\lambda^2, \sigma^2)$, the likelihood for the unknown parameters λ^2 and σ^2,

$$L(\lambda^2, \sigma^2) = \int_{-\infty}^{\infty} \pi(\theta \mid y, \lambda^2, \sigma^2)\, d\theta. \tag{2.8}$$

Good (1965) referred to the maximization of (2.8) as a type II maximum likelihood method. Since $\pi(\theta \mid y, \lambda^2, \sigma^2)$ is normally distributed, (2.8) can be expressed in the closed form (Akaike, 1980)

$$L(\lambda^2, \sigma^2) = (2\pi\sigma^2)^{-N/2}|\lambda^2 D^T D|^{1/2}|X^T X + \lambda^2 D^T D|^{-1/2}\exp\left\{\frac{-1}{2\sigma^2}\text{SSE}(\hat{\theta}, \lambda^2)\right\}. \tag{2.9}$$

The maximum likelihood estimator of σ^2 is

$$\hat{\sigma} = \frac{\text{SSE}(\hat{\theta}, \lambda^2)}{N}. \tag{2.10}$$

It is convenient to work with -2 log likelihood. Using (2.10) in (2.9) yields

$$-2 \log L(\lambda^2, \hat{\sigma}^2) = N \log 2\pi + N \log \left(\frac{\text{SSE}(\hat{\sigma}, \lambda^2)}{N} \right) + \log |X^T X + \lambda^2 D^T D| - \log |\lambda^2 D^T D| + N. \tag{2.11}$$

A practical way to determine the value of λ^2 for which the -2 log likelihood is minimized is to compute the likelihood for discrete values of λ^2 and search the discrete -2 log likelihood–hyperparameter space for the minimum. Akaike (1980) is very likely the first practical use of the likelihood of the Bayesian model and the use of the likelihood of the hyperparameters, as a measure of the goodness of fit of a model to data.

Estimating a trend

Here we return to the original problem posed in Section 1. We use the notation $f_n = t_n$, where t_n is the trend at time n, to emphasize the fact that we are estimating the mean of a nonstationary mean time series. A critically important observation is that from the stochastic regression or Bayesian point of view, the difference equation constraints in the Whittaker problem are stochastic. That is, $\nabla^k t_n = w_n$, with w_n assumed to be a normally distributed zero-mean sequence with unknown variance τ^2. For example, for $k = 1$ and $k = 2$ those constraints are

$$t_n = t_{n-1} + w_n;$$
$$t_n = 2t_{n-1} - t_{n-2} + w_n. \tag{2.12}$$

A parameterization that relates the trend estimation problem to the earlier development in this section is $\tau^2 = \sigma^2/\lambda^2$. Corresponding to the matrix D in (2.2), for $k = 1$ and $k = 2$, the smoothness constraints can be expressed in terms of the following $N \times N$ constraint matrices:

$$D_1 = \begin{bmatrix} \alpha & & & & & \\ -1 & 1 & & & & \\ & -1 & 1 & & & \\ & & \cdot & & & \\ & 0 & & \cdot\cdot & & \\ & & & \cdot\cdot & \cdot & \\ & & & & -1 & 1 \end{bmatrix},$$

$$D_2 = \begin{bmatrix} \alpha & & & & & & \\ -\beta & \beta & & & & & \\ 1 & -2 & 1 & & & & \\ & 1 & -2 & 1 & 0 & & \\ & & \cdot & \cdot & \cdot & & \\ & & & \cdot & \cdot & \cdot & \\ & & & & \cdot & & \cdot \\ & & & & 1 & -2 & 1 \end{bmatrix}. \tag{2.13}$$

In (2.13) α and β are small numbers that are chosen to satisfy initial conditions.

For fixed k and fixed λ^2, the least squares solution can be simply expressed in the form of (2.3). For example, with $k = 2$, the solution $\{t_n, n = 1, \ldots, N\}$ satisfies

$$\left\| \begin{pmatrix} y \\ 0 \end{pmatrix} - \begin{pmatrix} I \\ \lambda D_2 \end{pmatrix} t \right\|^2. \tag{2.14}$$

Note that the problem in (2.14) is a version of the Bayesian linear stochastaic regression problem in (2.3) with $\theta = t = (t_1, \ldots, t_N)^T$, $X = I$, the $N \times N$ identity matrix, and $D = D_1$ or D_2. From (2.3), the solution to (2.14), with $D = D_2$, is

$$\hat{t} = [I + \lambda^2 D_2^T D_2]^{-1} y \tag{2.15}$$

and the value of $\mathrm{SSE}(\hat{\theta}, \lambda^2)$ is given by (2.5) with $\hat{\theta} = \hat{t}$, $X = I$, $D = D_2$. The smoothing problem expression of (2.15) is that the solution vector is $\hat{t} = (t_{1|N}, \ldots, t_{N|N})^T$. The least squares problem in (2.14) is solved for discrete values of λ and the -2 log likelihood–hyperparameter space is searched for a minimum. From (2.11) the minimized value of -2 log likelihood for this problem is

$$-2 \log L(\hat{\lambda}^2, \hat{\sigma}^2) = N \log 2\pi + N \log \left(\frac{\mathrm{SSE}(\hat{t}^2, \hat{\lambda}^2)}{N} \right) + \log |\hat{\lambda}^2 D_2^T D_2 + I| - \log |\hat{\lambda}^2 D_2^T D_2| + N. \tag{2.16}$$

The numerical values of $\mathrm{SSE}(\hat{t}^2, \lambda^2)$ and of the determinants in (2.16) are transparent in a least squares algorithm analysis of (2.14). Since $\lambda = \sigma/r$, λ^2 has a noise-to-signal ratio interpretation. Smaller λ corresponds to smoother trends.

Example of trend estimation

We consider the example of an asymmetrically truncated normal densitylike function in the presence of additive noise, $N(0, \sigma^2)$. Figure 1(A) shows the

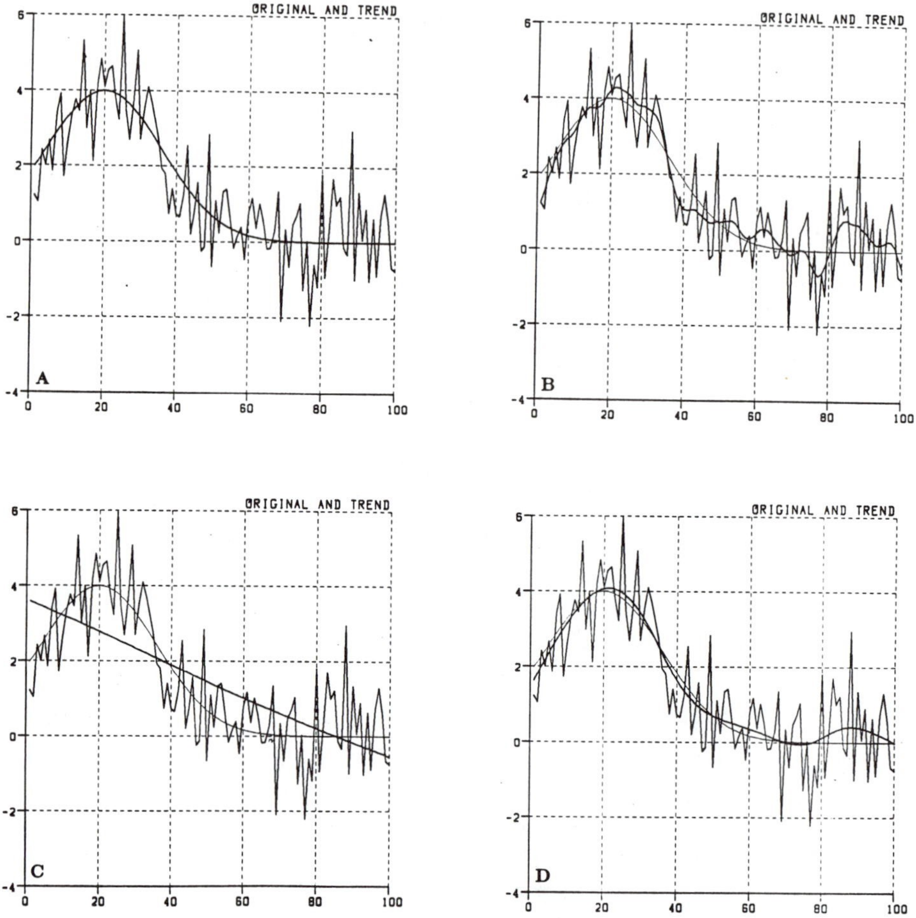

Figure 1. Trend estimation: (A) truncated Gaussian signal and signal plus noise; (B) signal plus noise plus smoothed trend with a too large hyperparameter; (C) signal plus noise plus smoothed trend with a too small hyperparameter; (D) signal plus noise plus smoothed trend with optimum hyperparameter.

smooth function t_n, $n = 1, \ldots, N$ and the superposition of t_n and the additive noise. The problem is: Given the noisy observations y_n, estimate the unknown smooth function that is in the noise (i.e., specify $\hat{t}_{n|N}$, $n = 1, \ldots, N$). We solved the least squares computational problem in (2.14) using the Householder transformation method. $-2 \log$ likelihood of the hyperparameter model is computed from (2.16).

The critical role of the hyperparameter is transparent in this example. Figure 1(B)–(D) shows the estimated trend for values of the hyperparameters that are too small ($\lambda^2 = 0.00001$) and too large ($\lambda = 0.1$) as well as the hyperparameter for which $-2 \log$ likelihood is minimized ($\lambda = 0.00136$). As anticipated, with the hyperparameter defined as indicated above, the estimated trend for a too large value of the hyperparameter is too bumpy and the estimated trend for a too small value of the hyperparameter is too smooth.

It is important to note that in this example, although the truncated Gaussian satisfies $\nabla^2 \log t_n = 0$, we estimate the trend with the "incorrect" model $\nabla^2 t_n = w_n$, the stochastically perturbed second-order difference equation. The point is that a priori we do not know a correct expression for the underlying smooth function in (1.1). Different hyperparameter values result in solutions of the stochastically perturbed second-order difference equation with very different smoothness properties. The best of those solutions yields a very good approximation to the original unknown smooth function. This key observation was referred to by Shiller (1973) as the "flexible ruler approach."

2.2 Smoothness Priors in the Frequency Domain

The smoothness priors in the estimation of the mean value of a nonstationary time series was expressed as a time domain, stochastically perturbed difference equation constraint on the evolution of the trend. Smoothness priors constraints can also be expressed in the frequency domain. In this section we illustrate the use of frequency-domain priors for the estimation of the power spectral density of a stationary time series.

Long AR model for spectral estimation

A smoothness priors–long autoregressive (AR) model approach is used here for spectral density estimation. The classical windowed periodogram method of spectral estimation is satisfactory for spectral analysis when the data set is "long." The alternative of spectral estimation via the fitting of parametric models to moderate length data spans became popular in the last decade (Kesler, 1986). When the data span is relatively short, three facts render parametric modeling methods of spectral estimation statistically unreliable. One is the instability or small-sample variability of whatever statistic is used for determining the best order of parametric model fitted to the data. The second is that usually the "parsimonious" parametric model is not a very good replica of the system that generated the data. The third is that the spectral density of the fitted parametric model cannot possibly be correct.

Independent of which parametric model order is selected, there is information in the data to select models of different orders. A Bayesian estimate of power spectral density requires that the spectral density of parametric models of different model orders be weighted in accordance with the likelihood and the prior of the model order of different models.

The smoothness priors AR model of spectral estimation alleviates this problem. A particular class of frequency-domain smoothness priors is assumed for the coefficients of AR model order M, with M relatively large. The likelihood of the hyperparameters that characterize the class of smoothness priors is maximized to yield the best AR model of order M with the best data dependent priors. [A more complete treatment of the modeling discussed here is in Kitagawa and Gersch (1985a).]

Smoothness priors long AR model

Consider the autoregressive model of order M,

$$y_n = \sum_{m=1}^{M} a_m y_{n-m} + \varepsilon_n. \tag{2.17}$$

In (2.17) $\{\varepsilon_n\}$ is a Gaussian white noise with mean zero and variance σ^2. A least squares fit of the AR model to the data, $y_1, \ldots, y_N$, with the first M observations $y_{1-M}, y_{2-M}, \ldots, y_0$ treated as given constants, leads to the minimization of

$$\sum_{n=1}^{N} \left[y_n - \sum_{m=1}^{M} a_m y_{n-m} \right]^2. \tag{2.18}$$

If M is comparable to N, the result of the least squares computation can be meaningless. The smoothness priors solution mitigates this difficulty by considering the solution of the constrained least squares problem. We consider a frequency-domain smoothness priors constraint on the distribution of the AR model parameters. The frequency-response function of the whitening filter of the AR process is given by

$$A(f) = 1 - \sum_{m=1}^{M} a_m \exp(-2\pi imf). \tag{2.19}$$

Let a measure of the smoothness of the frequency-response function be

$$R_k = \int_{-1/2}^{1/2} \left| \frac{d^k A(f)}{df^k} \right|^2 df = (2\pi)^{2k} \sum_{m=1}^{M} m^{2k} a_m^2. \tag{2.20}$$

From the definition in (2.20), a large value of R_k means an unsmooth (in the sense of differential) frequency-response function. We also use the zero derivative smoothness constraint,

$$R_0 = \int_{-1/2}^{1/2} |A(f)|^2 \, df = 1 + \sum_{m=1}^{M} a_m^2 \tag{2.21}$$

as a penalty to the whitening filter.

With these constraints, and with λ^2 and v^2 fixed, the AR model coefficients $\{a_m, m = 1, \ldots, M\}$, minimize

$$\sum_{n=1}^{N} \left[y_n - \sum_{m=1}^{M} a_m y_{n-m} \right]^2 + \lambda^2 \sum_{m=1}^{M} m^{2k} a_m^2 + v^2 \sum_{m=1}^{M} a_m^2. \tag{2.22}$$

In (2.22), λ^2 and v^2 are the hyperparameters. By a proper choice of these parameters, our estimates of the AR model coefficients balance the trade-off between the infidelity of the model to the data and the infidelity of the model to the frequency-domain smoothness constraints. For completeness, to within a constant, the Gaussian priors on the AR model coefficients corresponding to the R_0 and R_k constraints are

$$\exp\left\{ \left(\frac{-\lambda^2}{2\sigma^2} \right) \sum_{m=1}^{M} m^{2k} a_m^2 \right\} \exp\left\{ \left(\frac{-v^2}{2\sigma^2} \right) \sum_{m=1}^{M} a_m^2 \right\}. \tag{2.23}$$

Following our earlier discussion, define the matrices D and a and the matrices X and y by

$$D = \begin{bmatrix} (v^2 + \lambda^2)^{1/2} & & & \\ & (v^2 + 2^{2k}\lambda^2)^{1/2} & & \\ & & \ddots & \\ & & & (v^2 + M^{2k}\lambda^2)^{1/2} \end{bmatrix}, \quad a = \begin{bmatrix} a_1 \\ a_2 \\ \vdots \\ a_M \end{bmatrix},$$

$$X = \begin{bmatrix} y_0 & \cdots & y_{1-M} \\ y_1 & \cdots & y_{2-M} \\ \vdots & & \vdots \\ y_{N-1} & \cdots & y_{N-M} \end{bmatrix}, \quad y = \begin{bmatrix} y_1 \\ y_2 \\ \vdots \\ y_N \end{bmatrix}. \tag{2.24}$$

Then the AR model coefficients satisfy

$$\hat{a} = (X^T X + D^T D)^{-1} X^T y, \tag{2.25}$$

and the residual sum of squares is

$$S(\lambda^2, v^2) = y^T y - \hat{a}^T (X^T X + D^T D) \hat{a}. \tag{2.26}$$

The likelihood of the hyperparameter model is

$$L(\lambda^2, v^2, \sigma^2) = \left(\frac{1}{2\pi\sigma^2} \right)^{N/2} |D^T D|^{1/2} |X^T X + D^T D|^{-1/2} \exp\left\{ \frac{-1}{2\sigma^2} S(\lambda^2, v^2) \right\}. \tag{2.27}$$

Given λ^2 and v^2, the maximum likelihood estimate of σ^2 is, $\hat{\sigma}^2 = S(\lambda^2, v^2)/N$. The ML estimates of λ^2 and v^2 are obtained by minimizing

$$-2 \log L(y|\lambda^2, v^2, \hat{\sigma}^2) = N \log 2\pi\hat{\sigma}^2 + \log |D^T D| - \log |X^T X + D^T D| + N \tag{2.28}$$

with respect to λ^2 and v^2. Computation of the likelihood over a discrete k, λ^2, v^2 parameter grid and searching over the resulting discrete likelihood-hyperparameter space for the minimum of $-2 \log$ likelihood yields the desired smoothness priors long AR model.

The frequency-domain smoothness priors constraint used here has an interpretation as a constraint on the smoothness of the whitening filter of the AR model. (The zeroth derivative has an energy constraint interpretation.) An important facet of our computations is that they are computationally tractable. That allows us to remain within the framework of the general linear model.

Example: analysis of Canadian lynx data

The data example discussed here is the analysis of the Canadian lynx data ($n = 114$). Other examples are shown in Kitagawa and Gersch (1985a).

First, AR models of order up to 20 were fitted by a least squares algorithm–AIC criterion method. The AIC best order model was 11. Smoothness priors AR model orders up to 20 and up to the fourth-order smoothness prior constraint were tried. The hyperparameters λ and v were searched over the discrete values $\lambda = 2^{j-3k}\sigma_0$, $j = 1, \ldots, 10$, where σ_0^2 is the sample variance of the data and $v = i^2\sigma_0$, $i = 0, 1, \ldots, 4$, for each value of the order of the smoothness prior constraint.

The AIC overall best model was model order $M = 14$, $k = 1$, $v = 0$, $\lambda = 0.173$. The smoothness priors AR model estimate of the spectrum is shown in Figure 2(A). The AIC criterion-AR modeled spectrum is shown in Figure 2(B). In the Bayesian model spectral estimate, the peaks at the high frequencies are significantly reduced compared to the AR model spectrum estimate, while the ones in the lower frequencies remain unchanged. Figure 2(C) shows the superimposed estimated spectra obtained from AR models with different model orders. From Figure 2(B) and (C) we see that the shape of the two rightmost peaks of Figure 2.1(B) vary considerably with model order. Thus they are not estimated reliably by fixed-order models. That is typical of the problem of estimating spectral density by fixed-order parametric models. If the model fitted to the data is not in the class of models that generated the data, the model fitting is only approximate. The selection of the best non-Bayesian parametric model ignores the evidence, in the Bayesian sense, for other parametric models when in fact it should be taken into account. The

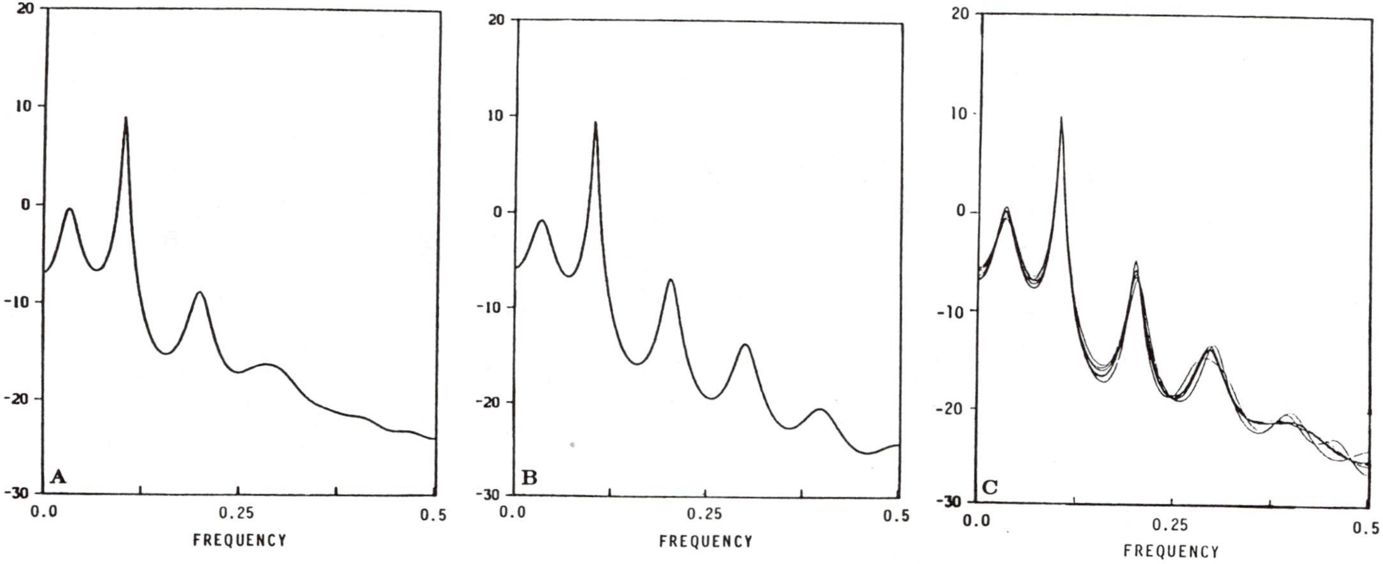

Figure 2. Spectral densities from Canadian lynx data example: (A) spectral density versus frequency, smoothness priors model; (B) spectral density versus frequency, AIC-AR model; (C) superposition of spectral densities versus frequency, AR models.

suppression of those peaks by the smoothness priors–long AR model method, shown in Figure 2(A), therefore seems quite reasonable.

2.3 Discussion

The variation of the behavior of the solution, from very rough to very smooth, in the trend estimation problem under the smoothness priors constraints for different values of the hyperparameters, is characteristic of the profound effect of the hyperparameter. The log likelihood of the hyperparameter versus the hyperparameter changes gradually in the vicinity of the maximum log likelihood. That fact permits a discrete likelihood-hyperparameter search procedure to be used in conjunction with a Householder transformation algorithm to realize a reasonable computational procedure.

The stochastically perturbed difference equation constraint in the trend estimation problem is a time-domain smoothness priors constraint. Akaike (1980) considered other time-domain smoothness constraint problems, including the Shiller distributed lag problem and the seasonal adjustment of time series. Ishiguro et al. (1981) used time-domain smoothness constraints and fixed-effects regression in the analysis of earth tide data. The Householder transformation least squares algorithm FORTRAN programs BAYSEA and BAYTAP-G in TIMSAC-84 (Akaike et al., 1985) are suitable for seasonal adjustment and tidal analyses, respectively.

Akaike (1979) illustrated a frequency-domain smoothness prior for the solution of the Shiller (1973) impulse response estimation problem. We used frequency-domain smoothness priors here for spectral density estimation. Gersch and Kitagawa (1984) is an application of the frequency domain–smoothness priors method to transfer function estimation. Our smoothness priors method is particularly suited for the situation in which only a short span of data is available for analysis. In that situation, the results of conventional parametric model analysis methods are particularly sensitive to the choice of model order. We circumvent that problem, with the Bayesian smoothness priors method, by tending to fit models that are "too long." The model parameters are specified as the solution to a constrained least squares problem in which the constraints are expressed in the frequency domain. The likelihood of the hyperparameters is readily computable in a least squares framework with the frequency-domain priors.

The goodness of the choice of the frequency-domain smoothness priors can be appraised by evaluating its performance in various conceivable situations. The smoothness priors-long AR model gives reasonable results in the analysis of the real Lynx data in comparison with the minimum AIC-AR model method. Kitagawa and Gersch (1985a) show smoothness priors-long AR model results that were superior to the minimum AIC-AR model method in a simulated two-sine-waves-in-noise case, when the data actually correspond to an ARMA model. This flexibility of performance is what is desired from a Bayesian model.

Also in Kitagawa and Gersch (1985a), a Monte Carlo study of the expected entropy experiment was done to appraise the performance of the smoothness priors–long AR model for spectral estimation against performance of parametric AR models whose order was determined by Akaike's AIC criterion. The smoothness priors–long AR model method was superior to the minimum AIC-AR model method in the two simulation model cases studied. In one case, the simulation model was in the AR model class. In the other case, the simulation model was not in the AR model class. Thus the example shown here and the Monte Carlo study reported in Kitagawa and Gersch (1985a) are evidence to support the soundness of our empirical frequency-domain smoothness priors approach.

3. STATE-SPACE GAUSSIAN SMOOTHNESS PRIORS MODELING

A state-space modeling approach for the linear model with Gaussian system and observation noise that is the equivalent of the least squares computational approach to smoothness priors modeling was shown in Brotherton and Gersch (1981) and Kitagawa (1981). The state-space smoothness priors modeling method was applied to the modeling of nonstationary mean and nonstationary covariance time series by Gersch and Kitagawa (1983, 1985) and Kitagawa and Gersch (1984, 1985b).

In the modeling of nonstationary time series discussed below, there tends to be more parameters than data. In that case, attempts to fit the parameters by least squares or any other ordinary means will yield poor parameter estimates. The smoothness priors permit the model parameters to be expressed implicitly as the solution of zero-mean unknown variance stochastically perturbed difference equations. The variances are hyperparameters of the prior distribution of the parameters. One interpretation of the role of the smoothness priors is that they permit a realization of a computational procedure to estimate the model parameters.

In this section, computational procedures for the modeling of nonstationary mean and nonstationary covariance time series are discussed that are variations of the procedures discussed in our previous papers. Examples are shown in Section 3.4. A discussion of other problems treated by the smoothness priors–state space-linear–Gaussian model and comments appear in Section 3.5. Kalman filter, prediction and smoothing formulas, and computation of the likelihood of the linear Gaussian model are shown in Section 3.2.

3.1 Nonstationary Mean Smoothness Priors State-Space Modeling

Time series with trend and seasonal components occur for example in meteorological, oceanographic, and econometric studies. Here we consider a complex nonstationary mean time series problem motivated by economic time series considerations. The economic time series nonstationary mean can be decomposed into a trend t_n, a globally stationary component v_n, a seasonal component s_n, a trading day factor d_n, and an observation noise component ε_n,

$$y_n = t_n + s_n + v_n + d_n + \varepsilon_n. \tag{3.1}$$

Each of the aforementioned components can be modeled as a stochastically perturbed difference equation. The generic state-space model for this decomposition can be expressed by

$$\begin{aligned} x_n &= F x_{n-1} + G w_n \\ y_n &= H_n x_n + \varepsilon_n, \end{aligned} \tag{3.2}$$

where F, G, and H_n are $M \times M$, $M \times L$, and $1 \times M$ matrices, respectively. w_n and ε_n are each assumed to be zero-mean independent normally distributed random variables. x_n is the state vector at time n and y_n is the observation at time n. For any particular model of the time series, the matrices F, G, and H_n are known and the observations are generated recursively starting from an initial state that is assumed to be normally distributed with mean x_0 and covariance matrix V_0.

The state-space model that includes the local polynomial trend, stationary AR coefficient, trading-day effects, and observation error components can be written in the orthogonal decomposition form

$$x_n = \begin{bmatrix} F_1 & 0 & 0 & 0 \\ 0 & F_2 & 0 & 0 \\ 0 & 0 & F_3 & 0 \\ 0 & 0 & 0 & F_4 \end{bmatrix} x_{n-1} + \begin{bmatrix} G_1 & 0 & 0 & 0 \\ 0 & G_2 & 0 & 0 \\ 0 & 0 & G_3 & 0 \\ 0 & 0 & 0 & G_4 \end{bmatrix} w_n \tag{3.3}$$

$$y_n = [H_1 \quad H_2 \quad H_3 \quad H_{4,n}] x_n + \varepsilon_n.$$

The component models (F_j, G_j, H_j), in order ($j = 1, \ldots, 4$), represent the trend, stationary AR, seasonal, and trading-day effects components, respectively. Some of the particular trend, AR, seasonal, and trading-day difference equation constraints that we have employed and that have representations in the (F_j, G_j, H_j) matrices in (3.3) are shown below.

The trend component t_n satisfies a kth-order stochastically perturbed difference equation

$$\nabla^k t_n = w_{1,n} \tag{3.4}$$

where $w_{1,n}$ is an independent, identically distributed (i.i.d.) sequence with $w_{1,n} \sim N(0, \tau_1^2)$ [see (2.12)]. The stationary AR component v_n is assumed to satisfy an AR model of order p. That is given by

$$v_n = a_1 v_{n-1} + \cdots + a_p v_{n-p} + w_{2,n}. \tag{3.5}$$

In (3.5) $w_{2,n}$ is an i.i.d. sequence with $w_{2,n} \sim N(0, \tau_2^2)$. The seasonal component of the period L difference equation is

$$s_n = -s_{n-1} - s_{n-2} - \cdots - s_{n-L+1} + w_{3,n}. \tag{3.6}$$

In (3.6), $w_{3,n}$ is an i.i.d. sequence with $w_{3,n} \sim N(0, \tau_3^2)$.

The trading-day effect model is

$$d_n = \beta_{1,n} d_{1,n} + \cdots + \beta_{6,n} d_{6,n}, \tag{3.7}$$

where $\beta_{i,n}$ denotes the trading-day effect factor and $d_{i,n}$ corresponds to the number of the ith day of the week at time n. Implicit in (3.7) is the constraint $\sum_{i=1}^{7} \beta_{i,n} = 0$. There is no stochastic component in (3.7).

For a general model including local polynomial trend, AR component trend, local seasonal component and trading-day effect components, the state or system noise vector, and observation noise ε_n are assumed to i.i.d. with zero mean and diagonal covariance matrix

$$\begin{pmatrix} w_n \\ \varepsilon_n \end{pmatrix} \sim N\left(\begin{pmatrix} 0 \\ 0 \\ 0 \\ 0 \end{pmatrix}, \begin{pmatrix} \tau_1^2 & 0 & 0 & 0 \\ 0 & \tau_2^2 & 0 & 0 \\ 0 & 0 & \tau_3^2 & 0 \\ 0 & 0 & 0 & \sigma^2 \end{pmatrix} \right). \tag{3.8}$$

An example of a state-space model that incorporates each of the components with trend order 2, AR model order 2, and seasonal component with period L is

$$x_n = \begin{bmatrix} t_n \\ t_{n-1} \\ v_n \\ v_{n-1} \\ s_n \\ \cdot \\ \cdot \\ \cdot \\ \cdot \\ \cdot \\ s_{n+L-2} \\ \beta_{1,n} \\ \cdot \\ \cdot \\ \cdot \\ \beta_{6,n} \end{bmatrix} = \begin{bmatrix} 2 & -1 & 0 & 0 & 0 & \cdot & \cdot & \cdot & 0 & 0 & \cdot & 0 \\ 1 & 0 & 0 & 0 & 0 & \cdot & \cdot & \cdot & 0 & 0 & \cdot & 0 \\ 0 & 0 & \alpha_1 & \alpha_2 & 0 & \cdot & \cdot & \cdot & 0 & 0 & \cdot & 0 \\ 0 & 0 & 1 & 0 & 0 & \cdot & \cdot & \cdot & 0 & 0 & \cdot & 0 \\ 0 & 0 & 0 & 0 & -1 & \cdot & \cdot & \cdot & -1 & 0 & 0 & 0 \\ \cdot & \cdot & \cdot & \cdot & 1 & \cdot & \cdot & & 0 & \cdot & \cdot & \cdot \\ \cdot & \cdot & \cdot & \cdot & 0 & 1 & \cdot & & \cdot & \cdot & \cdot & \cdot \\ \cdot & \cdot & \cdot & \cdot & \cdot & \cdot & \cdot & & \cdot & \cdot & \cdot & \cdot \\ \cdot & \cdot & \cdot & \cdot & \cdot & & & & & \cdot & \cdot & \cdot \\ \cdot & \cdot & \cdot & \cdot & \cdot & & & & & \cdot & \cdot & \cdot \\ 0 & 0 & 0 & 0 & 0 & \cdot & \cdot & 1 & 0 & 0 & & 0 \\ 0 & 0 & 0 & 0 & 0 & 0 & 0 & 0 & 0 & 1 & \cdot & 0 \\ \cdot & \cdot & \cdot & \cdot & \cdot & \cdot & \cdot & \cdot & \cdot & \cdot & 1 & \cdot \\ \cdot & \cdot & \cdot & \cdot & \cdot & \cdot & \cdot & \cdot & \cdot & \cdot & \cdot & \cdot \\ \cdot & \cdot & \cdot & \cdot & \cdot & \cdot & \cdot & \cdot & \cdot & \cdot & \cdot & \cdot \\ 0 & 0 & 0 & 0 & 0 & 0 & 0 & 0 & 0 & 0 & \cdot & 1 \end{bmatrix} x_{n-1} + \begin{bmatrix} 1 & 0 & 0 \\ 0 & 0 & 0 \\ 0 & 1 & 0 \\ 0 & 0 & 0 \\ 0 & 0 & 1 \\ \cdot & \cdot & \cdot \\ \cdot & \cdot & \cdot \\ \cdot & \cdot & \cdot \\ \cdot & \cdot & \cdot \\ \cdot & \cdot & \cdot \\ 0 & 0 & 0 \\ 0 & 0 & 0 \\ \cdot & \cdot & \cdot \\ \cdot & \cdot & \cdot \\ \cdot & \cdot & \cdot \\ 0 & 0 & 0 \end{bmatrix} w_n$$

$$y_n = [1 \quad 0 \quad 1 \quad 0 \quad 1 \quad \cdots \quad 0 \quad d_{1,n} \quad \cdots \quad d_{6,n}] x_n + \varepsilon_n. \tag{3.9}$$

The smoothness priors problem that includes all of the components in the decomposition identified above corresponds to the maximization of

$$\exp\left\{-\frac{1}{2\sigma^2}\sum_{n=1}^{N}(y_n - t_n - s_n - d_n)^2\right\}\exp\left\{-\frac{\tau_1^2}{2\sigma^2}\sum_{n=1}^{N}(\nabla^2 t_n)^2\right\}$$
$$\cdot\exp\left\{-\frac{\tau_2^2}{2\sigma^2}\sum_{n=1}^{N}\left(v_n - \sum_{i=1}^{2} a_i v_{n-i}\right)^2\right\}\exp\left\{-\frac{\tau_3^2}{2\sigma^2}\sum_{n=1}^{N}\left(\sum_{i=0}^{L-1} s_{n-i}\right)^2\right\}. \tag{3.10}$$

The first term in (3.10) corresponds to the conditional data distribution. The remaining terms in (3.10), in order, correspond to the priors on the trend, the globally stochastic component, and the seasonal component.

The role of the hyperparameters τ_1^2 and τ_3^2 as a measure of the uncertainty in the belief of the priors is clear from (3.10). Relatively small τ_1^2 (τ_3^2) imply relatively wiggly trend (seasonal) components. Relatively large τ_1^2 (τ_3^2) imply relatively smooth trend (seasonal) components. Correspondingly, the ratio of τ_j^2/σ^2, $j = 1$ or 3, can be interpreted as signal-to-noise ratios. [The value of σ^2 in (3.10) is essentially estimated free of computational cost in the Kalman filter algorithm.]

3.2 Recursive Estimation of State and Likelihood Computation

Let a state-space model be given by

$$x_n = F_n x_{n-1} + G_n w_n$$
$$y_n = H_n x_n + \varepsilon_n, \tag{3.11}$$

where $w_n \sim N(0, Q_n)$ and $\varepsilon_n \sim N(0, R_n)$. Given the observations $y_1, \ldots, y_N$ and the initial conditions $x_{0|0}$, $V_{0|0}$, the one-step-ahead predictor and the filter are obtained from the Kalman filter algorithm:

Time update (prediction):

$$x_{n|n-1} = F_n x_{n-1|n-1}$$
$$V_{n|n-1} = F_n V_{n-1|n-1} F_n^T + G_n Q_n G_n^T. \tag{3.12}$$

Observation update (filtering):

$$K_n = V_{n|n-1} H_n^T [H_n V_{n|n-1} H_n^T + R_n]^{-1}$$
$$x_{n|n} = x_{n|n-1} + K_n [y_n - H_n x_{n|n-1}]$$
$$V_{n|n} = [I - K_n H_n] V_{n|n-1}. \tag{3.13}$$

Using these estimates, the smoothed value of the state x_n given the entire observation set, $y_1, \ldots, y_N$, is obtained by the fixed-interval smoothing algorithm (Anderson and Moore, 1979),

$$
\begin{aligned}
A_n &= V_{n|n} F_n^T V_{n+1|n}^{-1} \\
x_{n|N} &= x_{n|n} + A_n[x_{n+1|N} - x_{n+1|n}] \\
V_{n|N} &= V_{n|n} + A_n[V_{n+1|N} - V_{n+1|n}]A_n^T.
\end{aligned}
\tag{3.14}
$$

The state-space representation and the Kalman filter yield an efficient algorithm for the likelihood of a time series model. The joint distribution of $y_1, \ldots, y_N$ is

$$
f(y_1, \ldots, y_N) = \prod_{n=1}^{N} f(y_n | y_1, \ldots, y_{n-1}),
\tag{3.15}
$$

with

$$
\begin{aligned}
f(y_n | y_1, \ldots, y_{n-1}) &= (2\pi v_n)^{-1/2} \exp\left\{ \frac{-1}{2v_n} (y_n - H_n x_{n|n-1})^2 \right\}, \\
v_n &= H_n V_{n|n-1} H_n^T + R_n.
\end{aligned}
\tag{3.16}
$$

Then the log likelihood l of the model is obtained by

$$
l = -\frac{1}{2}\left[N \log 2\pi + \sum_{n=1}^{N} \log v_n + \sum_{n=1}^{N} \frac{-1}{2v_n} (y_n - H_n x_{n|n-1})^2 \right].
\tag{3.17}
$$

The maximum or likelihood estimate of the model parameters are obtained by maximizing (3.17) with respect to those parameters. The AIC is defined by

$$
\text{AIC} = -2(\text{maximum log} - \text{likelihood}) + 2(\text{number of parameters}).
\tag{3.18}
$$

Alternative models for time series might be models with and without trading-day effects or models with and without AR component effects. In each case, when we consider alternative models, the model with the smallest value of the AIC statistic is selected as the AIC best model.

In fitting a stationary model, we can utilize the theoretical mean and the theoretical covariance of the state vector as the initial values $x_{0|0}$ and $V_{0|0}$. In the nonstationary case we consider the initial vector, $x_{0|0}$, as an unknown parameter and estimate it by using the entire set of data. The log likelihood obtained by estimating the initial state vector is a natural estimate of the expected log likelihood of the predictive distribution (Akaike, 1980b; Gersch and Kitagawa, 1983).

3.3 Nonstationary Covariance Modeling

Here time series with nonstationary covariances are modeled by a time-varying autoregressive (AR) model with smoothness constraints on the AR parameters. Time-varying AR coefficient models have been a topic of research for some time. For example, see Whittle (1965), Kozin (1977), and Nicholls and Quinn (1985) and the extensive references therein, particularly to the use of random coefficients in econometric modeling. Earlier, in engineering applications, the modeling of nonstationary covariance time series was done by fitting locally stationary models, and by orthogonal polynomial expansions of AR coefficient models. Astrom and Wittenmark (1971, Theorem 5) express a time-varying AR coefficient model that includes the possibility of random AR coefficients. Bohlin (1976) is an early application of the analysis of time series models with time-dependent coefficients. The concept of the likelihood of the Bayesian model or of hyperparameters or anything related to a smoothness prior do not appear in the earlier papers. Those are key concepts here. Kitagawa (1983) is a precedent to the material discussed in this section.

The problem in modeling nonstationary covariance time series is to achieve an efficient parameterization to capture the local and global statistical relationships in the time series. That objective is achieved here by imposing smoothness priors constraints in the form of stochastically perturbed difference equations on the evolution of the AR coefficients. The variances of the white noise stochastic perturbations are the hyperparameters of the AR coefficient distribution. The difference equations are embedded into a state-space representation. A relatively large AR model order is chosen, the AR coefficients at each time instant are also smoothed using the frequency-domain differential constraints on AR coefficients, as in the smoothness priors–long AR model for spectral estimation (Section 2.2). For each order of the differential constraint, the Kalman filter yields the likelihood of the hyperparameters. The smoothed estimate of the nonstationary innovations series variance is also computed. That is used in the computation of an instantaneous spectral density which is defined in terms of the instantaneous AR model coefficients and the innovations series variance.

Time-varying AR coefficient model

A time-varying AR coefficient model is given by

$$y_n = \sum_{i=1}^{m} a_{i,n} y_{n-i} + \varepsilon_n. \tag{3.19}$$

In (3.19), the coefficients $a_{i,n}$ are assumed to change "gradually" with time and ε_n is assumed to be a normally distributed white noise sequence with variance σ^2. Since there are $m \times N$ AR coefficients in the model in (3.19), an attempt to fit the parameters by least squares or any other ordinary means to the N observations $y_1, \ldots, y_N$, will yield poor parameter estimates. We consider the unknown AR coefficients to be random variables and impose stochastic constraints on those coefficients. Those constraints define a Gaussian smoothness prior distribution on the time history of the AR coefficients and on the spectrum at each time instant. A simple and useful model for a time-varying AR coefficient model is obtained by the stochastically perturbed difference equation constraint model

$$\nabla^{k_1} a_{i,n} = \delta_{i,n}, \qquad i = 1, \ldots, m. \tag{3.20}$$

For convenience, in (3.20) $\delta_{i,n}$ is assumed to be a zero-mean Gaussian white noise sequence with variance τ_i^2 independent of i and n. That is, $\tau_i^2 = \tau^2$, $i = 1, \ldots, m$.

The smoothness priors constraints on the AR coefficients mitigate the problem of overparameterization by permitting the AR coefficients to be expressed as the solution of the constrained least squares problem

$$\sum_{n=1}^{N} \left[y_n - \sum_{i=1}^{m} a_{i,n} y_{n-i} \right]^2 + \tau^2 \sum_{n=1}^{N} \sum_{i=1}^{m} [\nabla^{k_1} a_{i,n}]^2 + \lambda^2 \sum_{n=1}^{N} \sum_{i=1}^{m} i^{2k_2} a_{i,n}^2 + v^2 \sum_{n=1}^{N} \sum_{i=1}^{m} a_{i,n}^2. \tag{3.21}$$

In (3.21) m and k_1, k_2 are assumed known and τ^2, λ^2, v^2 are the trade-off parameters which balance the infidelity of the model to the data and the infidelity of the model to the smoothness constraints.

Similar to the analysis in Section 2.1, (3.21) yields a Bayesian interpretation of the least squares problem. Multiply (3.21) by $-1/2\sigma^2$ and exponentiate. Then, to within a constant term,

$$\exp\left\{ -\frac{\tau^2}{2\sigma^2} \sum_{n=1}^{N} \sum_{i=1}^{m} (\nabla^{k_1} a_{i,n})^2 \right\} \exp\left\{ -\frac{\lambda^2}{2\sigma^2} \sum_{i=1}^{m} i^{2k_2} a_{i,n}^2 \right\} \cdot \exp\left\{ -\frac{v^2}{2\sigma^2} \sum_{i=1}^{m} a_{i,n}^2 \right\} \tag{3.22}$$

expresses the product of the prior distribution on the smoothness of the spectrum and the prior distribution on the smoothness of the AR parameters. The trade-off parameters τ^2, λ^2, v^2 are the hyperparameters of the prior distribution. As in the development in Section 2.1, the product of the conditional data distribution [proportional to the leftmost term in (3.21)], and the prior distribution in (3.22) yields the posterior distribution for the AR parameters. As in (2.8), integration of the posterior distribution for the AR parameters yields the likelihood for the smoothness trade-off parameters.

State-space time-varying AR coefficient model

Define the $km \times 1$ state vector at time n to be $x_n = (a_{1,n}, \ldots, a_{m,n}, \ldots, a_{1,n-k+1}, \ldots, a_{m,n-k+1})^T$. The state-space time-varying AR coefficient model is

$$x_n = Fx_{n-1} + Gw_n$$
$$y_n = H_n x_n + v_n. \tag{3.23}$$

In (3.23), H_n is a $(m+1) \times km$ observation matrix, w_n is the m vector, $w_n = (\delta_{1,n}, \ldots, \delta_{m,n})^T$, and the $m+1$ vector $v_n = (\varepsilon_n, \xi_n)^T$ is defined in (3.25). For the difference equation orders $k_1 = 1$ and $k_1 = 2$, the matrices F, G, and H are

$$k_1 = 1: F = (I_m), \quad G = (I_m), \quad H_n = H_{1,n} = [y_{n-1}, \ldots, y_{n-m}]$$

$$k_1 = 2: \ F = \begin{bmatrix} 2I_m & -I_m \\ I_m & 0 \end{bmatrix}, \quad G = \begin{bmatrix} I_m \\ 0 \end{bmatrix}, \quad H_n = H_{2,n} = [H_{1,n} \quad 0 \quad \cdots \quad 0]. \tag{3.24}$$

The input process noise w_n, the observation noise ε_n, and the spectrum smoothness priors noise form the $2km + 1$ vector $(w_n^T, \varepsilon_n, \xi_n)^T$ that is assumed to be normally distributed and independent with time with the mean and covariance matrix.

$$\begin{pmatrix} w_n \\ \varepsilon_n \\ \xi_n \end{pmatrix} \sim N\left(\begin{pmatrix} 0 \\ 0 \\ 0 \end{pmatrix}, \begin{pmatrix} Q & 0 & 0 \\ 0 & \sigma^2 & 0 \\ 0 & 0 & S \end{pmatrix}\right). \tag{3.25}$$

In (3.25), the $m \times m$ diagonal matrix Q has the element $1/\tau^2$ on the diagonal and for $k_2 = 2$ [in (3.21)], the $m \times m$ diagonal matrix S has diagonal elements $(1/(\lambda^2 + v^2), \ldots, 1/(m^4\lambda^2 + v^2))$ [see (2.22)].

For a fixed difference order k_1, the best fit of the state-space smoothness priors constraints-time-varying AR coefficient model to the data $y_1, \ldots, y_N$ is the one for which the likelihood of the hyperparameters τ^2, λ^2, v^2 are maximized. The likelihood is computed using the recursive formulas indicated in Section 3.2.

Instantaneous variance and instantaneous spectrum

In many practical data analysis situations, the relatively fast wiggles of a nonstationary covariance time series appears to be modulated by a relatively slowly changing envelope function. The envelope function has an interpretation as a change of scale of the time-varying AR coefficient model or equivalently as the smoothed (trend) value of the instantaneous variance (Section 3.1). A key idea in that modeling is to find a variance stabilizing transformation of the innovations that yields the instantaneous trend in an additive zero-mean constant variance observation noise (Wahba, 1980). Let s_n, $n = 1, \ldots, N$ be a realization of a zero-mean normally distributed white noise with unknown variance σ_n^2. Then if $\sigma_{2n}^2 = \sigma_{2n-1}^2$, $\chi_m^2 = [s_{2m-1}^2 + s_{2m}^2]/2$

is an independent sequence of chi-square random variables with two degrees of freedom. From Wahba (1980), the transformation $t_m = \log[\chi_m^2] + \gamma$, where $\gamma = 0.57721$ is the Euler constant, leaves the independent random variable t_m with a distribution that is almost normal with the moments $E[t_m] = \log \sigma_m^2$, $\text{var}[t_m] = \pi^2/6$. That idea is exploited here.

Consider a second-order difference equation constraint on the log variance defined by

$$\nabla^2 t_m = w_m. \tag{3.26}$$

In (3.26), $\{w_m\}$ is an independent zero-mean normally distributed sequence with unknown variance τ^2. Define a state vector by $x_m = [t_m \; t_{m-1}]^T$. Then in state-space form the constraint model in (3.26) and the observation model are given by

$$x_m = \begin{bmatrix} -2 & 1 \\ 1 & 0 \end{bmatrix} x_{m-1} + \begin{bmatrix} 1 \\ 0 \end{bmatrix} w_m$$

$$y_m = [1 \quad 0]x_m + \varepsilon_m. \tag{3.27}$$

Application of the Kalman filter, prediction, and smoothing algorithms described earlier yield the smooth value $t_{n|N}$, the logarithm of the smoothed estimate of the changing variance. Our estimate of the changing variance is $\sigma^2_{2m|N} = \sigma^2_{2m-1|N} = \exp(t_{m|N} + \gamma)$.

Motivated by earlier work on spectrum estimation, we define the instantaneous spectrum of a time-varying coefficient AR process by

$$S_n(f) = \frac{\sigma_n^2}{\left|1 - \sum_{k=1}^{m} a_{k,n} \exp(-2\pi i k f)\right|^2}; \qquad -\tfrac{1}{2} \leqslant f \leqslant \tfrac{1}{2}. \tag{3.28}$$

The value of the instantaneous spectrum is obtained by substituting the smoothed estimates of the time-varying AR coefficients and the smoothed estimate of the innovations variance σ_n^2 into (3.28).

3.4 Examples

Nonstationary mean time series example

The RSWOMEN series of the Bureau of the Census data (Zellner, 1983) is analyzed here. The series consists of the retail sales of women's apparel, reported in millions of dollars. The sales for each month are affected by the number of times each day of the week occurs, the trading-day effect, because buying behavior differs for each day of the week. (The sales are also affected by holidays.) We are interested in determining whether or not it is appropriate to include a trading-day effect in the model and whether or not the globally stochastic AR component should be included in the model. The AIC statistic is used to determine the best of the alternative models.

The computations were done using the DECOMP.FORT program in TIMSAC-84 (Akaike et al., 1985). The model was fitted to the data $y_1, \ldots, y_{139}$ and 24 data points were withheld. The AICs for AR model orders $p = 0, 1, 2, 3$ [as in (3.5)], respectively, for the non-trading-day effect and trading-day-effect models are (111.51, 96.96, 98.80, 98.65) and (88.07, 68.34, 67.23, 68.91). An interpretation of those results is that models with an AR component, $p \neq 0$, are superior to models without AR components both with and without trading-day effects and that the AIC best model (AIC = 67.23) is the trading-day-effect model with AR model order $p = 2$. Figure 3(A) to (E) shows selected computational responses for the non-AR component–trading-day-effect model. Figure 3(F) to (J) shows selected computational results for the AR component–trading-day-effects model. The seasonal components, residual noise, and trading-day effects and seasonal plus trading effects are quite similar in appearance for both models.

Several aspects of the modeling results are noteworthy. The trend of the trend plus AR component model is smoother than the trend–non-AR component model and the trend plus AR component is almost indistinguishable from the trend in the non-AR component model. Also, the seasonal component is very regular, whereas the seasonal plus trading-day component reveals the expected slight irregularities.

An important property of the AIC best trend plus seasonal plus AR component model, instead of the trend plus seasonal model, can be seen in the (out-of-sample) forecasts for these models as shown in Figure 3(E) and (J). In those illustrations we show the true series, the forecasted series, and $\pm 1\sigma$ of the forecast random variable. The $\pm 1\sigma$ prediction intervals of the trend plus AR plus seasonal plus trading-day components model is much tighter than the same quantity for the non-AR component model. The increase in prediction variance per step in increasing horizon forecast grows in accordance with the variance component terms in the matrix, in (3.8). The variance of the (wiggly) trend term in the non-AR component model is larger than the sum-of-variance terms of the (smooth) trend and AR component in the AR component model. That larger variance is reflected into the larger $\pm 1\sigma$ prediction interval of the non-AR component model.

These results illustrate the flexibility of the decomposition of the nonstationary mean concept via smoothness priors modeling and the importance of the role of the AIC in selecting the best of alternative models.

Time-varying AR coefficient modeling, nonstationary covariance time series

The computations were realized using the TVCAR.FORT program in TIMSAC-84 (Akaike et al., 1985). Figure 4(A) shows a seismic data event, $y_1, \ldots, y_N$. The stochastic "background noise," P wave (the first abrupt

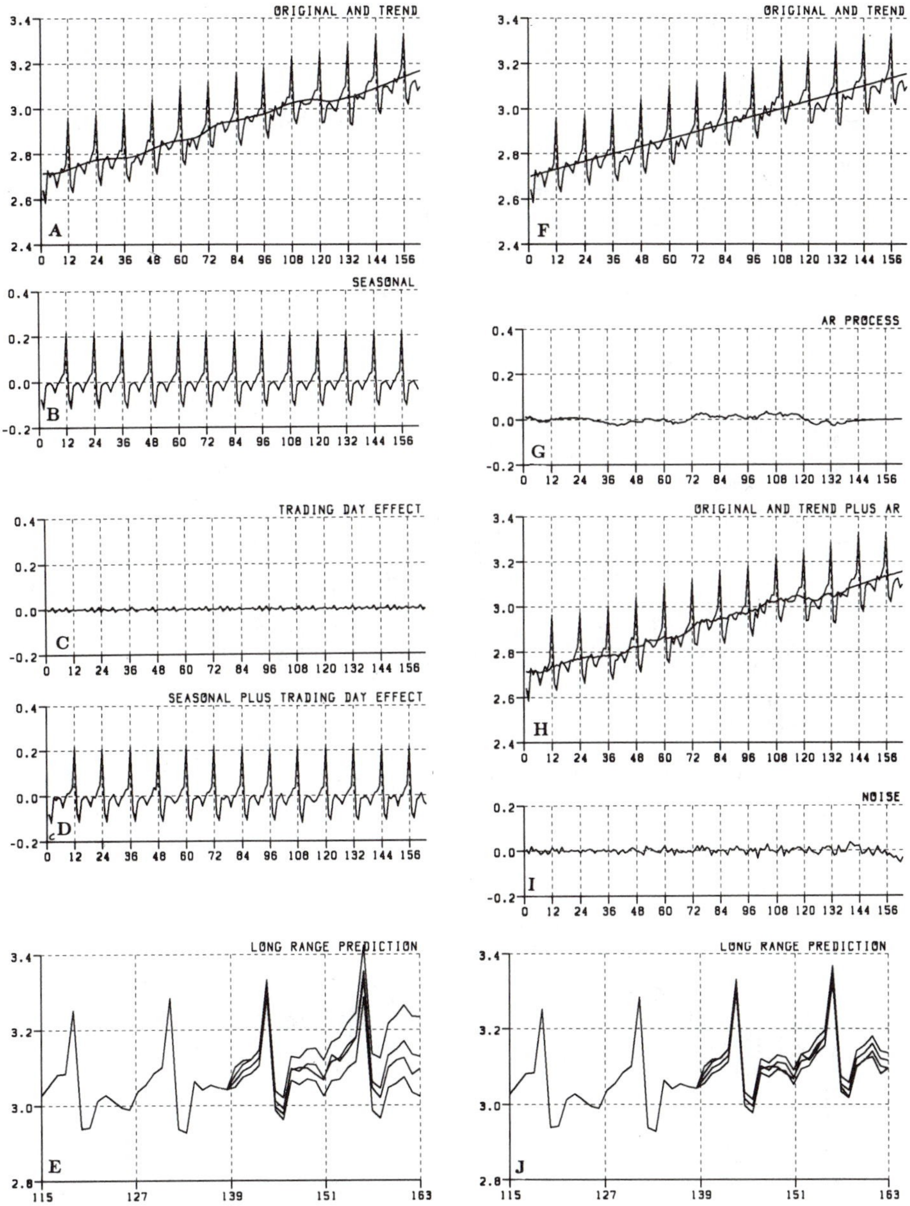

Figure 3. Nonstationary mean, RSWOMEN data. Trend plus seasonal plus trading component model: (A) original data and trend; (B) seasonal component; (C) trading-day effect; (D) seasonal plus trading-day effect; (E) true, predicted and $\pm 1\sigma$ plus predicted. Trend plus seasonal plus AR plus trading component model: (F) original data and trend; (G) AR component; (H) original plus trend plus AR component; (I) residual noise; (J) true, predicted and $\pm 1\sigma$ plus predicted.

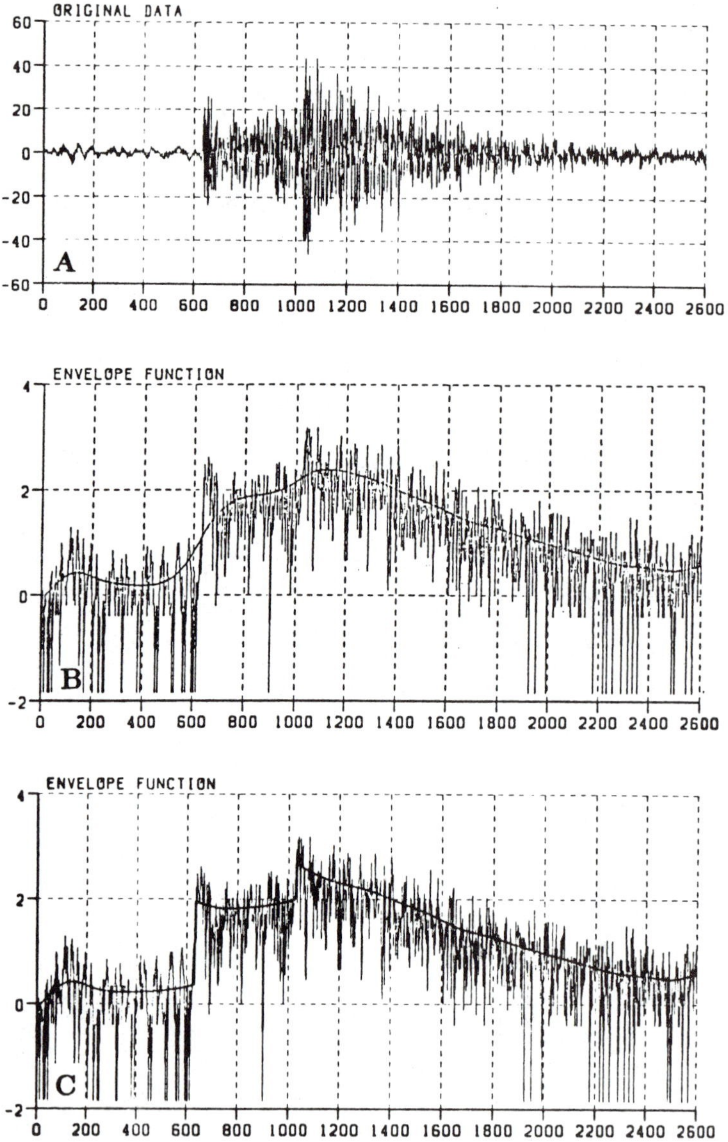

Figure 4. Nonstationary covariance, seismic data: (A) Seismic data, $y_1, \ldots, y_N$ ("ordinary model," B,D,F; "intervention model," C,E,G); (B, C) log $((y_{2m}^2 + y_{2m-1}^2)/2)$, $m = 1, \ldots, N/2$ data and smoothed envelope; (D, E) instantaneous power spectral density; (F, G) parcors.

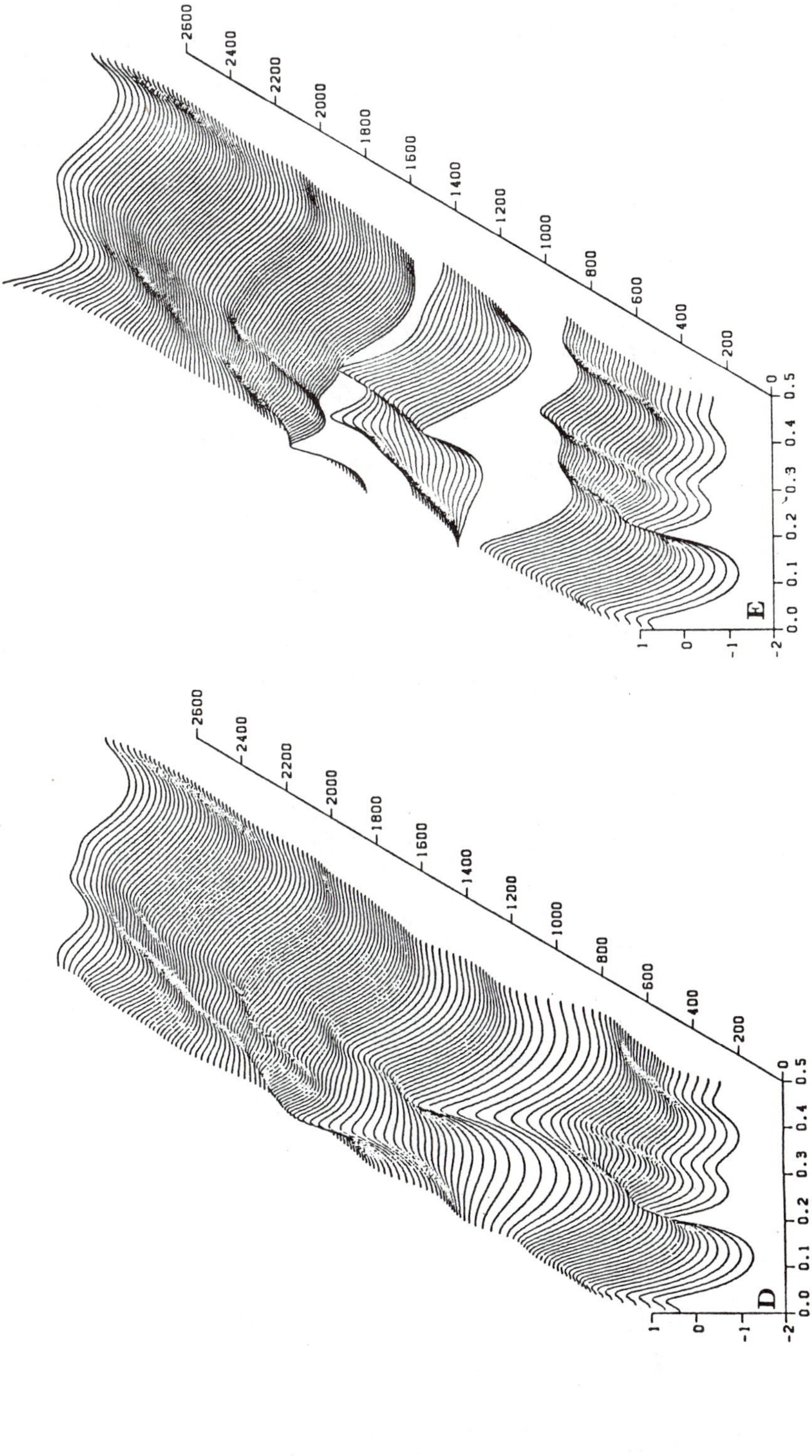

Figure 4. (Continued)

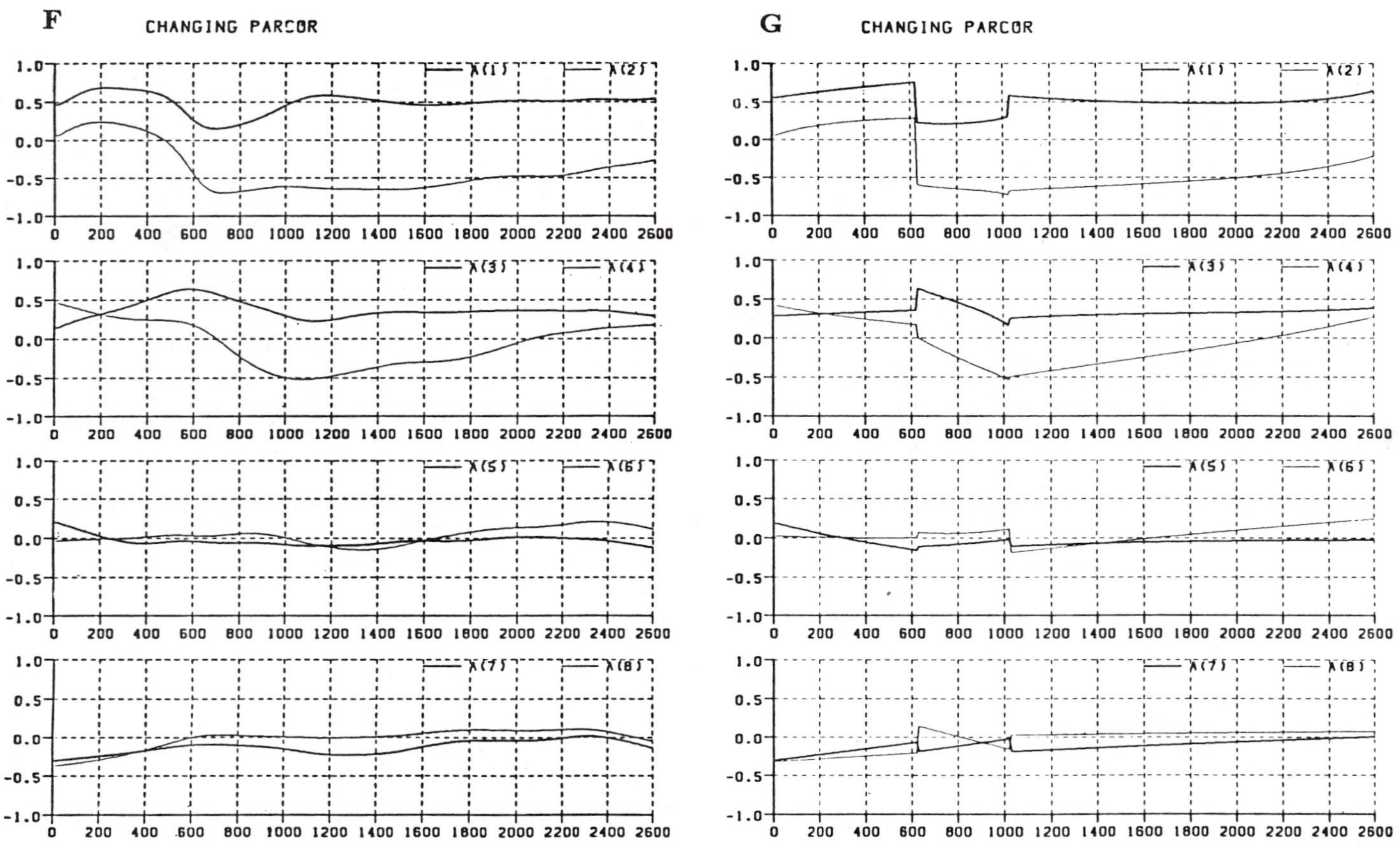

Figure 4. (Continued)

change in the signal) and S wave (the second abrupt change in the signal) are clearly discernible. Figure 4(B) to (F) show graphs of computational results from the time-varying AR coefficient model described in Section 3.3. Respectively they show the $\log((y_{2m}^2 + y_{2m-1}^2),\ m = 1, \ldots, N/2$ "unsmoothed envelope" data and the superimposed estimate of the envelope (changing variance), the evolution of the instantaneous power spectral density, and the evolution of the partial correlation coefficients (parcors) of the fitted, AR order $m = 8$ time-varying AR coefficient model. Figure 4(C), (E), and (G) show the corresponding computational results from an "intervention analysis" model that is part of TVCAR.FORT.

The smoothed envelope for the nonintervention model is in fact quite smooth. Similarly, the instantaneous spectrum and partial correlation coefficients (parcors) reflect the smooth transition the time-varying AR coefficients model, (3.2). Two visual-inspection-determined "outlier" events occur at $n = 635$ and $n = 1030$. They correspond to the arrival times of the P and S waves and are identified to the program, by human intervention, as large observation variance events. The large observation variance relaxes the priors constraint and permits the AR coefficients and subsequently derived quantities to change abruptly at those instants. The "validity" of this intervention-type analysis is suggested by comparison of the results of the intervention-type and nonintervention-type analysis. The properties of the latter "drift" towards the former. The abrupt changes in the appearance of the envelope function and in the instantaneous spectra and parcors correspond to the physical interpretation that the P waves and S waves are different sources of energy at the observing seismometer.

3.5 Comments and Discussion

The paper by Akaike (1980) motivated our work in smoothness priors. In Akaike (1980), computations were done by a Householder transformation least squares algorithm of computational complexity $O(N^3)$. Brotherton and Gersch (1981) and Kitagawa (1981) demonstrated an equivalent state-space modeling approach for the linear model with Gaussian system and observation noise. In the vicinity of the maximized likelihood, the likelihood is a rather flat function of the hyperparameters. This fact permits a relatively coarse grid discrete hyperparameter-likelihood search procedure to determine the values of the hyperparameters that tend to maximize the likelihood. Such a procedure preserves the $O(N)$ computational complexity inherent in the Kalman filter computations. A computational complexity of $O(N)$ version of that method was subsequently applied to a variety of nonstationary mean and nonstationary covariance time series modeling problems (Gersch and Kitagawa, 1983, 1985; Kitagawa and Gersch, 1984, 1985b). Variations of the procedures in those papers expressed in computer programs DECOMP.FORT and TVCAR.FORT (TIMSAC-84, Akaike et al., 1985) yielded the computational results shown here.

Potentially, many more combinations and extensions of the models shown here for the modeling of nonstationary mean and nonstationary covariance time series by smoothness priors methods are possible. For example, a generalization of the regression on trading-days components, in the nonstationary mean decomposition of time series modeling, could take into account constant-coefficient and/or time-varying coefficient regression on other time series. A time-varying partial AR coefficients variation of the present time-varying AR coefficient model for the modeling of nonstationary covariance time series has already been implemented. Another potential variation on the time-varying AR coefficient model would be to estimate the full nondiagonal $m \times m$ system noise covariance matrix (the matrix of hyperparameters). Gersch and Kitagawa (1985), an application of the time-varying AR coefficient model, includes computation of the time-varying covariance function. That computation is useful to permit computation of the mean square response to nonstationary excitation of building structures to single realizations of seismic event data.

Some other linear model–Gaussian disturbances–state space smoothness priors models have been implemented. Kitagawa and Takanami (1985) show a smoothness priors modeling method for the extraction of seismic signals from correlated background noise. The smoothness priors innovation in that work is the implementation of a nonconstant or time-varying hyperparameter. That hyperparameter achieves a time-varying balance of the trade-off between the variances of the seismic signal and the background noise. The modeling of continuous-model time series with discrete-time observations is another domain where smoothness priors state-space modeling has been exhibited. Kitagawa (1984) includes a smoothness priors variation of the Jones (1980) continuous-time AR process–discrete time observations modeling. The application in Kitagawa (1984) is to irregularly spaced or missing data time series modeling.

4. STATE-SPACE NON-GAUSSIAN MODELING OF NONSTATIONARY MEAN TIME SERIES

A non-Gaussian state-space approach to the modeling of nonstationary time series is shown. Neither the system noise nor the observation noise need be Gaussian. Recursive formulas for the prediction, filtering, and smoothing of the state are given. A numerical method, based on a piecewise linear approximation to the density functions for realizing these formulas, is also given. The merits and potential wide applicability of this approach to non-

Gaussian modeling are illustrated by some numerical examples. Extention of this method to the state-space modeling of nonlinear systems is straightforward.

Earlier in this chapter we demonstrated the wide range of applicability of the linear model with Gaussian system and observation noise. There are numerous problems for which Gaussian modeling is inadequate. For example, the problem of trend estimation becomes difficult when the trend has discontinuities as well as smooth changes and when there are observation outliers. A simple linear Gaussian model with small process noise variance does not track jumps or discontinuities very well. A model with large process noise variance will respond to sudden changes in the trend, but it will also be inappropriately wiggly where the trend is quite smooth. The treatment of such trend discontinuities with the included possibility of observation outliers in the linear Gaussian model framework requires a complicated model. Heavy-tailed distributions for process and observation noise can cope with these problems with a simple model. Also, smoothing problems in which there is a time-varying and/or a nonhomogeneous binomial or Poisson mean require a non-Gaussian system noise model formulation. Similarly, nonlinear models such as storage models for river flow and a ship's nonlinear maneuverability require non-Gaussian distribution models.

Thus the development of methods for treating systems with non-Gaussian distributions is well motivated. In earlier attempts, systems with non-Gaussian distributions were approximated by the use of extended Kalman filters, sums of Gaussian distributions, by Edgeworth or GramCharlier expansions, and so on (see, e.g., Alspach and Sorenson, 1972). Here we approximate the probability density functions directly by a piecewise linear function. The recursive prediction, filtering, and smoothing computation required by the state-space modeling are realized by numerical integration. A similar approach was considered by Bucy and Senne (1971) and de Figueiredo and Jan (1971) in the context of nonlinear filtering problems. Such an approach is more feasible now than it was several years ago because of the development and proliferation of fast computational facilities. In Section 4.1 the state-space prediction, filtering, and smoothing formula aspects of the numerical computations are derived and the computation of the likelihood for the not necessarily Gaussian distribution model are shown. Numerical examples are shown in Section 4.2 and a discussion and comments are given in Section 4.3.

4.1 Non-Gaussian State-Space Model

Consider the stationary state-space system described by

$$\begin{aligned} x_n &= Fx_{n-1} + Gw_n \\ y_n &= Hx_n + \varepsilon_n, \end{aligned} \tag{4.1}$$

where, as before, F, G, and H are linear transformations. The independent and independent of each other, but not necessarily Gaussian process and observation noises are w_n and ε_n, respectively. The initial state vector x_0 is distributed in accordance with $p(x_0)$ and the conditional density of the state at time n given the observations $(y_1, \ldots, y_m) = Y_m$ is denoted by $p(x_n|Y_m)$. Then the recursive formulas for the one-step-ahead prediction, filtering, and smoothing densities are derived as follows:

One-step-ahead prediction (time update):

$$\begin{aligned} p(x_n|Y_{n-1}) &= \int_{-\infty}^{\infty} p(x_n, x_{n-1}|Y_{n-1})\, dx_{n-1} \\ &= \int_{-\infty}^{\infty} p(x_n|x_{n-1})p(x_{n-1}|Y_{n-1})\, dx_{n-1}. \end{aligned} \tag{4.2}$$

Filtering (observation update):

$$\begin{aligned} p(x_n|Y_n) = p(x_n|y_n, Y_{n-1}) &= \frac{p(x_n, y_n|Y_{n-1})}{p(y_n|Y_{n-1})} \\ &= \frac{p(y_n|x_n)p(x_n|Y_{n-1})}{p(y_n|Y_{n-1})}, \end{aligned} \tag{4.3}$$

where $p(x_n|x_{n-1})$ is the density of x_n given the previous state vector x_{n-1}, $p(y_n|x_n)$ is the density of y_n given x_n and $p(y_n|Y_{n-1})$ is obtained by $\int p(y_n|x_n)p(x_n|Y_{n-1})\, dx_n$.

Similarly, consider the expression for the joint density of x_n and x_{n+1}, given the entire observation sequence Y_N,

$$\begin{aligned} p(x_n, x_{n+1}|Y_N) &= p(x_{n+1}|Y_N)p(x_n|x_{n+1}, Y_n) \\ &= \frac{p(x_{n+1}|Y_N)p(x_n, x_{n+1}|Y_n)}{p(x_{n+1}|Y_n)} \\ &= \frac{p(x_{n+1}|Y_N)p(x_{n+1}|x_n)p(x_n|Y_n)}{p(x_{n+1}|Y_n)}. \end{aligned} \tag{4.4}$$

From (4.4) we obtain the formula for smoothing:

$$\begin{aligned} p(x_n|Y_N) &= \int_{-\infty}^{\infty} p(x_n, x_{n+1}|Y_N)\, dx_{n+1} \\ &= p(x_n|Y_n) \int_{-\infty}^{\infty} \frac{p(x_{n+1}|Y_N)p(x_{n+1}|x_n|}{p(x_{n+1}|Y_n)}\, dx_{n+1}. \end{aligned} \tag{4.5}$$

In the linear Gaussian case, the conditional densities $p(x_n|Y_{n-1})$, $p(x_n|Y_N)$, and $p(x_n|Y_N)$ are characterized by mean vectors and covariance matrices. Correspondingly, (4.2), (4.3), and (4.5) lead to the Kalman filter and the fixed-interval smoothing algorithm. In the non-Gaussian or nonlinear case, however, it is necessary to evaluate the non-Gaussian densities explicitly at each step. The algorithms above, (4.3)–(4.5), can be realized numerically by piecewise linear approximations to the density functions, transformation of densities, convolution of densities, and Bayes theorem (product of two densities and normalization). Details of the numerical computations are in Kitagawa (1987).

In general, the non-Gaussian model has some unknown parameters. The best choice of the parameters can be found by maximizing the log likelihood defined by

$$l(\theta) = \log p(y_1, \ldots, y_N) = \sum_{n=1}^{N} \log p(y_n|y_1, \ldots, y_{n-1}) = \sum_{n=1}^{N} \log p(y_n|Y_{n-1}). \tag{4.6}$$

The term $p(y_n|Y_{n-1})$ appears in (4.3) and can be evaluated numerically. If we have several candidate models, including models with different types of system noise or observation noise density functions, we choose the model for which the AIC is minimum.

4.2 Numerical Examples

Estimation of a shifting mean value

Consider the data simulated from the following model,

$$Y_n \sim N(\mu_n, 1)$$

$$\mu_n = \begin{cases} 0 & n = 1, \ldots, 100 \\ 1 & n = 101, \ldots, 200 \\ -1 & n = 201, \ldots, 300 \\ 2 & n = 301, \ldots, 400. \end{cases} \tag{4.7}$$

The data are shown in Figure 5(A). The problem is to estimate the abruptly changing mean value function μ_n.

For this type of data we used the model

$$\begin{aligned} \nabla^k t_n &= w_n \\ y_n &= t_n + \varepsilon_n. \end{aligned} \tag{4.8}$$

As before, ∇ is the difference operator defined by $\nabla t_n = t_n - t_{n-1}$ and w_n and ε_n are white noise sequences that are not necessarily normally distributed. For simplicity we assume that the difference order k is 1. Equation (4.8) is a special form of the state-space model (Section 3.1), with $x_n = t_n$, $F = G = H = 1$. We

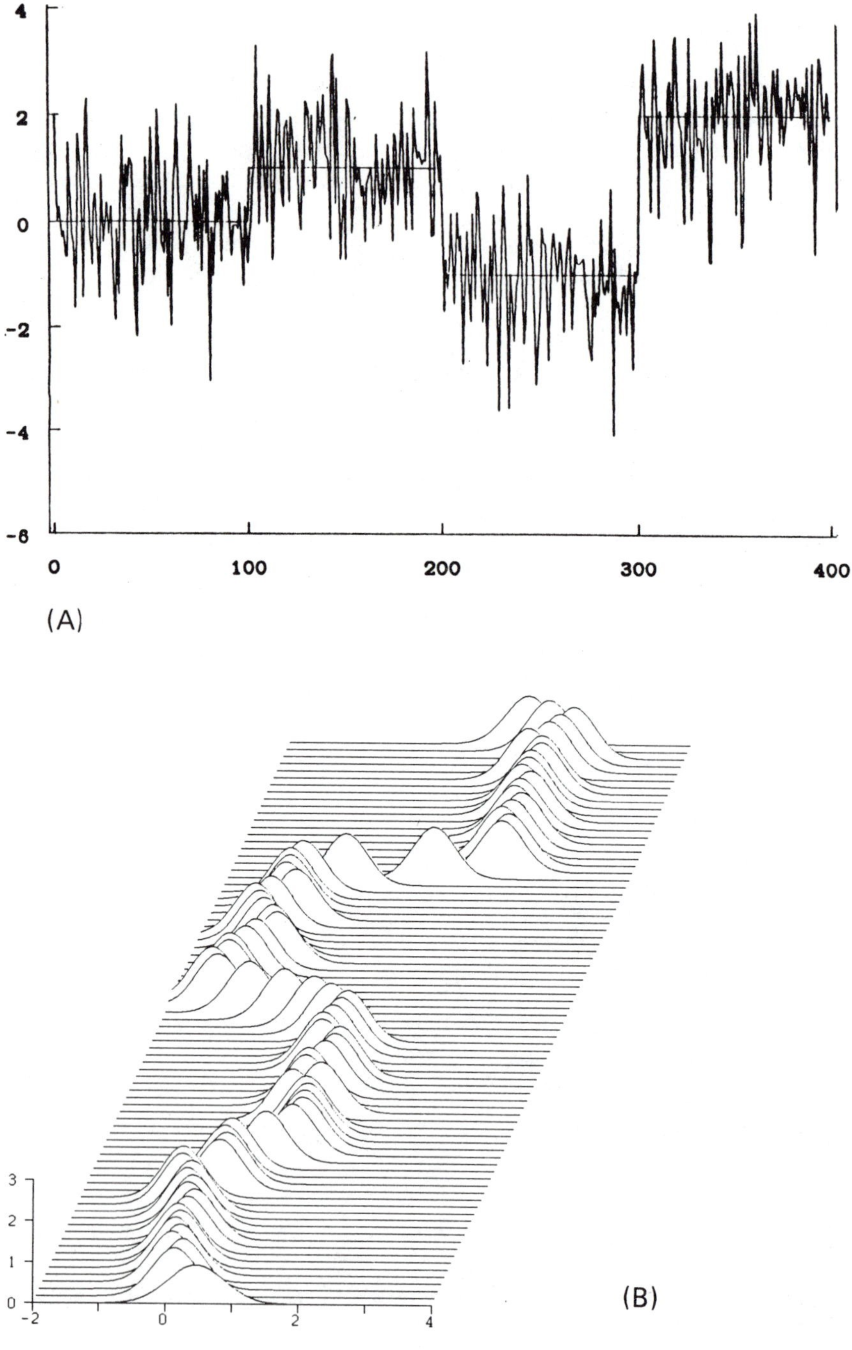

Figure 5.

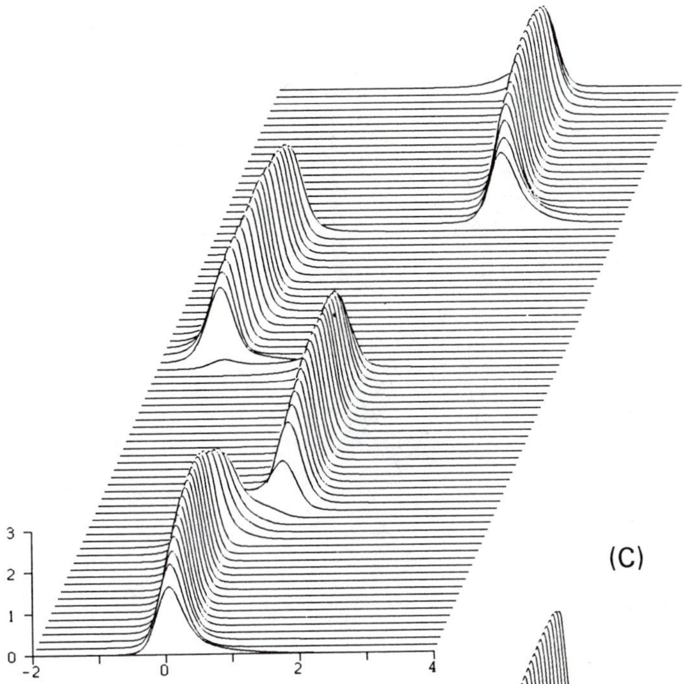

(C)

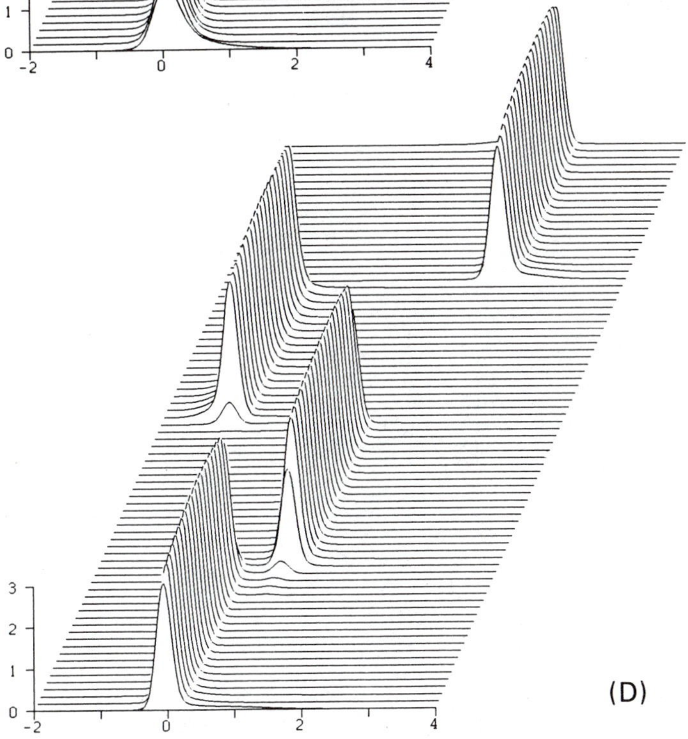

(D)

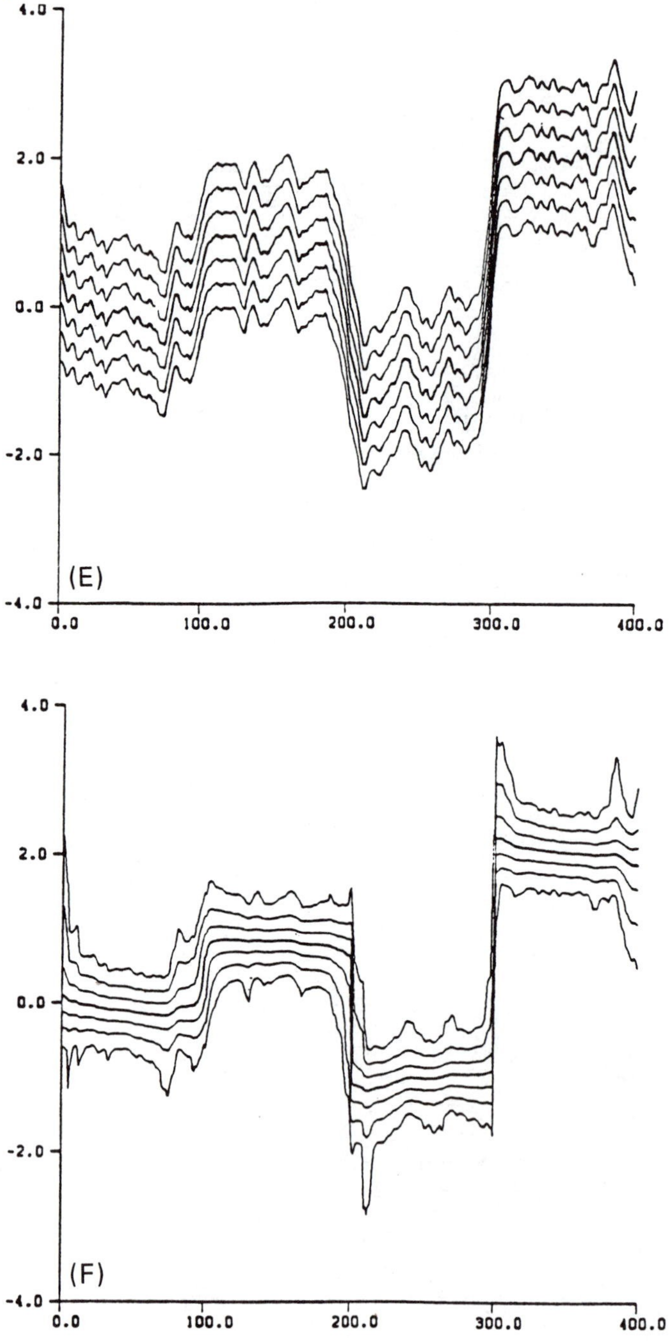

Figure 5. (Continued)

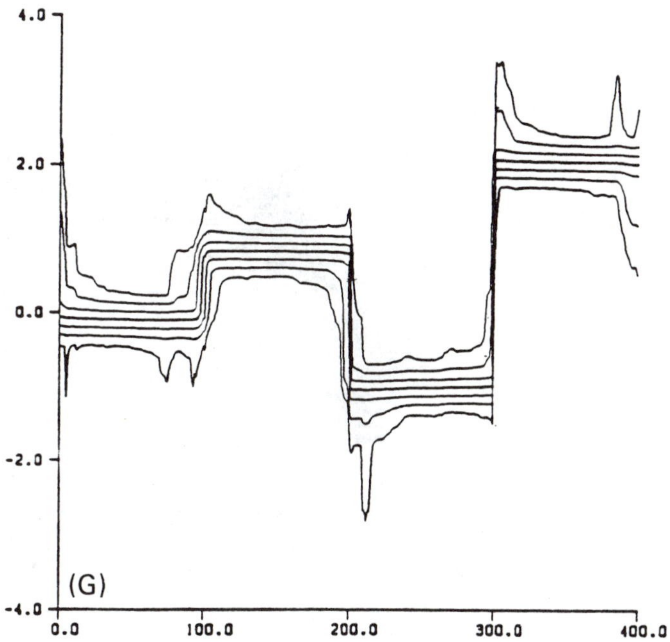

Figure 5. State-space-model non-Gaussian discontinuous trend example: (A) abruptly changing trend data; (B) smoothed state estimate, Gaussian system noise model; (C) smoothed state estimate, Pearson system-system noise model; (D) smoothed state estimate, Gaussian mixture system noise model; (E) posterior mean, $\pm 1,2,3\sigma$, Gaussian system noise model; (F) posterior median and (0.13, 2.27, 15.87, 84.13, 97.73) percentage points, Pearson system-system noise model; (G) posterior median and (0.13, 2.27, 15.87, 84.13, 97.73) percentage points, Gaussian system noise model.

considered the following model classes:

$$\begin{aligned} &\text{Model (a):} \quad w_n \sim aN(0, \tau^2) + (1-a)N(0, \tau_x^2), \qquad \varepsilon_n \sim N(0,1) \\ &\text{Model (b):} \quad w_n \sim Q(b, \tau^2), \qquad \varepsilon_n \sim N(0,1). \end{aligned} \tag{4.9}$$

Model (a) denotes a mixture of Gaussian system noises. In (4.9), for models (a) and (b), $N(0, r)$ denotes the Gaussian distribution with mean 0 and variance r. In model (b), $Q(b, \tau^2)$ denotes the distribution of the Pearson system with density $q(x; b, \tau^2) = C(\tau^2 + x^2)^{-b}$ with $\frac{1}{2} < b \leqslant \infty$ and $C = \tau^{2b-1}\Gamma(b)/\Gamma(\frac{1}{2})$. This family links the Cauchy distribution ($b = 1$) and the Gaussian distribution ($b = \infty$). In model (a), τ_x^2 was arbitrarily set to 4.0, approximately the sample variance of the simulated data. The maximum likelihood estimate of

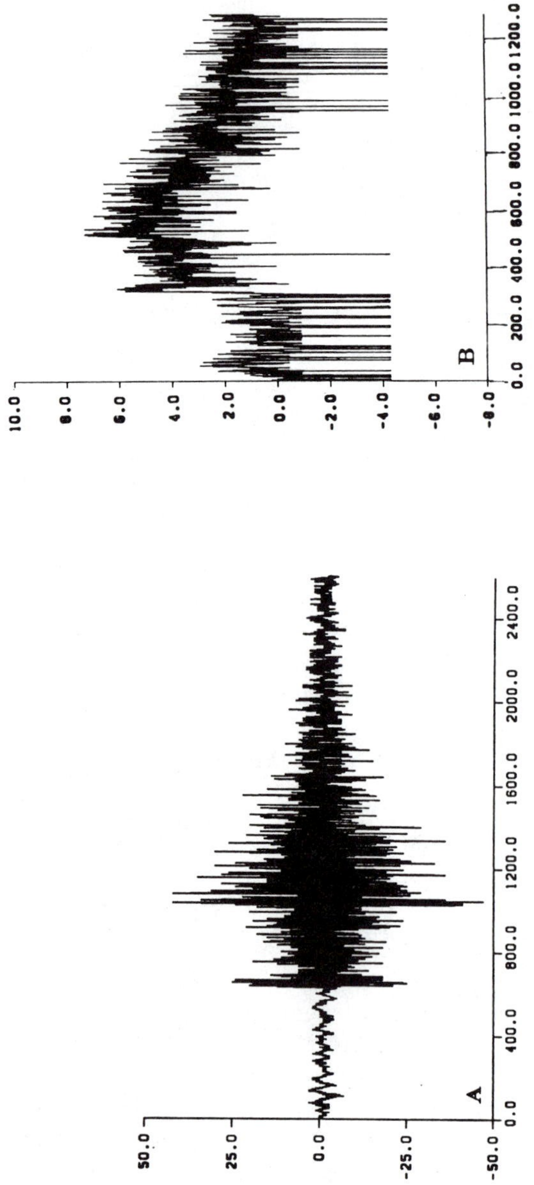
B
10.0
8.0
6.0
4.0
2.0
0.0
-2.0
-4.0
-6.0
-8.0
0.0
200.0
400.0
600.0
800.0
1000.0
1200.0
A
50.0
25.0
0.0
-25.0
-50.0
0.0
400.0
800.0
1200.0
1600.0
2000.0
2400.0

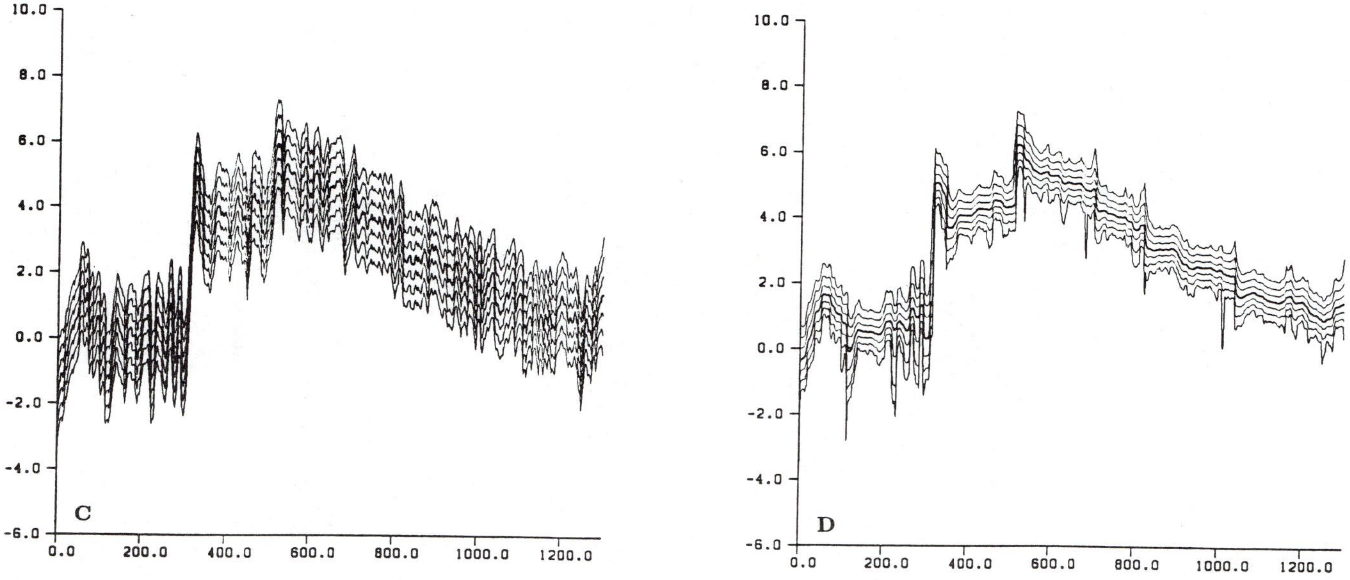

Figure 6. State-space model non-Gaussian envelope of seismic signal example: (A) seismic data, $y_1, \ldots, y_N$; (B) log "envelope" data; (C) posterior mean $\pm 1, 2, 3\sigma$, Gaussian disturbances model; (D) posterior mode (0.13, 2.27, 15.87, 84.13, 97.73) percentage points, non-Gaussian disturbances model.

τ^2 for the Gaussian model, model (a), with $a = 1.0$ or equivalently, model (b) with $b = \infty$, was $\hat{\tau}^2 = 0.0429$. The AIC of the model was 1240.33. For the mixture of Gaussian system noises model, $\hat{a} = 0.989$, $\hat{\tau}^2 = 0.0000014$, and AIC = 1212.48. We tried four Pearson family models: $b = 0.6, 0.8, 1.0$, and ∞. $b = 0.80$ is the AIC best Pearson family model with $\hat{\tau}^2 = 0.000002$ and AIC = 1215.20. The AIC best model is the mixture of Gaussian system noises model.

Figure 5(B) to (D) shows the marginal posterior density $p(x_n | Y_N)$ versus time n for the Gaussian model, the best Pearson system model and the mixture of Gaussians model, respectively. For the Gaussian model [Figure 5(B)], the densities obtained have identical shape except for the ends of the time interval where the densities become slightly broader. In Figure 5(C), the shape of the posterior density varies with time. When the mean value shifts, the density becomes heavy tailed on one side. The Gaussian mixtures model also exhibits the latter behavior.

Figure 5(E) to (G) shows the mean (bold) and $\pm 1, 2, 3\sigma$ intervals of the $p(x_n | Y_N)$ versus n for the Gaussian model and the median (bold) and corresponding 0.13, 2.27, 15.87, 84.13, 97.73, and 99.87 percentage points of the Pearson system model and the Gaussian mixtures model, respectively. For the Gaussian model, the estimated mean value function becomes a wiggly curve and does not reflect the sudden change of the mean value. The estimated median of the Pearson system and the Gaussian mixtures models do capture the sudden change of the signal mean value. The multimodal or skewed distribution and jumps of the mean value are typical of the phenomenon that are seen in non-Gaussian modeling.

Estimation of changing variance

The estimation of changing variance was discussed in Section 3.3 in the context of fitting a time-varying AR coefficient model to a seismic signal. The same idea is exploited here to estimate the changing variance of the same seismic signal with the state-space non-Gaussian modeling method. A first-order difference model for the trend of the log of the sum of the squares of successive observations was used, $\nabla t_m = t_m - t_{m-1} \sim q(b, \tau^2)$, $y_m = t_m + \varepsilon_m$ with $m = 1, \ldots, N/2$. Two models were considered, the Gaussian system noise-Gaussian observation noise and the Cauchy system noise with an $r(x) = \exp(x - \exp x)$ observation noise model. The corresponding AICs were 4778.94 and 4222.84. The latter model was the AIC best model. The original seismic wave $y_n, n = 1, \ldots, N$ and the $\log((y_{2m}^2 + y_{2m-1}^2)/2)$, $m = 1, \ldots, N/2$ signals are shown in Figure 6(A) and (B), respectively. The Gaussian model smoothed mean and $\pm 1, 2, 3\sigma$ and non-Gaussian model smoothed median and corresponding probability point curves are shown in Figure 6(C) and (D), respectively. Those illustrations indicate that the Cauchy system noise model yields better estimates of the smooth mean and abrupt changes of the mean innovations variance than the simple Gaussian system noise model. Modeling the real data changing variance seismic signal with the state-space non-Gaussian system noise method automatically yields the location of abrupt changes in the mean of the signal.

4.3 Comments and Discussion

The results shown here were obtained using a simple one-dimensional state vector. In principle, it is straightforward to extend the computational formulas to higher-dimensional state systems. The resulting increase in the computational burden required to compute the convolution of density functions becomes quite severe. A variety of numerical techniques have been investigated to cope with this problem. Very likely the use of more powerful computers rather than the increase of effort in numerical analysis methods will be more expeditious in the development of the non-Gaussian state-space smoothness priors method.

Several other problems lend themselves to the application of the one-dimensional non-Gaussian modeling shown here. Kitagawa (1987) shows an application to the handling of discrete distributions. The time-varying mean of a real, nonstationary (nonhomogeneous) binary process is estimated. Also, smoothing of the log periodogram using a state-space model with $w_n \sim \log \chi^2$ and $\varepsilon_n \sim$ Cauchy or $\varepsilon_n \sim$ Gaussian is an alternative to the Gaussian distributions approach in Wahba (1980).

Our procedure extends quite naturally to the analysis of nonlinear systems. The one-step-ahead prediction formula (4.2) and the filtering formula (4.3) are applicable even for nonlinear systems.

Time series with nonstationarities, nonlinearities, and outliers that have been difficult to analyze by conventional linear Gaussian models can be quite simply analyzed with the non-Gaussian model. The computational burden using the non-Gaussian filter and smoother is substantial. The development of faster algorithms and the use of faster computers will alleviate this burden.

5. SUMMARY

The ingredients for the smoothness priors analysis of time series are the model, the prescription of the priors, the criterion for goodness of model fit, and the computational method. Initially, smoothness priors modeling of stationary, nonstationary mean, and nonstationary covariance time series was demonstrated in the context of the linear model with Gaussian distributed system noise and with Gaussian distributed observation noise. Both time-domain and frequency-domain specifications of the prior distribution of the model parameters were considered. A hyperparameter specifies the degree of belief in the prior distribution. The smoothness priors method of analysis derives its unity from the fact that the likelihood of the Bayesian model [the likelihood of the hyperparameter(s)] is the single criterion by which the goodness of fit of the model is determined. The maximization of the likelihood of a small number of hyperparameters permits the modeling of time series with complex structure and a large number of implicitly inferred parameters. When there are alternative candidate smoothness priors models, we use Akaike's AIC to determine the best of alternative models (the likelihood of the model has a central role in the AIC). Householder transformation least squares and Kalman filter algorithms were the means for the realization of the smoothness priors time series modeling.

Finally, we demonstrated a state-space representation not necessarily Gaussian not necessarily linear model method of smoothness priors modeling. In that method, piecewise linear approximation to densities and numerical integration computations were employed. Conceptually, all the possible combinations of models and smoothness priors computations could be realized with this method.

The extensive applicability of smoothness priors modeling methods in time series modeling was demonstrated. A large number of other problems that have not been well solved by more traditional time series methods remain to be solved by that method.

ACKNOWLEDGMENT

The authors gratefully acknowledge the reviewers' careful reading and constructive suggestions. The first author was supported by an NSF grant.

REFERENCES

Akaike, H. (1973). Information Theory and an Extension of the Maximum Likelihood Principle, in *2nd International Symposium in Information Theory* (B. N. Petroc and F. Caski, eds.), Akadémiai Kiadó, Budapest, Hungary, pp. 267–281.

Akaike, H. (1979). *Smoothness Priors and the Distributed Lag Estimator*, Technical Report 40 (T. A. Anderson, Project Director), Department of Statistics, Stanford University, Standford, Calif.

Akaike, H. (1980a). Likelihood and the Bayes Procedure, in *Bayesian Statistics, Proceedings of the First International Meeting, Valencia* (J. M. Bernardo, M. H. De Groot, D. V. Lindley, and A. F. M. Smith, eds.), University Press, Valencia, Spain, pp. 143–166.

Akaike, H. (1980b). Seasonal adjustment by a Bayesian modeling, *J. Time Ser. Anal.* 1, 1–13.

Akaike, H., T. Ozaki, M. Ishiguro, Y. Ogata, G. Kitagawa, Y-H. Tamura, E. Arahata, K. Katsura, and Y. Tamura (1985). TIMSAC-84, Part 1 and Part 2, Vols. 22 and 23, Computer Science Monographs, The Institute of Statistical Mathematics, Tokyo.

Alspach, D. L., and H. W. Sorenson (1972). Nonlinear Bayesian estimation using Gaussian sum approximations, *IEEE Trans. Autom. Control* AC-17, 439–448.

Anderson, B. D. O., and J. B. Moore (1979). *Optimal Filtering*, Prentice-Hall, Englewood Cliffs, N.J.

Askar, M., and H. Derin (1981). A recursive algorithm for the Bayes solution of the smoothing problem, *IEEE Trans. Autom. Control* AC-26, 558–561.

Astrom, K. J. A., and B. Whittenmark (1971). Problems of identification and control, *J. Math. Anal. Appl.* 34, 90–113.

Berger, J. O. (1985). *Statistical Decision Theory, Foundations, Concepts and Methods, 2nd ed.*, Springer-Verlag, New York.

Bohlin, T. (1976). Four Cases of Identification of Changing Systems, in *System Identification Advances and Case Studies* (R. Mehra and D. G. Lainotis, eds.), Academic Press, New York.

Box, G. E. P., and G. M. Jenkins (1970). *Time Series Analysis: Forecasting and Control*, Holden-Day, San Francisco.

Brotherton, T., and W. Gersch (1981). A Data Analytic Approach to the Smoothing Problem and Some of its Variations, in *Proceedings of the 20th IEEE Conference on Decision and Control*, pp. 1061–1069.

Bucy, R. S., and K. D. Senne (1971). Digital synthesis of nonlinear filters, *Automatica* 7, 287–289.

Campbell, M. J., and A. M. Walker (1977). A survey of statistical work on the McKenzie River series of annual Canadian lynx trappings for the years 1821–1934 and a new analysis, *J. R. Stat. Soc. Ser. A* 140, 411–431.

de Figueiredo, J. R. P., and Y. G. Jan (1971). Spline Filters, in *Proceedings of the 2nd Symposium on Nonlinear Estimation Theory and Its Applications*, San Diego, Calif., pp. 127–141.

Gersch, W., and G. Kitagawa (1983). The prediction of time series with trends and seasonalities, *J. Bus. Econ. Stat.* 1, 253–264.

Gersch, W., and G. Kitagawa (1984). A Smoothness Priors Method for Transfer Function Estimation, in *Proceedings of the 23rd IEEE Conference on Decision and Control*, pp. 363–367.

Gersch, W., and G. Kitagawa (1985). A time varying AR coefficient model for modeling and simulating earthquake ground motion, *Earthquake Eng. Struct. Dyn.* 13, 243–254.

Good, I. J. (1965). *The Estimation of Probabilities*, MIT Press, Cambridge, Mass.

Good, I. J., and J. R. Gaskins (1980). Density estimation and bump hunting by the penalized likelihood method exemplified by scattering and meteorite data, *J. Am. Stat. Assoc.* 75, 42–73.

Ishiguro, M., and Y. Sakamoto (1983). A Bayesian approach to binary response curve estimation, *Ann. Inst. Stat. Math. Part B* 35, 115–137.

Ishiguo, M., H. Akaike, M. Doe, and S. Nakai (1981). A Bayesian Approach to the Analysis of Earth Tides, *Proceedings of the 9th International Conference on Earth Tides.*

Jones, R. H. (1980). Maximum likelihood fitting of ARMA models to time series with missing observations, *Technometrics* 22, 389–395.

Kalman, R. E. K. (1960). A new approach to linear filtering and prediction problems, *Trans. ASME J. Basic Eng.* 82D, 35–45.

Kesler, S. B., ed., (1986). *Modern Spectrum Analysis II*, IEEE Press, New York.

Kitagawa, G. (1981). A nonstationary time series model and its fitting by a recursive filter, *J. Time Ser. Anal.* 2, 103–116.

Kitagawa, G. (1983). Changing spectrum estimation, *J. Sound Vib.* 89(4), 443–445.

Kitagawa, G. (1984). State Space Modeling of Nonstationary Time Series and Smoothing of Unequally Spaced Data, in *Time Series Analysis of Irregularly Observed Data*, Vol. 25 of *Lecture Notes in Statistics* (E. Parzen, ed.), New York, Springer-Verlag, pp. 189–210.

Kitagawa, G. (1987). Non-Gaussian state space modeling of nonstationary time series (with discussion), *J. Am. Stat. Assoc.* 82, 1032–1063.

Kitagawa, G., and W. Gersch (1984). A smoothness priors-state modeling of time series with trend and seasonality, *J. Am. Stat. Assoc.* 79, 378–389.

Kitagawa, G., and W. Gersch (1985a). A smoothness priors long AR model method for spectral estimation, *IEEE Trans. Autom. Control* AC-30, 57–65.

Kitagawa, G., and W. Gersch (1985b). A smoothness priors time varying AR coefficient modeling of nonstationary time series, *IEEE Trans. Autom. Control* AC-30, 48–56.

Kitagawa, G., and T. Takanami (1985). Extraction of signal by a time series model and screening out micro earthquakes, *Signal Process.* 8, 303–314.

Kozin, F. (1977). Estimation and Modeling of Nonstationary Time series, in *Proceedings of the Symposium in Applied Computational Methods in Engineering*, University of Southern California, Los Angeles, pp. 603–612.

Lindley, D. V., and A. F. M. Smith (1972). Bayes estimate for the linear model, *J. R. Stat. Soc. Ser. B* 34, 1–41.

Nakamura, T. (1986). Bayesian cohort models for general cohort table analysis, *Ann. Inst. Stat. Math. Part B* 38, 353–370.

Naniwa, S. (1986). Trend Estimation via Smoothness Priors-State Space Modeling, in *Monetary and Economic Studies*, Vol. 4, Institute for Monetary and Economic Studies, Bank of Japan, pp. 79–112.

Nicholls and Quinn (1985). Time Series in the Time Domain, in *Handbook of Statistics* (E. J. Hannan and P. R. Krishnaiah, eds.), North-Holland, Amsterdam.

O'Sullivan, F., B. S. Yandell, and W. J. Raynor, Jr. (1986). Automatic smoothing of regression functions in generalized linear models, *J. Am. Stat. Assoc.* 81, 96–103.

Shiller, R. (1973). A distributed lag estimator derived from smoothness priors, *Econometrica* 41, 775–778.

Tanabe, K., and Sagae (1987). Smoothness priors density estimation, in preparation.

Tikhonov, A. N. (1965). Incorrect problems of linear algebra and a stable method for their solution, *Sov. Math. Dokl.* 6, 988–991.

Titterington, D. M. (1985). Common structure of smoothing techniques in statistics, *Int. Stat. Rev.* 53, 141–170.

Vinod, H. D., and J. L. Ullah (1981). *Recent Advances in Regression Methods*, Marcel Dekker, New York.

Wahba, G. (1977). A Survey of Some Smoothing Problems and the Method of Generalized Cross-Validation for Solving Them, in *Applications of Statistics* (P. R. Krishnaih, ed.), North-Holland, Amsterdam, pp. 507–524.

Wahba, G. (1980). Automatic smoothing of the log periodogram, *J. Am. Stat. Assoc.* 75, 122–132.

Wahba, G. (1982). Constrained Regularization for Ill Posed Linear Operator Equations with Applications in Meteorology and Medicine, in *Statistical Theory and Related Topics III*, Vol. 2 (S. S. Gupta and J. O. Berger, eds.), Academic Press, New York, pp. 383–418.

Wahba, G. (1983). Bayesian confidence intervals for the cross-validated smoothing spline, *J. R. Stat. Soc. Ser. B* 45, 133–150.

Wecker, W. E., and C. R. Ansley (1983). The signal extraction approach to nonlinear regression and spline smoothing, *J. Am. Stat. Assoc.* 78, 81–89.

Whittaker, E. T. (1923). On a new method of graduation, *Proc. Edinborough Math. Assoc.* 78, 81–89.

Whittaker, E. T., and G. Robinson (1924). *Calculus of Observations, A Treasure on Numerical Calculations*, Blackie & Son, Ltd., London. pp. 303–306.

Whittle, P. (1965). Recursive relations for predictors of non-stationary processes, *J. R. Stat. Soc. Ser. B* 27, 523–532.
Zellner, A. ed. (1983). *Applied Time Series Analysis of Economic Data*, U.S. Department of Commerce, Bureau of the Census, Washington, D.C.

17

Noninformative Priors and Restricted Maximum Likelihood Estimation in the Kalman Filter

WILLIAM M. SALLAS IMSL Inc., Houston, Texas

DAVID A. HARVILLE Department of Statistics, Iowa State University, Ames, Iowa

1. INTRODUCTION

In many applications, data accumulate over time. In such applications, the state-space model—the model underlying the Kalman filter—is a very general model to use. The model has been applied extensively by engineers and scientists in conjunction with linear and nonlinear stochastic differential equations (see, e.g., Maybeck, 1979, Chap. 4; 1982, Chap. 11). The model has also been used in the statistical and econometric literature in conjunction with time series models (see, e.g., Harvey, 1981, Chaps. 4 and 5), with regression models having time series errors (see, e.g., Harvey, 1981, Chap. 7), with some time-varying parameter models (see, e.g., Belsley and Kuh, 1973; Cooley and Prescott, 1976), with Bayesian models, with multivariate models, and with data containing missing values (Harvey and McKenzie, 1984) or with data collected at unequally spaced time points (Jones, 1984).

The model is specified stage by stage, where the stages generally correspond to the time points at which observations become available. The approach is essentially Bayesian. At stage k, the model for the new observations includes the prior information that is available at stage k for estimation of the state vector at stage $k + 1$.

Initially, for estimation at stage 1, a priori information for the state vector may not be available. In this case, Kalman (1961), Albert and Sittler (1965), Kaminski et al. (1971), Young (1974), and others have indicated that a zero prior mean and infinite prior variance (i.e., a noninformative prior) for each element of the state vector is appropriate. In this case, at the completion of stage 1, the Kalman filter produces (in the limit) the generalized least squares estimates given by Aitken (1934). Then, when the state vector is modeled as unchanging in subsequent stages, the formulas for the Kalman filter reduce to the formulas for recursive least squares estimators given by Plackett (1950) and Albert and Sittler (1965) for the "fixed-effects" linear regression model.

When there is a priori information for some elements of the state vector and not for some others, Sallas and Harville (1981) showed that a noninformative prior for each element of the state vector without a priori information is also appropriate. In this case, at the completion of stage 1, the Kalman filter produces (in the limit) estimators given by Henderson (1973) and Harville (1976a) for the mixed linear model—a model that incorporates both fixed and random effects. Then, when those elements of the state vector without a priori information are modeled as unchanging in subsequent stages, the formulas for the Kalman filter can be used to produce recursive formulas for the estimators in mixed linear models. These ideas are reviewed in the remaining parts of Section 1 and in Section 2.

In practice, unknown parameters can enter into elements of the state transition matrix and state- and observational-error variance-covariance matrices. Maximum likelihood techniques have been suggested in the engineering literature for estimating these unknown parameters. Mehra (1970), Kashyap (1970), Sage and Melsa (1971), Gupta and Mehra (1974), and Maybeck (1982) discuss the problem of estimation of the elements of the state transition matrix and/or state- and observation-error variance-covariance matrices. However, these techniques do not account for the fact that some elements of the initial state vector may have arbitrarily large variances. In this case, we show that the maximum likelihood approach (in the limit) is equivalent to the restricted maximum likelihood approach proposed by Patterson and Thompson (1971) in the context of the mixed linear model. In their approach the likelihood is taken to be that associated with linearly independent error contrasts (linear combinations of the observations with zero means). By a careful choice of these error contrasts, we derive a computationally useful representation for this likelihood in terms of quantities computed by the Kalman filter. This technique and the computing formulas are discussed in Section 3.

In Section 4 statistical inference and diagnostics for use in conjunction with REML and the Kalman filter are discussed. In Section 5 some extensions to the model are discussed. In Section 6 concluding remarks are given. The prediction of National Football League games is used as an example throughout our discussion.

1.1 Mixed Linear Model

The model we adopt is a mixed linear model, incorporating fixed and random effects. The model includes a $p \times 1$ parameter vector α corresponding to the fixed effects and a $q \times 1$ vector b_k corresponding to random effects that may change with $k = 1, 2, \ldots$. We shall assume that the fixed effects are estimable in the model for the data available at $k = 1$. The notation used here follows that of Sallas and Harville (1981). The subscript k is used here rather than t, which is more customary in time series, to emphasize that the model is expressed in stages $k = 1, 2, \ldots$ and that these stages need not correspond to equally spaced time points. In fact, they need not correspond to time points of any kind.

Let y_k be the $n_k \times 1$ vector of observations that first become available at stage k. The *observation equation* for the mixed linear model is

$$y_k = X_k\alpha + Z_k b_k + e_k, \qquad k = 1, 2, \ldots. \tag{1.1}$$

Here, the $n_k \times p$ matrix X_k and the $n_k \times q$ matrix Z_k are known. The random effects b_k are allowed to change in accordance with the formula

$$b_{k+1} = T_{k+1} b_k + w_{k+1}, \qquad k = 1, 2, \ldots \tag{1.2}$$

starting with $b_1 = w_1$. In the engineering literature, b_k is called the *state vector*, and (1.2) is called the *state equation.*

The change in the random effects from stage k to $k + 1$ is explained in part by the *transition matrix* T_{k+1}. It is assumed that the q-dimensional state-error vectors w_k ($k = 1, 2, \ldots$) are independently distributed multivariate normal with mean vector 0 and variance-covariance matrix $\sigma^2 Q_k$, that the n_k-dimensional observational-error vectors e_k ($k = 1, 2, \ldots$) are independently distributed multivariate normal with mean vector 0 and variance-covariance matrix $\sigma^2 R_k$, and that the w_k's and e_k's are independent of each other. Here, σ^2 is an unknown positive scalar. Often, the matrices R_k, T_{k+1}, and Q_{k+1} are assumed known. Here, we regard elements of these matrices as functions of unknown parameters that have to be estimated.

When $p = 0$ (i.e., with no fixed effects) (1.1) and (1.2) reduce to the state-space model—a completely random model. When $q = 0$, (1.1) and (1.2) reduce to the fixed-effects model. Engineers frequently use the state-space model given by (1.1)–(1.2) with $p = 0$. However, because of some bias in the measurement process they may subtract a constant from all their data. Rather than treat this constant as known, the mixed linear model allows it to be treated as a fixed effect to be estimated. Similarly, in analyzing time series, practitioners frequently remove the mean before applying some type of ARMA model. Instead, a mixed linear model could be used in which the mean is treated as a fixed effect to be estimated. Mixed linear model (1.1)–

(1.2) has a Bayesian interpretation. The state vector is divided into two parts: elements corresponding to the fixed effects, and elements corresponding to the random effects. The fixed effects are modeled with noninformative priors on the initial state and are unchanging in time.

1.2 Football Example

Harville (1980) used mixed linear model (1.1)–(1.2) for making week-ahead predictions for National Football League games. Harville's model used the difference in points scored (home team's point total minus visiting team's point total) as the dependent variable. It incorporated one fixed effect reflecting home field advantage (in points) and random effects for each NFL team, which reflect its yearly performance levels (in points) relative to that of the "average" team. We shall use an extended version of Harville's model to motivate the methodology to be discussed.

Number the NFL teams $i = 1, 2, \ldots, 28$; number the games in each season $j = 1, 2, \ldots$; and number the seasons $s = 1, 2, \ldots$ with the first season in the data set coded $s = 1$. Let h_{sj} and v_{sj} denote the home and visiting team, respectively, for the jth game in season s. (If the game was played on a neutral field, h_{sj} and v_{sj} are taken to be the two participating teams in either order.)

Let y_{1sj} represent the home team's score minus the visiting team's score for the jth game in season s. Let y_{2sj} represent the corresponding difference in the teams' net yards offense. (Net yards offense equals net yards rushing, plus gross yards passing, minus yards lost to "sacks.") The model is a bivariate model with y_{1sj} and y_{2sj} as the dependent variables. The observation equation incorporates fixed effects for the home field advantage in points (α_1) and in yards (α_2). Furthermore, the observation equation incorporates random effects that represent the performance level in points (b_{1is}) and in yards offense (b_{2is}) of team i in season s. The observation equation for the jth game in the sth season is

$$y_{1sj} = \alpha_1 x_{sj} + b_{1h_{sj}s} - b_{1v_{sj}s} + e_{1sj} \tag{1.3}$$

$$y_{2sj} = \alpha_2 x_{sj} + b_{2h_{sj}s} - b_{2v_{sj}s} + e_{2sj}. \tag{1.4}$$

Here x_{sj} equals 0 if the sjth game is played at a neutral site (e.g., a Super Bowl game); otherwise, x_{sj} equals 1. The errors e_{1sj} and e_{2sj} have mean zero. The observational-error variance-covariance matrix is

$$\operatorname{var}\begin{pmatrix} e_{1sj} \\ e_{2sj} \end{pmatrix} = R.$$

The yearly performance levels of team i as measured by points and yards offense are assumed to obey a bivariate first order autoregressive process. Let

T be the 2×2 transition matrix. We have for $i = 1, 2, \ldots, 28$, the state equation

$$\begin{pmatrix} b_{1,i,s+1} \\ b_{2,i,s+1} \end{pmatrix} = T \begin{pmatrix} b_{1is} \\ b_{2is} \end{pmatrix} + \begin{pmatrix} w_{1,i,s+1} \\ w_{2,i,s+1} \end{pmatrix} \tag{1.5}$$

starting with

$$\begin{pmatrix} b_{1i1} \\ b_{2i1} \end{pmatrix} = \begin{pmatrix} w_{1i1} \\ w_{2i1} \end{pmatrix}.$$

The errors w_{1is} and w_{2is} have mean zero. Take the state-error variance-covariance matrices to be Q_1 for $s = 1$ and Q_2 for $s = 2, 3, \ldots$. We assume the eigenvalues of T are less than one in absolute value. Further, we assume that Q_1 satisfies $Q_1 - TQ_1T^T = Q_2$, so that the yearly performance levels of the 28 NFL teams follow a stationary bivariate first-order autoregressive process. The common variance-covariance matrix of the yearly performance levels is Q_1. The autocorrelation matrix between performance levels of the same team in seasons s and $r = s - \tau$ is

$$\begin{pmatrix} \sqrt{E(b_{1is}^2)} & 0 \\ 0 & \sqrt{E(b_{2is}^2)} \end{pmatrix}^{-1} \begin{pmatrix} E(b_{1is}b_{1ir}) & E(b_{1is}b_{2ir}) \\ E(b_{2is}b_{1ir}) & E(b_{2is}b_{2ir}) \end{pmatrix} \begin{pmatrix} \sqrt{E(b_{1is}^2)} & 0 \\ 0 & \sqrt{E(b_{2is}^2)} \end{pmatrix}^{-1}$$

$$= [\operatorname{diag}(Q_1)]^{-1/2} T^{\tau} Q_1 [\operatorname{diag}(Q_1)]^{-1/2}.$$

Furthermore, the yearly performance levels of any given team are independent of those of any other team.

The results of 1050 NFL games played during the regular and playoff seasons for 1979–1983 gave the estimates listed in Table 1. The estimates of the fixed effects α_1 and α_2 are 2.69 (0.39) and 16.1 (3.4), respectively. Standard errors for the estimates are given in parentheses. The estimates of the 56 random effects (performance levels in points and yards for each of the 28 teams) are given at the conclusion of the Super Bowl for the 1983 season (actually played in January 1984). Although the Los Angeles Raiders defeated the Washington Redskins in the Super Bowl 38 to 9, Washington was rated higher by 1.14 points and 8.1 yards.

Estimates of the elements of the state transition matrix and state- and observational-error variance-covariance matrices are as follows:

$$T = \begin{pmatrix} 1.80 & -0.16 \\ 19.24 & -1.81 \end{pmatrix}, \quad Q_2 = \begin{pmatrix} 13.2 & 93.5 \\ 93.5 & 661.5 \end{pmatrix}, \quad R = \begin{pmatrix} 166 & 708 \\ 708 & 12{,}386 \end{pmatrix}.$$

The standard deviations of the observational errors are $\sqrt{166} = 12.9$ points and $\sqrt{12{,}386} = 111$ yards. The common variance-covariance matrix for the

Table 1 Fixed and Random Effect Estimates

Effect	Points	(Std. Error)	Yards	(Std. Error)
Home field advantage	2.69	(0.39)	16.1	(3.4)
Atlanta Falcons	−0.45	(2.50)	−5.3	(19.9)
Indianapolis Colts	−5.09	(2.51)	−42.9	(20.0)
Buffalo Bills	−3.58	(2.51)	−28.3	(20.0)
Chicago Bears	0.47	(2.51)	2.9	(20.0)
Cincinnati Bengals	1.99	(2.50)	16.2	(19.9)
Cleveland Browns	−0.34	(2.50)	−4.1	(19.9)
Dallas Cowboys	4.80	(2.46)	41.7	(19.5)
Denver Broncos	−3.26	(2.47)	−32.4	(19.7)
Detroit Lions	1.19	(2.46)	7.1	(19.6)
Green Bay Packers	−0.48	(2.50)	−0.1	(19.9)
Houston Oilers	−6.55	(2.50)	−57.1	(19.9)
Kansas City Chiefs	1.40	(2.50)	10.4	(20.0)
Los Angeles Rams	0.08	(2.41)	−2.1	(19.3)
Miami Dolphins	3.49	(2.46)	30.2	(19.5)
Minnesota Vikings	−1.90	(2.50)	−14.7	(19.9)
New England Patriots	−1.31	(2.50)	−12.3	(19.9)
New Orleans Saints	−0.11	(2.50)	0.1	(19.9)
New York Giants	−0.49	(2.51)	−2.2	(20.0)
New York Jets	0.14	(2.50)	8.3	(19.8)
Los Angeles Raiders	7.00	(2.38)	60.5	(19.0)
Philadelphia Eagles	−3.18	(2.51)	−26.9	(20.0)
Pittsburg Steelers	0.49	(2.45)	5.4	(19.6)
San Diego Chargers	−0.80	(2.50)	0.0	(19.9)
San Francisco 49ers	4.65	(2.41)	33.9	(19.3)
St. Louis Cardinals	−0.25	(2.50)	−1.4	(19.9)
Washington Redskins	8.14	(2.38)	68.6	(18.9)
Tampa Bay Buccaneers	−4.99	(2.50)	−39.1	(19.9)
Seattle Seahawks	−0.81	(2.39)	−12.4	(19.1)

yearly performance levels of the teams in points and in yards, Q_1, obtained by solving $Q_1 - TQ_1T^T = Q_2$ is

$$Q_1 = \begin{pmatrix} 19.6 & 154 \\ 1545 & 1240 \end{pmatrix}.$$

The standard deviations of the team effects are $\sqrt{19.6} = 4.4$ points and

$\sqrt{1240} = 35$ yards. The autocorrelation matrix between the yearly performance levels of the same team $\tau = 1$ year apart is

$$\begin{pmatrix} 0.54 & 0.51 \\ 0.63 & 0.58 \end{pmatrix}.$$

In Sections 2 and 3, we discuss the procedures used to compute these estimates.

2. ESTIMATION OF THE EFFECTS

With the infusion of new data at each stage, computation of revisions of estimates of the fixed and random effects in model (1.1)–(1.2) and of elements of T_k, Q_k, and R_k becomes an ongoing problem. Generally, revisions of estimates of the random effects are most likely to change substantially from stage to stage, while revisions of the other estimates are likely to change very slowly. The estimates of the effects (fixed, as well as random) are of primary interest. The estimates of T_k, Q_k, and R_k are generally of interest only in that they are needed to refine the estimates of the effects.

In the case of the football data, weekly updates of the fixed and random effects are needed. One could consider weekly updates of estimates of the elements of T, Q_2, and R, but the changes would be so small that the resulting changes in the estimates of the effects would be negligible. Yearly updates of the elements of T, Q_2, and R seem more appropriate and will be considered in Section 3.

In Section 2.1 we consider estimation of the effects using a recursive algorithm given by Kalman (1960), which has come to be known as the Kalman filter. The model on which the Kalman filter is based is the state-space model, that is, model (1.1)–(1.2) with $p = 0$. The theory assumes that the elements of the T_k's, Q_k's, and R_k's are all known. In practice, if good estimates of the elements are available, we could use these instead. The computations and storage requirements required by the Kalman filter are much less than those that would be necessary if one were to process all the data at the end of each stage. This is accomplished by taking advantage of previous computations and retaining in storage only those items essential for the processing of future observations.

In Section 2.2 we consider the modifications to the Kalman filter required to incorporate the fixed effects in model (1.1)–(1.2).

2.1 Kalman Filter

Suppose that the model is the state-space model, given by (1.1) and (1.2) with $p = 0$. Denote the estimator of the realization of the state vector b_k given the observations $y_1, y_2, \ldots, y_j$ by $\hat{\beta}_{k|j}$. The mean square error matrix for $\hat{\beta}_{k|j}$ is

$$\sigma^2 C_{k|j} \equiv E(\hat{\beta}_{k|j} - b_k)(\hat{\beta}_{k|j} - b_k)^T.$$

The Kalman filter, proceeds in stages. The quantities $\hat{\beta}_{k|k-1}$ and $C_{k|k-1}$ are available at the kth stage, having been computed during the $(k-1)$st stage. During the kth stage the filtered estimate $\hat{\beta}_{k|k}$, along with $C_{k|k}$, are computed by the *update equations*:

$$\hat{\beta}_{k|k} = \hat{\beta}_{k|k-1} + C_{k|k-1} Z_k^T H_k^{-1} v_k \tag{2.1}$$

$$C_{k|k} = C_{k|k-1} - C_{k|k-1} Z_k^T H_k^{-1} Z_k C_{k|k-1}, \tag{2.2}$$

where $v_k = y_k - Z_k \hat{\beta}_{k|k-1}$ and $H_k = R_k + Z_k C_{k|k-1} Z_k^T$. Here, v_k is the one-step-ahead prediction error, and $\sigma^2 H_k$ is the variance-covariance matrix for v_k. The "startup values" needed at the first stage are the prior mean of b_1, $\hat{\beta}_{1|0} = 0$, and the prior variance of b_1 (modulo σ^2), $C_{1|0} = Q_1$.

The kth stage is completed by computing the one-step-ahead estimate $\hat{\beta}_{k+1|k}$, along with $C_{k+1|k}$, in accordance with the *prediction equations*:

$$\hat{\beta}_{k+1|k} = T_{k+1} \hat{\beta}_{k|k} \tag{2.3}$$

$$C_{k+1|k} = T_{k+1} C_{k|k} T_{k+1}^T + Q_{k+1}. \tag{2.4}$$

If the estimate of the state vector is needed for more than one step into the future, that is, an estimate $\hat{\beta}_{k|j}$ is needed where $k > j + 1$, then (2.3) and (2.4) can be used recursively to compute $\hat{\beta}_{j+2|j}, \hat{\beta}_{j+3|j}, \ldots, \hat{\beta}_{k|j}$ along with their associated mean square error matrices. If several stages correspond to a single time point, that is, there is no change in the effects from one stage to the next, the prediction equations can be skipped.

The update equations are the conditional mean and variance of a multivariate normal distribution. The estimator $\hat{\beta}_{k|j}$ produced by the Kalman filter is the conditional mean of b_k given the observations $y_1, y_2, \ldots, y_j$. Each component of the estimator (and more generally, each linear combination of the components) has minimum mean square error. The mean square error matrix for $\hat{\beta}_{k|j}$ is $\sigma^2 C_{k|j}$, which is the variance-covariance matrix in the conditional multivariate normal distribution for b_k given $y_1, y_2, \ldots, y_j$. The matrix $\sigma^2 H_k$ is the conditional variance-covariance matrix of y_k given $y_1, y_2, \ldots, y_{k-1}$.

The IMSL subroutine KALMN (IMSL, 1987b, Chap. 8) can be used to compute the quantities in (2.1) to (2.4). As options, one can input no observations, which has the effect of skipping the update equations, or one can indicate a transition matrix equal to the identity along with no state equation error, which has the effect of skipping the prediction equations.

2.2 Modifications to the Kalman Filter for Mixed Model Estimation

The Kalman filter is directly applicable to the state-space model, a completely random model. However, a fixed-effects linear model can also be used via a Bayesian approach. The fixed effects are modeled as unchanging in time, with prior means of zero and arbitrarily large prior variances, that is, with noninformative (diffuse) priors (see, e.g., Kaminski et al., 1971; Young, 1974; Harvey, 1981, p. 197). This approach gives generalized least squares estimates. Two implementations have been used: (1) the use of the Kalman filter with large prior variances, and (2) the use of an alternative filter algorithm known as the information filter which propagates $C_{k|k}^{-1}$, the information matrix modulo σ^{-2}, in place of $C_{k|k}$ and $C_{k|k}^{-1}\hat{\beta}_{k|k}$ in place of $\hat{\beta}_{k|k}$ (see e.g., Anderson and Moore, 1979, Sec. 6.3).

A numerical problem is introduced by the first implementation. How large should the prior variances be? One choice for the prior variance-covariance matrix of the fixed effects is (say) $10^6 I$. However, too large a value can produce a loss of precision through rounding errors. Kaminski et al. (1971) suggested the use of the information filter with $C_{1|0}^{-1} = 0$ to avoid these numerical problems. In fact, use of the information filter for the "fixed-effects" linear regression model is just updating sums of squares and cross-product matrices in the Aitken equations. If estimates are needed along with mean square error matrices at each stage, this algorithm fails to address the problem of efficiently updating the estimates and the mean square error matrices. If its use is confined to some initial startup period, a switch to the Kalman filter algorithm for computing updates of estimates along with mean square error matrices can be made after the mean square error matrix for the estimates is nonsingular. Then this algorithm is identical to the recursive least squares algorithm given by Plackett (1950) and Albert and Sittler (1965) for the fixed-effects linear regression model.

In the case when some, but not all, of the elements of the state vector, are fixed and some random, Sallas and Harville (1981) showed that mixed model (1.1)–(1.2) can be regarded as a limiting version of the following two-part random model:

$$y_k = X_k a + Z_k b_k + e_k, \qquad k = 1, 2, \ldots .$$

The vector a is of length p and contains random effects that are "to become fixed." The elements of the vector a are modeled as unchanging in time with a common large prior variance. More specifically, it is assumed that

$$\begin{pmatrix} a \\ b_{k+1} \end{pmatrix} = \begin{pmatrix} I & 0 \\ 0 & T_{k+1} \end{pmatrix} \begin{pmatrix} a \\ b_k \end{pmatrix} + \begin{pmatrix} 0 \\ w_{k+1} \end{pmatrix}, \qquad k = 1, 2, \ldots$$

and

$$\begin{pmatrix} a \\ b_1 \end{pmatrix} = \begin{pmatrix} d \\ w_1 \end{pmatrix},$$

where

$$\text{var} \begin{pmatrix} d \\ w_k \end{pmatrix} = \begin{pmatrix} \varepsilon^{-1} I & 0 \\ 0 & \sigma^2 Q_k \end{pmatrix}.$$

Note that the elements of a are independently and identically normally distributed with mean 0 and variance $1/\varepsilon$ and are independent of w_k and e_k.

By applying the Kalman filter to this two-part random model and then taking limits with $\varepsilon \to 0^+$, Sallas and Harville (1981) derived recursive formulas for the mixed model. These recursive formulas are modifications of the Kalman filter that make a distinction between the fixed and random effects, but only at stage 1. After stage 1, the minimum mean square estimators can be computed by the Kalman filter.

At stage 1 estimators obtained are the mixed model formulas as given by Henderson (1973, 1975) and Harville (1976a). They yield, for $k = 1$, fixed- and random-effect estimators

$$\hat{\alpha}_1 = (X_1^T H_1^{-1} X_1)^{-1} X_1^T H_1^{-1} y_1 \tag{2.5}$$

$$\hat{\beta}_{1|1} = Q_1 Z_1^T H_1^{-1} (y_1 - X_1 \hat{\alpha}_1), \tag{2.6}$$

where $H_1 = R_1 + Z_1 Q_1 Z_1^T$, along with

$$\begin{aligned} C_{1|1} &= \sigma^{-2} \begin{pmatrix} E(\hat{\alpha}_1 - \alpha)(\hat{\alpha}_1 - \alpha)^T & E(\hat{\alpha}_1 - \alpha)(\hat{\beta}_{1|1} - b_1)^T \\ E(\hat{\beta}_{1|1} - b_1)(\hat{\alpha}_1 - \alpha)^T & E(\hat{\beta}_{1|1} - b_1)(\hat{\beta}_{1|1} - b_1)^T \end{pmatrix} \\ &= \begin{pmatrix} (X_1^T H_1^{-1} X_1)^{-1} & -(X_1^T H_1^{-1} X_1)^{-1} X_1^T H_1^{-1} Z_1 Q_1 \\ \text{(symmetric)} & Q_1 - Q_1 Z_1^T H_1^{-1} Z_1 Q_1 \\ & \quad + Q_1 Z_1^T H_1^{-1} X_1 (X_1^T H_1^{-1} X_1)^{-1} X_1^T H_1^{-1} Z_1 Q_1 \end{pmatrix}. \end{aligned} \tag{2.7}$$

A less general form for (2.5) to (2.7) was given by Henderson (1963). Henderson's equations, which are applicable when Q_1 is nonsingular, are

$$\begin{pmatrix} X_1^T R_1^{-1} X_1 & X_1^T R_1^{-1} Z_1 \\ Z_1^T R_1^{-1} X_1 & Z_1^T R_1^{-1} Z_1 + Q_1^{-1} \end{pmatrix} \begin{pmatrix} \hat{\alpha}_1 \\ \hat{\beta}_{1|1} \end{pmatrix} = \begin{pmatrix} X_1^T R_1^{-1} y_1 \\ Z_1^T R_1^{-1} y_1 \end{pmatrix}. \tag{2.8}$$

Equation (2.8) is identical to that obtained by application of the update equations in the information filter after first assigning each element of $C_{1|0}^{-1}$ corresponding to a fixed effect (both the rows and columns) the value zero. In particular, the information filter update equations add the generalized sums of squares and cross-products of X_1, Z_1, and y_1 to the startup values

$$C_{1|0}^{-1} = \begin{pmatrix} 0 & 0 \\ 0 & Q_1^{-1} \end{pmatrix} \quad \text{and} \quad C_{1|0}^{-1}\begin{pmatrix} \hat{\alpha}_1 \\ \hat{\beta}_{1|0} \end{pmatrix} = \begin{pmatrix} 0 \\ 0 \end{pmatrix}.$$

After this update, the information filter gives $C_{1|1}^{-1}$ given by the coefficient matrix on the left-hand side of (2.8) and $C_{1|1}^{-1}(\hat{\alpha}_1^T, \hat{\beta}_{1|1}^T)^T$ given by the right-hand side of (2.8).

The continued use of the information filter does not provide an efficient means for computing the estimates and the mean square error matrix at each stage, and it requires restrictive assumptions—the prediction equations require that the $C_{k|k}$'s and the T_k's are nonsingular. Given $\hat{\alpha}_1$, $\hat{\beta}_{1|1}$, and $C_{1|1}$, the Kalman filter can be efficiently used to compute the mixed model estimators $\hat{\alpha}_k$, $\hat{\beta}_{k|k}$, along with $C_{k|k}$.

The Kalman filter can be used by making the following adjustments:

1. Concatenate the fixed and random effects into a single vector of length $p + q$, and take $U_k = (X_k, Z_k)$, so that (1.1) and (1.2) can be rewritten as

$$y_k = U_k \begin{pmatrix} \alpha \\ b_k \end{pmatrix} + e_k$$

$$\begin{pmatrix} \alpha \\ b_{k+1} \end{pmatrix} = \begin{pmatrix} I & 0 \\ 0 & T_{k+1} \end{pmatrix}\begin{pmatrix} \alpha \\ b_k \end{pmatrix} + \begin{pmatrix} 0 \\ w_{k+1} \end{pmatrix}.$$

2. Apply the Kalman filter [(2.1) to (2.4)] with $(\hat{\alpha}_k^T, \hat{\beta}_{k|k}^T)^T$ in place of $\hat{\beta}_{k|k}$, $(\hat{\alpha}_{k-1}^T, \hat{\beta}_{k|k-1}^T)^T$ in place of $\hat{\beta}_{k|k-1}$, U_k in place of Z_k, and

$$\begin{pmatrix} I & 0 \\ 0 & T_{k+1} \end{pmatrix} \quad \text{and} \quad \begin{pmatrix} 0 & 0 \\ 0 & Q_{k+1} \end{pmatrix}$$

in place for T_{k+1} and Q_{k+1}, respectively.

In summary, once $\hat{\alpha}_1$, $\hat{\beta}_{1|1}$, and $C_{1|1}$ are obtained via mixed model formulas, the revision of estimates for the mixed model proceeds as for a state-space model in which the fixed effects are treated as states that are unchanging in time.

2.3 Football Example

In this section we return to the football model discussed in Section 1.2. Using the effect estimates and the mean square error matrix as of the end of the 1983 NFL season (Table 1), weekly revisions of estimates of the effects were computed during the 1984 NFL season using the modifications to the Kalman filter discussed in Section 2.2. The elements of T, Q_2, and R given in Section 1.2 were used. Each game was regarded as a stage. In applying the filter to successive games within the same season, the prediction equations can be skipped because the performance levels of the teams do not change. The estimates of the effects as of the end of each of weeks 7 to 14 are given in Table 2. Only the effect estimates for the scores are given in Table 2. Note that the effect estimates for home field advantage change very little from week to week. This is because the home field advantage is modeled as unchanging in time so the current observations have essentially the same affect on its estimation as observations in the past. Also, the estimates of the home field advantage are based on 1134 to 1232 games (week 1, 1979 through weeks 6 to 13 of 1984), so that the estimates are very precise. In contrast, note the

Table 2 Estimates of Effects Prior to Games Played on the Week Indicated

Effect	7	8	9	10	11	12	13	14
Home field advantage	2.74	2.71	2.65	2.66	2.66	2.64	2.65	2.65
Atlanta Falcons	1.42	0.42	−0.41	−1.23	−1.16	−1.30	−2.37	−2.94
Indianapolis Colts	−5.73	−5.84	−5.76	−6.54	−7.63	−7.02	−8.03	−8.01
Buffalo Bills	−5.50	−4.51	−5.68	−6.32	−6.55	−7.32	−6.61	−7.31
Chicago Bears	3.06	2.41	3.89	4.31	4.73	4.09	4.15	5.07
Cincinnati Bengals	−1.77	−1.77	−1.75	−1.38	−1.26	−1.19	−1.43	−0.72
Cleveland Browns	−1.33	−1.43	−1.25	−1.63	−1.46	−2.80	−1.86	−1.53
Dallas Cowboys	2.27	1.88	1.58	2.35	1.49	1.66	0.87	0.93
Denver Broncos	−0.25	−1.16	0.25	0.57	0.22	0.08	0.86	0.55
Detroit Lions	−0.75	−1.42	−0.85	−2.26	−1.99	−1.74	−1.78	−1.26
Green Bay Packers	−3.18	−2.45	−2.39	−0.89	−0.19	0.95	1.64	1.00
Houston Oilers	−7.95	−8.56	−8.60	−9.13	−9.51	−8.98	−8.42	−8.70
Kansas City Chiefs	−0.78	0.42	−0.58	−0.84	−2.05	−2.66	−2.63	−2.62
Los Angeles Rams	−0.50	0.50	1.28	−0.34	−0.17	0.57	−0.25	−0.28
Miami Dolphins	7.28	7.85	8.52	9.19	9.69	8.90	8.47	8.03
Minnesota Vikings	−2.45	−2.70	−3.54	−3.88	−3.71	−4.75	−5.19	−6.21
New England Patriots	−0.11	0.08	−0.70	−0.27	0.07	0.38	1.38	1.30
New Orleans Saints	0.59	−0.54	−0.07	0.39	−0.30	0.13	−0.08	−1.13
New York Giants	−0.94	0.04	−0.71	0.45	1.19	0.94	0.80	1.00
New York Jets	−1.09	−0.75	0.24	−0.25	−1.01	−1.85	−2.48	−2.06
Los Angeles Raiders	4.45	4.30	4.52	4.11	3.79	3.98	4.36	4.22
Philadelphia Eagles	−2.54	−2.16	−1.75	−2.37	−2.61	−1.83	−1.86	−1.32
Pittsburg Steelers	0.21	0.86	0.97	1.59	1.82	1.73	1.68	2.42
San Diego Chargers	2.44	1.07	0.72	−0.57	0.46	0.54	1.18	0.39
San Francisco 49ers	5.31	4.79	4.81	6.39	6.14	7.52	7.06	8.25
St. Louis Cardinals	2.25	2.79	3.40	4.24	3.94	4.02	3.79	3.31
Washington Redskins	5.49	6.19	5.59	4.55	4.48	4.44	4.09	4.85
Tampa Bay Buccaneers	−3.49	−3.16	−4.46	−4.23	−4.18	−3.79	−3.46	−3.34
Seattle Seahawks	2.77	1.77	1.75	2.95	4.30	3.88	4.50	4.64

estimates of the random effects fluctuate considerably from week to week. As discussed in Section 1.2, a team's performance level in any particular year has a correlation of 0.54 with that in the preceding year. In the case of the random effect estimates, the 1984 data are weighted more heavily, relative to the data from earlier years, than in the case of the fixed-effect estimates.

The effect estimates in the last column of Table 2 were used to predict the outcomes of football games scheduled in week 14 of the 1984 season. For example, the Washington Redskins were to play the Minnesota Vikings in Minnesota during week 14 (Thursday, November 29). The prediction for that game, as computed from (1.3), was $2.65 - 6.21 - 4.85 = -8.41$, that is, the visiting team, Washington, was predicted to win by 8.41 points. The standard error of this prediction, which is given by the square root of the first entry in the matrix H_k, was 13.4 points. The actual outcome of the game was Washington 31, Minnesota 17, so the prediction error was 5.59 points.

3. MAXIMUM LIKELIHOOD ESTIMATION

In discussing (in Section 2) the Kalman filter and its modifications for mixed model estimation, it was assumed that the matrices R_k, Q_k, and T_k ($k = 1, 2, \ldots$) are known. Now this assumption is relaxed and it is assumed that these matrices are known functions of the elements of the vector θ contained in some subset of u-dimensional Euclidean space. We begin by giving formulas for the log likelihood function, for its first- and second-order partial derivatives and for the information matrix. The formulas are useful in the estimation of θ. The formulas are presented first for the state-space model. Then modifications of these results are considered for mixed model (1.1)–(1.2). Finally, we discuss some methods for maximizing the likelihood function and schemes for computing revised parameter estimates.

3.1 Log Likelihood Function for the State-Space Model

Take the model to be the state-space model given by (1.1)–(1.2) with $p = 0$. Consider the observations accumulated through time m, and put $y_m^* = (y_1^T, y_2^T, \ldots, y_m^T)^T$ and $N = \sum_{k=1}^m n_k$. The conditional distribution of the n_k-dimensional y_k given $y_1, y_2, \ldots, y_{k-1}$ is multivariate normal with mean vector $Z_k \hat{\beta}_{k|k-1}$ and variance-covariance matrix $\sigma^2 H_k$. The joint probability density function of y_m^* is the product of these conditional densities for $k = 1, 2, \ldots, m$ and is given by

$$f(y_m^*; \theta, \sigma^2) = (2\pi\sigma^2)^{-N/2} \prod_{k=1}^{m} [\det(H_k)]^{-1/2} \exp\left(-\frac{1}{2\sigma^2} \sum_{k=1}^{m} v_k^T H_k^{-1} v_k \right)$$

(Schweppe, 1965). Regarded as a function of θ and σ^2, $f(y_m^*; \theta, \sigma^2)$ is the likelihood function. The natural logarithm of the likelihood function differs by no more than an additive constant from

$$L(\theta, \sigma^2; y_m^*) = -\frac{N}{2} \ln(\sigma^2) - \frac{1}{2} \sum_{k=1}^{m} \ln[\det(H_k)] - \frac{1}{2\sigma^2} \sum_{k=1}^{m} v_k^T H_k^{-1} v_k. \quad (3.1)$$

The first-order partial derivatives of $L(\theta, \sigma^2; y_m^*)$ are as follows:

$$\frac{\partial L}{\partial \theta_i} = -\frac{1}{2} \sum_{k=1}^{m} \operatorname{tr}\left[H_k^{-1}\left(\frac{\partial H_k}{\partial \theta_i}\right)\right] + \frac{1}{2\sigma^2} \sum_{k=1}^{m} v_k^T H_k^{-1} \left(\frac{\partial H_k}{\partial \theta_i}\right) H_k^{-1} v_k$$
$$- \frac{1}{\sigma^2} \sum_{k=1}^{m} v_k^T H_k^{-1} \left(\frac{\partial v_k}{\partial \theta_i}\right), \qquad i = 1, 2, \ldots, u \quad (3.2)$$

$$\frac{\partial L}{\partial \sigma^2} = -\frac{N}{2\sigma^2} + \frac{1}{2\sigma^4} \sum_{k=1}^{m} v_k^T H_k^{-1} v_k \quad (3.3)$$

(see, e.g., Maybeck, 1972, eq. 4.13; Gupta and Mehra, 1974, eq. 11). The second-order partial derivatives of $L(\theta, \sigma^2; y_m^*)$ are

$$\frac{\partial^2 L}{\partial \theta_i \partial \theta_j} = -\frac{1}{2} \sum_{k=1}^{m} \operatorname{tr}\left\{H_k^{-1}\left[\left(\frac{\partial^2 H_k}{\partial \theta_i \theta_j}\right) - \left(\frac{\partial H_k}{\partial \theta_i}\right) H_k^{-1} \left(\frac{\partial H_k}{\partial \theta_j}\right)\right]\right\}$$
$$+ \frac{1}{2\sigma^2} \sum_{k=1}^{m} v_k^T \left\{H_k^{-1}\left[\left(\frac{\partial^2 H_k}{\partial \theta_i \theta_j}\right) - 2\left(\frac{\partial H_k}{\partial \theta_i}\right) H_k^{-1} \left(\frac{\partial H_k}{\partial \theta_j}\right)\right] H_k^{-1}\right\} v_k$$
$$+ \frac{1}{\sigma^2} \sum_{k=1}^{m} v_k^T H_k^{-1} \left(\frac{\partial H_k}{\partial \theta_i}\right) H_k^{-1} \left(\frac{\partial v_k}{\partial \theta_j}\right)$$
$$+ \frac{1}{\sigma^2} \sum_{k=1}^{m} v_k^T H_k^{-1} \left(\frac{\partial H_k}{\partial \theta_j}\right) H_k^{-1} \left(\frac{\partial v_k}{\partial \theta_i}\right)$$
$$- \frac{1}{\sigma^2} \sum_{k=1}^{m} \left(\frac{\partial v_k}{\partial \theta_i}\right)^T H_k^{-1} \left(\frac{\partial v_k}{\partial \theta_j}\right)$$
$$- \frac{1}{\sigma^2} \sum_{k=1}^{m} v_k^T H_k^{-1} \left(\frac{\partial^2 v_k}{\partial \theta_i \theta_j}\right), \qquad i, j = 1, 2, \ldots, u \quad (3.4)$$

$$\frac{\partial^2 L}{\partial \theta_i \partial \sigma^2} = -\frac{1}{2\sigma^4} \sum_{k=1}^{m} v_k^T H_k^{-1} \left(\frac{\partial H_k}{\partial \theta_i}\right) H_k^{-1} v_k$$
$$+ \frac{1}{\sigma^4} \sum_{k=1}^{m} v_k^T H_k^{-1} \left(\frac{\partial v_k}{\partial \theta_i}\right), \qquad i = 1, 2, \ldots, u \quad (3.5)$$

$$\frac{\partial^2 L}{\partial (\sigma^2)^2} = \frac{N}{2\sigma^4} - \frac{1}{\sigma^6} \sum_{k=1}^{m} v_k^T H_k^{-1} v_k, \quad (3.6)$$

as can be derived by using matrix differentiation rules (see, e.g., Nering, 1970, p. 280; Graybill, 1969, p. 266).

The elements of the information matrix are

$$-E\left(\frac{\partial^2 L}{\partial\theta_i \partial\theta_j}\right) = \frac{1}{2}\sum_{k=1}^{m} \operatorname{tr}\left[H_k^{-1}\left(\frac{\partial H_k}{\partial\theta_i}\right) H_k^{-1}\left(\frac{\partial H_k}{\partial\theta_j}\right)\right]$$
$$+\frac{1}{\sigma^2}\sum_{k=1}^{m} E\left[\left(\frac{\partial v_k}{\partial\theta_i}\right)^T H_k^{-1}\left(\frac{\partial v_k}{\partial\theta_j}\right)\right], \qquad i, j = 1, 2, \ldots, u \tag{3.7}$$

$$-E\left(\frac{\partial^2 L}{\partial\theta_i \partial\sigma^2}\right) = \frac{1}{2\sigma^2}\sum_{k=1}^{m} \operatorname{tr}\left[H_k^{-1}\left(\frac{\partial H_k}{\partial\theta_i}\right)\right], \qquad i = 1, 2, \ldots, u \tag{3.8}$$

$$-E\left(\frac{\partial^2 L}{\partial(\sigma^2)^2}\right) = \frac{N}{2\sigma^4} \tag{3.9}$$

(Maybeck 1972, eq. 4.1.21; Goodwin and Payne 1977, p. 158, eq. 6.5.6). These expressions are derived on the basis of the following considerations: (1) v_k has mean zero and variance-covariance matrix $H_k\sigma^2$; (2) $\partial v_k/\partial\theta_i$ and $\partial^2 v_k/\partial\theta_i\partial\theta_j$ which are linear functions of y^*_{k-1} are uncorrelated with e_k, w_k and $\hat{\beta}_{k-1|k-1} - b_{k-1}$ and thus are uncorrelated with v_k; and (3) if x and y are random vectors and $E(x) = 0$, then $E(x^T A y) = \operatorname{tr}(\operatorname{cov}(x, y)A)$.

The computations required for the evaluation of the likelihood at a parameter point are essentially the same as those that arise in estimating the state vector of the state-space model and can be carried out by, for example, using the IMSL routine KALMN (IMSL, 1987b, Chap. 8).

To compute the first-order partial derivatives (3.2) recursively, u parallel Kalman filters are required. The recursive computations of second-order partial derivatives (3.4) requires an additional $u(u + 1)/2$ parallel Kalman filters. For computational purposes, in (3.7) it is useful to invoke the approximation

$$E\left[\left(\frac{\partial v_k}{\partial\theta_i}\right)^T H_k^{-1}\left(\frac{\partial v_k}{\partial\theta_j}\right)\right] \approx \left(\frac{\partial v_k}{\partial\theta_i}\right)^T H_k^{-1}\left(\frac{\partial v_k}{\partial\theta_j}\right). \tag{3.10}$$

The approximate information matrix involves only quantities obtainable from the first-order derivative filters.

3.2 Modifications for REML in the Mixed Linear Model

Suppose that the approach of using large prior variances for elements of the state vector in the state-space model is used now in conjunction with maximum likelihood estimation in order to estimate T_k, Q_k, and R_k. In a manner similar to that of Dempster et al. (1981), we discuss here how this leads to the use of the restricted maximum likelihood (REML) estimator of Patterson and Thompson (1971) for mixed linear models. Then we indicate how to modify formulas (3.1) to (3.9) for purposes of obtaining REML estimates of θ and σ^2 under model (1.1)–(1.2).

Consider the random linear model $y = Xa + e$ where the $N \times p$ matrix X has full column rank, $\text{var}(a) = \varepsilon^{-1}I$, and $\text{var}(e) = V$. The matrix V depends on an unknown parameter vector θ. The log likelihood is given by

$$\begin{aligned} L(\theta; y) &= -\tfrac{1}{2}\ln[\det(\varepsilon^{-1}XX^T + V)] - \tfrac{1}{2}y^T(\varepsilon^{-1}XX^T + V)^{-1}y \\ &= -\tfrac{1}{2}\ln[\varepsilon^{-p}\det(V)\det(X^TV^{-1}X + \varepsilon I)] \\ &\quad - \tfrac{1}{2}y^T[V^{-1} - V^{-1}X(X^TV^{-1}X + \varepsilon I)^{-1}X^TV^{-1}]y, \end{aligned}$$

where the second equality follows from well-known lemmas on matrix inverses (see, e.g., Duncan and Horn, 1972, Lemma 3.3).

The term $(p/2)\ln\varepsilon$ is free of θ and can be dropped from the function to be maximized. The limit of the remaining function as $\varepsilon \to 0^+$ (Dempster et al., 1981, Section 5.4; Sallas and Harville, 1981, Theorem 2) yields

$$\begin{aligned} L_1(\theta; y) &= -\tfrac{1}{2}\ln[\det(V)] - \tfrac{1}{2}\ln[\det(X^TV^{-1}X)] \\ &\quad - \tfrac{1}{2}(y - X\hat{\alpha}^T)V^{-1}(y - X\hat{\alpha}), \end{aligned}$$

where $\hat{\alpha}$ satisfies $X^TV^{-1}X\hat{\alpha} = X^TV^{-1}y$. The function $L_1(\theta; y)$ is the restricted maximum likelihood (REML) function introduced by Patterson and Thompson (1971). Harville (1977) derived the same function by writing the likelihood for a set of linearly independent functions of y with mean zero with respect to the associated model with the random effects a replaced by the fixed effects α. Ansley and Kohn (1985) also demonstrated the equivalence of placing a noninformative prior on part of the initial state vector and of the application of maximum likelihood to a set of transformed observations. They did not use the terminology "REML"; however, if part of the initial state (η in their notation) is regarded as a parameter vector, the transformations they describe are exactly the same as those employed in REML.

Here, under model (1.1)–(1.2) we display a set of $N - p$ linearly independent error contrasts represented by the elements of $c_1, c_2, \ldots, c_m$, where for $k = 1, 2, \ldots, m$ the elements of c_k are linearly independent functions of $y_k^* = (y_1^T, y_2^T, \ldots, y_k^T)^T$ (free of θ and σ^2) with means of zero. Let $X_k^* = (X_1^T, X_2^T, \ldots, X_k^T)^T$. Let $\tilde{\alpha}_k = (X_k^{*T}X_k^*)^{-1}X_k^{*T}y_k^*$ represent the ordinary least squares estimate of α, which is functionally independent of θ and σ^2. Define A_1 to be any matrix such that $A_1^TA_1$ equals the $(n_1 - p)$-order identity matrix and $A_1A_1^T = I - X_1(X_1^TX_1)^{-1}X_1$. Then the elements of $c_1 = A_1^Ty_1$, $c_2 = y_2 - X_2\tilde{\alpha}_1, \ldots, c_m = y_m - X_m\tilde{\alpha}_{m-1}$ form a set of $N - p$ linearly independent error contrasts. For $k = 1, 2, \ldots, m$, let $c_k^* = (c_1^T, c_2^T, \ldots, c_k^T)^T$. The distribution of the $(n_1 - p)$- dimensional c_1 is multivariate normal with mean vector 0 and variance-covariance matrix $\sigma^2 A_1 H_1 A_1^T$. For $k = 2, 3, \ldots, m$, the conditional distribution of the n_k-dimensional c_k given c_{k-1}^* is multivariate normal with mean vector $Z_k \hat{\beta}_{k|k-1} - X_k \tilde{\alpha}_{k-1} + X_k \hat{\alpha}_{k-1}$ and variance-covariance matrix $\sigma^2 H_k$.

The natural logarithm of the restricted likelihood function differs by no more than an additive constant from

$$L_1(\theta, \sigma^2; c_m^*) = -\frac{N-p}{2} \ln(\sigma^2) - \frac{1}{2} \ln[\det(X_1^T H_1^{-1} X_1)]$$
$$- \frac{1}{2} \sum_{k=1}^{m} \ln[\det(H_k)]$$
$$- \frac{1}{2\sigma^2} (y_1 - X_1 \hat{\alpha}_1)^T H_1^{-1} (y_1 - X_1 \hat{\alpha}_1)$$
$$- \frac{1}{2\sigma^2} \sum_{k=2}^{m} v_k^T H_k^{-1} v_k. \qquad (3.11)$$

Equation (3.11) is in terms of quantities that can be computed by the modified version of the Kalman filter discussed in Section 2.2. In particular, for stage 1, $H_1 = R_1 + Z_1 Q_1 Z_1^T$; and for stages $k = 2, 3, \ldots, m$, $v_k = y_k - X_k \hat{\alpha}_{k-1} - Z_k \hat{\beta}_{k|k-1}$, and $H_k = R_k + U_k C_{k|k-1} U_k^T$ where $U_k = (X_k, Z_k)$.

Equation (3.11) is derived by writing the joint probability function for c_m^* as a product of the conditional densities and then, using at stage 1, results given by Searle (1979, eqs. 2.35 and 2.75) to obtain

$$A_1(A_1^T H_1 A_1)^{-1} A^T = H_1^{-1} - H_1^{-1} X_1 (X_1^T H_1^{-1} X_1)^{-1} X_1^T H_1^{-1}$$

and

$$\det(A_1^T H_1 A_1) = \det(H_1) \det(X_1 H_1^{-1} X_1) [\det(X_1^T X_1)^{-1}].$$

Modifications to (3.2) to (3.9) for REML estimation in mixed linear models are as follows: (1) replace N by $N - p$; (2) replace H_1^{-1} by $H_1^{-1} - H_1^{-1} X_1 (X_1^T H_1^{-1} X_1)^{-1} X_1^T H_1^{-1}$; and (3) for $k = 2, 3, \ldots, m$ re-define v_k and H_k in terms of the modified version of the Kalman filter discussed in Section 2.2 (i.e., $v_k = y_k - X_k \hat{\alpha}_{k-1} - Z_k \hat{\beta}_{k|k-1}$ and $H_k = R_k + U_k C_{k|k-1} U_k^T$).

3.3 Optimization Methods for REML Estimation

The REML estimates of θ and σ^2 are obtained by minimizing $-L_1(\theta, \sigma^2; c_m^*)$. In this section we briefly outline two general numerical methods that can be used to solve this minimization problem—quasi-Newton methods and successive quadratic programming (SQP) methods. These methods are iterative. Any particular method may work for some problems, but not for others. Consequently, a variety of methods are needed.

The quasi-Newton approach is discussed by Dennis and Schnabel (1983, pp. 111–129). Let $\theta^* = (\theta^T, \sigma^2)$, and let $f(\theta^*)$ represent a function of θ^* which is to be minimized. Take $\nabla f(\theta^*)$ to be the vector of first-order partial derivatives (the gradient vector), and $\nabla^2 f(\theta^*)$ the matrix of second-order partial derivatives (the Hessian matrix). Starting with the current value of θ^*, say θ_c^*, each iteration of the quasi-Newton algorithm consists of three steps. First, the gradient of the function $f(\theta^*)$ at θ_c^*, say $\nabla f(\theta_c^*)$, and an approximation to the Hessian at θ_c^*, say $B_c \approx \nabla^2 f(\theta_c^*)$, are computed.

Second, the equation

$$(B_c + \mu_c I)s_c = -\nabla f(\theta_c^*), \qquad \mu_c \geqslant 0$$

is solved for s_c. Here, $\mu_c = 0$, if B_c is positive definite; otherwise, a μ_c is taken to be a positive number such that $B_c + \mu_c I$ is positive definite. Finally, a line search is used to find a new value of θ^* of the form

$$\theta_+^* = \theta_c^* + \lambda_c s_c, \qquad 0 < \lambda_c \leqslant 1.$$

The line search is begun by attempting a full quasi-Newton step, corresponding to $\lambda_c = 1$. If this step is not acceptable, λ_c is decreased until an acceptable value is found.

SQP methods are frequently employed for minimization problems that include nonlinear inequality constraints, say $g_j(\theta^*) \geqslant 0$ $(j = 1, 2, \ldots, r)$. At each iteration, a quadratic programming problem with linear inequality constraints is formed by a quadratic approximation to the Lagrangian function $l(\theta^*, \phi) = f(\theta^*) - \phi^T g(\theta^*)$ and by linearizing the constraints. Starting with the current value of θ^*, say θ_c^*, each iteration of the SQP algorithm consists of three steps. First, the gradient of the function $f(\theta^*)$ at θ_c^*, say $\nabla f(\theta_c^*)$, the gradient of the constraints $g(\theta^*)$ at θ_c^*, say $\nabla g(\theta_c^*)$, and an approximation to the Hessian of the Lagrangian function $l(\theta^*, \phi) = f(\theta^*) - \phi^T g(\theta^*)$ at (θ_c^*, ϕ_c), say B_c, are computed. (The Hessian contains only the second-order partial derivatives with respect to θ^*.)

Second a solution, say s_c, is obtained to the quadratic programming problem

$$\begin{aligned} &\min && \nabla f(\theta_c^*)s + \tfrac{1}{2}s^T B_c s \\ &\text{subject to} && g_j(\theta_c^*) + \nabla g_j(\theta_c^*)^T s \geqslant 0, \qquad j = 1, 2, \ldots, r. \end{aligned}$$

Finally, a line search is then employed to find a new value $\theta_+^* = \theta_c^* + \lambda_c s_c$, where $0 < \lambda_c \leqslant 1$.

The *IMSL MATH/LIBRARY User's Manual* (IMSL, 1987a, Chap. 8) describes general-purpose optimization routines for these methods. Optimization software is also available in the following subroutine libraries: NAG (Oxford, United Kingdom and Downers Grove, Illinois), PORT (AT&T Bell Labs), MINPACK (Argonne Lab), Harwell (Harwell Laboratory, United Kingdom), and NPSOL (Systems Optimization Laboratory, Stanford).

Routines that require only function values [of $-L_1(\theta, \sigma^2; c_m^*)$] are the easiest to use. These routines use finite differences to approximate first-order partial derivatives. Routines that require exact first-order partial derivatives can also be used, and generally give more satisfactory results. However, their use may require extensive programming effort. Routines that require the Hessian matrix as well as the first-order partial derivatives of $-L_1(\theta, \sigma^2; c_m^*)$ can also be used. With these routines convergence can be quadratic, rather than linear. Approximations to the Hessian are sometimes used. The information matrix is sometimes used in place of the Hessian (Rao, 1973, pp. 366–374). Alternatively, approximation (3.10) might be used.

The requirement that the Q_k's and R_k's are variance-covariance matrices, i.e., that they are nonnegative definite, and restrictions placed on the T_k's, frequently lead to nonlinear inequality constraints on θ. SQP methods assume the objective function exists outside the region specified by the constraints and evaluate the function during intermediate iterations outside that region. For purposes of applying SQP methods, $-L_1(\theta, \sigma^2; c_m^*)$ can be set equal to a large positive number for values of θ and σ^2 for which $-L_1(\theta, \sigma^2; c_m^*)$ is undefined. However, when this is done, it is often found that the algorithm produces successive iterates along the boundary that are very close together, leading to slow convergence or false convergence. In these cases, it may be best to restart the algorithm at another interior point.

There is an analytical result that can be used to simplify the minimization of the function $-L_1(\theta, \sigma^2; c_m^*)$. Let

$$\hat{\sigma}^2(\theta) = \frac{(y_1 - X_1\hat{\alpha}_1)^T \mathrm{H}_1^{-1}(y_1 - X_1\hat{\alpha}_1) + \sum_{k=2}^{m} v_k^T H_k^{-1} v_k}{N - p}. \tag{3.12}$$

Minimization of

$$-L_2(\theta; c_m^*) = \frac{N-p}{2}\ln[\hat{\sigma}^2(\theta)] + \frac{1}{2}\ln[\det(X_1^T H_1^{-1} X_1)] + \frac{1}{2}\sum_{k=1}^{m}\ln[\det(H_k)] \tag{3.13}$$

gives the REML estimate of θ, and the REML estimate of σ^2 can be computed as $\hat{\sigma}^2 = \hat{\sigma}^2(\hat{\theta})$. By minimizing $-L_2(\theta; c_m^*)$ instead of $-L_1(\theta, \sigma^2; c_m^*)$ the dimension of the minimization problem is reduced by one. The two minimization problems are equivalent since $\hat{\sigma}^2(\theta)$ minimizes $-L_1(\theta, \sigma^2; c_m^*)$

for all θ; consequently, $\hat{\sigma}^2(\theta)$ can be substituted for σ^2 in $-L_1(\theta, \sigma^2; c_m^*)$ giving $-L_2(\theta; c_m^*)$. Formulas for the first- and second-order partial derivatives of $L_2(\theta; c_m^*)$ can be obtained from (3.2) to (3.5) by the chain rule.

3.4 Football Example

In this section we return to the football model (1.3)–(1.4) and discuss the methods employed for REML estimation and for on-line estimation of the parameters θ and σ^2. The estimates of T, Q_2, and R given in Section 1.2 were obtained by minimizing $-L_1(\theta, \sigma^2; c_m^*)$ given by (3.11). The constrained minimization problem was solved using the SQP method. Subroutine NLPQL, a FORTRAN code developed by Schittkowski (1986), was used to carry out the minimization. Forward finite differences were used to compute the gradient numerically. [An implementation of NLPQL using a finite-difference gradient is available in the IMSL MATH/LIBRARY; see subroutine NCONF (IMSL, 1987a, Chap. 8).] Schittkowski's code computes an approximation to the Hessian of the Lagrangian using a modified Broyden-Fletcher-Goldfarb-Shanno (BFGS) formula (Powell, 1978). This approximation makes use of the gradient vector from the previous and current iteration.

Seven nonlinear constraints were imposed as follows:

1. To ensure that R is nonnegative definite, the smallest eigenvalue of R is required to be nonnegative.
2. To ensure that Q_2 is nonnegative definite, the smallest eigenvalue of Q_2 is required to be nonnegative.
3. To ensure stationarity of the yearly performance levels of the teams and the existence of Q_1, the modulus of the largest eigenvalue of T is required to be less than one.
4. To ensure that the autocorrelations between the yearly performance levels of the teams are nonnegative, the four elements of TQ_1 are required to be nonnegative.

REML estimation for model (1.3)–(1.4) required extensive computations. The computation time was dominated by the time required to evaluate the objective function at various parameter values. With each iteration, this function had to be evaluated once at the current iterate, and (to obtain the finite difference approximation to the gradient vector) at 10 other points. Occasionally, a line search had to be carried out, in which case additional function evaluations were required. The CPU time for a single evaluation of the function $-L_1(\theta, \sigma^2; c_m^*)$ and the constraints when the computations were carried out in double precision for 840 NFL games during the 1979–1982 seasons was 63 seconds on a Data General MV/10000. (In double precision, the MV/10000 has a rate of execution of 0.4 million floating-point multiplications per second.) Consequently, only five iterations could be obtained per hour of CPU time. The constraints caused the optimization code to take small steps, and the algorithm had to be stopped before the convergence criteria were met.

The univariate model (1.3) was applied in connection with an examination of the amount of data to use when revising parameter estimates. To compare the performance of these approaches over many years, they were applied to a data set consisting of the results of NFL games from 1969 to 1983. The offensive yardage statistics were not available for this data set. This model was parameterized in terms of three scalars: (1) the first-order autocorrelation for a team's yearly performance levels, T; the variance of the yearly performance levels, Q_1; and the observational-error variance, σ^2. A modified version of the method of scoring, based on approximation (3.10) was used to compute the REML estimates of the parameters. Equations (3.2) and (3.3) were used to compute the gradient vector. The parameters were reestimated at the end of each year. Three different approaches to reestimation were tried:

1. The estimates were revised at the end of each year by using a one-pass, "on-line" algorithm.
2. The most recent four years of data were used to recompute the REML estimates of the parameters.
3. All available data was used to recompute the REML estimates of the parameters.

The on-line algorithm consists of an adaptation of the method of scoring in which:

1. At stage k the mixed-model versions of (2.1) to (2.4) and (3.2) are used to obtain first-order partial derivatives with respect to T and Q_1. In doing so, T and Q_1 are taken to be the estimates from the conclusion of the previous year.
2. An approximate information matrix is accumulated over all previous data, so that in some sense, it is a mixture with groups of terms evaluated at different year's estimates of Q_1 and T. The approximation (3.10) is used.
3. At the conclusion of each year σ^2 is set as defined by (3.12) where the generalized sum of squares for estimating σ^2 is taken to be a running sum over all the previous data. A full quasi-Newton step is taken at the end of each year with B_c taken to be the approximate information matrix that is evaluated at the value of σ^2 just set. The revised estimates of T and Q_1 are used for the next year. The revised estimate of σ^2 is used to reset the generalized sum of squares used for estimating σ^2 in (3.12).

Table 3 Parameter Estimates at End of Each Season

	$\hat{Q}_1$			$\hat{T}$			$\hat{\sigma}^2$		
Season	On-line	4 yr	All	On-line	4 yr	All	On-line	4 yr	All
1969			0.329			0.771			185
1970	0.283		0.286	0.790		0.787	186		186
1971	0.228	0.259	0.259	0.828	0.829	0.829	185	183	183
1972	0.250	0.222	0.271	0.809	0.812	0.811	182	179	181
1973	0.275	0.229	0.282	0.811	0.812	0.815	184	182	183
1974	0.239	0.217	0.258	0.779	0.691	0.776	178	169	176
1975	0.249	0.260	0.268	0.796	0.767	0.797	177	170	176
1976	0.251	0.271	0.272	0.790	0.786	0.789	175	166	174
1977	0.222	0.217	0.252	0.778	0.729	0.775	173	154	171
1978	0.189	0.195	0.231	0.755	0.665	0.753	171	157	169
1979	0.165	0.209	0.221	0.753	0.765	0.764	171	156	169
1980	0.159	0.124	0.217	0.753	0.858	0.766	170	158	168
1981	0.156	0.104	0.208	0.737	0.645	0.746	171	164	169
1982	0.152	0.119	0.206	0.735	0.508	0.744	169	164	168
1983	0.145	0.125	0.199	0.733	0.586	0.746	170	166	168

4. First-order partial derivatives for Q_1 and T are accumulated only for the current year. The first-order partial derivatives are reset to zero at the end of each year.

The parameter estimates computed by the three approaches—on-line, REML based on four years of data, and REML based on all the data—are given in Table 3.

From Table 3, note that all three parameters—Q_1 (variation in the performance levels of the teams), T (the correlation between performance levels of a team in two consecutive years), and σ^2 (the variance of the observational error)—have all decreased since 1969. Note that the magnitude of the changes in the on-line estimates decreases with time. This is primarily because the amount of past information increases with time. At some point in time, one should consider fixing the approximate Hessian used to generate the next step so that the steps do not become too small.

4. CONFIDENCE SETS, TESTS, AND DIAGNOSTICS

4.1 Methods

Approximate confidence sets for θ and σ^2 can be based on asymptotic normality of θ and σ^2. The variance-covariance matrix can be approximated as the inverse of the information matrix, or as the inverse of a finite-difference approximation to the Hessian of $-L_1(\theta, \sigma^2; c_m^*)$ (see, e.g., Kendall and Stuart, 1973, Chap. 18). Alternatively, approximate tests of hypotheses can be performed using generalized likelihood ratio methods (see e.g., Kendall and Stuart, 1973, Chap. 24).

For the case where θ is known, Harville (1976b) gave confidence sets for linearly independent combinations of the fixed effects and realized values of the random effects in mixed linear models. Take $\hat{\sigma}^2$ to be the REML estimate of σ^2 based on $v = N - p$ linearly independent error contrasts. The confidence sets are modifications of the usual formulas for confidence sets on linearly independent combinations of regression coefficients in a linear regression model. For model (1.1)–(1.2) a $(1 - \delta)$ 100% confidence set at stage $k - 1$ for the observations at stage k $(k \geqslant 2)$ consists of those values of y_k that satisfy the inequality

$$(y_k - X_k\hat{\alpha}_{k-1} - Z_k\hat{\beta}_{k|k-1})^T H_k^{-1}(y_k - X_k\hat{\alpha}_{k-1} - Z_k\hat{\beta}_{k|k-1}) < n_k\hat{\sigma}^2 F_{\delta;n_k,v},$$

where $F_{\delta;n_k,v}$ is the upper δ fractile of the F distribution with n_k and v degrees of freedom.

For θ unknown, the traditional approach to estimation of the fixed and random effects is to simply replace θ by its estimate and to proceed as though θ is known. This approach tends to produce overly optimistic confidence sets. The approximation of Kackar and Harville (1984), which accounts to some degree for the unknown θ, can be useful in obtaining more satisfactory confidence sets. Let $\tilde{\tau}(y;\theta)$ be the mixed model estimator of τ supposing θ were known. The mean square error of the estimator obtained by replacing θ by its estimator is approximated by the adding a correction term, $\mathrm{tr}(AB^{-1})$ where $A = \mathrm{var}[\partial\tilde{\tau}(y;\theta)/\partial\theta]$ and B^{-1} is the mean square error matrix $E[(\hat{\theta}-\theta)(\hat{\theta}-\theta)]^T$. [The information matrix or Hessian of $-L_1(\theta, \sigma^2; c_m^*)$, that is, the inverse of the asymptotic variance-covariance matrix of the REML estimator $\hat{\theta}$, can be used in place of B here.]

The one-step-ahead prediction errors (for θ known) are independently distributed normal with mean vector 0. A plot of these errors over time can be useful in detecting outliers or invalid model assumptions.

4.2 Football Example

In addition to model (1.3)–(1.4), several other models were applied to data for 1979–1982 NFL seasons. Univariate models using (1.3) and (1.4) separately, rather than jointly, were fitted. The univariate models were used to compute initial values for the parameters in the bivariate model. Also, variations on model (1.3)–(1.4) were examined in an attempt to find better models. Some of the models included additional fixed effects. Five additional fixed effects were examined—grass field advantage, artificial turf advantage, home crowd advantage, time zone advantage, and preparation time advantage. With the exception of preparation time advantage, the other fixed effects represent some part of the total home field advantage reflected by α_1 and α_2 in model (1.3)–(1.4). Also, models containing additional random effects were fitted. These random effects reflected weekly performance levels of the teams. Equations (1.3) and (1.4) contained a difference in weekly performance levels of the teams in addition to the difference in their yearly performance level. The state equation used for the weekly performance levels of the teams was similar to that for the yearly performance levels.

The inclusion of these additional fixed or random effects did not improve the model. An F test for the additional fixed effects under a modified univariate model (1.3) to incorporate the five additional fixed effects gave an $F_{5,834} = 0.77$ (p-value $= 0.57$). A similar result was obtained with univariate model (1.4). A generalized likelihood ratio test was used to test the importance of additional random effects reflecting the weekly performance levels. Univariate model (1.3) was modified to incorporate the additional random effects. The likelihood ratio test of the null hypothesis that the first-order autocorrelation for the weekly performance levels is zero and the error variance of the process is zero gave a $\chi_2^2 = 0.2$ (p-value $= 0.90$).

Taking the parameter values to be those estimated under model (1.3)–(1.4) for the 1979–1982 NFL seasons, the model was applied to 1983 data. Some 220 of the 233 (94.4%) standardized week-ahead prediction errors for points were between -1.96 and 1.96. That is, 94.4% of the 95% prediction intervals captured the actual outcome of the game. A chi-squared test for normality based on eight cells of equal probability gave a $\chi_7^2 = 7.24$ (p-value $= 0.404$). Figure 1 is a plot of the points component of the standardized week-ahead prediction errors. (In order to improve resolution of the plot, the playoff games have been excluded.)

The week-ahead errors were also examined by teams. Three diagnostic tests were performed for each team. Regular season and playoff games were used. The most significant results of each of these three tests is as follows: (1) a negative first-order autocorrelation from week to week for San Diego (p-value $= 0.002$), Green Bay (p-value $= 0.010$), Washington (p-value $= 0.014$), Buffalo (p-value $= 0.033$), and Houston (p-value $= 0.057$); (2) a New York Jet's home field advantage less than that for other teams (p-value $= 0.073$); and (3) a Washington team effect better than predicted (p-value $= 0.008$) and a San Diego team effect worse (p-value$=0.046$) than predicted. The p-values reported here are two-sided and for one test at a time. Figure 2 shows the

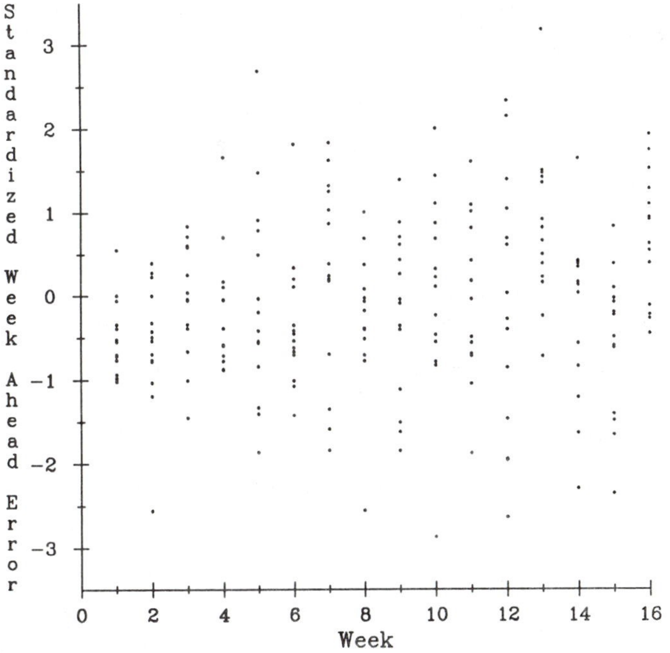

Figure 1. Standardized week-ahead prediction errors.

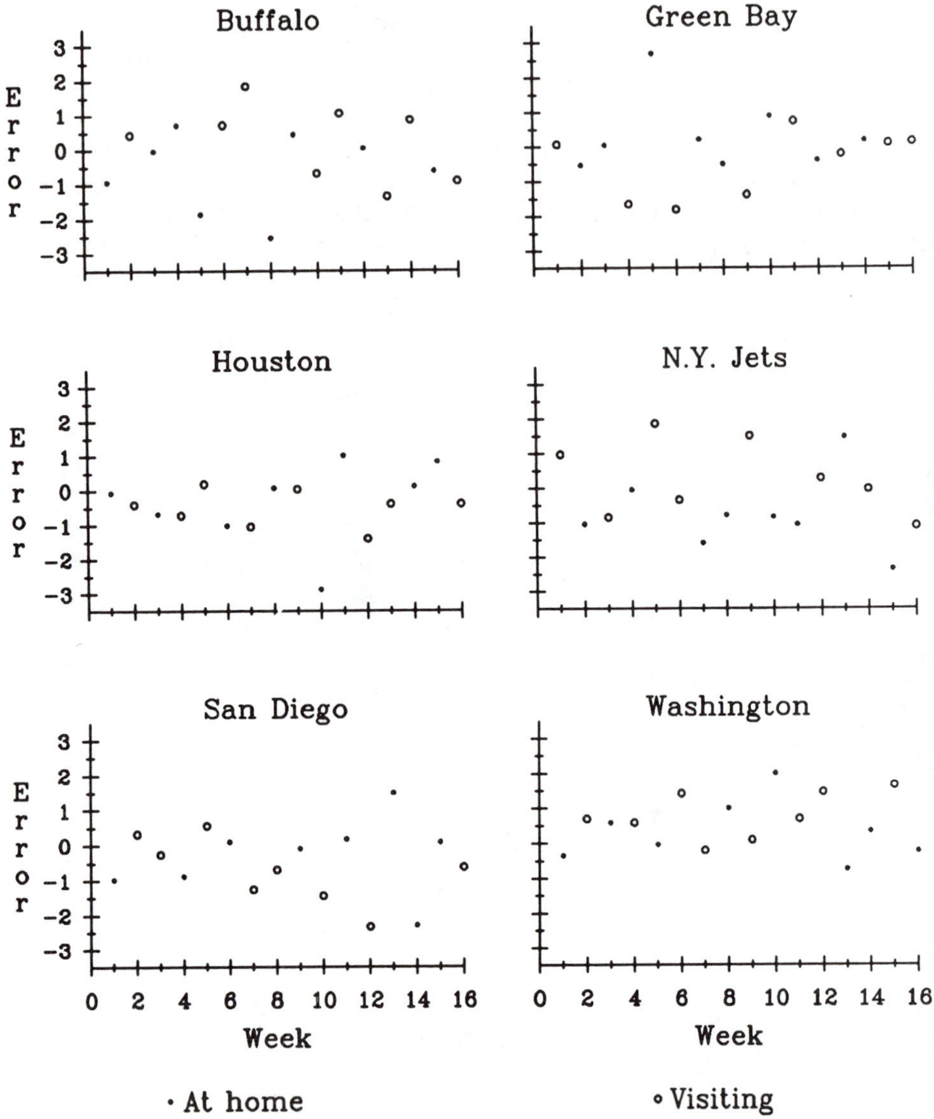

Figure 2. Standardized week-ahead prediction errors by team.

Table 4 Accuracy of Statistical Predictions and the Betting Line for the 1983 NFL Season

	On-line univariate	REML univariate	REML bivariate	Line
Winner (%)	57.3	56.9	59.5	58.6
Mean absolute error	11.13	11.12	11.02	11.10
Mean square error	195.7	197.1	196.5	198.8
Bets won (%)	52.7	51.4	56.7	

week-ahead errors for each of the six teams with the most significant results for the diagnostic tests. In these plots the sign of the error has been reversed when the team is visiting so that positive errors always reflect an observed performance for that team that was better than was predicted and negative errors always reflect an observed performance that was poorer than was predicted. (Washington's playoff games have been excluded in order to improve the resolution of the plot.) In view of the fact that the significance of the diagnostic tests as a whole is diminished by applying so many tests, there is very little evidence in these plots to suggest inadequacy of model (1.3)–(1.4).

A comparison in 1983 of the accuracy of various predictions was performed. The predictions were obtained as follows: (1) from univariate model (1.3) using the on-line implementation and data from 1968–1982 for the parameter estimates, (2) from univariate model (1.3) using REML estimates based on NFL seasons 1979–1982, (3) from bivariate model (1.3)–(1.4) using REML estimates based on NFL seasons 1979–1982, and (4) from the betting line. The primary source for the betting line was the *Houston Chronicle*, which, in its Wednesday editions, reported the opening line listed in *Harrah's Reno-Tahoe Sports Book*. Four methods of comparison were used for comparing accuracy: (1) percentage of winners correctly predicted, (2) mean absolute error, (3) mean square error, and (4) percentage of bets won versus the betting line. The results are summarized in Table 4.

The bivariate model (1.3)–(1.4) was generally best and could have been used profitably in 1983 to beat the line. (One needs to beat the line at least 52.4% of the time to gain a profit in Las Vegas. Here the one-sided p-value for the test that we can beat the line more than 52.4% of the time is 0.095.*)

*Preliminary findings in 1984 were less encouraging.

5. EXTENSIONS

To simplify the notation, model (1.1)–(1.2) assumed the fixed-effects vector α and the random-effects vector b_k to be of fixed length p and q, respectively. In some cases, additional fixed effects and random effects may enter the model at some stage. No modifications to formulas are needed to handle the additional random effects. The random-effects vector in model (1.1)–(1.2) at stage k can be taken of length q_k and then the transition matrix T_k has dimensions q_{k-1} by q_k. For the football model additional random effects in 1976 had to be added for the expansion teams Seattle and Tampa Bay.

When additional fixed effects enter the model at stage k, modifications to the Kalman filter and REML function at stage k must be made that are similar to the modifications indicated in Sections 2.2 and 3.2 for stage 1. The modifications necessary are described in detail in Sallas and Harville (1981) and Sallas (1981).

6. CONCLUDING REMARKS

When some elements of the state vector have noninformative priors and others do not, the state-space model can be viewed as mixed linear model (1.1)–(1.2) in which the elements of the state vector with noninformative priors correspond to fixed effects in the mixed model and the remaining elements in the state vector correspond to the random effects. At stage 1, mixed-model formulas can be used for estimating the effects. After stage 1, the Kalman filter can be used to update the mixed model estimates for the fixed and random effects. At that time, the algorithm for computing mixed-model estimates makes no distinction between the fixed and random effects.

When a parameter vector θ associated with the matrices T_k, Q_k, and R_k needs to be estimated in addition to the parameter vector α for the fixed effects, REML estimation under the assumption of normal errors should be considered. The REML approach corresponds to maximum likelihood when a Bayesian interpretation of the fixed effects is given. The restricted likelihood function can be evaluated using (3.11) together with the mixed-model formulas (2.5) to (2.7), the Kalman filter (2.1)–(2.4), and the modifications given at the end of Section 2.2. Various optimization methods can be used to carry out the maximization of (3.11) or equivalently the minimization of (3.13).

If periodic revisions of T_k, Q_k, and R_k are needed, a choice of the frequency of the revisions must be made. If estimates are revised too frequently undesirable oscillations in the estimates of θ may occur. On the other hand, if estimates are updated infrequently, parameter changes may be overlooked.

The choice of one year in the univariate football model (1.3) worked well.

Choices concerning the amount of past data to incorporate and how to incorporate it must also be made when periodic revisions of T_k, Q_k, and R_k are needed. An on-line method that makes use of REML formulas but makes a single pass through the data revising estimates of T_k, Q_k, and R_k periodically has the advantages over REML using all the data in that it can allow different estimates of T_k, Q_k, and R_k in different time periods for estimating the effects and more quickly adapt to parameter changes without requiring extensive amounts of computer time. If revisions of these matrices are not needed too frequently, such as in the football example, a method that uses REML with some recent part of the data can work, too.

ACKNOWLEDGMENTS

The authors are grateful to K. Schittkowski for providing the NLPQL code and to James C. Spall and the two referees for indicating references to the engineering literature and for many other helpful comments. Harville's work was supported in part by Office of Naval Research Contract N0014-85-K-0418.

REFERENCES

Aitken, A. C. (1934). On least squares and linear combination of observations, *Proc. R. Soc. Edinburgh* 55, 42–47.

Albert, A., and R. W. Sittler (1965). A method for computing least squares estimators that keep up with the data, *J. Soc. Ind. Appl. Math. Ser. A* 3, 384–417.

Anderson, B. D. O., and J. B. Moore (1979). *Optimal Filtering*, Prentice-Hall, Englewood Cliffs, N.J.

Ansley, C. F., and R. Kohn (1985). Estimation, filtering, and smoothing in state space models with incompletely specified initial conditions, *Ann. Stat.* 13, 1286–1316.

Belsley, D. A., and E. Kuh (1973). Time-varying parameter structures: an overview, *Ann. Econ. Soc. Meas.* 2, 375–379.

Cooley, T. F., and E. C. Prescott (1976). Estimation in the presence of stochastic parameter variation, *Econometrica* 44, 167–184.

Dempster, A. P., D. B. Rubin, and R. K. Tsutakawa (1981). Estimation in covariance components models, *J. Am. Stat. Assoc.* 76, 341–353.

Dennis, J. E., and R. B. Schnabel (1983). *Numerical Methods for Unconstrained Optimization and Nonlinear Equations*, Prentice-Hall, Englewood Cliffs, N.J.

Duncan, D. B., and S. D. Horn (1972). Linear dynamic recursive estimation from the viewpoint of regression analysis, *J. Am. Stat. Assoc.* 67, 815–821.

Goodwin, G. C., and R. L. Payne (1977). *Dynamic System Identification: Experimental Design and Data Analysis*, Academic Press, New York.

Graybill, F. A. (1969). *Introduction to Matrices with Applications in Statistics*, Wadsworth, Belmont, Calif.

Gupta, N. K., and R. K. Mehra (1974). Computational aspects of maximum likelihood estimation and reduction in sensitivity function calculations, *IEEE Trans. Autom. Control* AC-19, 774–783.

Harvey, A. C. (1981). *Time Series Models*, Halsted Press, New York.

Harvey, A. C., and C. R. McKenzie (1984). Missing Observations in Dynamic Econometric Models: A Partial Synthesis, in *Time Series Analysis of Irregularly Observed Data*, Vol. 25 of *Lecture Notes in Statistics* (Emanuel Parzen, ed.), Springer-Verlag, New York, pp. 108–133.

Harville, D. A. (1976a). Extension of the Gauss-Markov theorem to include the estimation of random effects, *Ann. Stat.* 4, 384–395.

Harville, D. A. (1976b). Confidence intervals and sets for linear combinations of fixed and random effects, *Biometrics* 32, 403–407.

Harville, D. A. (1977). Maximum likelihood approaches to variance component estimation and to related problems, *J. Am. Stat. Assoc.* 72, 320–338.

Harville, D. A. (1980). Predictions for National Football League games via linear-model methodology, *J. Am. Stat. Assoc.* 75, 516–524.

Henderson, C. R. (1963). Selection index and expected genetic advance, in *Statistical Genetics and Plant Breeding*, National Research Council Publication 982, National Academy of Sciences, Washington, D.C., pp. 141–163.

Henderson, C. R. (1973). Sire Evaluation and Genetic Trends, in *Proceedings of the Animal Breeding and Genetics Symposium in Honor of Dr. Jay L. Lush*, American Society of Animal Science, Champaign, Ill., pp. 10–41.

Henderson, C. R. (1975). Best linear unbiased estimation and prediction under a selection model, *Biometrics* 31, 423–447.

IMSL (1987a). *MATH/LIBRARY User's Manual*, version 1.0, IMSL, Houston, Tex.

IMSL (1987b). *STAT/LIBRARY User's Manual*, version 1.0, IMSL, Houston, Tex.

Jones, R. H. (1984). Fitting Multivariate Models to Unequally Spaced Data, in *Times Series Analysis of Irregularly Observed Data*, Vol. 25 of *Lecture Notes in Statistics* (Emanuel Parzen, ed.), Springer-Verlag, New York, pp. 158–188.

Kackar, R. N., and D. A. Harville (1984). Approximations for standard errors of estimators of fixed and random effects in mixed linear models, *J. Am. Stat. Assoc.* 79, 853–862.

Kalman, R. E. (1960). A new approach to linear filtering and prediction problems, *J. Basic Eng.* (*Transactions of the ASME, Series D*) 82, 35–45.

Kalman, R. E. (1961). *New Methods and Results in Linear Prediction and Filtering Theory*, Technical Report 61-1, RIAS, Baltimore, Md.

Kaminski, P. G., A. E. Bryson, and S. F. Schmidt (1971). Discrete square root filtering: a survey of current techniques, *IEEE Trans. Autom. Control* AC-16, 727–736.

Kashyap, R. L. (1970). Maximum likelihood identification of stochastic linear systems, *IEEE Trans. Autom. Control* AC-15, 25–34.

Kendall, M. G., and A. Stuart (1973). *The Advanced Theory of Statistics*, Vol. 2; *Inference and Relationship*, 3rd ed., Charles Griffin, London.

Maybeck, P. S. (1972). Combined state and parameter estimation for on-line applications, Ph.D. dissertation, Massachusetts Institute of Technology.

Maybeck, P. S. (1979). *Stochastic Models, Estimation, and Control*, Vol. 1, Academic Press, New York.

Maybeck, P. S. (1982). *Stochastic Models, Estimation, and Control*, Vol. 2, Academic Press, New York.

Mehra, R. K. (1970). On the identification of variances and adaptive Kalman filtering, *IEEE Trans. Autom. Control* AC-15, 175–184.

Nering, E. D. (1970). *Linear Algebra and Matrix Theory*, 2nd ed., Wiley, New York.

Patterson, H. D., and R. Thompson (1971). Recovery of inter-block information when block sizes are unequal, *Biometrika* 58, 545–554.

Plackett, R. L. (1950). Some theorems in least squares, *Biometrika* 37, 149–157.

Powell, M. J. D. (1978). A Fast algorithm for Nonlinearly Constrained Optimization Calculations, in *Numerical Analysis Proceedings, Dundee 1977*, Vol. 630 of *Lecture Notes in Mathematics* (G. A. Watson, ed.), Springer-Verlag, Berlin, pp. 144–157.

Rao, C. R. (1973). *Linear Statistical Inference and Its Applications*, 2nd ed., Wiley, New York.

Sage, A. P., and J. L. Melsa (1971). *System Identification*, Academic Press, New York.

Sallas, W. M. (1981). On-Line Estimation in Mixed Linear Models, in *Proceedings of the 6th Annual SAS Users Group International Conference*, SAS, Cary, N.C., pp. 57–65.

Sallas, W. M., and D. A. Harville (1981). Best linear recursive estimation for mixed linear models, *J. Am. Stat. Assoc.* 76, 860–869.

Schittkowski, K. (1986). NLPQL: a FORTRAN subroutine solving constrained nonlinear programming problems, *Ann. Oper. Res.* 5, 485–500.

Schweppe, F. C. (1965). Evaluation of likelihood functions for Gaussian signals, *IEEE Trans. Inf. Theory* IT-11, 61–70.

Searle, S. R. (1979). *Notes on Variance Component Estimation: A Detailed Account of Maximum Likelihood and Kindred Methodology*, Biometrics Unit, Cornell University, Ithaca, N.Y.

Young, P. (1974). Recursive approaches to time series analysis, *Bull. Inst. Math. Its Appl.* 10, 209–224.

18

Shannon Information-Theoretic Priors for State-Space Model Parameters

STACY D. HILL and JAMES C. SPALL The Johns Hopkins University, Applied Physics Laboratory, Laurel, Maryland

1. INTRODUCTION

Much effort has been devoted to developing Bayesian approaches to constructing and testing models of dynamic systems. The Bayesian framework provides an appealing formal approach to dynamic system analysis and frequently reduces the need to develop specialized ad hoc statistical and numerical techniques. A main difficulty in implementing a Bayesian technique is that of choosing the prior distribution, particularly when there is little prior information available from which to derive this distribution. This chapter presents a technique for determining noninformative priors—priors that contain little information relative to that provided in the data—for dynamic models such as AR, ARMA, and state space. We determine the prior that, subject to certain constraints, maximizes the difference between the information in the posterior and that in the prior. In this way, we let the information contained in the data dominate the posterior.

There is no one accepted criterion for choosing a noninformative prior, nor is there even one accepted definition of the word "noninformative." We will use a Shannon information-theoretic approach in deriving a noninformative prior. This approach has a sound axiomatic basis and is consistent with the Bayesian subjective interpretation of probability (Bernardo, 1979a, b; Good, 1966; Zellner, 1971, pp. 50–51).

Let us now consider some of the existing approaches to specifying noninformative priors. A common approach is to assign a uniform (either proper or improper) distribution to the parameters of interest. Although simple to implement, uniform priors are deficient in at least two respects: (1) in general, they have no sound information-theoretic justification, and (2) they are not transform invariant [i.e., if a uniform prior is assigned to θ, the resulting prior for $f(\theta)$ is not, in general, going to be uniform although there is no more information about $f(\theta)$ than there is about θ]. Another common noninformative prior is the Jeffreys' prior. The principal virtue of this prior is that it is transform invariant. As is the case with the uniform prior, however, there is generally no information-theoretic basis for the Jeffreys' prior. [An exception to this is discussed in Kashyap (1971).]

In the general statistics literature, both the uniform and the Jeffreys' prior have been widely discussed; in the context of dynamic models, uniform priors are used in Peterka (1981), Sage and Melsa (1971, p. 58), Spall and Koch (1985), and Spall (1988) while Jeffreys' priors are used in Poskitt and Tremayne (1983). Other noninformative priors have been proposed by Novick (1969), Akaike (1983), Bernardo (1979a), and Zellner (1977). Novick's procedure relies on developing a limiting form for the conjugate prior. As with the uniform and Jeffreys' priors, it lacks a formal information-theoretic basis; it may also be difficult to implement because of problems in identifying a conjugate prior. The Akaike, Bernardo, and Zellner approaches are based on information-theoretic considerations. The Akaike procedure is restricted to discrete priors and also deals with a prediction problem that is not appropriate in our framework. Bernardo's technique, which is based on some rather informal mathematics, produces noninformative prior distributions under assumptions different from those assumed here. The Zellner approach is computationally efficient but is difficult to interpret in relation to the key quantity in a Bayesian analysis—the posterior distribution. These existing information-theoretic approaches are discussed in slightly more detail in Section 2.

This chapter presents a computationally tractable procedure, based on an appealing information measure presented in Lindley (1956) for assessing noninformative priors. The exact procedure for determining noninformative priors, as described by Lindley, will generally be intractable since it involves an optimization problem over an infinite-dimensional function space. The tractability of the procedure presented here results from the fact that, instead of seeking the optimal prior distributions over the class of all probability distributions, we seek the optimal prior only over the class of convex combinations of a prespecified selection of distributions. In this way, we are able to convert the intractable infinite-dimensional functional optimization problem into a conventional (and feasible) finite-dimensional optimization problem in Euclidean space. Dyer and Chiou (1984) consider the information (in the sense of Lindley) associated with individual priors as opposed to convex combinations; thus their technique can be considered a special case of ours.

Since the (tractable) prior obtained here is an approximation to the (intractable) optimal noninformative prior, it is also of interest to examine the sensitivity of the posterior to the use of a prior other than the "true" prior. Edwards et al. (1963), Davis (1979), and Maryak and Spall (1987) are three references that present results that might apply in examining the effect of using the approximation.

The models considered here have the general discrete-time state-space form for $k = 1, 2, \ldots, n$:

$$\begin{aligned} x_k &= \Phi_{k-1} x_{k-1} + w_{k-1} \\ z_k &= H_k x_k + v_k, \end{aligned} \tag{0}$$

where $x_k \in R^p$ is the unobserved state, $z_k \in R^m$ is the noisy measurement of the state, and the random terms w_k, v_k, and x_0 are distributed $N(0, Q_k)$, $N(0, R_k)$, and $N(\mu, P_0)$, respectively. Except for allowing dependence between w_{k-1} and v_k for all k [i.e., $\operatorname{cov}(w_{k-1}, v_k) \equiv S_k$] we assume that $\{x_0, w_0, w_1, \ldots, w_{n-1}, v_1, v_2, \ldots, v_n\}$ are mutually independent. Model (0) is in state-space form; when $H_k = I$ and $v_k \equiv 0$ for all k, (0) reduces to the multivariate AR process. Because we are allowing dependence between w_{k-1} and v_k, model (0) is also a representation of an ARMA process (see Anderson and Moore, 1979, pp. 236–237). The model parameters μ, P_0, $\{\Phi_{k-1}, H_k, Q_{k-1}, R_k, S_k\}_{k=1}^n$ are assumed to be functions of an underlying parameter vector θ. It is about θ that we assume the analyst is interested in making inference.

The key quantity in a Bayesian analysis of the parameters θ in model (0) is the posterior density for θ given the data. We will let Y_n be a generic symbol to denote the data. In its most straightforward representation, Y_n simply represents the raw system measurements, that is, $Y_n = (z_1^T, z_2^T, \ldots, z_n^T)^T$. In some applications it is advantageous to work with a set of transformed system measurements, such as the sequence of Kalman filter estimates or residuals; Y_n will then denote this transformed sequence. With this notation, the posterior density can be expressed as

$$p(\theta \mid Y_n) = \frac{p(Y_n \mid \theta) p(\theta)}{\int_\Omega p(Y_n \mid \theta) p(\theta)\, d\theta},$$

where $p(\theta)$ is the prior on θ, and Ω is such that $\int_\Omega p(\theta) = 1$. This paper focuses on the determination of $p(\theta)$.

The remainder of this chapter is organized as follows: Section 2 presents the procedure for calculating noninformative priors, $p(\theta)$, using the Lindley information measure; Section 3 demonstrates how the recursive nature of the Kalman filter can be employed to enhance efficiency; Section 4 presents a numerical study; and Section 5 presents some concluding remarks.

2. NONINFORMATIVE PRIORS

As stated earlier, our goal is to find a prior density that, in some sense, contains much less information about θ than do the observation data, a prior density that is "noninformative." We take as our measure of information the Kullback-Leibler-Lindley (KLL) information measure, which expresses the information as the difference between the information in the posterior density and that in the prior density. The noninformative prior is the density that maximizes the information measure. This problem being an infinite-dimensional optimization problem, the noninformative prior is not obtainable. So we seek a suboptimal noninformative prior, a density that nearly maximizes the information measure and that is relatively easy to compute by some numerical procedure. In fact, we show here how to turn this problem into a *finite-dimensional* convex optimization problem—a problem for which readily available numerical techniques exist—and devise an efficient algorithm for obtaining a suboptimal noninformative prior. Note that the procedure described in this chapter is oriented towards finite samples ($n < \infty$), in contrast to the procedure of Hill and Spall (1987), which is oriented towards the asymptotic ($n \to \infty$) case.

2.1 Efficient Formulation for Calculating Noninformative Priors

To begin, we define the information measure. As mentioned in Section 1, we follow Lindley (1956) and define the measure $I_n(\cdot)$ to be

$$I_n(p(\theta)) = \int_{\Omega \times R^q} p(\theta | Y_n) \log \{p(\theta | Y_n)\} p(Y_n)\, d\theta\, dY_n - \int_{\Omega} p(\theta) \log \{p(\theta)\}\, d\theta, \tag{2.1}$$

where $Y_n \in R^q$. For ease of presentation we have abused notation: in the left side of (2.1) $p(\theta)$ denotes the prior density *function*, while in the right side of (2.1) it denotes the *value* of the prior density at θ. This functional, called the KLL information measure, is based on the notion of Shannon information (Shannon and Weaver, 1949); it expresses the information provided about θ as the difference between the expected information about θ in the posterior and that in the prior density $p(\theta)$. A noninformative prior density is defined to be any density that maximizes (2.1) over the set of all densities defined on Ω.

Of course, there are other ways of defining information that produce other noninformative priors. An expression in Zellner (1977) is the obvious analog to (2.1); it defines information to be the difference between the expected information in the data distribution $p(Y_n|\theta)$ and that in the prior density. However, the exact information interpretation of that measure is not clear since inference in a Bayesian analysis is based on the posterior $p(\theta | Y_n)$, not

$p(Y_n|\theta)$. Bernardo (1979a) assumes the observation data are i.i.d. But our observation data are not i.i.d because of the inherent sequential dependence associated with dynamic models. The approach taken by Akaike (1983) is similar to that taken here in that he defines information using the KLL information measure. His approach is a prediction analysis approach and requires, as in Bernardo's, independent observations, an assumption that does not hold for the system (0). Dyer and Chiou (1984) take as their optimal prior the density that maximizes the KLL measure over a certain restricted class of probability densities (e.g., the class of beta density functions).

Let us turn now to the problem of maximizing (2.1). The first step toward solving this problem is to recast it as a finite-dimensional problem. To accomplish this, we simply constrain the maximizing prior to be a convex combination of a finite set of prespecified priors, which we will call the base priors. Our solution, the convex combination that maximizes I_n, is then guaranteed to be more nearly noninformative than the standard noninformative priors (such as the uniform, Jeffreys', etc.) mentioned in Section 1 and the priors computed in Dyer and Chiou (1984) when those priors are among the base priors. This follows immediately since each base prior is itself a convex combination of the base priors. There is another reason for constraining the suboptimal prior to be a mixture distribution. Dalal and Hall (1983) show that bounded priors (and posteriors) can be well approximated by a mixture of conjugate priors.

Let the base priors $p_1(\theta), \ldots, p_l(\theta)$ be given and let $\alpha = (\alpha_1, \ldots, \alpha_l)$ be such that $0 \leqslant \alpha_j \leqslant 1$ and $\sum_{j=1}^{l} \alpha_j = 1$. Our goal is to maximize $I_n(\cdot)$ over all $p(\theta)$ of the form $p(\theta) = \sum_{j=1}^{l} \alpha_j p_j(\theta)$. More precisely, our goal is to find that convex combination $p^*(\theta)$ such that

$$I_n(p^*(\theta)) = \max_{\alpha} \left\{ I_n(p(\theta)) : p(\theta) = \sum_{j=1}^{l} \alpha_j p_j(\theta) \right\}.$$

It follows from Lindley (1956) that $I_n(\cdot)$ is a concave function of α, so our problem is a finite-dimensional convex optimization problem and therefore can be solved by one of the well known algorithms for finding the maxima of concave functions (see, e.g., Zangwill, 1967). It is even more desirable to have $I_n(\cdot)$ strictly concave, since then we are guaranteed the uniqueness of the maximum (and hence the noninformative prior) and assured of convergence of the numerical algorithm to this maximum. The next result gives a condition under which strict concavity holds.

Let A denote the set consisting of all convex combinations of the $p_j(\theta)$'s, and A^0 the set of all strict convex combinations of the $p_j(\theta)$'s (i.e., the α_j's in $\sum_{j=1}^{l} \alpha_j p_j(\theta)$ each satisfies $0 < \alpha_j < 1$). We have:

THEOREM 2.1 $I_n(\cdot)$ is strictly concave on A^0 (as a function of α) if and only if for each $p'(\theta), p''(\theta) \in A^0$ such that $p' \neq p''$, we have that $p'(Y_n) \neq p''(Y_n)$ on a set of Y_n's of positive Lebesgue measure.*

Proof. Strict concavity holds (see, e.g., Ortega and Rheinboldt, 1970, p. 84) if and only if

$$I_n(p''(\theta)) + \frac{d}{dt}\{I_n(tp'(\theta) + (1-t)p''(\theta))\}|_{t=0}$$

$$> I_n(p'(\theta)), \qquad \forall p', \quad p'' \in A^0, \quad p' \neq p''. \tag{2.2}$$

Fix $p(\theta)$ and note that $I_n(p(\theta))$ can be written equivalently as†

$$\int p(Y_n|\theta)p(\theta) \log \left\{\frac{p(Y_n|\theta)}{p(Y_n)}\right\} dY_n\, d\theta, \tag{2.3}$$

where

$$p(Y_n) = \int p(Y_n|\theta)p(\theta)\, d\theta.$$

A straightforward differentiation yields

$$\frac{d}{dt}\{I(tp'(\theta) + (1-t)p''(\theta))\}|_{t=0}$$

$$= \int p(Y_n|\theta)\{p'(\theta) - p''(\theta)\} \log \{p(Y_n|\theta)/p''(Y_n)\}\, dY_n\, d\theta,$$

which gives us, after some manipulation,

$$I(p''(\theta)) + \frac{d}{dt}\{I(tp'(\theta) + (1-t)p''(\theta))\}|_{t=0}$$

$$= \int p(Y_n|\theta)p'(\theta) \log \left\{\frac{p(Y_n|\theta)}{p''(Y_n)}\right\} dY_n\, d\theta.$$

Thus (2.2) holds if and only if

$$\int p'(Y_n) \log \left\{\frac{p'(Y_n)}{p''(Y_n)}\right\} dY_n > 0,$$

which equals zero if and only if $p'(Y_n) = p''(Y_n)$ a.s. (see, e.g., Lindley, 1956, proof of Theorem 1).

*Note that $p'(Y_n)$ and $p''(Y_n)$ differ on a set of positive measure if, for example, they differ a.s. as expected in many applications.

†Here, and in the remainder of this section, we suppress the subscript denoting the range of integration whenever an integral is taken over the entire range space.

2.2 Numerical Search Algorithm

In this section we present an algorithm for finding the optimal mixture prior p^*. The main benefit of this procedure is its ease of implementation: given a finite set of points, it searches for a maximum value by maximizing the information measure $I_n(\cdot)$ over the lines connecting each pair of points. In this sense, this procedure is similar to multidimensional line-search techniques of Bazaraa and Shetty (1979, pp. 270–289).

We begin with some notation and definitions. Identify the convex combination $\sum_{j=1}^{l} \alpha_j p_j(\theta)$ with the point $\alpha = (\alpha_1, \ldots, \alpha_{l-1})$ in R^{l-1} (recall that $\sum_{j=1}^{l} \alpha_j = 1$; $\sum_{j=1}^{l} \alpha_j p_j(\theta) = \sum_{j=1}^{l-1} \alpha_j(p_j(\theta) - p_l(\theta)) + p_l(\theta)$). This identification sets up a one-to-one correspondence between the set A of convex combinations of base priors and the set $S = \{\alpha = (\alpha_1, \ldots, \alpha_{l-1}) \in R^{l-1} : \alpha_j \geqslant 0, \sum_{j=1}^{l-1} \alpha_j \leqslant 1\}$ in R^{l-1}. We will suppress the subscript n in $I_n(\cdot)$ and abuse notation by writing $I(\alpha)$ for $I(p(\theta))$ where $p(\theta) = \sum_{j=1}^{l} \alpha_j p_j(\theta)$. It is assumed throughout that $I(\cdot)$ is strictly concave on S. A *direction* is any unit vector in R^{l-1}. A direction d is *feasible* for a set $S' \subset S$ at a point $\alpha' \in S'$ if there exists a $\delta > 0$ such that $\alpha + \lambda d \in S'$ whenever $0 < \lambda < \delta$. The derivative of $I(\cdot)$ at α in the direction d is denoted $I'(\alpha; d)$. This derivative can, of course, be evaluated directly; Section 3 shows how it can be obtained recursively.

The algorithm generates at each iteration k a finite set of candidate optimal priors, obtained by maximizing $I(\cdot)$ over line segments joining the pairs of points obtained at iteration $k - 1$. The algorithm attempts to find a sufficiently small region that contains the unique optimal prior p^*. The procedure is outlined below, and the justification for it is given in the proposition.

Choose $\varepsilon > 0$ to terminate the algorithm.

1. Choose as starting points for the optimal α, say α^*, the points $e_0^{(1)}, e_1^{(1)}, \ldots, e_{l_1}^{(1)}$ in R^{l-1}, where $e_0^{(1)}$ is the origin and the $e_i^{(1)}$'s are the standard basis vectors. Let $k = 1$, $l_1 = l - 1$; go to step 2.
2. If $\max_i |I(e_i^{(k)}) - I(e_{l_k}^{(k)})| < \varepsilon$, then stop; take the optimal α to be the point in the interior of the span of the $e_i^{(k)}$'s; otherwise, go to step 3.
3. Let $\alpha_{ij}^{(k)}$ be the point in R^l that maximizes $I(\cdot)$ over the line $\lambda e_i^{(k)} + (1 - \lambda)e_j^{(k)}$, $0 \leqslant \lambda \leqslant 1$; there are m_{k+1} such points. Label these points (order is not important) $e_0^{(k+1)}, \ldots, e_{m_{k+1}}^{(k+1)}$, and go to step 2.

At iteration k the algorithm attempts to find the optimal α^* somewhere in the convex region spanned by the $e_i^{(k)}$'s. Starting with this region, the algorithm reduces the region of uncertainty to a smaller region. It is natural to ask then under what conditions does α^* belong to the smaller region, given that it belongs to the larger one? This is answered by the proposition below. For each k, $k = 1, 2, \ldots$, let S_k denote the convex span of the $e_i^{(k)}$'s. We shall assume for mathematical convenience that each S_k has l distinct vertices, which implies that open sets in S_k are also open in S_1.

PROPOSITION 2.1 Suppose that the optimal value α^* lies in the interior of S_k. If at each vertex $e_j^{(k+2)}$ $(j = 0, \ldots, l-1)$ of S_{k+2} there exists at least one feasible direction d for S_{k+1} such that $I'(e_j^{(k+2)}; d) > 0$, then $\alpha^* \in S_{k+1}$ (see Figure 1 for an illustration).

Proof. Since $I(\cdot)$ is strictly concave, for each j there is a point $x_j \in S'_{k+1}$ such that $I(e_j^{(k+2)}) < I(x_j)$. Since the $e_j^{(k+2)}$'s are points on the boundary of S_{k+1}, the points x_j must lie in the interior of S_{k+1}. Also,

$$\max_j I(e_j^{(k+2)}) < \max_j I(x_j) \leqslant \max_{\alpha \in S_{k+1}} I(\alpha),$$

which implies that $I(\cdot)$ has a local maximum at some point α^{**} in the interior of S_{k+1}. Since $I(\cdot)$ is strictly concave, the point at which it attains its maximum is unique; consequently, $\alpha^* = \alpha^{**}$.

Figure 1 illustrates the algorithm and the conditions of the proposition for the case of three base priors ($l = 3$). The points $e_i^{(k)}$, $i = 1, 2, 3$, are obtained after the kth iteration. The points $e_i^{(k+1)}$, $i = 1, 2, 3$, are obtained by maximizing $I(\cdot)$ along the lines connecting pairs of the $e_i^{(k)}$'s. To check that α^*

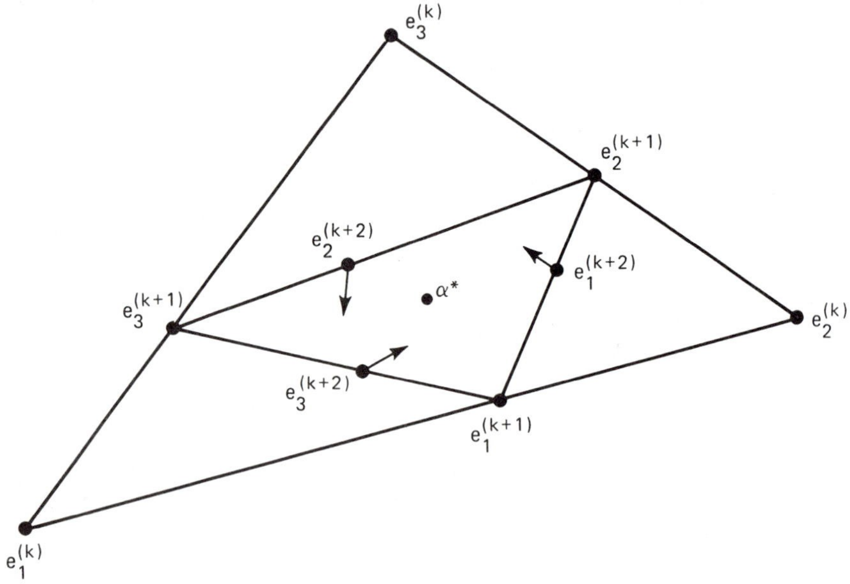

Figure 1. Illustration of the numerical search algorithm.

belongs to the convex span of the $e_i^{(k+1)}$'s requires checking the gradient of $I(\cdot)$ at the points $e_i^{(k+2)}$, $i = 1, 2, 3$, that are obtained after the $(k+2)$th iteration. The arrows at each $e_i^{(k+2)}$ indicate the direction in which the gradient of $I(\cdot)$ is positive.

3. RECURSIVE FORMULA FOR $I_n(\cdot)$

To find the suboptimal prior $p^*(\theta)$, the maximum of $I_n(\cdot)$ over A, via a numerical algorithm such as one given in Zangwill (1967) or Ortega and Rheinboldt (1970), we need to compute the derivatives of $I_n(\cdot)$ with respect to the α_j's. We can, of course, compute these derivatives directly. But we will show how, by taking advantage of the sequential structure of (0), to obtain them iteratively. Before doing this we convert the system measurements to a more convenient form.

Consider the Kalman filter state estimate $\hat{x}_k = E(x_k | z_1, \ldots, z_k, \theta)$ of x_k and take Y_k to be $Y_k = (\hat{x}_1^T, \ldots, \hat{x}_k^T)^T$, which is sufficient for $z_1, \ldots, z_k$. The $\hat{x}_k$'s satisfy (for $k = 1, \ldots, n$)

$$\hat{x}_k = \hat{x}_{k|k-1} + \Sigma_{k|k-1} H_k^T (H_k \Sigma_{k|k-1} H_k^T + R_k)^{-1} (z_k - H_k \hat{x}_{k|k-1}), \tag{3.1a}$$

where

$$\hat{x}_{k|k-1} = \Phi_{k-1} \hat{x}_{k-1} + S_{k-1} R_{k-1}^{-1} (z_{k-1} - H_{k-1} \hat{x}_{k-1}) \tag{3.1b}$$

$$\Sigma_k = \Sigma_{k|k-1} - \Sigma_{k|k-1} H_k^T (H_k \Sigma_{k|k-1} H_k^T + R_k)^{-1} H_k \Sigma_{k|k-1} \tag{3.2a}$$

$$\begin{aligned} \Sigma_{k|k-1} = {} & (\Phi_{k-1} - S_{k-1} R_{k-1}^{-1} H_{k-1}) \Sigma_{k-1} (\Phi_{k-1} - S_{k-1} R_{k-1}^{-1} H_{k-1})^T \\ & + (Q_{k-1} - S_{k-1} R_{k-1}^{-1} S_{k-1}^T) \end{aligned} \tag{3.2b}$$

with

$$\hat{x}_{0|-1} = \mu, \qquad \Sigma_{0|-1} = P_0. \tag{3.3}$$

To simplify the equations that follow, we will assume that $S_k = 0$, $k = 1, \ldots, n$; similar results hold for the more general case. The equations above then reduce to

$$\hat{x}_k = \Phi_{k-1} \hat{x}_{k-1} + K_k (z_k - H_k \Phi_{k-1} \hat{x}_{k-1}) \tag{3.4a}$$

$$K_k = \Sigma_{k|k-1} H_k^T (H_k \Sigma_{k|k-1} H_k^T + R_k)^{-1}. \tag{3.4b}$$

Since $\{\hat{x}_k\}$ is a Markov process, it immediately follows that

$$p(\hat{x}_k | Y_{k-1}, \theta) = p(\hat{x}_k | \hat{x}_{k-1}, \theta) \tag{3.5}$$

and $\hat{x}_k$ conditioned on $\hat{x}_{k-1}$, θ is distributed as $N(\Phi_{k-1} \hat{x}_{k-1}, W_k)$, where

$$W_k = K_k (H_k \Phi_{k-1} \Sigma_{k-1} \Phi_{k-1}^T H_k^T + H_k Q_{k-1} H_k^T + R_k) K_k^T. \tag{3.6}$$

We proceed now to develop the recursion for I_k. Our goal here is to show that

$$I_k(p(\theta)) = I_{k-1}(p(\theta)) + G_{k-1}(p(\theta)) + F_{k-1}(p(\theta)), \tag{3.7}$$

for appropriately defined G_k, F_k's.

From (2.3) and (3.5), I_k can be written equivalently as

$$\begin{aligned}&\int p(Y_{k-1}|\theta)p(\theta)[\log\{p(Y_{k-1}|\theta)\} - \log\{p(Y_{k-1})\}]\,dY_{k-1}\,d\theta\\ &\quad+\int p(\hat{x}_k|\hat{x}_{k-1},\theta)\log\{p(\hat{x}_k|\hat{x}_{k-1},\theta)\}p(\hat{x}_{k-1}|\theta)p(\theta)\,d\hat{x}_k\,d\hat{x}_{k-1}\,d\theta\\ &\quad-\int p(\hat{x}_k|\hat{x}_{k-1},\theta)\log\{p(\hat{x}_k|\hat{x}_{k-1})\}p(\hat{x}_{k-1}|\theta)p(\theta)\,d\hat{x}_k\,d\hat{x}_{k-1}\,d\theta.\end{aligned} \tag{3.8}$$

The first term of (3.8) is I_{k-1}. Consider the second term. Using (3.5), we obtain

$$G_{k-1}(p(\theta)) \equiv \int \log\{((2\pi e)^p|W_k|)^{-1/2}\}p(\theta)\,d\theta, \tag{3.9}$$

where $|\cdot|$ denotes the determinant and e the base of log. Next, we obtain the last term in (3.8), F_{k-1}. Note that

$$\begin{aligned}p(\hat{x}_k|\hat{x}_{k-1}) &= \int p(\hat{x}_k|\hat{x}_{k-1},\theta)p(\theta)\,d\theta\\ &= \int \{(2\pi)^p|W_k|\}^{-1/2}\\ &\qquad\times \exp\{-\tfrac{1}{2}(\hat{x}_k-\Phi_{k-1}\hat{x}_{k-1})^T W_k^{-1}(\hat{x}_k-\Phi_{k-1}\hat{x}_{k-1})\}p(\theta)\,d\theta,\end{aligned} \tag{3.10}$$

the last equality following from (3.6). Also, by (3.1) to (3.4), $\{\hat{x}_k|\theta\} \sim N(m_k, B_k)$, where $B_k = \Phi_{k-1}B_{k-1}\Phi_{k-1}^T + K_kH_k\Sigma_{k|k}H_k^TK_k^T + K_kR_kK_k^T$, $B_0 = 0$, and $m_k = \Phi_{k-1}m_{k-1}$, $1 \leqslant k \leqslant n$, $m_0 = \mu$; thus

$$p(\hat{x}_k) = \int \{(2\pi)^p|B_k|\}^{-1/2}\exp\{-\tfrac{1}{2}(\hat{x}_k-m_k)^TB_k^{-1}(\hat{x}_k-m_k)\}p(\theta)\,d\theta. \tag{3.11}$$

Hence

$$F_{k-1}(p(\theta)) = -\int p(\hat{x}_k|\hat{x}_{k-1})\log\{p(\hat{x}_k|\hat{x}_{k-1})\}p(\hat{x}_{k-1})\,d\hat{x}_k\,d\hat{x}_{k-1}, \tag{3.12}$$

where $p(\hat{x}_k|\hat{x}_{k-1})$ and $p(\hat{x}_{k-1})$ are obtained from (3.10) and (3.11), respectively.

The derivatives of I_n can now be easily obtained from (3.7). To compute these derivatives, rewrite $p(\theta) = \Sigma_{j=1}^{l}\alpha_jp_j(\theta)$ as $\Sigma_{j<l}\alpha_j(p_j - p_l) + p_l$ (remem-

ber that $\Sigma_{j=1}^{l} \alpha_j = 1$). For the term $G_{k-1}(p(\theta))$ we have, using (3.9),

$$\frac{\partial G_{k-1}}{\partial \alpha_j} = G_{k-1}(p_j(\theta)) - G_{k-1}(p_l(\theta)) \tag{3.13}$$

for $1 \leqslant j \leqslant l-1$. Consequently,

$$\frac{\partial^2 G_{k-1}}{\partial \alpha_i \partial \alpha_j} = 0 \qquad \forall\, i,\, j, \quad 1 \leqslant i, j \leqslant l-1.$$

Now consider the term $F_{k-1}(p(\theta))$. Using (3.12), we have

$$\begin{aligned}\frac{\partial F_{k-1}}{\partial \alpha_j} = &- \int \frac{\partial}{\partial \alpha_j} \{p(\hat{x}_k|\hat{x}_{k-1})\} \log\{p(\hat{x}_k|\hat{x}_{k-1})\} p(\hat{x}_{k-1})\, d\hat{x}_k\, d\hat{x}_{k-1} \\ &- \int p(\hat{x}_k|\hat{x}_{k-1}) \left\{ \frac{\frac{\partial}{\partial \alpha_j} p(\hat{x}_k|\hat{x}_{k-1})}{p(\hat{x}_k|\hat{x}_{k-1})} \right\} p(\hat{x}_{k-1})\, d\hat{x}_k\, d\hat{x}_{k-1} \\ &- \int p(\hat{x}_k|\hat{x}_{k-1}) \log\{p(\hat{x}_k|\hat{x}_{k-1})\} \frac{\partial}{\partial \alpha_j} p(\hat{x}_{k-1})\, d\hat{x}_k\, d\hat{x}_{k-1}.\end{aligned} \tag{3.14}$$

But $\partial p(\hat{x}_k|\hat{x}_{k-1})/\partial \alpha_j = p_j(\hat{x}_k|\hat{x}_{k-1}) - p_l(\hat{x}_k|\hat{x}_{k-1})$, where

$$p_i(\hat{x}_k|\hat{x}_{k-1}) = \int p(\hat{x}_k|\hat{x}_{k-1}, \theta) p_i(\theta)\, d\theta, \qquad i = 1, \ldots, l,$$

and $\partial p(\hat{x}_{k-1})/\partial \alpha_j = p_j(\hat{x}_{k-1}) - p_l(\hat{x}_{k-1})$, with $p_j(\hat{x}_{k-1}) = \int p(\hat{x}_{k-1}|\theta) p_j(\theta)\, d\theta$. Hence (3.14) becomes

$$\begin{aligned}\frac{\partial F_{k-1}}{\partial \alpha_j} = &\int \{p_l(\hat{x}_k|\hat{x}_{k-1}) - p_j(\hat{x}_k|\hat{x}_{k-1})\} \\ &\cdot \log\{p(\hat{x}_k|\hat{x}_{k-1})\} p(\hat{x}_{k-1})\, d\hat{x}_{k-1}\, d\hat{x}_k \\ &+ \int \{p_l(\hat{x}_{k-1}) - p_j(\hat{x}_{k-1})\} p(\hat{x}_k|\hat{x}_{k-1}) \\ &\times \log\{p(\hat{x}_k \hat{x}_{k-1})\}\, d\hat{x}_{k-1}\, d\hat{x}_k.\end{aligned}$$

Hence

$$\begin{aligned}\frac{\partial^2 F_{k-1}}{\partial \alpha_i \partial \alpha_j} = &\int \frac{\{p_l(\hat{x}_k|\hat{x}_{k-1}) - p_j(\hat{x}_k|\hat{x}_{k-1})\}}{p(\hat{x}_k|\hat{x}_{k-1})} [\{p_i(\hat{x}_k|\hat{x}_{k-1}) - p_l(\hat{x}_k|\hat{x}_{k-1})\} \\ &+ \log\{p(\hat{x}_k \hat{x}_{k-1})\}\{p_i(\hat{x}_{k-1}) - p_l(\hat{x}_{k-1})\}] p(\hat{x}_{k-1})\, d\hat{x}_k\, d\hat{x}_{k-1} \\ &+ \int [\{p_i(\hat{x}_k|\hat{x}_{k-1}) - p_l(\hat{x}_k|\hat{x}_{k-1})\} \log\{p(\hat{x}_k|\hat{x}_{k-1})\} + \{p_i(\hat{x}_k|\hat{x}_{k-1}) \\ &- p_l(\hat{x}_k|\hat{x}_{k-1})\}]\{p_l(\hat{x}_{k-1}) - p_j(\hat{x}_{k-1})\}\, d\hat{x}_k\, d\hat{x}_{k-1}.\end{aligned} \tag{3.15}$$

Although these derivatives appear cumbersome, they are more efficient than those obtained by differentiating I_n directly, which entails keeping track of the correlations between the $\hat{x}_k$'s. The recursive form has another advantage: When the system (0) is constant coefficient (the system matrices do not depend on k), many of the terms simplify considerably.

To sum up, we seek the suboptimal prior $p^*(\theta)$ within the convex span of the prespecified base priors, say $p_1(\theta), \ldots, p_l(\theta)$. To carry out this search, we apply a numerical algorithm, which requires the derivatives of the information functional $I_n(\cdot)$. The expression (3.7) for $I_n(\cdot)$ allows us to obtain these derivatives in an efficient recursive manner.

4. NUMERICAL STUDY

This section illustrates how the information measure in (2.1) is increased for optimal convex combinations of base priors in comparison to the value of the measure associated with the individual base priors. We consider a simple state-space model with scalar state and observation. Working with two base priors, as we do in this section, is illustrative of what would be done in implementing the search algorithm of Section 2.2.

Figures 2 and 3 present the main results of this section. Both figures illustrate the concave shape of $I_n(\cdot)$ as a function of α, where $p^*(\theta) = \alpha p_1(\theta) + (1 - \alpha)p_2(\theta)$ and p_1 and p_2 (the base priors) are described below.

The state-space model corresponding to Figure 2 (hereafter model 1) is constant coefficient with

$$\mu = 0, \qquad P_0 = \frac{0.0025}{1 - \theta^2}, \qquad \Phi = \theta,$$

$$Q = 0.0025, \qquad H = 1, \qquad R = 0.0025,$$

and $n = 4$. The base priors for θ are both truncated Gaussian distributions: p_1 represents an $N(-\frac{1}{4}, \frac{1}{16})$ distribution and p_2 represents an $N(\frac{1}{4}, \frac{1}{18})$ distribution with both p_1 and p_2 truncated at $\pm\frac{1}{2}$ (and appropriately normalized, of course). Note that $\{z_k\}$ is a stationary process.

The state-space model for Figure 3 (i.e., model 2) is also constant coefficient, but with the parameters

$$\mu = 0, \qquad P_0 = \frac{0.005}{1 - \theta^2}, \qquad \Phi = \theta,$$

$$Q = 0.005, \qquad H = 1, \qquad R = 0,$$

and $n = 2$. Since there is no measurement noise, this case corresponds to an

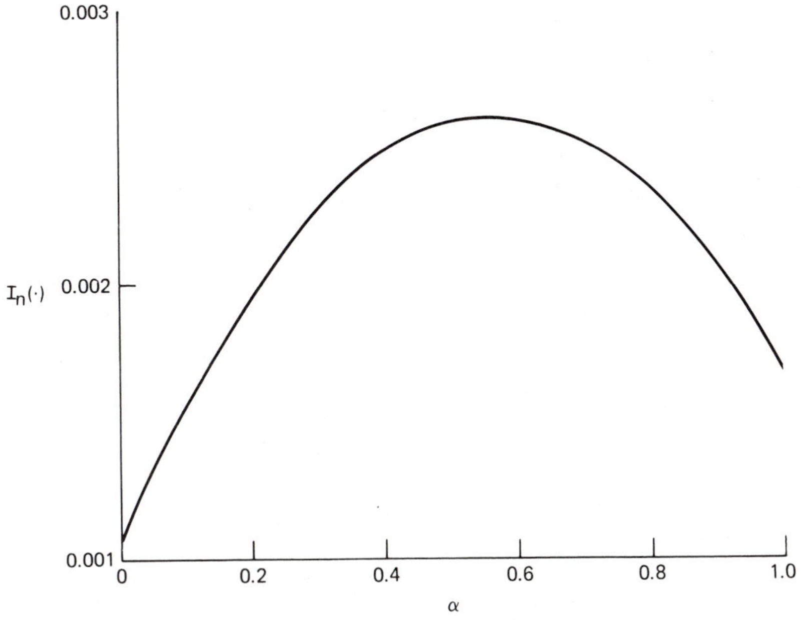

Figure 2. Values for $I_n(\alpha p_1 + (1 - \alpha)p_2)$ for model 1.

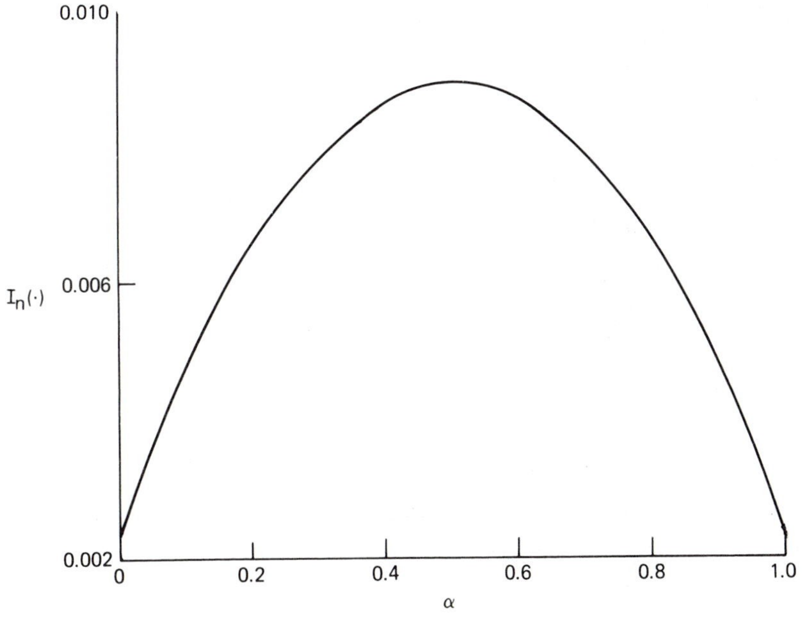

Figure 3. Values for $I_n(\alpha p_1 + (1 - \alpha)p_2)$ for model 2.

AR(1) model. The base priors are two uniform distributions, one over the interval $(-\frac{1}{4}, 0)$ and the other over the interval $(0, \frac{1}{4})$. As with model 1, $\{z_k\}$ is a stationary process.

Figure 2 illustrates that as one would expect, the base prior that is more noninformative is given greater weight in the convex combination. (*Note:* $I_n(p_1) = 0.0169$, $I_n(p_2) = 0.0107$.) A bisection search method was applied to find the optimal value of α, which was found to be $0.56(\pm 0.01)$ with a corresponding value of $I_n(p^*)$ of 0.00261. In Figure 3 both base priors had the same value of $I_n(=0.00210)$, and the optimal value of α was 0.5 with $I_n(p^*) = 0.00878$. Note that this corresponds to the case where p^* represents a uniform $(-\frac{1}{4}, \frac{1}{4})$ distribution.

Several other studies were performed with the state-space models of Figures 2 and 3 to gain additional insight into the methodology. First, since the model of Figure 3 is stationary, a result in Kashyap (1971, Theorem 4 for Markov processes) applies, which indicates that asymptotically $(n \to \infty)$ a prior that is proportional to the square root of the determinant of the Fisher information matrix is the optimal prior (with respect to maximizing $I_n(\cdot)$). For the state-space model of Figure 3 (which is Markov, since $R = 0$), we have that the value of I_n with the Kashyap prior is 0.00886, which exceeds the value of $I_n(p^*)(=0.00878)$. It was also found that when the Kashyap prior was combined with, say, one of several other priors (e.g., uniform, truncated Gaussian, ramp), 100% weighting was given to the Kashyap prior, as one would expect if the above-mentioned Kashyap theorem is applicable. Although a realization of length 2 can hardly be considered asymptotic, it may be true that the Kashyap result is applicable (or at least provides a good approximation) since with $|\Phi| < \frac{1}{4}$, the system is highly stable. Interestingly, we also found that in the stationary state-space model of Figure 2 (which is not Markov since $R \neq 0$), the Kashyap prior was optimal among those considered (uniform, truncated Gaussian, and p^*), yielding a value of I_n of 0.0344. As with the model of Figure 3, this state-space model is highly stable ($|\Phi| < \frac{1}{2}$ with $n = 4$). The studies above suggest that it might be a good idea to include the Kashyap prior among the base priors.

Finding p^* is a computationally demanding procedure, even for fairly small-dimensional systems, as can be seen from the form for I_n with its multifold integrals. An evaluation of I_n for the models above with a fixed value of α took between 13 and 120 CPU seconds on an IBM 3090 (the integrals were evaluated using the IMSL subroutine DMLIN). We would expect reductions in run times with more efficiently designed software and advances in computer technology.

5. CONCLUDING REMARKS

We have presented a computationally tractable method for assessing noninformative priors for AR, ARMA, and state space models. The method is a constrained optimization of an intuitively attractive information measure over a class of mixtures of distributions from a given collection. When the information measure is strictly concave, convergence to a unique solution is assured. The computational tractability is due to the constraint on the solution space and to an efficient recursion for computing the derivatives necessary for the optimization algorithms.

The technique presented here can be computationally intensive. This is apparent by examining the form for the information measure with its (generally) high-dimensional multifold integral. Although intensive, the method is, at least in principal, tractable (and will become more so with advances in computer technology), whereas the exact solution to maximizing the information measure over all possible densities is, in general, intractable.

ACKNOWLEDGMENTS

J. C. Spall received support from a JHU/APL Stuart S. Janney Fellowship for the preparation of this chapter. Both authors received additional support from U.S. Navy Contract N00039-87-C-5301.

REFERENCES

Akaike, H. (1983). On minimum information prior distributions, *Ann. Inst. Stat. Math. Part A* 35, 139–149.

Anderson, B. D. O., and J. B. Moore (1979). *Optimal Filtering*, Prentice-Hall, Englewood Cliffs, N.J.

Bazaraa, M., and C. C. Shetty (1979). *Nonlinear Programming: Theory and Methods*, Wiley, New York.

Bernardo, J. M. (1979a). Reference posterior distributions for Bayesian inference, *J. R. Stat. Soc. Ser. B* 41, 113–147.

Bernardo, J. M. (1979b). Expected information as expected utility, *Ann. Stat.* 7, 686–690.

Dalal, S. R., and W. J. Hall (1983). Approximating priors by mixtures of natural conjugate priors, *J. R. Stat. Soc. Ser. B* 45, 278–286.

Davis, W. W. (1979). Approximate Bayesian predictive distributions and model selection, *J. Am. Stat. Assoc.* 74, 312–317.

Dyer, D., and P. Chiou (1984). An information-theoretic approach to incorporating prior information in binomial sampling, *Commun. Stat. Theory Methods* 13, 2051–2083.

Edwards, W., H. Lindman, and L. J. Savage (1963). Bayesian statistical inference for psychological research, *Psychol. Rev.* 70, 193–242.

Good, I. J. (1966). A derivation of the probabilistic explication of information, *J. R. Stat. Soc. Ser. B* 28, 578–581.

Hill, S. D. and J. C. Spall (1987). Noninformative Bayesian priors for large samples based on Shannon information theory, in *Proceedings of the IEEE Conference on Decision and Control*, Los Angeles, December 9–11, pp. 1690–1693.

Kashyap, R. L. (1971). Prior probability and uncertainty, *IEEE Trans. Inf. Theory*, IT-17, 641–650.

Lindley, D. V. (1956). On a measure of the information provided by an experiment, *Ann. Math. Stat.* 27, 986–1005.

Maryak, J. L., and J. C. Spall (1987). Conditions for the insensitivity of the Bayesian posterior distribution to the choice of prior distribution, *Stat. Probab. Lett.* 5, 399–405.

Novick, M. R. (1969). Multiparameter Bayesian indifference procedures, *J. R. Stat. Soc. Ser. B* 31, 29–64.

Ortega, J. M., and W. C. Rheinboldt (1970). *Iterative Solution of Nonlinear Equations in Several Variables*, Academic Press, New York.

Peterka, V. (1981). Bayesian system identification, *Automatica* 17, 41–53.

Poskitt, D. S., and A. R. Tremayne (1983). On the posterior odds of time series models, *Biometrika* 70, 157–162.

Sage, A. P., and J. L. Melsa (1971). *System Identification*, Academic Press, New York.

Shannon, C. E., and W. Weaver (1949). *The Mathematical Theory of Communication*, University of Illinois Press, Urbana, Ill.

Spall, J. C., and M. I. Koch (1985). An Approach to Isolating Sources of Errors in Invalid State Space Models Based on Stochastic Approximation, in *Proceedings of the American Control Conference*, Boston, June 19–21, pp. 1350–1356.

Spall, J. C. (1988). Bayesian error isolation for models of large-scale systems, *IEEE Trans. Autom. Control* AC-33, 341–347.

Zangwill, W. I. (1987). Nonlinear programming via penalty functions, *Manage. Sci.* 13, 344–358.

Zellner, A. (1971). *An Introduction to Bayesian Inference in Econometrics*, Wiley, New York.

Zellner, A (1977). Maximal Data Information Prior Distributions, in *New Developments in the Applications of Bayesian Methods* (A. Aykac and C. Brumat, eds.), North-Holland, Amsterdam, pp. 211–232.

Referee Acknowledgment

Every chapter in this book was reviewed by at least two referees. The editor is most grateful to the following individuals for serving as referees.

David J. Abeshouse The Johns Hopkins University, Applied Physics Laboratory, Laurel, Maryland,

Gerald W. Bodoh Hadron, Inc. , Fairfax, Virginia

Lyle D. Broemeling Office of Naval Research, Arlington, Virginia

Peyton Cook University of Tulsa, Tulsa, Oklahoma

Frederick E. Daum Raytheon Company, Wayland, Massachusetts

William W. Davis U.S. Bureau of the Census, Washington, District of Columbia

Franklin Dellon The Johns Hopkins University, Applied Physics Laboratory, Laurel, Maryland

Joaquin Diaz University of Houston, Houston, Texas

James T. Everett The Johns Hopkins University, Applied Physics Laboratory, Laurel, Maryland

Robert C. Ferguson The Johns Hopkins University, Applied Physics Laboratory, Laurel, Maryland

Catherine R. Foster The Johns Hopkins University, Applied Physics Laboratory, Laurel, Maryland

Will Gersch University of Hawaii, Honolulu, Hawaii

Paul Gilbert Bank of Canada, Ottawa, Ontario

James D. Hamilton University of Virginia, Charlottesville, Virginia

Francis T. Heuring Old Dominion Systems, Inc., Germantown, Maryland

Robert L. Hickerson The Johns Hopkins University, Applied Physics Laboratory, Laurel, Maryland

Stacy D. Hill The Johns Hopkins University, Applied Physics Laboratory, Laurel, Maryland

Robert P. Judd Oakland University, Rochester, Michigan

Matthew I. Koch* The Johns Hopkins University, Applied Physics Laboratory, Laurel, Maryland

Robert Kohn University of Chicago, Chicago, Illinois

Charles G. Markon The Johns Hopkins University, Applied Physics Laboratory, Laurel, Maryland

George W. Martin The Johns Hopkins University, Applied Physics Laboratory, Laurel, Maryland

John L. Maryak* The Johns Hopkins University, Applied Physics Laboratory, Laurel, Maryland

John R. McNellis The Johns Hopkins University, Applied Physics Laboratory, Laurel, Maryland

Wanda M. Mercer Independent Consultant, Baltimore, Maryland

David H. Moen University of South Dakota, Vermillion, South Dakota

Mehdi Mostaghimi* Southern Connecticut State University, New Haven, Connecticut

Fahimeh Rezayat University of Virginia, Charlottesville, Virginia

Robert H. Shumway University of California, Davis, California

Malcolm D. Shuster Business and Technological Systems, Inc., Laurel, Maryland

Lee S. Simkins The Johns Hopkins University, Applied Physics Laboratory, Laurel, Maryland

Richard H. Smith The Johns Hopkins University, Applied Physics Laboratory, Laurel, Maryland

*Reviewed more than one chapter

Jacqueline K. Telford The Johns Hopkins University, Applied Physics Laboratory, Laurel, Maryland

Kent D. Wall Naval Postgraduate School, Monterey, California

James V. White The Analytic Sciences Corporation, Reading, Massachusetts

David A. Whitney The Analytic Sciences Corporation, Reading, Massachusetts

Paul A. Zucker The Johns Hopkins University, Applied Physics Laboratory, Laurel, Maryland

Index

V

W

Z